Lecture Notes in Economics and Mathematical Systems

Springer
Berlin
Heidelberg
New York
Barcelona
Budapest
Hong Kong
London
Milan
Paris
Santa Clara
Singapore
Tokyo

Kurt Marti   Peter Kall (Eds.)

Stochastic Programming Methods and Technical Applications

Proceedings of the 3rd GAMM/IFIP-Workshop
on "Stochastic Optimization:
Numerical Methods and Technical Applications"
held at the Federal Armed Forces University Munich,
Neubiberg/München, Germany, June 17–20, 1996

 Springer

Editors

Prof. Dr. Kurt Marti
Universität der Bundeswehr München
Fakultät für Luft- und Raumfahrttechnik
D-85577 Neubiberg/München, Germany

Prof. Dr. Peter Kall
Universität Zürich
Institut für Operations Research
Moussonstraße 15
CH-8044 Zürich, Switzerland

Library of Congress Cataloging-in-Publication Data

GAMM/IFIP-Workshop on "Stochastic Optimization: Numerical Methods and
  Technical Applications" (3rd : 1996 : Federal Armed Forces
  University Munich)
  Stochastic programming methods and technical applications :
proceedings of the 3rd GAMM/IFIP-Workshop on "Stochastic
Optimization: Numerical Methods and Technical Applications", held at
the Federal Armed Forces University Munich, Neubiberg/München,
Germany, June 17-20, 1996 / Kurt Marti, Peter Kall, eds.
     p.   cm. -- (Lecture notes in economics and mathematical
systems, ISSN 0075-8442 ; 458)
  Includes bibliographical references and index.
  ISBN 3-540-63924-1
  1. Stochastic programming--Congresses. 2. Mathematical
optimization--Congresses. 3. Stochastic processes--Congresses.
I. Marti, Kurt, 1943- . II. Kall, Peter. III. Title.
IV. Series.
T57.79.G36 1996
519.7--dc21   97-48263
                                                      CIP

ISSN 0075-8442
ISBN 3-540-63924-1 Springer-Verlag Berlin Heidelberg New York

© Springer-Verlag Berlin Heidelberg 1998
Printed in Germany

Typesetting: Camera ready by author
SPIN: 10649589   42/3142-543210 - Printed on acid-free paper

PREFACE

This volume includes a selection of tutorial and technical papers presented at the 3rd GAMM/IFIP-Workshop on "Stochastic Optimization: Numerical Methods and Technical Applications", held at the Federal Armed Forces University Munich, Neubiberg/Munich, June 17-20, 1996.

Optimization problems arising in practice contain usually several random parameters. Hence, in order to get robust solutions with respect to random parameter variations, i.e., to reduce expensive online measurements and corrections, the mostly available statistical informations (samples, moments, etc.) about the random parameters should be considered already at the planning phase. Thus, the original problem with random coefficients must be replaced by an appropriate deterministic substitute problem, and efficient numerical solution/approximation techniques have to be developed for solving the resulting substitute problems.

E.g., evaluating the violation of the random constraints by means of penalty functions, or applying a reliability-based approach, one obtains a stochastic program with recourse, a chance-constrained stochastic program, respectively.

Solving the chosen deterministic substitute problem, one has then to deal with the numerical evaluation of probability and mean value functions (represented by certain multiple integrals) and its derivatives.

Therefore, also the aim of the 3rd GAMM/IFIP-Workshop on "Stochastic Optimization" was to bring together scientists from Stochastic Programming, Numerical Optimization and from Reliability-based Engineering Optimization, as e.g. Optimal Structural Design, Optimal Trajectory Planning for Robots, Optimal Power Dispatch, etc.

The following Scientific Program Committee was formed:

H.A. Eschenauer (Germany)

P. Kall (Switzerland)

K. Marti (Germany, Chairman)

J. Mayer (Switzerland)

F. Pfeiffer (Germany)

R. Rackwitz (Germany)

G.I. Schuëller (Austria).

The first day of the Workshop was devoted mainly to four one-hour tutorial papers on one of the main topics of the Workshop: Modelling aspects, approximation and numerical solution techniques, technical applications of stochastic optimization. The tutorials are contained in part I. TUTORIAL PAPERS; the technical contributions providing new theoretical results, numerical solution procedures and new applications to reliability-based optimization of technical structures/systems are divided into the following three parts: II. THEORETICAL MODELS AND CONCEPTUAL METHODS, III. NUMERICAL METHODS AND COMPUTER SUPPORT and IV. TECHNICAL APPLICATIONS.

In order to guarantee again a high scientific level of the Proceedings volume of the third workshop on this topic, all papers were refereed. We express our gratitude to all referees, and we thank all authors for delivering the final version of their papers in due time.

We gratefully acknowledge the support of the Workshop by GAMM (Society for Applied Mathematics and Mechanics), IFIP (International Federation of Information Processing), the Federal Armed Forces University Munich, the Friends of the University, and we thank the commander of the student division for the kind accommodational support.

Finally we thank Springer-Verlag for including the Proceedings in the Springer Lecture Notes Series.

München/Zürich

October 1997   K. Marti, P. Kall

TABLE OF CONTENTS

I. TUTORIAL PAPERS

Bounds for and Approximations to
Stochastic Linear Programs with Recourse - Tutorial -
P. Kall 1

Optimal Power Generation under Uncertainty
via Stochastic Programming
D. Dentcheva and W. Römisch 22

Position and Controller Optimization
for Robotic Parts Mating
G. Prokop and F. Pfeiffer 57

Some Basic Principles of Reliability-Based Optimization (RBO)
of Structures and Mechanical Components
M. Gasser and G.I. Schuëller 80

II. THEORETICAL MODELS AND CONCEPTUAL METHODS

Stochastic Optimization Approach to Dynamic Problems
with Jump Changing Structure
V.I. Arkin 104

Reflections on Robust Optimization
J. Dupačová 111

On Constrained Discontinuous Optimization
Y. Ermoliev and V. Norkin 128

On the Equivalence in Stochastic Programming with
Probability and Quantile Objectives
Yu.S. Kan and A.A. Mistryukov 145

A Note on Multifunctions in Stochastic Programming
V. Kaňková 154

Approximation to Extremum Problems
with Probability Cost Functionals
R. Lepp 169

Global Optimization of Probabilities
by the Stochastic Branch and Bound Method
V. Norkin 186

Robust Stability of Interval Matrices:
a Stochastic Approach
B.T. Polyak 202

A Note on Preprocessing via Fourier-Motzkin Elimination
in Two-Stage Stochastic Programming
R. Schultz 208

Bounds for the Reliability of
k-out-of-connected-(r,s)-from-(m,n): F Lattice Systems
T. Szántai 223

On a Relation between Problems of Calculus of Variations
and Mathematical Programming
T. Rapcsák and A. Vásárhelyi 238

III. NUMERICAL METHODS AND COMPUTER SUPPORT

Parameter Sensitivity of Deterministic and Stochastic Search Methods
K.-J. Böttcher 249

Regression Estimators related to Multinormal Distributions:
Computer Experiences in Root Finding
I. Deák 279

Some Aspects of Algorithmic Differentiation of
Ordinary Differential Equations
P. Eberhard and Ch. Bischof 294

Refinement Issues in Stochastic Multistage Linear Programming
K. Frauendorfer and Ch. Marohn 305

On Solving Stochastic Linear Programming Problems
P. Kall and J. Mayer 329

IV. TECHNICAL APPLICATIONS

On an ON/OFF Type Source with Long Range Correlations
R. Antkiewicz and A. Manikowski 345

Optimization Methods in Structural Reliability

K. Breitung, F. Casciati and L. Faravelli 359

Mathematical Aspects of the Boundary Initial Value Problems

for Thermoelasticity Theory of Non-simple Materials

with Control for Temperature

J. Gawinecki and L. Kowalski 370

Stochastic Trajectory Planning for Manutec r3

with Random Payload

S. Qu 382

Implementation of the Response Surface Method (RSM)

for Stochastic Structural Optimization Problems

J. Reinhart 394

Optimization of an Engine Air Intake System for Minimum

Noise Transmission using Function Approximation Concepts

M.H. van Houten, A.J.G. Schoofs and D.H. van Campen 410

Stochastic Structural Optimization of Powertrain

Mounting Systems with Dynamic Constraints

R. Deges and T. Vietor 424

Bounds for and Approximations to Stochastic Linear Programs with Recourse[1]
— Tutorial —

P. Kall
IOR, University of Zurich, Moussonstr. 15, CH–8044 Zürich

Abstract. The objective of stochastic linear programs with recourse contains a multivariate integral $\mathcal{Q}(x) = \int_\Xi Q(x,\xi)P(d\xi)$, in general. This, although having convenient properties under mild assumptions (like e.g. convexity, smoothness), causes difficulties in computational solution procedures. Therefore we usually replace $\mathcal{Q}(\cdot)$ by successively improved lower and upper bounding functions more amenable to optimization procedures, the involved bounding functions being solutions to various (generalized) moment problems.

Keywords. 90C15 (1991 MSC)

1 Stochastic Programs with Recourse

Consider a stochastic program of the type

$$(1.1) \qquad \min_{x \in X}\{c^{\mathrm{T}}x + \mathcal{Q}(x)\}$$

under the assumptions:

- $X \subset \mathbb{R}^n$ convex polyhedral,

- $\mathcal{Q}(x) := \displaystyle\int_\Xi Q(x,\xi)P(d\xi)$ with $\Xi \subset \mathbb{R}^K$ a convex polyhedron, P a probability measure on Ξ,

- $Q(x,\xi) := \min_y\{q^{\mathrm{T}}y \mid Wy = h(\xi) - T(\xi)x,\ y \geq 0\}$ with $h(\cdot), T(\cdot)$ linear affine in $\xi \in \Xi$, i.e. $h(\xi) := h^0 + \displaystyle\sum_{i=1}^K \xi_i h^i$, $T(\xi) := T^0 + \displaystyle\sum_{i=1}^K \xi_i T^i$, where

---

[1] Revised version of a paper presented at IFIP WG 7.7 Workshop —and Tutorial— on Stochastic Optimization; Tucson, AZ, Jan 15–19, 1996.

$h^i \in \mathbb{R}^m$ and T^i are $m \times n$–matrices, $i = 0, \cdots, K$.

Then we have

Proposition 1.1 $Q(x, \cdot) : \Xi \longrightarrow \mathbb{R}$ *is convex* $\forall x \in X$.

and

Proposition 1.2 *Provided the existence of the integral, then* $\mathcal{Q}(\cdot) : X \longrightarrow \mathbb{R}$ *is convex.*

In spite of the last fact, we may not solve (1.1) by just applying any iterative method for convex programming since the repeated evaluation of $\mathcal{Q}(x)$ (and possibly of its gradient) would involve repeated multivariate integration which in general cannot be efficiently done. Hence we usually try to approximate $\mathcal{Q}(x)$ by lower and upper bounding functions which are easier to deal with.

2   Bounds for Univariate Integrals

Consider

$$(2.1) \qquad \int_\Xi \varphi(\xi) P(d\xi).$$

With $\Xi := [\alpha_0, \alpha_1] \subset \mathbb{R}$, $\varphi : \Xi \longrightarrow \mathbb{R}$ convex,
we have

Proposition 2.1 (Jensen inequality [15]) *For* $\bar{\xi} := E\xi$ *holds*

$$(2.2) \qquad \varphi(\bar{\xi}) \leq \int_\Xi \varphi(\xi) P(d\xi),$$

reducing to an equality if φ *is linear on* Ξ.

Sketch of the proof: For any discrete distribution $\tilde{P}$,

$$\{\xi_i, p_i\}_{i=1,\cdots,r}, \ (\xi_i \in \Xi) : \ p_i > 0, \sum_{i=1}^r p_i = 1,$$

it follows from convexity that

$$\varphi(\sum_{i=1}^r \xi_i \cdot p_i) \leq \sum_{i=1}^r \varphi(\xi_i) \cdot p_i = \int_\Xi \varphi(\xi) \tilde{P}(d\xi).$$

Defining a sequence of *simple functions* $\{\xi^\nu\}$ on Ξ such that the sequences $\{\xi^\nu\}$ and $\{\varphi(\xi^\nu)\}$ are *mean fundamental* (w.r.t. P), that $\xi^\nu \xrightarrow{\text{in } P} \xi$, and that $\varphi(\xi^\nu) \xrightarrow{\text{in } P} \varphi(\xi)$ (see Halmos [13]), implies the hypothesis. $\qquad\square$

On the other hand, for $\varphi : \Xi \longrightarrow \mathbb{R}$ a convex function on $\Xi := [\alpha_0, \alpha_1] \subset \mathbb{R}$, we have

Proposition 2.2 (Edmundson–Madansky inequality [9, 26]) *With $\bar\xi := E\xi$, for the*
(E–M) distribution

$$(2.3) \qquad \left\{ \left(\alpha_0, p_{\alpha_0} = \frac{\alpha_1 - \bar\xi}{\alpha_1 - \alpha_0} \right), \left(\alpha_1, p_{\alpha_1} = \frac{\bar\xi - \alpha_0}{\alpha_1 - \alpha_0} \right) \right\}$$

holds the (E–M) inequality

$$(2.4) \qquad \int_\Xi \varphi(\xi) P(d\xi) \le \varphi(\alpha_0) \cdot p_{\alpha_0} + \varphi(\alpha_1) \cdot p_{\alpha_1}.$$

Proof: For $\xi \in [\alpha_0, \alpha_1]$ holds

$$\xi = \frac{\alpha_1 - \xi}{\alpha_1 - \alpha_0} \cdot \alpha_0 + \frac{\xi - \alpha_0}{\alpha_1 - \alpha_0} \cdot \alpha_1$$

and hence, due to convexity,

$$\varphi(\xi) \le \varphi(\alpha_0) \cdot \frac{\alpha_1 - \xi}{\alpha_1 - \alpha_0} + \varphi(\alpha_1) \cdot \frac{\xi - \alpha_0}{\alpha_1 - \alpha_0}.$$

Integrating this inequality yields

$$\int_\Xi \varphi(\xi) P(d\xi) \le \varphi(\alpha_0) \cdot \frac{\alpha_1 - \bar\xi}{\alpha_1 - \alpha_0} + \varphi(\alpha_1) \cdot \frac{\bar\xi - \alpha_0}{\alpha_1 - \alpha_0}.$$

and hence the hypothesis. $\qquad\square$

3 Bounds for Multivariate Integrals

Let $\Xi := \bigtimes_{i=1}^K [\alpha_{i0}, \alpha_{i1}] \subset \mathbb{R}^K$, $\varphi : \Xi \longrightarrow \mathbb{R}$ convex. Again we have

Proposition 3.1 (Jensen inequality [15]) *For $\bar\xi := E\xi$ holds*

$$(3.1) \qquad \varphi(\bar\xi) \le \int_\Xi \varphi(\xi) P(d\xi),$$

reducing to an equality if φ is linear affine on Ξ.

Proof: In analogy to Prop. 2.1 □

Consider now for $\xi \in \Xi$ the *one-point-distribution* P_ξ : $P_\xi(\xi) = 1$. The corresponding E–M distributions per component, $P_{\eta_i}^{\{\xi\}}, i = 1, \cdots, K$, are

$$p(\alpha_{i0}; \xi_i) := \quad P_{\eta_i}^{\{\xi\}}(\eta_i = \alpha_{i0}) \quad = \quad \frac{\alpha_{i1} - \xi_i}{\alpha_{i1} - \alpha_{i0}}$$

$$p(\alpha_{i1}; \xi_i) := \quad P_{\eta_i}^{\{\xi\}}(\eta_i = \alpha_{i1}) \quad = \quad (-1)\frac{\alpha_{i0} - \xi_i}{\alpha_{i1} - \alpha_{i0}}.$$

Denoting the vertices of Ξ by a^ν, with $\nu = (\nu_1, \cdots, \nu_K)^T$ and $\nu_i \in \{0, 1\}$, such that $a_i^\nu = \alpha_{i\nu_i}$, and assuming the components η_i of η to be stochastically independent, we get the joint distribution $P_\eta^{\{\xi\}}$ as

$$(3.2) \qquad p(a^\nu; \xi) := P_\eta^{\{\xi\}}(a^\nu) = \frac{\prod_{i=1}^K (-1)^{\nu_i} (\alpha_{i\bar\nu_i} - \xi_i)}{\prod_{i=1}^K (\alpha_{i1} - \alpha_{i0})},$$

where $\bar\nu_i = 1 - \nu_i$.
Observing that $E_{\eta_i}^{\{\xi\}}\eta_i = \xi_i$ and hence

$$(3.3) \qquad E_\eta^{\{\xi\}}\eta = \sum_\nu a^\nu p(a^\nu; \xi) = \xi,$$

we have from Jensen's inequality

Proposition 3.2 *For $\xi \in \Xi$ holds*

$$(3.4) \qquad \begin{aligned} \varphi(\xi) \le \int_\Xi \varphi(\eta) P_\eta^{\{\xi\}}(d\eta) \quad &= \sum_\nu \varphi(a^\nu) P_\eta^{\{\xi\}}(a^\nu) \\ &= \sum_\nu \varphi(a^\nu) p(a^\nu; \xi). \end{aligned}$$

If φ is linear on Ξ, then

$$(3.5) \qquad \begin{aligned} \varphi(\xi) \quad &= \sum_\nu \varphi(a^\nu) P_\eta^{\{\xi\}}(a^\nu) \\ &= \sum_\nu \varphi(a^\nu) p(a^\nu; \xi). \end{aligned}$$

From Prop. 3.2 follows immediately

Proposition 3.3 (Edmundson–Madansky inequality) *For*

$$(3.6) \qquad P^0(a^\nu) = \int_\Xi P_\eta^{\{\xi\}}(a^\nu) P(d\xi) = \int_\Xi p(a^\nu; \xi) P(d\xi)$$

holds the E–M inequality

(3.7)
$$\int_\Xi \varphi(\xi) P(d\xi) \le \sum_\nu \varphi(a^\nu) P^0(a^\nu).$$

If φ is linear on Ξ, then

(3.8)
$$\int_\Xi \varphi(\xi) P(d\xi) = \sum_\nu \varphi(a^\nu) P^0(a^\nu).$$

Observe that from (3.3) and (3.6) follows

(3.9)
$$\begin{cases} \displaystyle\sum_\nu a^\nu P^0(a^\nu) &= \displaystyle\int_\Xi \sum_\nu a^\nu p(a^\nu;\xi) P(d\xi) \\[3mm] &= \displaystyle\int_\Xi \xi P(d\xi) = \overline{\xi}. \end{cases}$$

To get the E–M distribution $P^0(a^\nu)$ explicitly we have to distinguish whether the components $\{\xi_1,\cdots,\xi_K\}$ of ξ are stochastically independent or not.

For the independent case we have due to Kall-Stoyan [21]

Proposition 3.4 *If the components of ξ are independent, then the E–M distribution is*

(3.10)
$$P^0(a^\nu) = \frac{\prod_{i=1}^K (-1)^{\nu_i}(\alpha_{i\overline{\nu}_i} - \overline{\xi}_i)}{\prod_{i=1}^K (\alpha_{i1} - \alpha_{i0})}.$$

Proof: Due to (3.2) and Prop. 3.3 we have according to the assumed independence

$$\begin{aligned} P^0(a^\nu) &= \int_\Xi p(a^\nu;\xi) P(d\xi) \\[3mm] &= \int_\Xi \frac{\prod_{i=1}^K (-1)^{\nu_i}(\alpha_{i\overline{\nu}_i} - \xi_i)}{\prod_{i=1}^K (\alpha_{i1} - \alpha_{i0})} P(d\xi) \\[3mm] &= \frac{\prod_{i=1}^K (-1)^{\nu_i}(\alpha_{i\overline{\nu}_i} - \overline{\xi}_i)}{\prod_{i=1}^K (\alpha_{i1} - \alpha_{i0})} \end{aligned}$$

$\square$

For the dependent case Frauendorfer [10] has derived

Proposition 3.5 *If the components of ξ are dependent, then the E–M distribution is*

$$P^0(a^\nu) = \frac{\prod_{i=1}^K (-1)^{\nu_i}(\alpha_{i\bar\nu_i} - \bar\xi_i)}{\prod_{i=1}^K (\alpha_{i1} - \alpha_{i0})}$$

(3.11)

$$+ \frac{\sum_{\Lambda \in \mathcal{B}} \delta_{\overline\Lambda}(\bar\nu) \prod_{i \in \overline\Lambda} \alpha_{i\bar\nu_i} \delta_\Lambda(\nu) \rho_\Lambda}{\prod_{i=1}^K (\alpha_{i1} - \alpha_{i0})},$$

where

$\mathcal{B}$ *the set of all subsets of* $\{1, \cdots, K\}$,
$\delta_\Lambda(\nu) := \prod_{i \in \Lambda} (-1)^{\bar\nu_i}$, $\Lambda \in \mathcal{B}$, *and* $\overline\Lambda := \{1, \cdots, K\}\backslash\Lambda$,
$m_\Lambda := \int_\Xi (\prod_{i \in \Lambda} \xi_i) P(d\xi)$, $\Lambda \in \mathcal{B}$,
$\rho_\Lambda := m_\Lambda - \prod_{i \in \Lambda} \bar\xi_i$, $\Lambda \in \mathcal{B}$.

Proof: see Frauendorfer [10] □

Remark 3.1 *For the components $\xi_1, \cdots, \xi_K$ being independent we have $\rho_\Lambda = 0\ \forall \Lambda \in \mathcal{B}$. Hence, in this case (3.11) coincides with (3.10).*

Defining the class $\mathcal{P}$ of probability measures on Ξ as the set of distributions having the joint moments m_Λ, $\Lambda \in \mathcal{B}$, i.e.

$$\mathcal{P} := \{P \mid \int_\Xi (\prod_{i \in \Lambda} \xi_i) P(d\xi) = m_\Lambda\ \forall \Lambda \in \mathcal{B}\},$$

and introducing for probability measures the partial ordering $<^{(c)}$ by

$$P <^{(c)} Q \Longleftrightarrow \int_\Xi \psi(\xi) P(d\xi) \le \int_\Xi \psi(\xi) Q(d\xi)\ \forall\ convex\ \psi : \Xi \longrightarrow \mathbb{R}$$

(Stoyan [31]), we see that $P^0 \in \sup^{(c)} \mathcal{P}$, i.e. P^0 solves the moment problem

$$\max\nolimits^{(c)}\{P \mid P \in \mathcal{P}\}.$$

Moreover it was shown (Kall [18]), that $\sup^{(c)} \mathcal{P}$ is a singleton, i.e. that $\max^{(c)}\{P \mid P \in \mathcal{P}\}$ is uniquely determined by (3.11).

□

4 Bounds on Simplices

If instead of $\Xi := \bigtimes_{i=1}^K [\alpha_{i0}, \alpha_{i1}]$ we have the simplex $\Delta = \text{conv}\{d_0, d_1, \cdots, d_K\}$ (with $d_0, d_1, \cdots, d_K$ being affine independent) containing supp P, then for any $\xi \in \Delta$ the system of linear equations

$$\begin{array}{ccccccc} p_0(\xi) & + & p_1(\xi) & + & \cdots & + & p_K(\xi) & = & 1 \\ d_0 p_0(\xi) & + & d_1 p_1(\xi) & + & \cdots & + & d_K p_K(\xi) & = & \xi, \end{array}$$

or briefly the system

$$Dp(\xi) = \begin{pmatrix} 1 \\ \xi \end{pmatrix}$$

with

$$D = \begin{pmatrix} 1 & 1 & \cdots & 1 \\ d_0 & d_1 & \cdots & d_K \end{pmatrix},$$

has the unique solution

$$p(\xi) = D^{-1} \begin{pmatrix} 1 \\ \xi \end{pmatrix} \geq 0.$$

Hence for $\varphi : \Delta \longrightarrow \mathbb{R}$ being convex we have

$$\varphi(\xi) \leq \sum_{i=0}^{K} p_i(\xi)\varphi(d_i),$$

yielding the following version of the E–M inequality:

Proposition 4.1 *For* $\Delta \supset \text{supp}\, P$ *and* $\varphi : \Delta \longrightarrow \mathbb{R}$ *being convex it holds*

$$(4.1) \qquad \int_\Delta \varphi(\xi) P(d\xi) \leq \sum_{i=0}^{K} \varphi(d_i) P_i^\Delta,$$

where the E–M distribution on the vertices of Δ *is given by*

$$(4.2) \qquad \begin{cases} P^\Delta &= \displaystyle\int_\Delta D^{-1} \begin{pmatrix} 1 \\ \xi \end{pmatrix} P(d\xi) \\[2em] &= D^{-1} \begin{pmatrix} 1 \\ \bar{\xi} \end{pmatrix}. \end{cases}$$

Therefore, to determine the E–M distribution requires just to solve the linear equations

$$DP^\Delta = \begin{pmatrix} 1 \\ \bar{\xi} \end{pmatrix}$$

involving $\bar{\xi}$ but not the higher order mixed moments (see Frauendorfer [11]). However, if $\Delta \supset \Xi = \text{supp}\, P$, we have to expect for the E–M bounds that

$$\overline{\varphi}^\Delta := \sum_{i=0}^{K} \varphi(d_i) P_i^\Delta \geq \overline{\varphi}^0 := \sum_\nu \varphi(a^\nu) P^0(a^\nu).$$

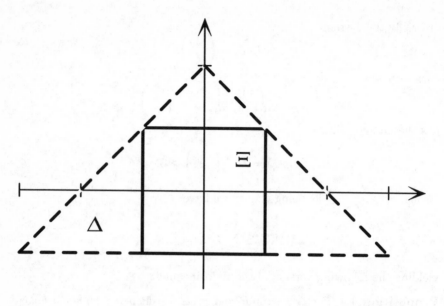

Figure 1: Cube vs. Simplex

Example 4.1 *Assume that*
$\Xi := [-1, 1] \times [-1, 1]$ *and* $\Delta := \operatorname{conv} \{(3, -1)^{\mathrm{T}}, (0, 2)^{\mathrm{T}}, (-3, -1)^{\mathrm{T}}\}$.
Assume further, that P *is the uniform distribution on* Ξ, *and that* $\varphi(\xi) = \xi_1^2 + \xi_2^2$.
Then we have

$$D = \begin{pmatrix} 1 & 1 & 1 \\ 3 & 0 & -3 \\ -1 & 2 & -1 \end{pmatrix} \quad and \quad D^{-1} = \begin{pmatrix} \frac{1}{3} & \frac{1}{6} & -\frac{1}{6} \\ \frac{1}{3} & 0 & \frac{1}{3} \\ \frac{1}{3} & -\frac{1}{6} & -\frac{1}{6} \end{pmatrix},$$

yielding

$$P^\Delta = \left(\frac{1}{3}, \frac{1}{3}, \frac{1}{3}\right)^{\mathrm{T}}.$$

On the other hand obviously $P^0(a^\nu) = \dfrac{1}{4} \ \forall \nu$. *Hence with*

$$\varphi(a^\nu) = 2 \ \forall \nu, \quad \varphi(d_0) = \varphi(d_2) = 10, \quad \varphi(d_1) = 4$$

we get $\overline{\varphi}^\Delta = 8$ *whereas* $\overline{\varphi}^0 = 2$. □

5 Generalized Moment Problems

Observe that for a K-dimensional interval Ξ as well as for a simplex Δ the E–M distribution does not depend on the particular convex function φ. However

- we need the first moments $\overline{\xi}_i$ to compute P^Δ on Δ;

- we need all mixed moments m_Λ to find P^0 on Ξ;

- for supp $P \subset \Xi \subset \Delta$ holds for the E–M upper bounds $\overline{\varphi}^0 \leq \overline{\varphi}^\Delta$ for any convex φ.

To avoid the computation of the 2^K mixed moments, we may find another upper bound as follows:

For a probability measure $\tilde{P}$ with $E_{\tilde{P}}\xi = \overline{\xi}$ and supp $\tilde{P}$ being contained in any convex polyhedron

$$\mathcal{B} := \text{conv}\{v_0, v_1, \cdots, v_r\} \subset \mathbb{R}^K \quad (v_i \text{ the vertices}),$$

such that $\overline{\xi} \in \text{int}\, \mathcal{B}$, consider the class of probability measures

$$\mathcal{P}_\mathcal{B} := \{P \mid P(\mathcal{B}) = 1,\ E_P\xi = \overline{\xi}\}.$$

Due to well known duality statements for semi-infinte LP's and their dual (generalized) moment problems (Karlin-Studden [23] and Krein-Nudel'man [25]) we have that for any convex continuous function $\varphi : \mathcal{B} \longrightarrow \mathbb{R}$

$$\max_{P \in \mathcal{P}_\mathcal{B}} E_P\varphi(\xi) = \inf\{\sum_{i=1}^{K} \overline{\xi}_i t_i + t_0 \mid \sum_{i=1}^{K} \xi_i t_i + t_0 \geq \varphi(\xi)\ \forall \xi \in \mathcal{B}\}.$$

Defining $t := (t_0, t_1, \cdots, t_K)^T$, we get due to the convexity of φ (see Dupačová [4, 5, 6])

Proposition 5.1 *An upper bound $\overline{\varphi}^\star := \max_{P \in \mathcal{P}} E_P\varphi$ and the corresponding extremal distribution $P^\star = \{p_0^\star, \cdots, p_r^\star\}$ (on the vertices of $\mathcal{B}$) result from solving*

(5.1) $\begin{aligned} &\min\{(1, \overline{\xi}^T)t \mid (1, v_j^T)t \geq \varphi(v_j),\ j = 0, \cdots, r\} = \\ &= \max\{\sum_{j=0}^{r} p_j\varphi(v_j) \mid \sum_{j=0}^{r} p_j v_j = \overline{\xi},\ \sum_{j=0}^{r} p_j = 1,\ p_j \geq 0\ \forall j\}. \end{aligned}$

$\overline{\varphi}^\star$ and $P^\star$ depend on $\overline{\xi}$ *and* on φ.

If in particular $\mathcal{B}$ is an interval $\Xi = \text{conv}\{v_0, \cdots, v_{2^K-1}\}$, then the E–M distribution P^0 is feasible in (5.1) and hence $E_P\varphi(\xi) \leq \overline{\varphi}^0 \leq \overline{\varphi}^\star$. But to solve the LP (5.1) may be cheaper than to compute all mixed moments.

Example 5.1 *As in Example 4.1 we consider*
$\Xi := [-1, 1] \times [-1, 1]$ *with the uniform distribution.*
Numbering the vertices v^j counter-clockwise, starting with $v^0 = (1, -1)$, we

know that $P_0^0 = \cdots = P_3^0 = \frac{1}{4}$.
Choosing $\varphi(\xi) := \xi_1^2 + \xi_1\xi_2 + \xi_2^2$ we have

$$\varphi(v^j) = \begin{cases} 1 & for \quad j = 0 \\ 3 & for \quad j = 1 \\ 1 & for \quad j = 2 \\ 3 & for \quad j = 3 \end{cases}$$

and hence $\overline{\varphi}^0 = 2$, whereas with $p_1^\star = p_3^\star = \frac{1}{2}$, $p_0^\star = p_2^\star = 0$ follows $\overline{\varphi}^\star = 3$. $\square$

The particular case, to construct an upper bound $\overline{\varphi}^\star$ for $E_P\varphi(\xi)$ by solving the moment problem $\max_{P \in \mathcal{P}_\mathcal{B}} E_P\varphi$ under the moment conditions $\mathcal{P}_\mathcal{B} := \{P \mid P(\mathcal{B}) = 1,\ E_P\xi = \overline{\xi}\}$ to get $P^\star$, can be generalized in principle. This involves the duality statements for semi-infinite LP's mentioned above.

For probability measures P with $\operatorname{supp} P \subset \Theta \subset \mathbb{R}^K$ let

$$a : \Theta \longrightarrow \mathbb{R}^N, \quad \varphi : \Theta \longrightarrow \mathbb{R}$$

be measurable and $D := \operatorname{conv} a(\Theta)$, $m \in \mathbb{R}^N$. Then for the pair of dual problems

$$(5.2) \quad \begin{cases} v_{\text{prim}} := \inf_{y_0,y}\{y_0 + m^{\mathrm{T}}y \mid y_0 + a(\xi)^{\mathrm{T}}y \geq \varphi(\xi)\ \forall \xi \in \Theta\}, \\ v_{\text{dual}} := \sup_P \left\{ \int_\Theta \varphi(\xi)P(d\xi) \mid \int_\Theta a(\xi)P(d\xi) = m,\ \int_\Theta P(d\xi) = 1 \right\} \end{cases}$$

we have due to Kemperman [24]

Theorem 5.1 *If $m \in \operatorname{int} D$ then $v_{\text{prim}} = v_{\text{dual}}$; and for $v_{\text{dual}} < \infty$, v_{prim} is attained for some $(y_0, y^{\mathrm{T}})^{\mathrm{T}}$. The conditions*

i) *v_{dual} is attained;*

ii) *$\exists$ a primal feasible $(y_0, y^{\mathrm{T}})^{\mathrm{T}}$ such that $\exists \xi^i \in \Theta$, $i = 1, \cdots, \nu \geq 1$, for which $y_0 + a(\xi^i)^{\mathrm{T}}y = \varphi(\xi^i)\ \forall i$ and $m \in \operatorname{conv}\{a(\xi^1), \cdots, a(\xi^\nu)\}$;*

are equivalent.

From Richter [27] and Rogosinski [30] (see also Kemperman [24]) we know

Theorem 5.2 *For $f_1, \cdots, f_N$ being integrable functions on the probability space $(\Omega, \mathcal{F}, P)$, there exists a probability measure $\tilde{P}$ with finite support in Ω such that*

$$\int_\Omega f_i(\omega)P(d\omega) = \int_\Omega f_i(\omega)\tilde{P}(d\omega),\ i = 1, \cdots, N,$$

with $\operatorname{card}(\operatorname{supp}\tilde{P}) \leq N + 1$.

Remark 5.1 *Due to Theor. 5.2 we may restrict ourselves to discrete distributions in Θ. If we can satisfy condition ii) of Theor. 5.1, we have solved both problems in (5.2) at once since*

$$m \in \text{conv}\,\{a(\xi_1), \cdots, a(\xi_\nu)\} \implies \exists p_i \geq 0 : m = \sum_{i=1}^{\nu} a(\xi_i)p_i, \; \sum_{i=1}^{\nu} p_i = 1,$$

i.e. $(\xi_1, \cdots, \xi_\nu; \; p_1, \cdots, p_\nu)$ *is feasible for the dual in (5.2), implying*

$$v_{\text{dual}} \geq \sum_{i=1}^{\nu} \varphi(\xi_i)p_i = \sum_{i=1}^{\nu} (y_0 + a(\xi_i)^{\mathrm{T}}y)p_i = y_0 + m^{\mathrm{T}}y \geq v_{\text{prim}},$$

whereas by weak duality $v_{\text{dual}} \leq v_{\text{prim}}$. $\qquad\qquad\qquad\qquad\qquad\qquad\square$

As an instance for this approach consider a distribution $\hat{P}$ with

$$\mu := \int_{\mathrm{IR}^K} \xi \hat{P}(d\xi) \quad \text{and} \quad \rho := \int_{\mathrm{IR}^K} \|\xi\|^2 \hat{P}(d\xi),$$

$\|\cdot\|$ the Euclidean norm. Then choose for (5.2)

$$a(\xi) := \begin{pmatrix} \xi \\ \|\xi\|^2 \end{pmatrix} \quad \text{and} \quad m := \begin{pmatrix} \mu \\ \rho \end{pmatrix},$$

these moment conditions having been considered first by Dulá [3] for simplicial functions φ. The assumption of Theor. 5.1 may be checked (see Kall [20]) by

Lemma 5.1 *With a, m as above and $D := \text{conv}\,a(\mathrm{IR}^{K+1})$, it holds*

$$m \in \text{int}\, D \iff \rho > \|\mu\|^2.$$

Hence Theor. 5.1 applies iff the distribution $\hat{P}$ is not completely degenerate, i.e. iff $\sum_{i=1}^{K} \text{var}\,(\xi_i) > 0$.

Assuming that φ is nonlinear and is determined by

$$(5.3) \qquad\qquad \varphi(\xi) = \max_{1 \leq j \leq r} (d_j^{\mathrm{T}}\xi - f_j),$$

as it holds for the recourse function $Q(x, \xi) := \min_y \{q^{\mathrm{T}}y \mid Wy = h(\xi) - T(\xi)x, \; y \geq 0\}$ for any fixed x, the primal problem in (5.2) can be shown to be the *convex*[2] solvable program

$$(5.4) \quad \begin{cases} \quad v_{\text{prim}} = \inf\{y_0 + \mu^{\mathrm{T}}y + \rho y_{K+1}\} \\ \text{s.t.} \quad 4y_0 y_{K+1} + 4f_j y_{K+1} - \|d_j - y\|^2 \geq 0, \; j = 1, \cdots, r, \\ \quad y_{K+1} > 0. \end{cases}$$

For $(\tilde{y}_0, \tilde{y}, \tilde{y}_{K+1})$ solving (5.4), with $J \subset \{1, \cdots, r\}$ denoting the active constraints, follows (see Kall [20])

---

[2] i.e. convex objective and convex feasible set

Proposition 5.2 *For $\hat{\xi}^{(j)}$ determined as*

$$\hat{\xi}^{(j)} := \frac{d_j - \tilde{y}}{2\tilde{y}_{K+1}}, \quad j = 1, \cdots, r$$

holds

$$\begin{pmatrix} \mu \\ \rho \end{pmatrix} \in \text{conv} \left\{ \begin{pmatrix} \hat{\xi}^{(j)} \\ \|\hat{\xi}^{(j)}\|^2 \end{pmatrix} \mid j \in J \right\}.$$

Hence, by Theor. 5.1, with p_j solving

$$\begin{aligned} \sum_{j \in J} p_j &= 1 \\ \sum_{j \in J} p_j \hat{\xi}^{(j)} &= \mu \\ \sum_{j \in J} p_j \|\hat{\xi}^{(j)}\|^2 &= \rho \\ p_j &\geq 0, \ j \in J \end{aligned}$$

we get the upper bound

$$\sum_{j \in J} p_j \varphi(\hat{\xi}^{(j)}) \geq \int_{\mathbb{R}^K} \varphi(\xi) \hat{P}(d\xi).$$

Example 5.2 *Let $\hat{P}$ be the uniform distribution on $\Xi := [-1, 1] \times [-1, 1]$ such that*

$$\mu = \begin{pmatrix} 0 \\ 0 \end{pmatrix} \quad \text{and} \quad \rho = \frac{1}{3}.$$

With

$$\varphi(\xi) = \max\{\xi_1, \xi_2, -\xi_1, -\xi_2\}$$

and the notation of (5.3) we have $f_i = 0 \ \forall i$ and

$$d_1 = \begin{pmatrix} 1 \\ 0 \end{pmatrix}, \quad d_2 = \begin{pmatrix} 0 \\ 1 \end{pmatrix}, \quad d_3 = \begin{pmatrix} -1 \\ 0 \end{pmatrix}, \quad d_4 = \begin{pmatrix} 0 \\ -1 \end{pmatrix}.$$

Now problem (5.4) reads as

$$\begin{aligned} v_{\text{prim}} = \inf\{y_0 + \tfrac{1}{3}y_3\} \\ s.t. \quad 4y_0 y_3 &\geq (1 - y_1)^2 + y_2^2 \\ 4y_0 y_3 &\geq y_1^2 + (1 - y_2)^2 \\ 4y_0 y_3 &\geq (1 + y_1)^2 + y_2^2 \\ 4y_0 y_3 &\geq y_1^2 + (1 + y_2)^2 \\ y_3 &> 0 \end{aligned}$$

having the solution

$$\tilde{y}_0 = \frac{1}{2\sqrt{3}}, \ \tilde{y}_1 = \tilde{y}_2 = 0, \ \tilde{y}_3 = \frac{1}{2}\sqrt{3}.$$

From Prop. 5.2 we get the 'tangent points'

$$\hat{\xi}^{(1)} = \frac{1}{\sqrt{3}} \begin{pmatrix} 1 \\ 0 \end{pmatrix}, \quad \hat{\xi}^{(2)} = \frac{1}{\sqrt{3}} \begin{pmatrix} 0 \\ 1 \end{pmatrix},$$

$$\hat{\xi}^{(3)} = \frac{1}{\sqrt{3}} \begin{pmatrix} -1 \\ 0 \end{pmatrix}, \quad \hat{\xi}^{(4)} = \frac{1}{\sqrt{3}} \begin{pmatrix} 0 \\ -1 \end{pmatrix},$$

and then we have to solve the linear system

$$\begin{aligned}
\sum_{j \in J} p_j &= 1 \\
\sum_{j \in J} p_j \hat{\xi}^{(j)} &= \begin{pmatrix} 0 \\ 0 \end{pmatrix} \\
\sum_{j \in J} p_j \|\hat{\xi}^{(j)}\|^2 &= \frac{1}{3} \\
p_j &\geq 0, \quad j \in J,
\end{aligned}$$

yielding $p_j = \frac{1}{4}$, $j = 1, \cdots, 4$.
  Hence we get the upper bound

$$\sum_{j=1}^{4} \varphi(\hat{\xi}^{(j)}) p_j = \frac{1}{\sqrt{3}} = 0.57735,$$

whereas the E–M bound would yield (with $\tilde{p}_i = \frac{1}{4} \; \forall i$, $\tilde{\xi}^i$ the vertices of Ξ)

$$\sum_{i=1}^{4} \varphi(\tilde{\xi}^i) \tilde{p}_i = 1.$$

$\square$

For more on bounds and related moment problems see e.g. Edirisinghe and Ziemba [7, 8], Gassmann and Ziemba [12], Huang et al. [14], and Kall [19, 20].

6   Improving Bounds

Consider again the interval $\Xi := \bigtimes_{i=1}^{K} [\alpha_{i0}, \alpha_{i1}] \subset \mathbb{R}^K$ containing supp P, and a convex function $\varphi : \Xi \longrightarrow \mathbb{R}$. Using the Jensen (3.1) and the E–M (3.7) inequalities we know that

$$\varphi(\bar{\xi}) \leq \int_{\Xi} \varphi(\xi) P(d\xi) \leq \sum_{\nu} \varphi(a^\nu) P^0(a^\nu),$$

where linearity of φ on Ξ implies equality on both sides. Dividing $[\alpha_{10}, \alpha_{11}]$ at some $\beta \in (\alpha_{10}, \alpha_{11})$ into the to disjoint parts $[\alpha_{10}, \beta)$ and $[\beta, \alpha_{11}]$, we get the two disjoint intervals $\Xi^1 := [\alpha_{10}, \beta) \times \mathsf{X}_{i=2}^K [\alpha_{i0}, \alpha_{i1}]$ and $\Xi^2 := [\beta, \alpha_{11}] \times \mathsf{X}_{i=2}^K [\alpha_{i0}, \alpha_{i1}]$ with $\Xi^1 \cup \Xi^2 = \Xi$, i.e. we have *partitioned* Ξ into two intervals.

Using the conditional distributions $P^{(i)} := P(\cdot \mid \Xi^i)$ and conditional expectations $\overline{\xi}^{(i)} := E(\xi \mid \Xi^i)$, $i = 1, 2$, we get from the Jensen and E–M inequalities

$$\varphi(\overline{\xi}^{(i)}) \leq \int_{\Xi^i} \varphi(\xi) P^{(i)}(d\xi) \leq \sum_\nu \varphi(a^{(i)\nu}) P^{(i)0}(a^{(i)\nu}), \quad i = 1, 2,$$

where $a^{(i)\nu}$ are the vertices of Ξ^i and $P^{(i)0}(a^{(i)\nu})$ are the E–M probabilities derived from $P(\cdot \mid \Xi^i)$. With $p_i := P(\Xi^i)$ we have $\overline{\xi} = p_1 \overline{\xi}^{(1)} + p_2 \overline{\xi}^{(2)}$ and hence due to Jensen

$$\begin{aligned} \varphi(\overline{\xi}) &\leq p_1 \varphi(\overline{\xi}^{(1)}) + p_2 \varphi(\overline{\xi}^{(2)}) \\ &\leq p_1 \int_{\Xi^1} \varphi(\xi) P^{(1)}(d\xi) + p_2 \int_{\Xi^2} \varphi(\xi) P^{(2)}(d\xi). \end{aligned}$$

Further it is obvious that

$$\begin{aligned} \int_\Xi \varphi(\xi) P(d\xi) &= p_1 \int_{\Xi^1} \varphi(\xi) P^{(1)}(d\xi) + p_2 \int_{\Xi^2} \varphi(\xi) P^{(2)}(d\xi) \\ &\leq p_1 \sum_\nu \varphi(a^{(1)\nu}) P^{(1)0}(a^{(1)\nu}) + p_2 \sum_\nu \varphi(a^{(2)\nu}) P^{(2)0}(a^{(2)\nu}). \end{aligned}$$

And finally, due to (3.9) we have $\sum_\nu a^{(i)\nu} P^{(i)0}(a^{(i)\nu}) = \overline{\xi}^{(i)}$ and hence, since $\overline{\xi} = p_1 \overline{\xi}^{(1)} + p_2 \overline{\xi}^{(2)}$, from E–M follows

$$\begin{aligned} p_1 \sum_\nu \varphi(a^{(1)\nu}) P^{(1)0}(a^{(1)\nu}) &+ p_2 \sum_\nu \varphi(a^{(2)\nu}) P^{(2)0}(a^{(2)\nu}) \\ &\leq \sum_\nu \varphi(a^\nu) P^0(a^\nu). \end{aligned}$$

Putting these together we have

Proposition 6.1 *Partitioning the interval Ξ into Ξ^1 and Ξ^2, the lower bound (Jensen) is increased and the upper bound (E–M) is decreased to*

$$(6.1) \quad \begin{cases} p_1 \varphi(\overline{\xi}^{(1)}) + p_2 \varphi(\overline{\xi}^{(2)}) \leq \int_\Xi \varphi(\xi) P(d\xi) \\ \\ \qquad \leq p_1 \sum_\nu \varphi(a^{(1)\nu}) P^{(1)0}(a^{(1)\nu}) + p_2 \sum_\nu \varphi(a^{(2)\nu}) P^{(2)0}(a^{(2)\nu}). \end{cases}$$

Example 6.1 *For $\Xi := [-1, 1] \times [-1, 1]$ with P the uniform distribution consider $\varphi(\xi) := \max[\{3\xi_1 - \xi_2\}; \{-2\xi_1 + \xi_2\}]$.*

Then $\overline{\xi} = (0; 0)^{\mathrm{T}}$ and the E–M probabilities for

$$\begin{aligned} a^1 &:= (1; -1)^{\mathrm{T}}, \quad a^2 := (1; 1)^{\mathrm{T}}, \\ a^3 &:= (-1; 1)^{\mathrm{T}}, \quad a^4 := (-1; -1)^{\mathrm{T}} \end{aligned}$$

are $P^0(a^\nu) = \frac{1}{4} \ \forall \nu$ with $\varphi(a^1) = 4$, $\varphi(a^2) = 2$, $\varphi(a^3) = 3$ and $\varphi(a^4) = 1$.
Hence we have

$$\varphi(\bar{\xi}) = 0 \le \int_\Xi \varphi(\xi)P(d\xi) \le \sum_{\nu=1}^4 \varphi(a^\nu)P^0(a^\nu) = \frac{5}{2}.$$

Partitioning Ξ at $\xi_1 = \bar{\xi}_1 = 0$ into Ξ^A and Ξ^B with $p_A = P(\Xi^A) = p_B = P(\Xi^B) = \frac{1}{2}$, we get

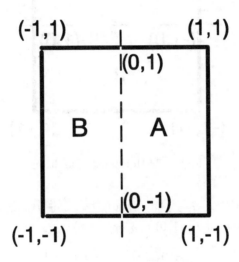

Figure 2: Vertical partition

$\bar{\xi}^{(A)} = (\frac{1}{2},0)^T$, $\bar{\xi}^{(B)} = (-\frac{1}{2},0)^T$, $P^{(A)0}(a^{(A)\nu}) = P^{(B)0}(a^{(B)\nu}) = \frac{1}{4} \ \forall \nu$, and
with the above values $\varphi(a^\nu)$ and $\varphi(0,1) = \varphi(0,-1) = 1$ follows

$$\varphi(\bar{\xi}^{(A)}) = \frac{3}{2}, \quad \varphi(\bar{\xi}^{(B)}) = 1;$$

and according to (6.1)

$$
\begin{aligned}
\frac{5}{4} &= p_A\varphi(\bar{\xi}^{(A)}) + p_B\varphi(\bar{\xi}^{(B)}) \\
&\le \int_\Xi \varphi(\xi)P(d\xi) \\
&\le p_A \sum_\nu \varphi(a^{(A)\nu})P^{(A)0}(a^{(A)\nu}) \\
&\quad + p_B \sum_\nu \varphi(a^{(B)\nu})P^{(B)0}(a^{(B)\nu}) = \frac{7}{4}.
\end{aligned}
$$

Partitioning Ξ at $\xi_2 = \bar{\xi}_2 = 0$ into Ξ^I and Ξ^{II} with $p_I = P(\Xi^I) = p_{II} =$

$P(\Xi^{II}) = \frac{1}{2}$, *we get*

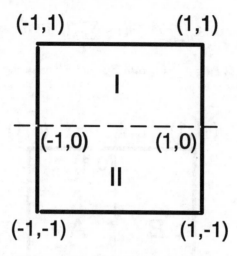

Figure 3: Horizontal partition

$\bar{\xi}^{(I)} = (0, \frac{1}{2})^{T}$, $\bar{\xi}^{(II)} = (0, -\frac{1}{2})^{T}$, $P^{(I)0}(a^{(I)\nu}) = P^{(II)0}(a^{(II)\nu}) = \frac{1}{4}$ $\forall \nu$, *and with* $\varphi(a^{\nu})$ *as above and* $\varphi(1, 0) = 3$, $\varphi(-1, 0) = 2$
follows

$$\varphi(\bar{\xi}^{(I)}) = \frac{1}{2}, \quad \varphi(\bar{\xi}^{(II)}) = \frac{1}{2}$$

and according to (6.1)

$$
\begin{aligned}
\frac{1}{2} &= p_{I}\varphi(\bar{\xi}^{(I)}) + p_{II}\varphi(\bar{\xi}^{(II)}) \\
&\leq \int_{\Xi} \varphi(\xi)P(d\xi) \\
&\leq p_{I} \sum_{\nu} \varphi(a^{(I)\nu})P^{(I)0}(a^{(I)\nu}) \\
&\quad + p_{II} \sum_{\nu} \varphi(a^{(II)\nu})P^{(II)0}(a^{(II)\nu}) = \frac{10}{4}.
\end{aligned}
$$

$\square$

If $\nabla\varphi(a^{\nu}) = \nabla\varphi(a^{\mu})$ $\forall \nu \neq \mu$, this implies linearity of φ on Ξ (by convexity). Due to Props. 3.1 and 3.3, then

$$\varphi(\bar{\xi}) = \int_{\Xi} \varphi(\xi)P(d\xi) = \sum_{\nu} \varphi(a^{\nu})P^{0}(a^{\nu}),$$

and hence we are done.

7 Convergence

We started with $\mathcal{S}_0 := \{\Xi_0\} := \{\Xi\}$ and constructed a partition $\mathcal{S}_1 := \{\Xi_1^1, \Xi_1^2\}$. We may continue by paritioning the same way Ξ_1^1 and/or Ξ_1^2 getting a refined partition $\mathcal{S}_2$ (i.e. for any element $\Xi_2^j \in \mathcal{S}_2$ exists an element $\Xi_1^l \in \mathcal{S}_1$ such that $\Xi_2^j \subset \Xi_1^l$). Thus we can construct a sequence of successive refinements $\{\mathcal{S}_k\}$. With $\delta(\Xi_k^j) := \sup\{\|\xi - \eta\| \mid \xi, \eta \in \Xi_k^j\}$ we define the width of the partition $\mathcal{S}_k$ as $\overline{\delta}(\mathcal{S}_k) := \max\{\delta(\Xi_k^j) \mid \Xi_k^j \in \mathcal{S}_k\}$. Hence $\{\overline{\delta}(\mathcal{S}_k)\}$ is a monotonically decreasing sequence.

Theorem 7.1 *Assume that*

- *supp* $P \subset \Xi$,

- *that* $\varphi : \Xi \longrightarrow \mathbb{R}$ *be continuous, convex and*

- *that* $\{\mathcal{S}_k\}$, $(\mathcal{S}_0 = \{\Xi\})$ *be a sequence of successively refined partitions such that* $\overline{\delta}(\mathcal{S}_k) \longrightarrow 0$.

Then for the corresponding sequences $\{\tilde{P}^{(k)}\}$ *of Jensen distributions and* $\{P^{(k)0}\}$ *of E–M distributions holds*

$$
(7.1) \quad \left\{
\begin{array}{ccc}
\displaystyle\int_\Xi \varphi(\xi)\tilde{P}^{(k)}(d\xi) & \longrightarrow & \displaystyle\int_\Xi \varphi(\xi)P(d\xi) \\[2mm]
\displaystyle\int_\Xi \varphi(\xi)P^{(k)0}(d\xi) & \longrightarrow & \displaystyle\int_\Xi \varphi(\xi)P(d\xi).
\end{array}
\right.
$$

Proof: Obviously φ is uniformly continuous on Ξ. Hence,

$$
\forall \varepsilon \;\; \exists \delta_\varepsilon : |\varphi(\xi) - \varphi(\eta)| < \varepsilon \;\; \forall \xi, \eta \in \Xi : \|\xi - \eta\| < \delta_\varepsilon.
$$

Since $\overline{\delta}(\mathcal{S}_k) \longrightarrow 0$, there

$$
\exists N(\delta_\varepsilon) : \overline{\delta}(\mathcal{S}_k) < \delta_\varepsilon \;\; \forall k \geq N(\delta_\varepsilon)
$$

such that for any $k \geq N(\delta_\varepsilon)$ and arbitrary $\Xi_k^j \in \mathcal{S}_k$ holds

$$
\sup_{\xi, \eta \in \Xi_k^j} |\varphi(\xi) - \varphi(\eta)| \leq \varepsilon.
$$

Observing that

$$
\int_{\Xi_k^j} \tilde{P}^{(k)}(d\xi) = \int_{\Xi_k^j} P(d\xi) = \int_{\Xi_k^j} P^{(k)0}(d\xi) \;\; \forall j
$$

and defining the simple functions $\underline{\varphi}_k, \overline{\varphi}_k : \Xi \longrightarrow \mathbb{R}$ by

$$
\left.
\begin{array}{rcl}
\underline{\varphi}_k(\xi) & := & \inf_{\eta \in \Xi_k^j} \varphi(\eta) \;\; \text{if } \xi \in \Xi_k^j \\[2mm]
\overline{\varphi}_k(\xi) & := & \sup_{\eta \in \Xi_k^j} \varphi(\eta) \;\; \text{if } \xi \in \Xi_k^j
\end{array}
\right\} \;\; \forall \xi \in \Xi,
$$

for which $0 \leq \overline{\varphi}_k(\xi) - \underline{\varphi}_k(\xi) \leq \varepsilon \ \forall \xi \in \Xi$ and $\forall k \geq N(\delta_\varepsilon)$, we get with $\underline{\xi}^{(k)} = E(\xi \mid S_k)$ and $\overline{\xi}^{(k)}$ attaining (only) vertices within the partition S_k:

$$
\begin{aligned}
\int_\Xi \underline{\varphi}_k(\xi) P(d\xi) &= \int_\Xi \underline{\varphi}_k(\xi) \tilde{P}^{(k)}(d\xi) &\leq& \int_\Xi \varphi(\underline{\xi}^{(k)}) \tilde{P}^{(k)}(d\xi) \\
&\leq \int_\Xi \varphi(\xi) P(d\xi) &\leq& \int_\Xi \varphi(\overline{\xi}^{(k)}) P^{(k)0}(d\xi) \\
&\leq \int_\Xi \overline{\varphi}_k(\xi) P^{(k)0}(d\xi) &=& \int_\Xi \overline{\varphi}_k(\xi) P(d\xi).
\end{aligned}
$$

Hence

$$
\left| \int_\Xi \varphi(\underline{\xi}^{(k)}) \tilde{P}^{(k)}(d\xi) - \int_\Xi \varphi(\xi) P(d\xi) \right| \text{ and } \left| \int_\Xi \overline{\varphi}_k(\xi) P^{(k)0}(d\xi) - \int_\Xi \varphi(\xi) P(d\xi) \right|
$$

are bounded by

$$
\int_\Xi \overline{\varphi}_k(\xi) P(d\xi) - \int_\Xi \underline{\varphi}_k(\xi) P(d\xi) =
$$

$$
= \int_\Xi |\overline{\varphi}_k(\xi) - \underline{\varphi}_k(\xi)| P(d\xi)
$$

$$
\leq \varepsilon
$$

$\square$

Theor. 7.1 states, in the terminology of Billingsley [1], the *weak convergence* of $\{\tilde{P}^{(k)}\}$ and $\{P^{(k)0}\}$ to P. For the recourse function $Q(x, \xi)$ as defined in Sec. 1 this implies the *epi-convergence* of $\underline{Q}_k(x) := \int_\Xi Q(x, \underline{\xi}^{(k)}) \tilde{P}^{(k)}(d\xi)$ and $\overline{Q}_k(x) := \int_\Xi Q(x, \overline{\xi}^{(k)}) P^{(k)0}(d\xi)$ to $Q(x) = \int_\Xi Q(x, \xi) P(d\xi)$, which in turn ensures that

$$
\begin{aligned}
\min_{x \in X} \{c^T x + \underline{Q}_k(x)\} &\longrightarrow \min_{x \in X} \{c^T x + Q(x)\}, \\
\min_{x \in X} \{c^T x + \overline{Q}_k(x)\} &\longrightarrow \min_{x \in X} \{c^T x + Q(x)\},
\end{aligned}
$$

and for arbitrary accumulation points $\underline{x}$ and $\overline{x}$ of $\{\underline{x}^{(k)} \mid \underline{x}^{(k)} \in \arg\min_{x \in X}[c^T x + \underline{Q}_k(x)]\}$ and $\{\overline{x}^{(k)} \mid \overline{x}^{(k)} \in \arg\min_{x \in X}[c^T x + \overline{Q}_k(x)]\}$, respectively, hold $\underline{x} \in \arg\min_{x \in X}[c^T x + Q(x)]\}$ and $\overline{x} \in \arg\min_{x \in X}[c^T x + Q(x)]\}$. (See e.g. Birge-Wets [2], Kall [16, 17], Robinson [28], Robinson-Wets [29] and Wets [32]).

References

[1] P. Billingsley. *Convergence of Probability Measures*. Wiley, New York, 1968.

[2] J.R. Birge and R.J.-B. Wets. Designing approximation schemes for stochastic optimization problems, in particular for stochastic programs with recourse. *Math. Prog. Study*, 27:54–102, 1986.

[3] J.H. Dulá. An upper bound on the expectation of simplicial functions of multivariate random variables. *Math. Prog.*, 55:69–80, 1992.

[4] J. Dupačová. Minimax stochastic programs with nonconvex nonseparable penalty functions. In A. Prékopa, editor, *Progress in Operations Research*, pages 303–316. North-Holland Publ. Co., 1976.

[5] J. Dupačová. Minimax stochastic programs with nonseparable penalties. In K. Iracki, K. Malanowski, and S. Walukiewicz, editors, *Optimization Techniques, Part I*, volume 22 of *Lecture Notes in Contr. Inf. Sci.*, pages 157–163, Berlin, 1980. Springer-Verlag.

[6] J. Dupačová. The minimax approach to stochastic programming and an illustrative application. *Stochastics*, 20:73–88, 1987.

[7] N.C.P. Edirisinghe and W.T. Ziemba. Tight bounds for stochastic convex programs. *Oper. Res.*, 40:660–677, 1992.

[8] N.C.P. Edirisinghe and W.T. Ziemba. Bounds for two-stage stochastic programs with fixed recourse. *Math. Oper. Res.*, 19:292–313, 1994.

[9] H.P. Edmundson. Bounds on the expectation of a convex function of a random variable. Technical Report Paper 982, The RAND Corporation, 1956.

[10] K. Frauendorfer. Solving SLP recourse problems with arbitrary multivariate distributions—the dependent case. *Math. Oper. Res.*, 13:377–394, 1988.

[11] K. Frauendorfer. *Stochastic Two-Stage Programming*, volume 392 of *Lecture Notes in Econ. Math. Syst.* Springer-Verlag, Berlin, 1992.

[12] H. Gassmann and W.T. Ziemba. A tight upper bound for the expectation of a convex function of a multivariate random variable. *Math. Prog. Study*, 27:39–53, 1986.

[13] P.R. Halmos. *Measure Theory*. D. van Nostrand, Princeton, New Jersey, 1950.

[14] C.C. Huang, W.T. Ziemba, and A. Ben-Tal. Bounds on the expectation of a convex function of a random variable: With applications to stochastic programming. *Oper. Res.*, 25:315–325, 1977.

[15] J.L. Jensen. Sur les fonctions convexes et les inégalités entre les valeurs moyennes. *Acta Math.*, 30:173–177, 1906.

[16] P. Kall. Approximation to optimization problems: An elementary review. *Math. Oper. Res.*, 11:9–18, 1986.

[17] P. Kall. On approximations and stability in stochastic programming. In J. Guddat, H. Th. Jongen, B. Kummer, and F. Nožička, editors, *Parametric Optimization and Related Topics*, pages 387–407. Akademie-Verlag, Berlin, 1987.

[18] P. Kall. Stochastic programs with recourse: An upper bound and the related moment problem. *ZOR*, 31:A119–A141, 1987.

[19] P. Kall. Stochastic programming with recourse: Upper bounds and moment problems—A review. In J. Guddat, B. Bank, H. Hollatz, P. Kall, D. Klatte, B. Kummer, K. Lommatzsch, K. Tammer, M. Vlach, and K. Zimmermann, editors, *Advances in Mathematical Optimization (Dedicated to Prof. Dr. Dr. hc. F. Nožička)*, pages 86–103. Akademie-Verlag, Berlin, 1988.

[20] P. Kall. An upper bound for SLP using first and total second moments. *Ann. Oper. Res.*, 30:267–276, 1991.

[21] P. Kall and D. Stoyan. Solving stochastic programming problems with recourse including error bounds. *Math. Operationsforsch. Statist., Ser. Opt.*, 13:431–447, 1982.

[22] P. Kall and S. W. Wallace. *Stochastic Programming*. John Wiley & Sons, Chichester, 1994.

[23] S. Karlin and W.J. Studden. *Tschebycheff Systems: With Applications in Analysis and Statistics*. Interscience Publ., New York, 1966.

[24] J.M.B. Kemperman. The general moment problem, a geometric approach. *Ann. Math. Statist.*, 39:93–122, 1968.

[25] M.G. Krein and A.A. Nudel'man. *The Markov Moment Problem and Extremal Problems*, volume 50 of *Transactions of Math. Mon.* AMS, 1977.

[26] A. Madansky. Bounds on the expectation of a convex function of a multivariate random variable. *Ann. Math. Statist.*, 30:743–746, 1959.

[27] H. Richter. Parameterfreie Abschätzung und Realisierung von Erwartungswerten. *Blätter Dt. Ges. Versicherungsmath.*, 3:147–161, 1957.

[28] S.M. Robinson. Local epi-continuity and local optimization. *Math. Prog.*, 37:208–222, 1987.

[29] S.M. Robinson and R.J.-B. Wets. Stability in two stage programming. *SIAM J. Contr. Opt.*, 25:1409–1416, 1987.

[30] W.W. Rogosinski. Moments of nonnegative mass. *Proc. Roy. Soc. London*, A 245:1–27, 1958.

[31] D. Stoyan. *Comparison Methods for Queues and other Stochastic Models*. John Wiley & Sons, New York, 1983. Revised English Edition by D.J. Daley of *Qualitative Eigenschaften und Abschätzungen stochastischer Modelle*, Oldenbourg-Verlag, München, 1977.

[32] R. Wets. Stochastic programming: Solution techniques and approximation schemes. In A. Bachem, M. Grötschel, and B. Korte, editors, *Mathematical Programming: The State-of-the-Art, Bonn 1982*, pages 566–603. Springer-Verlag, Berlin, 1983.

Optimal Power Generation under Uncertainty via Stochastic Programming *

Darinka Dentcheva and Werner Römisch
Humboldt-Universität zu Berlin, Institut für Mathematik
Unter den Linden 6, 10099 Berlin, Germany

Abstract: A power generation system comprising thermal and pumped-storage hydro plants is considered. Two kinds of models for the cost-optimal generation of electric power under uncertain load are introduced: (i) a dynamic model for the short-term operation and (ii) a power production planning model. In both cases, the presence of stochastic data in the optimization model leads to multi-stage and two-stage stochastic programs, respectively. Both stochastic programming problems involve a large number of mixed-integer (stochastic) decisions, but their constraints are loosely coupled across operating power units. This is used to design Lagrangian relaxation methods for both models, which lead to a decomposition into stochastic single unit subproblems. For the dynamic model a Lagrangian decomposition based algorithm is described in more detail. Special emphasis is put on a discussion of the duality gap, the efficient solution of the multi-stage single unit subproblems and on solving the dual problem by bundle methods for convex nondifferentiable optimization.

Keywords: hydro-thermal power system, uncertain load, stochastic
programming, multi-stage, two-stage, mixed-integer,
Lagrangian relaxation, bundle methods

MSC 1991: 90C15, 90C90, 90C11

1 Introduction

The efficient operation and planning of electric power generation systems play an important role for electric utilities as well as the whole human activity. On the one hand, the efficient use of the available fuel for the production of electrical energy is of growing importance, both monetarily and because most of the primary energy sources, which today's energy supply is based

---

*This research is supported by the Schwerpunktprogramm "Echtzeit-Optimierung großer Systeme" of the Deutsche Forschungsgemeinschaft

on, are not renewable and have limited scope. Savings of a small percentage in the operation of a moderately large power system represent a significant reduction in operation cost as well as in the quantities of fuel consumed. On the other hand, in the future, the human community and, in particular, the power supply industry will be confronted with general economic and ecological conditions that are partly contradictory and aggravating. Some of these conditions are the rise in global energy demand, the scarcity of essential resources and the limits to the local and global environmental damage. Another contemporary challenge for the electric utility industry arises from the changes of market structures for electric power. There has been a worldwide movement towards deregulation of the electric utility industry and an opening of the market to nonutility participants. Moreover, there are plans to open the use of the transmission system in the European Community. All this has led and will further lead to a growth of the number and size of energy transactions. This development raises questions about the prices involved which are based on market actions rather than on costs as in traditional delivery contracts.

These issues have motivated a growing interest in applying mathematical modelling and optimization techniques for optimal system operation. Indeed, there is already a long tradition for applying mathematical programming methods and software to the solution of many relevant engineering problems (e. g. economic dispatch and unit commitment; see [67], [69] and the references therein). The recent substantial progress in many areas of mathematical optimization (e. g. in linear, mixed-integer, nonlinear, nondifferentiable and stochastic programming) opens the road to solving more and more involved models (e. g. [22]). Such complex and large optimization models arise, for instance, for the optimal operation of a hydro-thermal system when including additional aspects like data uncertainty, other regenerative sources of energy, the mid-term management of reservoirs, electricity trading etc. Models of this type are usually characterized by a combination of several difficulties like continuous as well as binary decision variables, very large dimension, nonlinearities (e. g. in hydro modelling, fuel costs, price structures in fuel as well as in electricity purchases) and the uncertainty of problem data (e. g. uncertainty of load forecasts, streamflows to reservoirs, pricing schemes, generator failures etc.).

The present paper aims, in particular, at applying a mathematical methodology, called *stochastic programming*, for handling uncertain data in optimization models. Stochastic programming is mostly concerned with problems that require a here-and-now decision on the basis of given probabilistic information on random quantities, but without making further observations. Possible formulations of stochastic programming models depend on when decisions must be taken relative to the realization of the random variables (e. g. at several stages in a dynamic model), the degree to which the constraint structure must be satisfied (e. g. with some probability), and the choice of the (stochastic) objective function (e. g. expected costs).

Stochastic programming approaches for tackling models in electric power generation under uncertainty have already found considerable attention (cf. chapters 24-26 in [17] for earlier works). We briefly mention here some of the recent and relevant works in this direction. A multi-stage stochastic optimization model for the optimal scheduling of a hydro-thermal generation system with uncertain inflows is developed in [51]. The authors present a solution strategy based on Benders decomposition and test results for a system comprising 39 hydroelectric plants, one aggregate thermal unit and a yearly planning period with monthly stages. The paper [10] offers an augmented Lagrangian decomposition technique for scheduling power systems under random disturbances which are modelled by scenario trees. In [32] a multi-stage stochastic program for scheduling hydroelectric generation under uncertainty is described and solved by an enhanced version of nested Benders decomposition. The paper also reports on the generation of monthly streamflow scenario trees and on model validation in the user's environment of the Pacific Gas & Electric Company. In [11] stochastic programming techniques based on Benders decomposition and importance sampling are applied to the facility expansion planning of electric power systems under uncertainty of the availability of generators and transmission lines, and on the demand. Schemes for the pricing of electric power, which is subject to demand and supply uncertainties, are designed and compared in [31] by means of a two-stage stochastic recourse model. The following papers deal with power scheduling under uncertain load. A two-stage stochastic program with simple recourse for the daily economic dispatch in a thermal power system is developed and solved in [8] under the assumption that the marginal distributions of the load are normal. In [25] and [26], this model is extended to power systems comprising thermal and pumped-storage hydro units and general load distributions. The extended model is solved by combining a smooth nonparametric estimation procedure for the marginal load distributions with standard nonlinear programming methods and it is validated by solving the daily economic dispatch problem of a system involving 24 thermal and 5 pumped-storage plants. Further extensions of the latter model by allowing for more general dynamics between decision and observation and for more appropriate recourse cost functions are discussed in [23] and [58]. These models do not yet include start-up and shut-down decisions into the optimization process. This is realized in [65], where a stochastic unit commitment problem for a thermal power system and a corresponding solution technique based on progressive hedging are developed. The progressive hedging methodology (cf. [57]) leads to a successive decomposition into scenario subproblems, which are deterministic unit commitment problems, and solved by Lagrangian relaxation and by an adapted subgradient method for dual maximization. In [66], the authors report on encouraging test runs for large real-life models.

The present paper aims at the development of two kinds of models for the cost-optimal scheduling of electric power in a hydro-thermal generation system under uncertain load: a dynamic stochastic recourse model for the short-term operation and a two-stage stochastic production planning model. Both models are further extensions of the stochastic models described in [25], [23] and [58]. They represent mixed-integer stochastic optimization problems which are large-scale for moderately large power systems. The second aim of the present paper consists in designing Lagrangian decomposition procedures for the two models by exploiting the particular structure of coupling constraints.

The models arise from a cooperation with the electric utility VEAG Vereinigte Energiewerke AG, which supplies the Eastern part of Germany. The VEAG owned generation system (in 1995) consists of 25 (coal-fired or gas-burning) thermal units and 6 pumped-storage hydro plants. Its total capacity is about 13.150 megawatts (MW) including a hydrogeneration capacity of 1.700 MW; the systems peak load amounts to 8.620 MW (in 1995). Hence, optimal scheduling of the VEAG-system exhibits two special features: the simultaneous optimization of thermal and hydro capacity is indispensable and the model is more large-scale than ever when including stochasticity. This gives rise to the need of solution algorithms for large-scale stochastic optimization problems which allow for handling mixed-integer decisions.

Existing solution procedures for large-scale stochastic programs are mostly based on approximating the underlying probability distribution by a discrete measure having finite support and on utilizing decomposition techniques for solving the large-scale approximate (deterministic) programs. For an overview and a discussion of much of the work done in this direction we refer to [15], [17], [20], [33], [52], [68]. In addition, we mention some of the recent relevant papers on decomposition approaches in stochastic programming. Primal decomposition techniques are based on the L-shaped or Benders decomposition method ([63]), its nested extension for multi-stage models ([4], [24]), and on regularized decomposition ([60]). A second group of (sometimes called *dual* or *scenario*) decomposition methods relax nonanticipativity constraints by introducing Lagrangian terms. For instance, the progressive hedging algorithm ([57]) and the scenario decomposition methods in [46], [59] are based on introducing augmented Lagrangians. Another augmented Lagrangian method by relaxing the recourse constraints is developed in [12]. A third group of methods consists of algorithms that combine decomposition and sampling techniques in various ways. For instance, sampling techniques are used for the generation of cuts in stochastic decomposition methods ([28]), for the efficient calculation of multivariate expected values by importance sampling ([30]), and for reducing the large dimensionality via EVPI-sampling ([13]) within nested Benders decomposition. Methods of a fourth group combine decomposition schemes and iterated approximations via refinement strategies (cf. [20], [21] and chapt. 3.5 in [33]).

Most of these numerical methods cannot be applied directly to stochastic programs involving integrality constraints. Methods for solving (mixed-) integer stochastic programs are rather rare. We refer to [62] for a brief overview of some recent approaches to stochastic integer programming. Moreover, let us mention a recently developed stochastic branch and bound method ([61]) and a dual decomposition method based on relaxing the scenario constraints and on (deterministic) branch and bound techniques ([9]), which also applies to mixed-integer situations.

Our paper is organized as follows. We introduce and discuss the two stochastic power scheduling models in Section 2. In Section 3 we briefly recall the Lagrangian relaxation approach and review some recent progress in solving the nondifferentiable duals. In the remaining two sections we develop Lagrangian decomposition methods for the dynamic recourse as well as for the two-stage stochastic model by relaxing coupling constraints. The dualization argument and the duality gap, the separability structure and the solution of the stochastic single unit subproblems are discussed in more detail for the dynamic model.

2   Models

2.1   Modelling a Hydro-Thermal System

We consider a power generation system comprising (coal-fired and gas-burning) thermal units, pumped-storage hydro plants and interchange contracts between interconnected utilities. We will develop and describe a mathematical model for a power system of this kind which has its origin in the earlier papers [25], [26]. The models allow for the simultaneous scheduling of all units and contracts over a certain time horizon.

Let T denote the number of time intervals obtained by discretizing the operation horizon. This discretization may be chosen uniformly (e. g. hourly or half-hourly) or non-uniformly. Let I and J denote the number of thermal and pumped-storage hydro units in the system. Delivery contracts are regarded as particular thermal units, but may have cost functions that are essentially different (e. g. nonconvex) from typical thermal costs. The decision variables in the model correspond to the outputs of each unit, i. e., the electric power generated or consumed by each unit of the system. These decision variables are denoted by

$$u_i^t \quad , \quad p_i^t \quad , \quad i = 1, \ldots, I \quad , \quad t = 1, \ldots, T,$$
$$s_j^t \quad , \quad w_j^t \quad , \quad j = 1, \ldots, J \quad , \quad t = 1, \ldots, T,$$

where $u_i^t \in \{0, 1\}$ and p_i^t are the on/off decisions and the production levels of the thermal unit i during the time period t. Correspondingly, s_j^t, w_j^t are the generation and pumping levels of the pumped-storage plant j during the period t, respectively. Thus, $u_i^t = 0$ and $u_i^t = 1$ mean that unit i is offline and on-line during period t, respectively. Further, by ℓ_j^t we denote the storage volume in the upper reservoir of plant j at the end of interval t. All variables mentioned above have finite upper and lower bounds representing unit capacity limits and reservoir capacities of the generation system:

$$p_{it}^{\min}u_i^t \le p_i^t \le p_{it}^{\max}u_i^t \ , \ u_i^t \in \{0,1\}, \ i = 1,\dots,I, \ t = 1,\dots,T,$$
$$0 \le s_j^t \le s_{jt}^{\max}, \ 0 \le w_j^t \le w_{jt}^{\max},$$
$$0 \le \ell_j^t \le \ell_{jt}^{\max}, j = 1,\dots,J, \ t = 1,\dots,T. \tag{2.1}$$

The constants $p_{it}^{\min}, p_{it}^{\max}, s_{jt}^{\max}, w_{jt}^{\max}$, and $\ell_{jt}^{\max}$ denote the minimal/maximal outputs of the units and the maximal storage volumes in the upper reservoirs during period t, respectively. The dynamics of the storage volume, which is measured in electrical energy, is modelled by the equations:

$$\begin{aligned} \ell_j^t &= \ell_j^{t-1} - s_j^t + \eta_j w_j^t \ , & t &= 1,\dots,T, \\ \ell_j^0 &= \ell_j^{\text{in}}, \ell_j^T = \ell_j^{\text{end}} \ , & j &= 1,\dots,J. \end{aligned} \tag{2.2}$$

Here, ℓ_j^{in} and ℓ_j^{end} denote the initial and final volumes in the upper reservoir, respectively, and η_j is the cycle efficiency of plant j. The cycle efficiency is defined as the quotient of the generation and of the pumping load that correspond to the same volume of water. The equalities (2.2) show, in particular, that there occur no in- or outflows in the upper reservoirs and, hence, that the pumped storage plants of the system operate with a constant amount of water. Together with the upper and lower bounds for ℓ_j^t the equations (2.2) mean that certain reservoir constraints have to be maintained for all pumped-storage plants during the whole time horizon.

Further single-unit constraints are minimum up- and down-times and possible must-on/off constraints for each thermal unit. Minimum up- and down-time constraints are imposed to prevent the thermal stress und high maintenance costs due to excessive unit cycling. Denoting by τ_i the minimum down-time of unit i, the corresponding constraints are described by the inequalities:

$$u_i^{t-1} - u_i^t \le 1 - u_i^\tau \ , \ \tau = t+1,\dots,\min\{t + \tau_i - 1, \ T\}, \ t = 1,\dots,T. \tag{2.3}$$

Analogous constraints can be formulated describing minimum-up times. Note that further single-unit constraints could be added, such as generator fuel limit constraints or air quality constraints in the form of limits on emissions from fossil-fired units.

The next constraints are coupling across the units: the loading and reserve constraints. The first constraints are essential for the operation of the power system and mean that the sum of the output powers is greater than or equal to the load demand in each time period. Denoting by d_t the load demand during period t, the loading constraints are described by the inequalities:

$$\sum_{i=1}^{I} p_i^t + \sum_{j=1}^{J}(s_j^t - w_j^t) \ge d^t, \ t = 1,\dots,T. \tag{2.4}$$

In order to compensate unexpected events within a specified short time period, a spinning reserve, describing the total amount of generation available from all units synchronized on the system minus the present load, is prescribed. For instance, such events are sudden load increases and the outage of one or more units. Beyond spinning reserve various classes of off-line reserves may be involved. These include gas-turbine units and pumped-storage hydro plants that can quickly be brought on-line and up to full capacity. Hence, the spinning reserve constraints concern the synchronized thermal units and are given by the following inequalities:

$$\sum_{i=1}^{I} (p_{it}^{\max} u_i^t - p_i^t) \geq r^t, \ t = 1, ..., T, \tag{2.5}$$

where $r^t > 0$ is a specified spinning reserve in period t.
The objective function is given by the total costs for operating the thermal units. These costs consist of the sum of the costs of each individual unit over the whole time horizon, i. e.,

$$\sum_{i=1}^{I} \sum_{t=1}^{T} \left[FC_{it}(p_i^t, u_i^t) + SC_{it}(u_i(t)) \right], \tag{2.6}$$

where FC_{it} is the fuel cost function and SC_{it} are the start-up costs for the operation of the thermal unit i during period t. We make the natural assumption that $FC_{it}(0, 0) = 0$ and that $FC_{it}(\cdot, 1)$ is strictly monotonically increasing. Often fuel cost functions are piecewise linear-quadratic and convex, i. e., they are functions of the form

$$FC_{it}(p, u) = \max_{\ell=1,...,L} f_{i\ell}(p) + u \, c_i, \tag{2.7}$$

where $f_{i\ell}$ are linear or convex quadratic functions having the property $\max_{\ell=1,...,L} f_{i\ell}(0) = 0$ and c_i is a fixed cost term. Non-convex set-ups for fuel costs are also possible and of particular importance for modelling costs in delivery contracts including discounts. Typical cost functions of this kind are general piecewise linear functions. Note that such functions can be modeled using binary variables for selecting the correct line segment for a given value of p (see e. g. [47]).
The start-up costs $SC_{it}(u_i(t))$, where $u_i(t) = (u_i^1, ..., u_i^t)$, can vary from a maximum cold-start value to a much smaller value when the unit i is still relatively close to the operating temperature. A simple description for start-up costs is given by

$$SC_{it}(u_i(t)) = C_i^f \max \left\{ u_i^t - u_i^{t-1}, 0 \right\}, \ t = 2, ..., T,$$

where C_i^f are fixed costs. This description has the advantage that it can be expressed in linear terms. On the other hand, it does not reflect that the costs depend on the cooling time. Alternatively, a more involved start-up cost function, which is time-dependent, is given by

$$SC_{it}(u_i(t)) = \left(C_i^f + C_i^c \left(1 - \exp \left(-(t - t_{s_i})/\alpha_i \right) \right) \right) \max \left\{ u_i^t - u_i^{t-1}, 0 \right\},$$

where C_i^f are again fixed costs, C_i^c cold-start costs, α_i the thermal time constant for the unit i and $t - t_{s_i}$ the down-time of unit i until period t, i. e.,

$$s_i = \max\left\{s \in I\!N : u_i^{t-j} = u_i^{t-1}, j = 2, \ldots, s\right\}.$$

Altogether, minimizing the objective function (2.6) subject to the constraints (2.1)-(2.5) leads to a cost-optimal schedule for all units of the power system during the specified time horizon. It is worth mentioning that a cost-optimal schedule has the following two interesting properties, which are both a consequence of the strict monotonicity of the fuel costs. If a schedule (u, p, s, w) is optimal, then the loading constraints (2.4) are typically satisfied with equality and we have $s_j^t w_j^t = 0$ for all $j = 1, \ldots, J$, $t = 1, \ldots, T$, i. e., generation and pumping do not occur simultaneously (see [27]).

The minimization problem (2.1)-(2.6) represents a mixed-integer program with (possibly) nonlinear objective, linear constraints, and IT binary and $(I + 2J)T$ continuous variables, respectively. For a typical configuration of the VEAG owned generation system with $I = 22$ (thermal), $J = 6$ (hydro) and $T = 192$ (i. e., 8 days with hourly discretization), this amounts to 4224 binary and 6528 continuous variables.

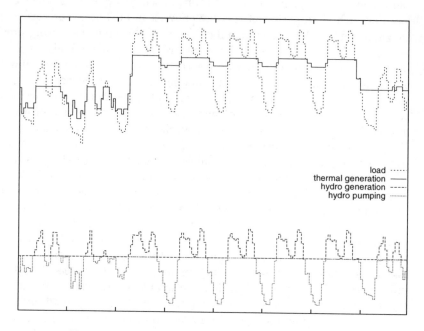

Fig. 1: load curve and hydro-thermal schedule

For this park of the power system and for a peak load week, Figure 1 shows a typical load curve and a corresponding cost-optimal hydro-thermal schedule. Note that the mixed-integer program is solved by the methods described in [14], which Figure 1 is taken from. The load curve in Figure 1 shows two types of cycles: In general, the load is higher during the morning and the early evening (peak), with a small valley during the early afternoon, and it is lower during the night. In addition, the consumption of electric power exhibits a weekly cycle, because the load is lower over weekend days than weekdays. The efficient operation of pumped-storage hydro plants exploits these two cycles. They are designed to save fuel costs by serving the peak load with hydro-energy and then pumping to refill the reservoir during off-peak periods, i. e., during the nights and weekends. The hydro schedule in Figure 1 reflects this typical operation of pumped-storage plants. They may, in fact, be operated on a daily or weekly cycle. Figure 1 records a schedule when operating on a weekly cycle. The remaining load, i. e., the difference between the original system load and the hydro schedule, shows a much more uniform structure than the original load. This portion of the load is covered by the total thermal output. Among the thermal plants of the system, the base-load units are loaded nearly 100% of the time horizon and the "cycling" units are loaded for periods depending on their costs and the shape of the load pattern.

So far we have tacitly assumed that the electrical load is deterministic over the whole time horizon. In electric utilities, schedulers forecast the electrical load for each time period of the day or week in advance. For this purpose they make use of historical load data (e. g. of the same week from previous years), of their personal experience and of statistical methods (e. g. time series or regression analysis). But, clearly, the actual load demand may deviate from the predicted load at any time period for various reasons. Usually electric utilities record the actual system load and save the data over several years. These statistical data provide a basis for the development of stochastic models for the load process and the optimization of power scheduling.

Next we decribe two stochastic models for the optimal scheduling of electric power which differ mainly in the quality of available information on the load stochasticity. The first one represents a model for the optimal on-line or short-term operation of a power system, where future consequences of actual scheduling decisions as well as the future load uncertainty are taken into account. In this model we assume that the load is completely known (i. e., deterministic) at the beginning of the time horizon and that the load uncertainty increases with the growing number of time periods. Secondly, a model for short- or mid-term power production planning is developed. The essential difference to the first model is that the quality of available information on the load uncertainty does not depend on time. It aims at determining (optimal) power production schedules for a future planning period (e. g. next week or month). The second model represents a two-stage stochastic program, whereas the first one is a dynamic (multi-stage) stochastic optimization problem. Both models involve mixed-integer decisions in all stages.

2.2 Dynamic Recourse Model

We assume that the load $\{d^t : t = 1, \ldots, T\}$ forms a (discrete-time) stochastic process on some probability space $(\Omega, \mathcal{A}, \mu)$, that the information on the load is complete for $t = 1$, and that the uncertainty increases with growing t. Let $\{\mathcal{A}_t : t = 1, \ldots, T\}$ be the filtration generated by the load process, where $\mathcal{A}_t$ is the μ-completed σ-field defined by the random vector $(d^1, \ldots, d^t)$. Hence, we have $\mathcal{A}_1 \subseteq \mathcal{A}_2 \subseteq \ldots \subseteq \mathcal{A}_t \subseteq \ldots \subseteq \mathcal{A}_T \subseteq \mathcal{A}$ and $\mathcal{A}_1$ is the μ-completion of $\{\emptyset, \Omega\}$. The sequence of scheduling decisions $\{(u^t, p^t, s^t, w^t) : t = 1, \ldots, T\}$ also forms a stochastic process on $(\Omega, \mathcal{A}, \mu)$, which is assumed to be *adapted* to the filtration of σ-fields, i.e., *nonanticipative*. The latter condition means that the decision (u^t, p^t, s^t, w^t) depends only on the data history $(d^1, \ldots, d^t)$ or, equivalently, that (u^t, p^t, s^t, w^t) is $\mathcal{A}_t$-measurable. We mention that this condition is often formulated in terms of a closed linear subspace that is determined by the conditional expectations with respect to the σ-fields $\mathcal{A}_t$ ([55], [12]). Since all decision variables are uniformly bounded, we may restrict our attention to decisions (u, p, s, w) belonging to $L^\infty(\Omega, \mathcal{A}, \mu; \mathbb{R}^m)$, where $m := 2(I + J)T$. Then the nonanticipativity condition can be formulated equivalently as

$$x = (u, p, s, w) \in \underset{t=1}{\overset{T}{\times}} L^\infty(\Omega, \mathcal{A}_t, \mu; \mathbb{R}^{m_t}), \tag{2.8}$$

where $m_t := 2(I + J)$, and the (stochastic) optimization problem consists in minimizing the expected cost (cf. (2.6))

$$F(x) = \mathbb{E}\left\{ \sum_{i=1}^{I} \sum_{t=1}^{T} \left[FC_{it}\left(p_i^t, u_i^t\right) + SC_{it}\left(u_i(t)\right) \right] \right\} \tag{2.9}$$

over all decisions (u, p, s, w) satisfying the nonanticipativity constraint (2.8) and μ-almost surely the constraints (2.1)-(2.5). Among the constraints (2.1)-(2.5), (2.2) and (2.3) reflect the dynamics of the model and (2.4), (2.5) are coupling across units. Altogether, the stochastic program involves $2(I + J)T$ stochastic decision variables and, hence, an enormous number of stochastic scheduling decisions for real-life power generation systems. It is a discrete time dynamic or multi-stage recourse problem, where the "stages" do not necessarily refer to time periods, but correspond to steps in the decision process where observations of the uncertain environment (i. e. the load) take place. The number K of stages of the dynamic model thus corresponds to the (maximal) number of time steps $t_1 = 1 < t_2 < \ldots < t_k < \ldots < t_{K+1} = T$ such that we have the strict inclusion $\mathcal{A}_{t_k} \subset \mathcal{A}_{t_{k+1}}$, $k = 1, \ldots, K - 1$, for the σ-fields belonging to the filtration.

For the numerical solution of the dynamic recourse model we now assume that a *discrete* multivariate probability distribution of the stochastic load vector $d = (d^1, \ldots, d^T)$, whose finite support consists of the atoms or *scenarios* $d_n = (d_n^1, \ldots, d_n^T)$, with the probabilities $\pi_n = \mu(d = d_n)$, $n = 1, \ldots, N$, is given. Let n_k, $k = 1, \ldots, K$, denote the number of atoms corresponding to the σ-field $\mathcal{A}_{t_k}$. Then we have $n_1 = 1 < n_2 < \ldots < n_k < \ldots < n_K = N$ and the following scenario constraints at each stage $k \in \{1, \ldots, K\}$:

$$d_n^{t_k} = d_{\hat{n}}^{t_k} \text{ implies } d_n^t = d_{\hat{n}}^t, \text{ for all } t = 1, \ldots, t_k. \tag{2.10}$$

Hence, the information on the load can be represented in the form of a *scenario tree*. Each path from the root to a leaf of the tree corresponds to one scenario; each branching node corresponds to a (decision) stage. Figure 2 shows an example of a load scenario tree over a weekly time horizon, where observations of the load are made every day, leading to one additional daily scenario.

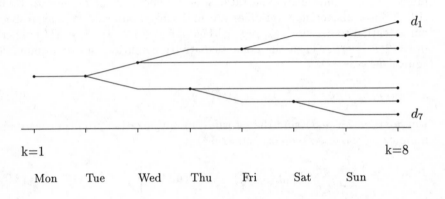

Fig. 2: Load scenario tree

The scenario information may have various origins. It can be obtained as an approximation of the multivariate load distribution, based on sampling from empirical data or on scenarios provided by experienced schedulers. We do not go into detail here, but refer to [16] (and the references therein) for a discussion of various approaches to the generation of scenarios that reflect the structure of the model as well as the information available on the underlying probability distribution. We also refer to [65] where several strategies for generating load scenarios (e. g. handling forecast uncertainty) are discussed. Although the primary aim of generating a scenario tree is to obtain a reasonable approximation for the underlying probability distribution, a compromise between the quality of approximation and the size of the approximate problem has to be taken into consideration, too. The size of the scenario based multi-stage model easily grows out of hand with increasing number of scenarios and stages. In order to illustrate this fact, let $u_{i,n}$, $p_{i,n}$, $s_{j,n}$, $w_{j,n}$, and $\ell_{j,n}$, denote the n-th scenario of the variables u_i, p_i, s_j, w_j, and ℓ_j. Then the scenario based model consists in minimizing the objective function

$$\sum_{n=1}^{N} \sum_{i=1}^{I} \sum_{t=1}^{T} \pi_n \left[FC_{it} \left(p_{i,n}^t, u_{i,n}^t \right) + SC_{it} \left(u_{i,n}(t) \right) \right] \qquad (2.11)$$

over all decisions $\{(u_n, p_n, s_n, w_n) : n = 1, \ldots, N\}$ satisfying the bound and integrality constraints (2.1), the system dynamics

$$\ell_{j,n}^t = \ell_{j,n}^{t-1} - s_{j,n}^t + \eta_j w_{j,n}^t, \; \ell_{j,n}^0 = \ell_j^{in}, \; \ell_{j,n}^T = \ell_j^{end}, j = 1, \ldots, J,$$

$$u_{i,n}^{t-1} - u_{i,n}^t \le 1 - u_{i,n}^\tau, \; \tau = t + 1, \ldots, \min\{t + \tau_i - 1, T\}, \qquad (2.12)$$

$$t = 1, \ldots, T, \; n = 1, \ldots, N,$$

the loading and reserve constraints

$$\sum_{i=1}^{I} p_{i,n}^t + \sum_{j=1}^{J} (s_{i,n}^t - w_{i,n}^t) \ge d_n^t, \; \sum_{i=1}^{I} \left(p_{it}^{\max} u_{i,n}^t - p_i^t \right) \ge r^t, \qquad (2.13)$$

$$t = 1, \ldots, T, \; n = 1, \ldots, N,$$

and the scenario nonanticipativity constraints, i.e., the equality

$$(u_n^t, p_n^t, s_n^t, w_n^t) = (u_{\hat{n}}^t, p_{\hat{n}}^t, s_{\hat{n}}^t, w_{\hat{n}}^t)$$

for $t = t_k$ implies that the same equality holds for all $t = 1, \ldots, t_k, \; k = 1, \ldots, K$.

When regarding the nonanticipativity constraints and introducing decision variables at each node of the scenario tree, the number of decisions in the (deterministic) optimization model (2.11)-(2.13) amounts to $2(I + J) \sum_{k=1}^{K} n_k (t_{k+1} - t_k)$. Hence, the model may easily become extremely large if the scenario tree contains too many paths. Even for the (very) small scenario tree in Figure 2 (i. e., with $K = 7$, $n_K = K$ and $t_{k+1} - t_k = 24$) the model involves $672 \cdot I$ binary and $672 \cdot (I + 2J)$ continuous variables and standard methods including those reviewed in Section 1, may not be able to solve the problem in reasonable time. This requires other techniques that exploit the underlying structure of the original stochastic model.

2.3   Two-Stage Stochastic Model

Again we assume the load $\{d^t : t = 1, \ldots, T\}$ to be given as a (discrete-time) stochastic process on some probability space $(\Omega, \mathcal{A}, \mu)$. However, this time the load process does not involve an information structure and the decision process consists of two stages where the first-stage decisions correspond to the here-and-now schedules for all power generation units over the whole time horizon. The second-stage decisions correspond to future compensation or recourse actions of each unit in each time period in response to the environment created by the chosen first-stage decision and the load realization in that specific time period. Hence, the aim of such a two-stage dynamic model can be formulated as follows: Find an optimal schedule for the whole power system and planning horizon such that the uncertain demand can be compensated by the system, all system constraints are satisfied and the sum of the total generation costs and the expected compensation costs is minimal.

In order to give a mathematical formulation of the model, let (u, p, s, w) denote the first-stage scheduling decisions as in Section 2.1 and $(\hat{u}, \hat{p}, \hat{s}, \hat{w})$ denote the stochastic compensation decisions having the components $\hat{u}_i^t$, $\hat{p}_i^t$, $\hat{s}_j^t$, $\hat{w}_j^t$, $i = 1, \ldots, I$, $j = 1, \ldots, J$, $t = 1, \ldots, T$, which correspond to the compensation actions of each unit at time period t.

In addition to the (non-stochastic) constraints for (u, p, s, w), (2.1) (capacity limits), (2.2) (storage dynamics), (2.3) (minimum down-time constraints) and (2.5) (reserve constraints), we have to require that the compensation actions also satisfy certain system constraints. These are the unit capacity limits, minimum-down time constraints and reservoir capacity bounds :

$$p_{it}^{\min} \hat{u}_i^t \le p_i^t \hat{u}_i^t + \hat{p}_i^t \le p_{it}^{\max} \hat{u}_i^t, \ \hat{u}_i^t \in \{0, 1\}, \ i = 1, \ldots, I, \tag{2.14}$$

$$\hat{u}_i^{t-1} - \hat{u}_i^t \le 1 - \hat{u}_i^\tau, \ \tau = t+1, \ldots, \min\{t + \tau_i - 1, T\}, i = 1, \ldots, I, \tag{2.15}$$

$$0 \le s_j^t + \hat{s}_j^t \le s_{jt}^{\max}, \qquad 0 \le w_j^t + \hat{w}_j^t \le w_{jt}^{\max},$$

$$0 \le \ell_j^t + \hat{\ell}_j^t \le \ell_j^{\max}, \qquad \hat{\ell}_j^0 = \hat{\ell}_j^T = 0 \tag{2.16}$$

$$\hat{\ell}_j^t = \hat{\ell}_j^{t-1} - \hat{s}_j^t + \eta_j \hat{w}_j^t, \qquad j = 1, \ldots, J, \ t = 1, \ldots, T, \ \mu - \text{a. s.}$$

In other words, the constraints (2.16) for the hydro scheduling decisions mean that the sum of first-stage decisions and recourse actions is feasible, too. The formulation (2.14) of the thermal unit capacity limits for the compensation stage becomes more involved because the term $p_i^t \hat{u}_i^t$ introduces a nonlinear constraint connecting first- and second-stage variables. The nonlinearity in (2.14) is avoided when requiring that a thermal unit, which is scheduled to be on-line in the first-stage, must not be off-line in the compensation action. In this case, (2.14) can be replaced by the (linear) constraints:

$$p_{it}^{\min} \hat{u}_i^t \le p_i^t + \hat{p}_i^t \le p_{it}^{\max} \hat{u}_i^t, \ u_i^t \le \hat{u}_i^t, \ \hat{u}_i^t \in \{0, 1\}, i = 1, \ldots, I. \tag{2.17}$$

This formulation of the thermal unit capacity limits seems to be quite natural and realistic because generation systems often possess sufficient flexibility to compensate load decreases by lowering output levels of thermal units. However, there might be a need for new on-line units in order to compensate unpredictable load increases. Another possible compensation strategy could be based on a subdivision of the set of available thermal units into two sets $\mathcal{I}_1$ and $\mathcal{I}_2$ such that $\mathcal{I}_1 \cup \mathcal{I}_2 = \{1, \ldots, I\}$ and the conditions

$$u_i^t = \hat{u}_i^t, \ i \in \mathcal{I}_1, \text{ and } u_i^t \leq \hat{u}_i^t, \ i \in \mathcal{I}_2, t = 1, \ldots, T, \ \mu - \text{ a. s. },$$

are satisfied. This means that only some of the available thermal units may change their on/off state when compensating uncertain load. From a modelling point of view this strategy would lead to a reduction of the number of binary variables.

In the following, we always assume that (2.17) instead of (2.14) is satisfied. Observe that the conditions (2.15) and (2.17) imply (2.3).

The loading constraints (2.4) are modified by requiring that the sum of the first-stage power outputs of all generation units satisfies the load with some probability $\pi_t \in (0, 1)$ in period t, $t = 1, \ldots, T$, and that the sum of the total power outputs satisfies the load with probability one. Denoting by F_{d^t} the distribution function of d^t, the (modified) loading constraints are given by the following inequalities:

$$\sum_{i=1}^{I} p_i^t + \sum_{j=1}^{J} (s_j^t - w_j^t) \geq F_{d^t}^{-1}(\pi_t), \ t = 1, \ldots, T, \tag{2.18 }$$

$$\sum_{i=1}^{I} (p_i^t + \hat{p}_i^t) + \sum_{j=1}^{J} (s_j^t + \hat{s}_j^t - (w_j^t + \hat{w}_j^t)) \geq d^t, \ t = 1, \ldots, T, \mu - \text{a.s.} \tag{2.19 }$$

A variant of (2.18), which will be considered in Section 5, is that the term $F_{d^t}^{-1}(\pi_t)$ is replaced by the expected load $\mathbb{E}(d_t)$, $t = 1, \ldots, T$. In both cases, the constraint (2.18) means that the sum of the first-stage output power satisfies a certain predicted or approximated load and the second-stage decisions take care of satisfying the stochastic load with probability one.

Since the real operation of the system takes place during the compensation action, the objective function corresponds to the total average costs for operating the thermal units, i. e.,

$$\mathbb{E} \left\{ \sum_{i=1}^{I} \sum_{t=1}^{T} \left[FC_{it} \left(p_i^t + \hat{p}_i^t, \hat{u}_i^t \right) + SC_{it} \left(\hat{u}_i(t) \right) \right] \right\} \tag{2.20 }$$

where FC_{it} and SC_{it} denote the fuel cost and start-up cost functions, respectively, for the operation of unit i during period t, and $\hat{u}_i(t) := (\hat{u}_i^1, \ldots, \hat{u}_i^t)$.

The stochastic power production planning model consists then in minimizing the objective function (2.20) over all deterministic decisions (u, p, s, w) and all stochastic decisions $(\hat{u}, \hat{p}, \hat{s}, \hat{w}) \in L^\infty(\Omega, \mathcal{A}, \mu; \mathbb{R}^m)$ satisfying the constraints (2.1), (2.2), (2.5), (2.15)-(2.19). The model represents a two-stage stochastic mixed-integer program involving $2(I+J)T$ deterministic and $2(I+J)T$ stochastic decision variables. Similar to the dynamic model in the previous section, only the loading constraints (2.18), (2.19) and the reserve constraints (2.5) are coupling across units.

3 Lagrangian Relaxation Approach

Lagrangian relaxation is a solution technique primarily for minimizing a non-smooth function. We would like to recall the basic ideas and some facts in order to clarify the reasons that make this approach appropriate for solving the problems introduced in the previous section. Our presentation is inspired by [40]. Let us consider an optimization problem

$$\min f(x) \text{ subject to } x \in C, \ g(x) \leq 0, \tag{3.1}$$

where $f : I\!\!R^n \to I\!\!R$, $C \subseteq I\!\!R^n$, $g : I\!\!R^n \to I\!\!R^m$. We make the general assumption that f and g_j, $j = 1, \ldots, m$ are convex functions and there exists an $\bar{x} \in C : g(\bar{x}) \leq 0$.

We suppose that the functions f and g and the set C have some special structure, which makes the Lagrangian problem

$$\min [L(x, \lambda) = f(x) + \lambda g(x)] \text{ subject to } x \in C \tag{3.2}$$

much easier to solve than the problem (3.1), where $\lambda \in I\!\!R_+^m$. Let us assume the following:

(A)   For all $\lambda \in I\!\!R_+^m$ there exists an element $x_\lambda \in C$ such that
$$\Theta(\lambda) = \min_{x \in C} L(x, \lambda) = L(x_\lambda, \lambda).$$

Be aware that $L(\cdot, \lambda)$ may have several minima for some λ, but $\Theta(\lambda)$ is well-defined, since the minimal value is non-ambiguous. By the weak duality theorem, we have

$$\Theta(\lambda) \leq \min_{x \in C} L(x, \lambda) \leq f(x)$$

for all feasible points x in (3.1). The following statement is straight-forward but important.

Proposition 3.1 *([18]) Any solution $\bar{x}$ of the Lagrangian problem (3.2) solves the perturbed problem (3.3):*

$$\min f(x) \text{ subject to } x \in C, \ -\tilde{g}(\lambda) \geq g(x), \tag{3.3}$$

where $\tilde{g}(\lambda) = -g(\bar{x})$.

Proof: For any feasible x in (3.3) and $\lambda \in I\!\!R_+^m$ we have

$$\begin{aligned}
f(x) &\geq f(x) + \lambda[g(x) + \tilde{g}(\lambda)] \\
&= L(x, \lambda) + \lambda\tilde{g}(\lambda) \geq L(\bar{x}, \lambda) - \lambda \cdot g(\bar{x}) \\
&= f(\bar{x})
\end{aligned}$$

$\square$

We conclude that if $\bar{x}$ is "almost feasible", it is "almost a solution" of (3.1). If we succeed in finding a solution to (3.2) which is also feasible for (3.1), then we have a solution to (3.1), because the inequality of (3.3) is satisfied. Having in mind the weak duality theorem, it is clear that any feasible point

$\bar{x}$ of (3.1) produces an upper bound $f(\bar{x})$ for $\Theta(\lambda)$. Hence, to solve (3.1) via (3.2) it is necessary to maximize Θ on $I\!\!R^m_+$.
We call $\Theta(\cdot)$ the *dual function*, λ the *dual variable*, and the problem

$$\max \Theta(\lambda) \text{ subject to } \lambda \in I\!\!R^m_+ \tag{3.4}$$

the *dual problem* to (3.1). We show that Θ is a concave function having subgradients at all λ by virtue of the assumption (A). Let us denote a solution of (3.2) for $\bar{\lambda}$ by $\bar{x}$.

$$\begin{aligned} \Theta(\lambda) &= \min_{x \in C} L(x, \lambda) \le L(\bar{x}, \lambda) \\ &= L(\bar{x}, \bar{\lambda}) + (\lambda - \bar{\lambda}) g(\bar{x}) \\ &= \Theta(\bar{\lambda}) - (\lambda - \bar{\lambda}) \tilde{g}(\bar{\lambda}) \end{aligned}$$

The latter inequality characterizes concavity and implies

$$\tilde{g}(\bar{\lambda}) \in \partial[-\Theta(\bar{\lambda})]$$

where $\partial[-\Theta(\bar{\lambda})]$ stands for the subdifferential of $-\Theta$ with respect to λ calculated at the point $\bar{\lambda}$.
Let us suppose that the problem under consideration has a separable structure, i. e., the problem is of the following form:

the variables $\qquad x = (x_1, \ldots, x_{\bar{n}})$ and $x_i \in I\!\!R^{n_i} \quad i = 1, \ldots, \bar{n}$,

the objective function $\qquad f(x) = \sum_{i=1}^{\bar{n}} f_i(x_i) + f_0$,

the related constraints $\qquad g_j(x) = \sum_{i=1}^{\bar{n}} g_j^i(x_i) + g_j^0, \quad j = 1, \ldots, m$,

where f_0 and g_j^0 $(j = 1, \ldots, m)$ are constants.
Let us further suppose some special structure of the set C. We assume the set C to be the following product

$$C = \left(\mathop{\times}_{i=1}^{i_0} \{0, 1\}^{n_i} \right) \times \left(\mathop{\times}_{i=i_0+1}^{\bar{n}} B_i \right),$$

where $B_i \subseteq I\!\!R^{n_i}$ are compact convex sets. This means that $x_1, \ldots, x_{i_0}$ are binary variables and we consider a mixed-integer problem.
Furthermore, let us assume the functions f_i and g^i to be convex piecewise linear or (piecewise) quadratic functions. Then $L(\cdot, \lambda)$ is a convex function, too.
The strong duality theorem does not apply due to the presence of integrality, i. e. the structure of the set C. However, we are in a favourable situation to have

- the assumption (A) is satisfied,

- decomposable structure of the relaxed problem,

- description of the subgradients of $\Theta(\lambda)$.

We call the following optimization problem a continuous relaxation of the problem (3.1)

$$\min f(x) \text{ subject to } x \in \tilde{C}, \ g(x) \leq 0,$$
$$\text{where } \tilde{C} = \left(\times_{i=1}^{i_0}[0,1]^{n_i}\right) \times \left(\times_{i=i_0+1}^{\bar{n}} B_i\right).$$

Proposition 3.2 *The Lagrangian relaxation provides a better lower bound of the optimal value of (3.1) than the continuous relaxation of the problem.*

Proof: The following sequence of inequalities holds true for each $\lambda \in I\!R_+^m$:

$$\min_{\substack{x \in C \\ g(x) \leq 0}} f(x) \geq \min_{x \in C} L(x, \lambda) \geq \min_{x \in \tilde{C}} L(x, \lambda)$$

This implies

$$\min_{\substack{x \in C \\ g(x) \leq 0}} f(x) \geq \max_{\lambda \in I\!R_+^m} \min_{x \in \tilde{C}} L(x, \lambda)$$

The maximum above is attained at some λ_0 since $\min_{x \in \tilde{C}} L(x, \cdot)$ is concave piecewise linear or (piecewise) quadratic function bounded from above on $I\!R_+^m$. Consequently, $L(x, \lambda)$ has a saddle point $(\lambda_0, x_{\lambda_0})$ and we obtain by virtue of the saddle point theorem:

$$\min_{\substack{x \in C \\ g(x) \leq 0}} f(x) \geq \max_{\lambda \in I\!R_+^m} \min_{x \in \tilde{C}} L(x, \lambda) = L(\lambda_0, x_{\lambda_0}) = \min_{\substack{x \in \tilde{C} \\ g(x) \leq 0}} f(x)$$

This proves the assertion. □

Observe that $L(x, \lambda)$ has a separable structure with respect to the components x_i, which together with the special structure of C leads to a decomposition of the problem 3.2 into $\bar{n}$ subproblems of dimension n_i each. The subproblems read

$$P_i(\lambda) : \min f_i(x_i) + \sum_{j=1}^{m} \lambda_j g_j^i(x_i) \text{ subject to } x_i \in C_i,$$

where:
$$C_i = \begin{cases} \{0,1\}^{n_i} & \text{if} \quad 1 \leq i \leq i_0 \\ B_i & \text{if} \quad i_0 \leq i \leq n_i \end{cases}$$

Denoting the marginal functions of the problems above by $\Theta_i(\lambda)$ $(i = 1, \ldots, \bar{n})$ we obtain for the dual function

$$\Theta_i(\lambda) = \sum_{i=1}^{\bar{n}} \Theta_i(\lambda) + f_0 + \sum_{j=1}^{m} \lambda_j g_j^0$$

Consequently, the dual problem has a separable structure, too. The latter observations make an approach to problems with decomposable structure via Lagrangian relaxation attractive. A solution procedure should include:

- a method for solving the non-smooth concave optimization problem (3.4).

- fast algorithms for minimizing the Lagrange-function $L(x, \lambda)$ at a given point λ, i. e., for solving the subproblems $P_i(\lambda), i = 1, \ldots, \bar{n}$. The solution provides then the value of Θ and its subgradients.

- a technique to obtain a primal feasible solution.

The latter point needs separate investigations. As already mentioned, a dual method does not provide a primal feasible solution due to the integrality conditions. Thus, we have to use the information on the dual solution to calculate a primal feasible point close to the dual solution efficiently. Due to the first proposition, such a procedure will obtain a fairly good point. In [2], it is shown that the relative duality gap for mixed integer problems with special structure becomes small under certain assumptions. We will see later how the estimate given there is modified for the dynamic recourse problem. Methods for nonsmooth optimization have been the subject of intensive development during the last 15 years. An algorithm for minimizing a convex function known for a long time is the cutting-plane method. It develops the natural idea to use subgradient-information and to generate a linear approximation of the function associated with it. Let us suppose that, at a certain moment, values $f(x_1), \ldots, f(x_k)$ and subgradients $z_1 \in \partial f(x_1), \ldots, z_k \in \partial f(x_k)$ are available. We define

$$\tilde{f}_k(x) = \max\{f(x_i) + < z_i, x - x_i >, \quad i = 1, \ldots, k\}$$

and, minimizing $\tilde{f}_k$, obtain a further point x_{k+1}. It is assumed that $\tilde{f}_k$ is bounded from below on C and we are able to compute values and subgradients of f.

However, this algorithm has some well-known drawbacks. The initial iterations are inefficient. The number of cuts increases after each iteration and there is no reliable rule for deleting them. The minimization of the approximate function is sensitive when approaching a point of nondifferentiability. Further developments have led to the so-called bundle methods which offer a stabilizing device based on the following ingredients:

- a sequence $\{x_n\}$ of stabilized iterates;

- a criterion (test) deciding whether a new iterate has been found and (or) whether the bundle of information, i.e., the approximation $\tilde{f}_k$, should be enriched;

- a sequence $\{M_n\}$ of positive definite matrices used for a stabilizing term.

Bundle methods are pioneered by Wolfe and Lemarechal. A detailed study on the subject can be found in [35] and [29]. A comprehensive review is given in [43]. One description of the main idea of (first-order) bundle methods is the following:

Suppose iterate x_n and a bundle of subgradients $z_k \in \partial f(y_k)$ have been computed. As above, we use the bundle of information to formulate a lower approximation of the function f, i. e.,
$\tilde{f}_n(x) = \max\{f(y_i)+ < z_i, x - y_i >, \quad i = 1, \ldots, k\}$, and

1. minimize $\tilde{f}_n(x) + \frac{1}{2} < M_n(x - x_n), x - x_n >$
   and let the point $\bar{x}$ be its minimal point.

2. compute a nominal decrease
   $\alpha_n = f(x_n) - \tilde{f}_k(\bar{x}) - \frac{1}{2} < M_n(\bar{x} - x_n), \bar{x} - x_n >$.
   A constant $c \in (0, 1)$ being chosen, we perform the descent test:
   $f(\bar{x}) \leq f(x_n) - c\alpha_n$
   If the inequality is satisfied we set $x_{n+1} = \bar{x}$; $y_{k+1} = \bar{x}$
   and increase n and k by 1.
   Otherwise, n is kept fixed, we set $y_{k+1} = \bar{x}$ and increase k by 1. In some versions (cf. [42]) an additional test is made before increasing k.

3. The choice of $\{M_n\}$ given in the literature is:
   - an abstract sequence, as in [39],
   - $M_n \equiv I$, as in [34],
   - $M_n = \mu_n I$ with heuristic rules for computing μ_n, in [36], [64],
   - solving a quasi-Newton equation in [42].

This description of the bundle methods corresponds to the proximal point concept (i. e., the Moreau-Yosida regularization). Recall that, given a positive semi-definite matrix M,

$$F(x) = \inf \left\{ f(y) + \frac{1}{2} < M(y - x), \ y - x > \right\} \qquad (3.5)$$

is the Moreau-Yosida regularization of the function f. In the classical framework M should be positive definite. In [42], it is suggested to allow a degenerate proximal term and it is shown there that the essential properties can be reproduced also in this case. A relationship between these concepts and certain first order bundle methods was observed by several authors, e.g. [29]. Methods of order higher than one are studied in [36] and [64] where a single stabilizing parameter is varied.

In [36] the choice of weights μ for updating the matrix in the proximal term is considered. The matrix M is intended to accumulate information about the curvature of f around the point $\bar{x}$. Safeguarded quadratic interpolation is proposed for choosing the weights μ_{n+1} so that the curvature of f between x_n and $\bar{x}$ is estimated. The algorithm computes a direction for the next iterate x_{n+1} by solving a quadratic program, then the descent test and the update of the bundle of subgradients are modified accordingly. The reported computational experiments indicate that this technique can decrease the number of objective evaluations necessary for reaching a desired accuracy in the optimal value significantly.

The algorithms presented in [7], [42], [44], referred to as variable metric bundle methods, make use of the Moreau-Yosida regularization of the objective function and develop some quasi-Newton formulas. Two strategies for updating the matrix M in the minimization procedure are suggested in [42]. In the first version, called diagonal quasi-Newton method, M is proportional to the identity matrix, while the second version uses a full quasi-Newton matrix. The matrix is updated at the end of a descent-step, when a new stabilizing iterate point is computed. The updating procedure corresponds to a regularizing scheme for the gradient of F.

In [44] M is a positive definite matrix and, thus, there is a unique solution of (3.5), which is denoted by $y(x)$. The main idea is to approximate $y(x)$ and to vary the matrix M in order to use the information gathered in finding one approximation to help in finding the next one. Let J be some approximation of the Jacobian $J(x)$ of $y(x)$. A Newton step $- \left[\nabla^2 F_\mu(x)\right]^{-1} \nabla F(x)$ is approximated there by

$$[M(I - J)]^{-1} M(y(x) - x) = [I - J]^{-1}(y(x) - y),$$

where I is the identity matrix. M could be fixed or updated by

$$M_n = \mu_n G_n,$$

where μ_n is some constant and G_n is an estimate of ∇F computed by information from previous iterations . How to compute the necessary estimate J of the Jacobian matrix of $y(x)$ is discussed in detail in [44]. The method developed there is called approximate Newton-method.

A precise study of the second-order properties of the Moreau-Yosida regularization is presented in [45] for the problem of minimizing a closed proper convex function, which is a selection of a finite number of twice continuously differentiable functions. It is proved that under certain constraint qualification the gradient ∇F_M is piecewise smooth. Further conditions are formulated that guarantee a superlinear (quadratic) convergence of an approximate Newton method for minimizing F.

Generally, one can consider any Newton-type method for nonsmooth equations in order to solve optimization problems. Newton-type methods in such a generality are considered in e. g. [38], [50], [53], [54]. The methods presented there are applied to solving optimization problems via augmented Lagrangians [54], via the Karush-Kuhn-Tucker equations [38], [53] or via the Moreau-Yosida regularization [5].

Our review is not an attempt to comment all recent developments of solution techniques for nonsmooth optimization problems. We only wish to present the main ideas of the well-established methods in order to clarify which of them are appropriate for solving the nonsmooth problems studied in the next two sections.

4 Lagrangian Relaxation for the Dynamic Recourse Problem

In this section, we consider the Lagrangian relaxation approach for the dynamic recourse model (2.1)-(2.9) in detail and sketch a conceptual algorithm for solving the problem. The decision variables are uniformly bounded functions $(u, p, s, w) \in \times_{t=1}^{T} L^{\infty}(\Omega, \mathcal{A}_t, \mu; \mathbb{R}^{m_t})$, $m_t = 2(I + J)$. The variables (u_i, p_i), $i = 1, \ldots, I$, and (s_j, w_j), $j = 1, \ldots, J$, are associated with one single operation unit i, and j, respectively. All constraints except for (2.4) and (2.5) are associated with a single operation unit. Thus, natural candidates for the relaxation are the coupling constraints (2.4) and (2.5). We associate Lagrange multipliers λ_1 and λ_2 with the load- and reserve-constraints, respectively. Setting $x = (u, p, s, w)$ and

$$
\begin{aligned}
L(x, \lambda) \;=\; & \mathbb{E}\Bigg\{ \sum_{i=1}^{I} \sum_{t=1}^{T} \left[FC_{it}\left(p_i^t, u_i^t\right) + SC_{it}\left(u_i(t)\right) \right] \\
& + \lambda_1^t \left(d^t - \sum_{i=1}^{I} p_i^t - \sum_{j=1}^{J} \left(s_j^t - w_j^t \right) \right) \\
& + \lambda_2^t \left(r^t - \sum_{i=1}^{I} \left(u_i^t \, p_{it}^{\max} - p_i^t \right) \right) \Bigg\},
\end{aligned}
\tag{4.1}
$$

we have to clarify what kind of objects λ_1 and λ_2 are. Duality theorems for dynamic models that are relevant for our setting are considered in [56], [58]. We utilize the results of [56]. For stating a duality result we neglect integrality and substitute $u_i^t \in \{0, 1\}$ by $u_i^t \in [0, 1]$ in (2.1) for a moment. We denote the modified constraint by (2.1)*.

First, let us recall that the dynamic recourse problem has *relatively complete recourse* if the following procedure leads to a choice of decisions x_k, $k = 1, \ldots, K$, almost surely for all stages k: Let x_1 be a feasible solution of the first stage. In the second stage (having a new observation of the load), we can choose x_2 satisfying the constraints and the dynamics of the system, i. e., in particular, (2.2) and (2.3) hold true with the corresponding components of x_1 and x_2. And so forth: In the k-th stage, we are able to choose a feasible decision x_k.

Nonanticipativity and relatively complete recourse provide sufficient conditions for considering L^1 to be the space of Lagrange multipliers λ, instead of working with esoteric objects from $(L^{\infty})^*$ (cf. [56]).

Suppose, additionally, that *strict feasibility* holds true. It means, that the feasible set determined by (2.1)*-(2.5) has a non-empty interior in $\times_{i=1}^{T} L^{\infty}(\Omega, \mathcal{A}_t, \mu; \mathbb{R}^{m_t})$, i. .e., there exists a positive real number ε, a point $\bar{x} \in \times_{i=1}^{T} L^{\infty}(\Omega, \mathcal{A}_t, \mu; \mathbb{R}^{m_t})$ and a neighbourhood U of $\bar{x}$ such that any point $x = (u, p, s, w) \in U$ satisfies (2.1)*-(2.3) and the inequalities:

$$
\sum_{i=1}^{I} p_i^t + \sum_{j=1}^{J} \left(s_j^t - w_j^t \right) \geq d^t + \varepsilon, \quad t = 1, \ldots, T,
$$

$$\sum_{i=1}^{I} \left(u_i^t p_{it}^{\max} - p_i^t \right) \geq r^t + \varepsilon, \quad t = 1, \ldots, T.$$

In terms of a power generation system, strict feasibility means that the generation system should have the capacity to produce power that satisfies every slightly changed demand and reserve-condition regarding the other constraints. This is a reasonable and acceptable restriction, which can be assumed to be satisfied.
We denote

$$X = \left\{ x \in \mathop{\times}_{i=1}^{T} L^{\infty} \left(\Omega, \mathcal{A}_t, \mu; I\!\!R^{m_t} \right) : \ (2.1\,)^* - (2.5\,) \text{ are fulfilled} \right\};$$

$$\Lambda = \left\{ \lambda \in \mathop{\times}_{i=1}^{T} L^{1} \left(\Omega, \mathcal{A}_t, \mu; I\!\!R^{2} \right) : \lambda_1^t, \lambda_2^t \geq 0 \ \mu - \text{ a.s. for } t = 1, \ldots, T \right\}.$$

The following duality statement holds true.

Proposition 4.1 *The Langrange function (4.1) has at least one saddle point $(\bar{x}, \bar{\lambda}) \in X \times \Lambda$ assuming (2.6), (2.7), relatively complete recourse and strict feasibility. In order that the function $\bar{x} \in X$ be an optimal solution of the problem (2.1) - (2.9) it is necessary and sufficient that the following conditions be satisfied a.s. for some $\bar{\lambda} \in \Lambda$:*

$$\begin{aligned} \bar{\lambda}_1^t \ & \left[d^t - \sum_{i=1}^{I} \bar{p}_i^t - \sum_{j=1}^{J} \left(\bar{s}_j^t - \bar{w}_j^t \right) \right] = 0 \\ \bar{\lambda}_2^t \ & \left[r^t - \sum_{i=1}^{I} \left(\bar{u}_i^t p_{it}^{\max} - \bar{p}_i^t \right) \right] = 0 \\ 0 \ & \in \partial_x L \left(\bar{x}^t, \bar{\lambda}^t \right), \quad t = 1, \ldots, T. \end{aligned}$$

(4.2)

Proof: The assertion follows by Theorem 1 and the arguments of Theorem 7 from [56]. □

Now we consider the relaxed problem:

$$\min_{(u, p, s, w)} \ L(u, p, s, w) \text{ subject to } (2.1\,) \text{ - } (2.3\,). \tag{4.3 }$$

Denoting the marginal function of the latter problem by $\Theta(\lambda)$, the dual problem reads

$$\max \Theta(\lambda) \text{ subject to } \lambda \in \Lambda. \tag{4.4 }$$

Now, we show that the dual problem is decomposable with respect to the single units. Using the notations of the previous section, we define

$$x_i = (u_i, p_i), \ i = 1, \ldots, I, \quad x_{I+j} = (s_j, w_j), \quad j = 1, \ldots, J, \quad \bar{n} = I + J,$$

and observe that all functions are separable with respect to x_i, $i = 1, \ldots, \bar{n}$. We define functions $\Theta_i(\cdot)$ and $\tilde{\Theta}_j(\cdot)$:

$$
\begin{aligned}
\Theta_i(\lambda) &= \min_{(u_i, p_i)} E \sum_{t=1}^{T} \Big[FC_{it}\left(p_i^t, u_i^t\right) + SC_{it}(u_i(t)) - \lambda_1^t p_i^t \\
&\qquad\qquad - \lambda_2^t \left(u_i^t p_{it}^{\max} - p_i^t\right) \Big] \\
&= \min_{u_i} E \sum_{t=1}^{T} \Big[\min_{p_i^t} \{ FC_{it}\left(p_I^t, u_i^t\right) - (\lambda_1^t - \lambda_2^t)\, p_i^t \} \\
&\qquad\qquad + SC_{it}(u_i(t)) - \lambda_2^t\, u_i^t\, p_{it}^{\max} \Big]
\end{aligned}
$$

The latter equality holds by the separable structure of the functions FC_{it} with respect to p_i^t and u_i^t (cf.(2.7)) and the possibility to exchange min and E in the above expression.

$$
\tilde{\Theta}_j(\lambda) = \min_{(s_j, w_j)} E \sum_{t=1}^{T} \left[-\lambda_1^t (s_j^t - w_j^t) \right]
$$

Consequently, the function $\Theta(\lambda)$ can be expressed as:

$$
\Theta(\lambda) = \sum_{i=1}^{I} \Theta_i(\lambda) + \sum_{j=1}^{J} \tilde{\Theta}_j(\lambda) + E \sum_{t=1}^{T} \left[\lambda_1^t\, d^t + \lambda_2^t\, r^t \right]
$$

It has a separable structure with respect to the single units as do the constraints (2.1) - (2.3), (2.6) - (2.8). Thus, the value and subgradients of $\Theta(\lambda)$ can be computed for a given argument λ by solving the subproblems $P_i(\lambda)$, $i = 1, \ldots, I$ and $\tilde{P}_j(\lambda)$ $j = 1, \ldots, J$:

$$
P_i(\lambda): \quad \min_{u_i} E \sum_{t=1}^{T} \Big[\min_{p_i^t} \{ FC_{it}\left(p_i^t, u_i^t\right) - (\lambda_1^t - \lambda_2^t)\, p_i^t \} + SC_{it}(u_i(t))
$$

$$
- \lambda_2^t\, u_i^t\, p_{it}^{\max} \Big] \text{ subject to (2.1), (2.3)}
$$

$$
\tilde{P}_j(\lambda): \quad \min_{(s_j, w_j)} E \sum_{t=1}^{T} \left[-\lambda_1^t \left(s_j^t - w_j^t \right) \right] \text{ subject to (2.1), (2.2), (2.6)}
$$

Note that these are dynamic recourse problems themselves associated with the single generation units. The subgradients of $\Theta(\lambda)$ with respect to λ_1 and λ_2 are given by

$$
d^t - \sum_{i=1}^{I} p_i^t - \sum_{j=1}^{J} \left(s_j^t - w_j^t \right) \text{ and}
$$

$$
r^t - \sum_{i=1}^{I} \left(u_i^t\, p_{it}^{\max} - p_i^t \right),
$$

where (u_i^t, p_i^t) and (s_j^t, w_j^t) are solutions of $P_i(\lambda)$, $i = 1 \ldots, I$, and $\tilde{P}_j(\lambda)$, $j = 1, \ldots, J$, respectively.

Now, we suppose that the measure μ has a finite support. Then the dynamic recourse problem can be viewed as a large-scale finite-dimensional optimization problem. Let us check the properties of Θ discussed in the previous section. The concavity of Θ follows immediately. The assumptions of Proposition 3.1 and Proposition 3.2 are satisfied for this problem. Observe that the assumption (A) of Section 3 is satisfied, too, i. e., the feasible set with respect to the continuous variables is a compact set because of (2.1). Therefore, the necessary properties for a nonsmooth optimization method of the kind discussed in Section 3 are at hand provided that efficient algorithms for solving the subproblems are available. Consequently, we shall have established an algorithm for solving the problem (2.1) - (2.9) if the following points are clarified:

- approximation of the stochastic process d^t by a scenario tree;

- choice of an appropriate method for solving the dual problem (4.4);

- efficient algorithms for solving the subproblems $P_i(\cdot)$, $\tilde{P}_j(\cdot)$,

- gaining information from the solution of the dual problem (4.4) for computing a primal feasible solution (Lagrangian heuristics) and providing an estimation of the occurring relative duality gap.

Let us comment on all of these points. The stochastic process d^t can be approximated by means of an analysis of statistical data using also expert knowledge. The first thing to clarify is the nature of the demand randomness. In order to estimate the load of the system one usually uses the data of the same week from previous years, data of days with similar weather conditions, and the experience of experts. The strategy of creating scenarios has to reflect truly all possible future demands. The number of scenarios that approximate the demand has to be chosen in such a way that a fairly good approximation is obtained but the speed of the optimization procedure is not affected critically since the execution time of the algorithm grows rapidly as the number of scenarios included increases. The probability assigned to each scenario can be calculated according to the likelihood of its occurrence.

The functions FC_{it} and SC_{it} $i = 1, \ldots, I$, $t = 1, \ldots, T$ are assumed to be piecewise linear or quadratic. Consequently, the function $\Theta(\lambda)$ is piecewise twice continuously differentiable. Therefore, any method of non-smooth optimization of those discussed in the previous section could be applied. The methods developed as bundle methods of order higher than one could be applied successfully, e. g. [36], [42], [44]. Unfortunately, for those guaranteeing superlinear convergence ([45]), no computational code is available up to now. The variable metric bundle methods [36], [42], [44] provide convergence but no estimate of the rate is given. We would like to emphasize that those methods are finite for piecewise linear convex functions. The published experience with NOA Version 3.0 ([37]) reports fast convergence in practice (cf. [36]). The efficiency of the optimization algorithm depends to great extent on the fast computation of the values and subgradients of the objective function $\Theta(\lambda)$. Therefore, the development of fast algorithms for solving the problems $P_i(\lambda)$ and $\tilde{P}_j(\lambda)$, $i = 1, \ldots, I$, $j = 1, \ldots, J$, is important. An algorithm for solving the problems $\tilde{P}_j(\lambda)$, $j = 1, \ldots, J$, has been developed in [48]. It regards $\tilde{P}_j(\lambda)$ as a network-flow problem and suggests a procedure adapted to the structure called EXCHA. The crucial point in this procedure is the selection of a proper direction from a prescribed subset of descent directions for minimizing the objective of $\tilde{P}_j(\lambda)$. Let us consider the problems $P_i(\lambda)$, $i = 1, \ldots, I$. The inner minimization (with respect to p_i) can be done explicitly or by one-dimensional optimization. Further, a dynamic programming procedure can be used to minimize the expected costs with respect to the integer variables u_i. A state transition graph of the unit to each scenario regarding the nonanticipativity constraint can be considered. Then the solution corresponds to a tree in this graph that has minimal weighted length. In order to reduce the number of nodes, we can include the constraints (2.3) into the process of generating the state transition graph by setting nodes "off" for at least τ_i periods.

Another substantial part of the solution procedure for the dynamic recourse problem consists in developing an algorithm for the determination of a primal feasible solution after one has found a solution of the dual problem. As already established, if we find an "almost" feasible point, it is "almost" a solution (Proposition 3.1). In addition, the optimal value $\Theta(\lambda)$ of the dual problem is a better lower bound of the objective function of the primal problem than the value of its continuous relaxation. It is possible to use some modification of the heuristic procedure presented for this purpose in [70] and further modified as in [14]. Recent publications [19], [41] suggest heuristics based on relaxed convexified primal problems These procedures are not directly applicable to the problem considered here due to the presence of pumped-storage hydro plants. An adaptation of these ideas to our setting needs further investigation.

In our case, the Lagrangian heuristics could work as follows:

- try to satisfy the reserve-constraints by using pumped-storage hydro plants in those time intervals, where the largest values of $d^t + r^t$ occur. If the reserve-constraints are still violated, use the procedure of [70].

- improve the feasible solution found at the end of the procedure above by solving the problem keeping the integer variables fixed. An algorithm for the latter problem is suggested in [49] that is a modification of the network-flow algorithm in [48]. The problem is considered as a network-flow problem again and the algorithm makes use of its special structure.

Summarizing, the presented solution technique includes the following basic steps:

- generation of a scenario tree (discrete approximation of d)

- solving the problem (4.4) e.g. by NOA Version 3.0,
  solving the problems $P_i(\lambda), i = 1, \ldots, I$, by dynamic programming and
  $\tilde{P}_j(\lambda), \; j = 1, \ldots, J$, by EXCHA.

- determination of a primal feasible solution by the procedure described
  above.

An illustrative example for an approximation of the load is given in Figure 3
and Figure 4 expresses the corresponding stochastic schedule for fixed binary
variables. The values of the approximative load are generated by using the
value of a given load, and a standard normal random variable (see [49] for
details).

A final remark is due. There is an estimate for the occurring duality gap. We
use the description of the problem (2.11)-(2.13) based on scenarios. At this
place, we incorporate the nonanticipativity condition into the representation
of the model. More precisely, we consider decisions $x_{in}^t = (u_{in}^t, p_{in}^t, s_{in}^t, w_{in}^t)$
and $x_{i\hat{n}}^t$ that correspond to scenarios n and $\hat{n}$ fulfilling $d_n^t = d_{\hat{n}}^t$ for all $t =
1, \ldots, t_k$ as indistinguishable up to the stage k. We use only one notation for
the decisions at stage k for all scenarios that are indistinguishable up to that
stage. Recall that the number of scenarios at the stage k is denoted by n_k and
the number of load and reserve-constraints amounts to $2 \sum_{k=1}^{K} n_k(t_{k+1} - t_k)$.

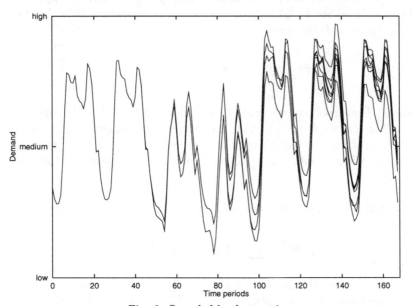

Fig. 3: Sampled load scenarios

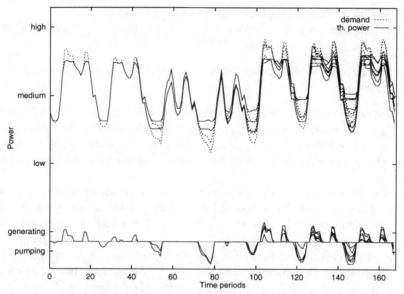

Fig. 4: Solution for the load given in Figure 3

Proposition 4.2 *Assume relatively complete recourse for the dynamic recourse problem. Let its optimal value be denoted by F^* and the optimal value of its dual problem by Θ^*. Then there exists a constant ρ such that the following estimate holds true:*

$$F^* - \Theta^* \leq (2 \sum_{k=1}^{K} n_k(t_{k+1} - t_k) + 1)\rho$$

Proof: The proof follows from Proposition 5.26 in [2]. We only have to show that the assumptions (A1)-(A3) made there are satisfied in our situation. (A1) is just the feasibility of the problem, which holds due to relatively complete recourse. (A2) and (A3) are easily checked specifying the required conditions. □

We consider the same dynamic recourse problem with a modified objective function:

$$\frac{1}{IN} \mathbb{E} \sum_{i=1}^{I} \sum_{t=1}^{T} \left[FC_{it}(p_i^t, u_i^t) + SC_{it}(u_i(t)) \right]$$

The objective function in this case represents the *average costs per scenario-term*. We have the same optimal solution for both problems and the duality gap becomes

$$F^* - \Theta^* \leq \frac{2 \sum_{k=1}^{K} n_k(t_{k+1} - t_k) + 1}{IN} \rho$$

The latter inequality implies that the duality gap goes to zero as $I \to \infty$. Consequently, the duality gap becomes small for large systems independently of making the discrete approximation of the load finer ($N \to \infty$).

5 Lagrangian Relaxation for the Two-Stage Model

We consider the two-stage stochastic power production planning model elaborated in Section 2.3 under the assumption that the fuel cost functions exhibit the form (2.7). Setting $x := (u, p, s, w)$ and $\hat{x} := (\hat{u}, \hat{p}, \hat{s}, \hat{w}) \in L^\infty(\Omega, \mathcal{A}, \mu; I\!\!R^m)$ the optimization problem consists in minimizing the objective function

$$F(x, \hat{x}) := I\!\!E \left\{ \sum_{i=1}^{I} \sum_{t=1}^{T} \left[FC_{it} \left(p_i^t + \hat{p}_i^t, \hat{u}_i^t \right) + SC_{it} \left(\hat{u}_i(t) \right) \right] \right\} \qquad (5.1\ )$$

over all decisions $x \in I\!\!R^m$ and $\hat{x} \in L^\infty(\Omega, \mathcal{A}, \mu, I\!\!R^m)$ such that the unit capacity limits (2.1), (2.2), (2.16), (2.17) the minimum down-time constraints (2.15) and the loading and reserve constraints

$$\sum_{i=1}^{I} p_i^t + \sum_{j=1}^{J} (s_j^t - w_j^t) \geq I\!\!E(d^t),$$

$$\sum_{i=1}^{I} (p_i^t + \hat{p}_i^t) + \sum_{j=1}^{J} \left(s_j^t + \hat{s}_j^t - (w_j^t + \hat{w}_j^t) \right) \geq d^t, \quad \mu - \text{ a. s.}, \qquad (5.2\ )$$

$$\sum_{i=1}^{I} (u_i^t p_{it}^{\max} - p_i^t) \geq r^t, \ t = 1, \dots T,$$

respectively, are satisfied. The constraints (5.2) are coupling across units while all remaining constraints are associated with the operation of single (thermal or hydro) units. With a similar argument based on a duality statement as in the previous section, we relax the constraints (5.2) by introducing Lagrange multipliers $\lambda = (\lambda_1, \lambda_2, \lambda_3)$, where $\lambda_1, \lambda_3 \in I\!\!R^T$ and $\lambda_2 \in L^1(\Omega, \mathcal{A}, \mu; I\!\!R^T)$. The dual problem is then of the following form:

$$\max \left\{ \Theta(\lambda) : \ \lambda \in I\!\!R^T \times L^1(\Omega, \mathcal{A}, \mu; I\!\!R^T) \times I\!\!R^T, \lambda \geq 0, \ \mu - \text{ a. s.} \right\} \qquad (5.3\ )$$

where

$$\Theta(\lambda) := \inf \left\{ L(x, \hat{x}; \lambda) : x \text{ and } \hat{x} \text{ satisfy } (2.1\ ), (2.2\ ) \text{ and } (2.15\ ) - (2.17\ ) \right\},$$

$L(x, \hat{x}; \lambda) :=$

$$F(x, \hat{x}) + \sum_{t=1}^{T} \lambda_1^t \left(I\!\!E(d^t) - \sum_{i=1}^{I} p_i^t - \sum_{j=1}^{J} (s_j^t - w_j^t) \right)$$

$$+ I\!\!E \left\{ \sum_{t=1}^{T} \lambda_2^t \left(d^t - \sum_{i=1}^{I} (p_i^t + \hat{p}_i^t) - \sum_{j=1}^{J} (s_j^t + \hat{s}_j^t - (w_j^t + \hat{w}_j^t)) \right) \right\}$$

$$+ \sum_{t=1}^{T} \lambda_3^t \left(r^t - \sum_{i=1}^{I} (u_i^t p_{it}^{\max} - p_i^t)) \right)$$

$$= \sum_{i=1}^{I} \sum_{t=1}^{T} [I\!\!E \{ FC_{it} (p_i^t + \hat{p}_i^t, \hat{u}_i^t) + SC_{it} (\hat{u}_i(t)) - \lambda_2^t (p_i^t + \hat{p}_i^t) \}$$

$$- (\lambda_1^t - \lambda_3^t) p_i^t - \lambda_3^t u_i^t p_{it}^{\max}]$$

$$- \sum_{j=1}^{J} \sum_{t=1}^{T} \left[\lambda_1^t (s_j^t - w_j^t) + I\!\!E \{ \lambda_2^t (s_j^t + \hat{s}_j^t - (w_j^t + \hat{w}_j^t)) \} \right]$$

$$+ \sum_{t=1}^{T} [\lambda_1^t I\!\!E(d^t) + I\!\!E(\lambda_2^t d^t) + \lambda_3^t r^t].$$

Hence, the dual function Θ decomposes into the form

$$\Theta(\lambda) = \sum_{i=1}^{I} \Theta_i(\lambda) + \sum_{j=1}^{J} \tilde{\Theta}_j(\lambda) + \sum_{t=1}^{T} [\lambda_1^t I\!\!E(d^t) + I\!\!E(\lambda_2^t d^t) + \lambda_3^t r^t]. \qquad (5.4)$$

Here $\Theta_i(\lambda)$ is the optimal value of a two-stage stochastic program for the (single) thermal unit i, which has the form:

$$\min \left\{ I\!\!E \left\{ \sum_{t=1}^{T} [FC_{it} (p_i^t + \hat{p}_i^t, \hat{u}_i^t) - \lambda_2^t (p_i^t + \hat{p}_i^t) + SC_{it} (\hat{u}_i(t)) \right. \right.$$

$$- (\lambda_1^t - \lambda_3^t) p_i^t - \lambda_3^t u_i^t p_{it}^{\max}] \} : \ p_{it}^{\min} \hat{u}_i^t \le p_i^t + \hat{p}_i^t \le p_{it}^{\max} \hat{u}_i^t, \qquad (5.5)$$

$$p_{it}^{\min} u_i^t \le p_i^t \le p_{it}^{\max} u_i^t, \text{ and minimum down-times (2.15)} \}$$

Introducing the optimal value function for the second-stage problem and taking into account the special form (2.7) of the fuel costs, the two-stage mixed-integer stochastic program (5.5) may be rewritten as

$$\min \{ \ \sum_{t=1}^{T} [(\lambda_3^t - \lambda_1^t) p_i^t - \lambda_3^t u_i^t p_{it}^{\max}] + I\!\!E \{ \Phi_i(u_i, p_i; \lambda_2) \} :$$

$$\qquad\qquad\qquad\qquad (5.6)$$

$$p_{it}^{\min} u_i^t \le p_i^t \le p_{it}^{\max} u_i^t, \ t = 1, \ldots, T, \text{ and (2.3) } \},$$

where

$$\Phi_i(u_i, p_i; \lambda_2) := \inf \Big\{ \sum_{t=1}^{T} \big[\max_{\ell=1,\dots,L} f_{i\ell}(p_i^t + \hat{p}_i^t) - \lambda_2^t(p_i^t + \hat{p}_i^t) + SC_{it}(\hat{u}_i(t)) + c_i \hat{u}_i^t \big] : p_{it}^{\min}\hat{u}_i^t \le p_i^t + \hat{p}_i^t \le p_{it}^{\max}\hat{u}_i^t,$$
$$t = 1, \dots, T, \text{ and (2.15)} \Big\}.$$

Since the minimization with respect to p_i and $\hat{p}_i$ (μ- a. s.) in (5.5) or (5.6) can be performed explicitly, the models represent two-stage stochastic combinatorial programs and can be solved by dynamic stochastic programming. Problem (5.6) simplifies essentially for the case of $\mathcal{I}_2 = \emptyset$, i. e., $\hat{u}_i^t = u_i^t$ ($i = 1, \dots, I$, $t = 1, \dots, T$), because the compensation program does not contain binary decisions, Φ_i enjoys a separability structure and can be computed explicitly. In the latter case (5.6) takes the form

$$\min \Big\{ \sum_{t=1}^{T} [SC_{it}(u_i(t)) + (\lambda_3^t - \lambda_1^t)p_i^t - (\lambda_3^t p_{it}^{\max} - c_i)u_i^t] +$$
$$\mathbb{E}\sum_{t=1}^{T} \hat{\Phi}_{it}(u_i^t, p_i^t; \lambda_2^t) : p_{it}^{\min}u_i^t \le p_i^t \le p_{it}^{\max}u_i^t, t = 1, \dots, T, \text{ and (2.3)}\Big\},$$

where $\hat{\Phi}_{it}(u_i^t, p_i^t; \lambda_2^t) := \inf \Big\{ \max_{\ell=1,\dots,L} f_{i\ell}(p_i^t + \hat{p}_i^t) - \lambda_2^t(p_i^t + \hat{p}_i^t) :$
$$p_{it}^{\min}u_i^t \le p_i^t + \hat{p}_i^t \le p_{it}^{\max}u_i^t \Big\}.$$

The term $\tilde{\Theta}_j(\lambda)$ in the representation (5.4) of the dual function Θ is the optimal value of the following stochastic pumped-storage subproblem for the plant j:

$$\min \Big\{ -\sum_{t=1}^{T} (\lambda_1^t + \mathbb{E}(\lambda_2^t))(s_j^t - w_j^t) + \mathbb{E}\Big[\sum_{t=1}^{T} \lambda_2^t (\hat{s}_j^t - \hat{w}_j^t)\Big] :$$

$$(s_j, w_j) \text{ and } (\hat{s}_j, \hat{w}_j) \text{ satisfy } 0 \le s_j^t \le s_{jt}^{\max}, \ 0 \le w_j^t \le w_{jt}^{\max}, \tag{5.7}$$

$$0 \le \ell_j^t \le \ell_{jt}^{\max}, \ t = 1, \dots, T, \text{ and (2.2), (2.16)}\Big\}.$$

Problem (5.7) represents a linear two-stage stochastic program, which can be solved by standard solution techniques (cf. [17], [20]).

These facts motivate a Lagrangian relaxation-based conceptual solution method for the two-stage stochastic model, which is similar to the algorithm developed in the previous section. Its basic steps are:

- Generation of scenarios $d_n, n = 1, \dots, N$, for the load process d and replacing d by this discrete approximation;

- solving the concave dual problem (5.3) by applying appropriate nondifferentiable optimization methods (cf. Section 3), where function values and subgradients of Θ are computed by solving the single unit subproblems (5.5) and (5.7). Note that (5.3) has dimension $2TN$ and Θ is piecewise linear or quadratic,

- determining a primal feasible solution for the first-stage variables by a procedure that is similar to the method described in Section 4.

Acknowledgements

The authors wish to thank G. Scheibner, G. Schwarzbach and J. Thomas (VEAG Vereinigte Energiewerke AG) for the fruitful and productive cooperation. Further thanks are due to M.P. Nowak (Humboldt-University Berlin) for valuable comments and to C. Carøe (University of Copenhagen), R. Gollmer (Humboldt-University Berlin) and R. Schultz (Konrad-Zuse-Zentrum für Informationstechnik Berlin) for helpful discussions on the two-stage model in Section 2.3.

References

[1] K. Aoki, M. Itoh, T. Satoh, K. Nara and M. Kanezashi: Optimal long-term unit commitment in large scale systems including fuel constrained thermal and pumped-storage hydro, IEEE Transactions on Power Systems 4(1989),1065-1073.

[2] D.P. Bertsekas: Constrained Optimization and Lagrange Multiplier Methods, Academic Press, New York 1982.

[3] D.P. Bertsekas, G.S. Lauer, N.R. Sandell Jr. and T.A. Posbergh: Optimal short-term scheduling of large-scale power systems, IEEE Transactions on Automatic Control AC-28(1983),1-11.

[4] J.R. Birge: Decomposition and partitioning methods for multi-stage stochastic linear programs, Operations Research 33(1985), 989-1007.

[5] J.R. Birge, L. Qi and Z. Wei: Two methods for nonsmooth optimization, Applied Mathematics Report AMR95/4, School of Mathematics, The University of New South Wales (Sydney, Australia), 1995.

[6] J.R. Birge and C.H. Rosa: Parallel decomposition of large-scale stochastic nonlinear programs, Annals of Operations Research 64(1996), 39-65.

[7] J.F. Bonnans, J.C. Gilbert, C. Lemaréchal and C.A. Sagastizábal: A family of variable metric proximal methods, Mathematical Programming 68(1995), 15-47.

[8] D.W. Bunn and S.N. Paschentis: Development of a stochastic model for the economic dispatch of electric power, European Journal of Operational Research 27(1986),179-191.

[9] C.C. Carøe and R. Schultz: Dual decomposition in stochastic integer programming, Preprint SC 96-46, Konrad-Zuse-Zentrum für Informationstechnik Berlin, 1996.

[10] P. Carpentier, G. Cohen, J.-C. Culioli and A. Renaud: Stochastic optimization of unit commitment: a new decomposition framework, IEEE Transactions on Power Systems 11(1996), 1067-1073.

[11] G.B. Dantzig and G. Infanger: Approaches to stochastic programming with application to electric power systems, in [22], 125-138.

[12] M.A.H. Dempster: On stochastic programming II: Dynamic problems under risk, Stochastics 25(1988), 15-42.

[13] M.A.H. Dempster and R.T. Thompson: EVPI-based importance sampling solution procedures for multistage stochastic linear programmes on parallel MIMD architectures, Annals of Operations Research (to appear).

[14] D. Dentcheva, R. Gollmer, A. Möller, W. Römisch and R. Schultz: Solving the unit commitment problem in power generation by primal and dual methods, in: Progress in Industrial Mathematics at ECMI 96 (M. Brøns, M.P. Bendsøe and M.P. Sørensen Eds.), Teubner, Stuttgart 1997, 332-339.

[15] J. Dupačová: Multistage stochastic programs: The state-of-the-art and selected bibliography, Kybernetika 31(1995), 151-174.

[16] J. Dupačová: Stochastic programming: Approximation via scenarios, Working Paper, Charles University Prague, 1996 and Aportaciones Matematicas (Mexican Society of Mathematics)(to appear).

[17] Y. Ermoliev and R.J.-B. Wets (Eds.): Numerical Techniques for Stochastic Optimization, Springer-Verlag, Berlin 1988.

[18] H. Everett: Generalized Lagrange multiplier method for solving problems of optimal allocation of resources, Operations Research 11(1963), 399-417.

[19] S. Feltenmark, K.C. Kiwiel and P.O. Lindberg: Solving unit commitment problems in power production planning, in: Operations Research Proceedings 1996 (U. Zimmermann et al. Eds.), Springer-Verlag, Berlin 1997, 236-241.

[20] K. Frauendorfer: Stochastic Two-Stage Programming, Lecture Notes in Economics and Mathematical Systems Vol. 392, Springer-Verlag, Berlin 1992.

[21] K. Frauendorfer: Barycentric scenario trees in convex multistage stochastic programming, Mathematical Programming 75(1996), 277-294.

[22] K. Frauendorfer, H. Glavitsch and R. Bacher (Eds.): Optimization in Planning and Operation of Electric Power Systems, Physica-Verlag, Heidelberg 1993.

[23] S. Früholz: Stochastische Optimierungsmodelle zu Lastverteilungsproblemen bei unsicherem Bedarf, Diplomarbeit, Humboldt-Universität Berlin, Institut für Mathematik 1995.

[24] H.I. Gassmann: MSLiP: A computer code for the multistage stochastic linear programming problem, Mathematical Programming 47(1990), 407-423.

[25] N. Gröwe and W. Römisch: A stochastic programming model for optimal power dispatch: Stability and numerical treatment; in: Stochastic Optimization (K. Marti, Ed.), Lecture Notes in Economics and Mathematical Systems Vol. 379, Springer-Verlag, Berlin 1992, 111-139.

[26] N. Gröwe, W. Römisch and R. Schultz: A simple recourse model for power dispatch under uncertain demand, Annals of Operations Research 59(1995), 135-164.

[27] J. Guddat, W. Römisch and R. Schultz: Some applications of mathematical programming techniques in optimal power dispatch, Computing 49(1992), 193-200.

[28] J.L. Higle and S. Sen: Stochastic Decomposition - A Statistical Method for Large Scale Stochastic Linear Programming, Kluwer, Dordrecht 1996.

[29] J.-B. Hiriart-Urruty and C. Lemaréchal: Convex Analysis and Minimization Algorithms I and II, Springer-Verlag, Berlin 1993.

[30] G. Infanger: Planning Under Uncertainty - Solving Large-Scale Stochastic Linear Programs, Boyd and Fraser, 1994.

[31] T. Ishikida and P.P. Varaiya: Pricing of electric power under uncertainty: information and efficiency, IEEE Transactions on Power Systems 10(1995), 884-890.

[32] J. Jacobs, G. Freeman, J. Grygier, D. Morton, G. Schultz, K. Staschus and J. Stedinger: SOCRATES: A system for scheduling hydroelectric generation under uncertainty, Annals of Operations Research 59 (1995), 99-133.

[33] P. Kall and S.W. Wallace: Stochastic Programming, Wiley, Chichester 1994.

[34] K.C. Kiwiel: An aggregate subgradient method for nonsmooth convex minimization, Mathematical Programming 27(1983), 320-341.

[35] K.C. Kiwiel: Methods of Descent for Nondifferentiable Optimization, Lecture Notes in Mathematics Vol. 1133, Springer-Verlag, Berlin 1985.

[36] K.C. Kiwiel: Proximity control in bundle methods for convex nondifferentiable minimization, Mathematical Programming 46(1990), 105-122.

[37] K.C. Kiwiel: User's Guide for NOA 2.0/3.0: A Fortran package for convex nondifferentiable optimization, Polish Academy of Sciences, Systems Research Institute, Warsaw (Poland), 1993/94.

[38] B. Kummer: Newton's method based on generalized derivatives for nonsmooth functions: convergence analysis, in: Advances in Optimization (W. Oettli and D. Pallaschke Eds.), Lecture Notes in Economics and Mathematical Systems Vol. 382, Springer-Verlag, Berlin 1992, 171-194.

[39] C. Lemaréchal: Bundle methods in nonsmooth optimization, in: C. Lemaréchal and R. Mifflin (Eds.), Nonsmooth Optimization, Pergamon Press, Oxford 1978.

[40] C. Lemaréchal: Lagrangian decomposition and nonsmooth optimization: Bundle algorithm, prox iteration, augmented Lagrangian, in: Nonsmooth Optimization Methods and Applications (F. Gianessi Ed.), Gordon and Breach, Amsterdam 1992, 201-216.

[41] C. Lemaréchal and A. Renaud: Dual equivalent convex and nonconvex problems, Research Report, INRIA, Rocquencourt, 1996.

[42] C. Lemaréchal and C.A. Sagastizábal: An approach to variable metric bundle methods, in: System Modelling and Optimization (J. Henry and J.P. Yvor Eds.), Lecture Notes in Control and Information Sciences Vol. 197, Springer-Verlag, New York 1994, 144-162.

[43] C. Lemaréchal and J. Zowe: A condensed introduction to bundle methods in nonsmooth optimization, DFG-Schwerpunktprogramm "Anwendungsbezogene Optimierung und Steuerung", Report No. 495, 1994.

[44] R. Mifflin: A quasi-second-order proximal bundle algorithm, Mathematical Programming 73(1996), 51-72.

[45] R. Mifflin, L. Qi and D. Sun: Properties of the Moreau-Yosida regularization of a piecewise C^2 convex function, manuscript, School of Mathematics, The University of New South Wales (Sydney, Australia), 1995.

[46] J.M. Mulvey and A. Ruszczyński: A new scenario decomposition method for large-scale stochastic optimization, Operations Research 43(1995), 477-490.

[47] G.L. Nemhauser and L.A. Wolsey: Integer and Combinatorial Optimization, Wiley, New York 1988.

[48] M.P. Nowak: A fast descent method for the hydro storage subproblem in power generation, International Institute for Applied Systems Analysis (Laxenburg, Austria), Working Paper WP-96-109, 1996.

[49] M.P. Nowak and W. Römisch: Optimal power dispatch via multistage stochastic programming, in: Progress in Industrial Mathematics at ECMI 96 (M. Brøns, M.P. Bendsøe and M.P. Sørensen Eds.), Teubner, Stuttgart 1997, 324-331.

[50] J.-S. Pang and L. Qi: Nonsmooth equations: Motivation and algorithms, SIAM Journal on Optimization 3(1993), 443-465.

[51] M.V.F. Pereira and L.M.V.G. Pinto: Multi-stage stochastic optimization applied to energy planning, Mathematical Programming 52 (1991), 359-375.

[52] A. Prékopa: Stochastic Programming, Kluwer, Dordrecht 1995.

[53] L. Qi and H. Jiang: Karush-Kuhn-Tucker equations and convergence analysis of Newton methods and Quasi-Newton methods for solving these equations, Applied Mathematics Report AMR94/5, School of Mathematics, The University of New South Wales (Sydney, Australia), 1994.

[54] L. Qi and J. Sun: A nonsmooth version of Newton's method, Mathematical Programming 58(1993), 353-367.

[55] R.T. Rockafellar and R.J.-B. Wets: Nonanticipativity and L^1-martingales in stochastic optimization problems, Mathematical Programming Study 6(1976), 170-187.

[56] R.T. Rockafellar and R.J.-B. Wets: The optimal recourse problem in discrete time: L^1-multipliers for inequality constraints, SIAM Journal on Control and Optimization 16(1978), 16-36.

[57] R.T. Rockafellar and R..J.-B. Wets: Scenarios and policy aggregation in optimization under uncertainty, Mathematics of Operations Research 16(1991), 119-147.

[58] W. Römisch and R. Schultz: Decomposition of a multi-stage stochastic program for power dispatch, ZAMM - Zeitschrift für Angewandte Mathematik und Mechanik 76(1996) Suppl.3, 29-32.

[59] C.H. Rosa and A. Ruszczyński: On augmented Lagrangian decomposition methods for multistage stochastic programs, Annals of Operations Research 64(1996), 289-309.

[60] A. Ruszczyński: A regularized decomposition method for minimizing a sum of polyhedral functions, Mathematical Programming 35(1986), 309-333.

[61] A. Ruszczyński, Y. Ermoliev and V.I. Norkin: On optimal allocation of indivisibles under uncertainty, Operations Research (to appear).

[62] R. Schultz: Discontinuous optimization problems in stochastic integer programming, ZAMM - Zeitschrift für Angewandte Mathematik und Mechanik 76(1996) Suppl.3, 33-36.

[63] R. Van Slyke and R. J.-B. Wets: L-shaped linear programs with applications to optimal control and stochastic programming, SIAM Journal on Applied Mathematics 17(1969), 638-663.

[64] H. Schramm and J. Zowe: A version of the bundle idea for minimizing a nonsmooth function: conceptual idea, convergence analysis, numerical results, SIAM Journal on Optimization 2(1992), 121-152.

[65] S. Takriti, J.R. Birge and E. Long: A stochastic model for the unit commitment problem, IEEE Transactions on Power Systems 11(1996), 1497-1508.

[66] S. Takriti, J.R. Birge and E. Long: Intelligent unified control of unit commitment and generation allocation, Technical Report 94-26, Department of Industrial and Operations Engineering, University of Michigan, Ann Arbor 1994.

[67] H. Wacker (Ed.): Applied Optimization Techniques in Energy Problems, Teubner, Stuttgart 1985.

[68] R. J.-B. Wets: Stochastic Programming, in: Handbooks in Operations Research and Management Science, Vol. 1, Optimization (G.L. Nemhauser, A.H.G. Rinnooy Kan, M.J. Todd Eds.), North-Holland, Amsterdam 1989, 573-629.

[69] A.J. Wood and B.F. Wollenberg: Power Generation, Operation, and Control (Second Edition), Wiley, New York 1996.

[70] F. Zhuang and F.D. Galiana: Towards a more rigorous and practical unit commitment by Lagrangian relaxation, IEEE Transactions on Power Systems 3(1988), 763-773.

Position and Controller Optimization for Robotic Parts Mating

G. Prokop and F. Pfeiffer

Lehrstuhl B für Mechanik; Technische Universität München

Tel.:   +49 (89) 289-15199
E-mail: prokop@lbm.mw.tu-muenchen.de
         pfeiffer@lbm.mw.tu-muenchen.de

Abstract. If a robot has to perform a specified manipulation task involving intentional environmental contacts, a certain response behavior is desired to reduce strains and ensure successful completion without damage of the contacting bodies. On the other hand, the dynamic behavior of a manipulator depends strongly on its position and the gains of its joint controllers. Hence, varying these parameters for an optimized performance during manipulation seems to be an obvious task. In order to deal with impacts, oscillations and constrained motion, a model-based optimization approach is suggested, which relies on a detailed dynamic model of the manipulator incorporating finite gear stiffnesses and damping. These models are used to define an optimization problem, which is then solved using numerical programming methods. It is illustrated with an assembly task, namely inserting a rigid peg into a hole with a PUMA 562 manipulator. The expected advantage in industrial applications is a comparatively easy implementation, because performance can be improved by simply adjusting 'external' parameters as mating position and coefficients of the standard joint controller. Particularly, no modifications of the control architecture and no additional hardware are required. Application of the proposed approach to a rigid peg-in-hole insertion under practical constraints can reduce the measure for impact sensitivity by 17 %, that for mating tolerances by 78 % and the damping of end-effector oscillations and motor torques by up to 79 %. These improvements are shown to be reproducable experimentally.

1   Introduction

A great deal of work has been done in recent years for enabling robots to perform complex manipulation tasks, where manipulation means that the robot interacts mechanically with its environment. High precision demands and the lack of sensoric capabilities, but also the deficiency of realistic models in the planning stage have to some extent prevented the automation of many technical applications such as assembly, grinding, burring or surface polishing.

In addition, handling devices specially designed for one single task are often too costly. Thus, programmable multipurpose robots are used to meet the increased demand for flexibility and the question arises how such devices can be efficiently programmed to perform a task 'optimally'. Several approaches have been developed in order to tune a robot to the specific properties of a task, especially with regard to environmental interaction, see among others [6, 16].

Elaborate control schemes have been implemented including nonlinear controllers, adaptive and robust control [7, 19], as well as hierarchical structures incorporating elaborate force control strategies, [14, 17]. There are promising approaches among them, which allow to perform a previously planned task fast and reliably, mainly because of their ability to detect possible errors during the task and take action to correct it.

However, in the planning stage the occurence of such errors can to a certain extent be avoided by the use of model-based analysis and optimization tools. The free parameters, which can be adjusted for this purpose are position and trajectory of the robot, its control coefficients and some design properties of the parts to be assembled. Depending on the properties of the task the effects of disturbances and finite tolerances can be predicted and systematically optimized.

Several measures have been developed for the judgement of the ability of a robot for manipulation, [1, 2, 9, 20]. In [20] a manipulability measure is defined, which gives an indication of how far the manipulator is from singularities and thus able to move and exert forces uniformly in all cartesian directions. These considerations were taken up in [2], where ellipsoids for force transmission, manipulability and impact magnitude are defined and visualized for a planar 4-DOF arm and a PUMA 560 robot. Asada [1] refers to the same effect as the virtual mass and pursues a concept, where the centroid of the end effector and its virtual mass are interactively optimized to achieve a desired dynamic behaviour.

To overcome problems stemming from uncertainties in the relative position of the mating parts to each other, Pfeiffer [9] uses a quasistatic force equilibrium between the robot's end effector and the environment for the different possible contact configurations. Tolerance areas are calculated, which show the permissible deviation from the ideal path to ensure successful completion, depending on the cartesian stiffness of the robot's tip.

Assembly strategies incorporating the process emerged first from purely geometric analyses, [3]. Such process investigations were extended to quasistatic investigations of the contact forces for different configurations, [15, 18] and interactions between process and robot were taken into account, [17]. With the ability of modeling structure-variant multibody systems incorporating unilateral contacts, complete dynamic models of robot and process including all interactions are available, [8, 9, 13]. These detailed models can be used to study the behavior of manipulators in connection with all effects emerging from environmental interaction, such as impacts, friction, possible stiction and constrained motion, see [17]. Therefrom, criteria and constraints have been evaluated to form an optimization problem for the automatic synthesis of a robotic assembly cell, [11]. It was soon clear that in optimizing manipulation tasks the problems shifted from difficulties in building realistic process models to computational expense and numerical convergence. Therefore, in the proposed approach computationally expensive process models are not part of the underlying model. They are rather contained in the formulation of the criteria by weighting factors for the different cartesian directions.

2 System Model and Problem Description

A typical feature in robotic manipulation tasks is the change in contact configuration between the corresponding workpieces. The resulting forces and moments acting on the end-effector influence the motion of the manipulator during the task. When modelling such processes, we can distinguish between the dynamic model of the robot and that of the process dynamics.

Industrial robots suitable for complex assembly tasks have to provide at least six degrees of freedom and – to ensure flexible operation – a large workspace. We will therefore focus on manipulators with 6 rigid links and 6 revolute joints, which are very common in industry. Such a robot can be modelled as a tree-structured multibody system, Fig. 1.

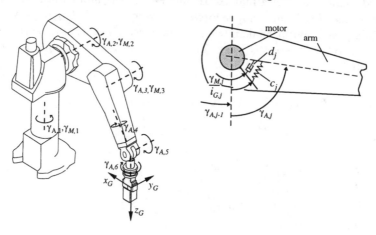

Fig. 1: Dynamic robot model

The joints of the first three axes are considered elastic in order to take the finite gear stiffnesses into account, which play an important role in precision assembly. For this purpose a linear force law consisting of a spring-damper combination c_j, d_j, combined with the gear ratio $i_{G,j}$, $j = 1, \ldots, 3$, is assumed. The gears of the hand axes are considered stiff and the motion of one arm and its corresponding motor is kinematically coupled.[1]

According to this, the robot possesses 9 degrees of freedom, 6 arm angles and 3 free motor angles connected with the respective joints by a linear force law.

$$\tau_{A,j} = c_j \left(\frac{\gamma_{M,j}}{i_{G,j}} - \gamma_{A,j} \right) + d_j \left(\frac{\dot{\gamma}_{M,j}}{i_{G,j}} - \dot{\gamma}_{A,j} \right) ; \qquad j = 1, \ldots, 3, \quad (1)$$

where $\gamma_{M,j}, \gamma_{A,j}$ denote the angle of the j-th motor and arm, respectively, relative to the previous body, see Fig. 1. c_j and d_j are stiffness and damping factors of the j-th gear and $i_{G,j}$ is the gear ratio.

Thus, for the vector of generalized coordinates

$$\bar{q} := [\gamma_{M,1}, \gamma_{M,2}, \gamma_{M,3}, \gamma_{A,1}, \ldots, \gamma_{A,6}]^T \qquad (2)$$

the equations of motion for the robot with forces acting on the gripper can be written as

$$M(\bar{q})\ddot{\bar{q}} + f(\bar{q}, \dot{\bar{q}}) = B\tau_C + W(\bar{q})\lambda \qquad (3)$$

with M being the inertia matrix, B and W are the input matrices for the motor torques (τ_C) and gripper forces (λ), respectively. $f(\bar{q}, \dot{\bar{q}})$ is a vector containing the gravitational and centrifugal forces. Let us assume the robot to be controlled by six PD joint controllers, one for each joint, which are represented by

$$\tau_C = -K_p \left(\bar{q}_M - \bar{q}_{Md} \right) - K_d \left(\dot{\bar{q}}_M - \dot{\bar{q}}_{Md} \right) , \qquad (4)$$

where

$$
\begin{aligned}
K_p &= \text{diag}\,[K_{p,1}, \ldots, K_{p,6}] , \\
K_d &= \text{diag}\,[K_{d,1}, \ldots, K_{d,6}] , \\
\bar{q}_{Md} &= [\gamma_{M1d}, \ldots, \gamma_{M6d}]^T , \\
\bar{q}_M &= B^T \bar{q}
\end{aligned}
$$

---

[1] In reality, the gears of the hand axes are elastic as well. However, the masses of the wrist bodies are comparatively small and thus the associated natural frequencies are out of the range of interest for our purposes.

domain	robot response behavior
criteria	impact sensitivity
	maximal robot force
	mating tolerance
	vibrational behavior
constraints	**robot design:**
	joint angle limitations
	joint torque limitations
	robot control:
	controller stability
	singularities
	practical demands in cell:
	workspace restrictions

Table 1: Optimization concept

with $K_{p,j}$, $K_{d,j}$ being the stiffness and damping control coefficients of the j-th axis referring to motor angles as inputs and motor torques as outputs. γ_{Mjd} is the motor angle of the j-th motor desired for a given position.

Supposing that the length of a trajectory for mating two parts together is small compared to the robot's characteristic measures, eq. (3) can be linearized around a working point $\overline{q} = q_0 + q, \dot{q}_0 = \ddot{q}_0 = 0$, which yields

$$
\begin{aligned}
M(q_0)\ddot{q} + \tilde{P}(q_0, K_d)\dot{q} + \tilde{Q}(q_0, K_p)q = \\
= h(q_0) + BK_p q_{Md} + BK_d \dot{q}_{Md} + W(q_0)\lambda(q, \dot{q}) \,,
\end{aligned} \tag{5}
$$

$$
\begin{aligned}
\tilde{P}(q_0, K_d) &= P(q_0) + BK_d B^T \\
\tilde{Q}(q_0, K_p) &= Q(q_0) + BK_p B^T
\end{aligned}
$$

with damping matrix $P(q_0)$ and stiffness matrix $Q(q_0)$. $h(q_0)$ contains only the gravitational forces.

With this model the parameters which affect the natural mode of vibrations of the manipulator, namely its position q_0 and control coefficients K_p and K_d, are considered, which – in mathematical terms – give an optimal linearization point for the equations of motion (5). The goal is to find q_0, K_p and K_d such that the system described by (5) behaves optimally with respect to the criteria relevant for a specific process. q_0, K_p and K_d can be optimized with only rough knowledge of the process to be carried out. The resulting optimization problem is described in table 1.

Optimization of robotic manipulation processes, particularly assembly, is essentially a trade-off between different, sometimes contradictory, aims. Optimizing for reduced sensitivity against gripper impacts, for example, may deteriorate the behavior with respect to the maximal applicable gripper force and vice versa. Thus, a set of criteria is established, with which a vector optimization problem can be stated. The specific needs of different processes are taken into account by the correct choice of criteria and weighting factors. For this purpose, the functional-efficient set of solutions is calculated, from which an optimal trade-off between the criteria can be chosen.

3 Criteria and Constraints

The effects, which influence the robot's behaviour during an assembly task have been worked out by a quasistatic and dynamic analysis of the robot dynamics in conjunction with a detailed modelling of different mating processes, see [8, 9, 13]. Scalar optimization criteria are derived from them, the minimization of which yields an improvement of the system's performance with regard to the respective effect.

3.1 Optimization Criteria

Impact sensitivity: When the mating parts are getting in contact with each other, impacts are unavoidable. However, their intensity is proportional to the effective mass $m_{red} = \left[w^T M^{-1} w \right]^{-1}$, reduced to the end effector, where w is the projection of the impact direction into the generalized coordinates and depends on the robot's position as well as on the cartesian impact direction. Fig. 2 shows an ellipsoid at the robot's gripper, from which

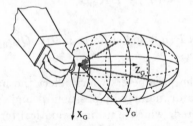

Fig. 2: Impact sensitivity in different cartesian directions

the reduced end-point inertia for each cartesian direction can be seen. In order to reduce the impact sensitivity, the volume of that ellipsoid must be minimized. Thus, we define the reduced endpoint inertia matrix M_{red} as

$$M_{red} = \left(\left[\begin{array}{c} J_{TG}(q_0) \\ J_{RG}(q_0) \end{array} \right] M(q_0)^{-1} \left[\begin{array}{c} J_{TG}(q_0) \\ J_{RG}(q_0) \end{array} \right]^T \right)^{-1}, \tag{6}$$

where $J_{TG}(q_0)$ and $J_{RG}(q_0)$ are the gripper's Jacobians of translation and rotation with respect to a coordinate frame fixed at the gripper. Depending on the specific needs of the mating process, a 6×6 diagonal positive semidefinite matrix g_M of weighting factors is introduced for the trade-off between the cartesian directions. In g_M the directions, in which impacts will occur during manipulation, can be emphasized. Thus, geometrically, the ellipsoid will be sqeezed or rotated during an optimization from directions, in which the mating process considered is sensitive against impacts into directions, in which impacts are not likely to occur. Therefore, as an optimization criterion for the minimization of impact intensities in the sensitive directions

$$G_1 = \left\| g_M^T M_{red} g_M \right\| \tag{7}$$

is stated, with $\|A\| = \sqrt{\text{trace}\left(A^T A\right)}$ being the Frobenius-norm of A.

Maximal applicable mating force λ_{max} in the direction of insertion: The upper bound for the applicable mating force λ is defined by the maximum torque of each motor multiplied by the resulting lever arms. λ_{max} can thus be evaluated from

$$\lambda_{max} = \min_i \left\{ \min \left\{ \begin{array}{c} \dfrac{h_i(q_0) - \tau_{i,max}}{J_{TG}^T(q_0)\,n} \\[2ex] \dfrac{h_i(q_0) - \tau_{i,min}}{J_{TG}^T(q_0)\,n} \end{array} \right\} \right\}, \\ i = 1, \ldots, n_M, \tag{8}$$

where n_M is the number of driven axes and n is a unit vector denoting the cartesian insertion direction. $h_i(q_0)$ is the torque necessary at joint i to balance the gravitational forces and is equal to the i-th component of $h(q_0)$. For a maximization of λ_{max}, its inverse is taken as the second criterion

$$G_2 = \frac{1}{\lambda_{max}}. \tag{9}$$

Mating tolerance: The deviation Δx_G from the desired path for a given static force depends on the endpoint stiffness, reduced to the cartesian gripper

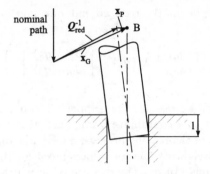

Fig. 3: Force equilibrium for mating tolerance

coordinates, [9], see Fig. 3. A quasistatical force equilibrium at the gripper
yields

$$\Delta x_G = Q_{red}^{-1}\lambda + \Delta x_P,$$
$$Q_{red} = \left(\left[\begin{array}{c} J_{TG} \\ J_{RG} \end{array}\right]\tilde{Q}^{-1}\left[\begin{array}{c} J_{TG} \\ J_{RG} \end{array}\right]^T\right)^{-1}. \tag{10}$$

Δx_P is the deviation resulting from the clearance between the two parts
and from the stiffness of the parts themselves. Therefore it depends only
on the mating process itself and needs not to be considered here. For a
maximization of Δx_G the reduced stiffnesses Q_{red} in the lateral directions
must be minimized. Together with a weighting factor g_Q, which contains
the cartesian directions, in which the tolerances are critical, this forms the
criterion for the maximization of the mating tolerance

$$G_3 = \|g_Q Q_{red}\|. \tag{11}$$

It should be noted that, for translational deviations, mainly the directions
perpendicular to the insertion direction should be emphasized by g_Q and all
rotational directions can possibly be contained, whereas the cartesian stiffness
in the insertion direction does not contribute to the mating tolerance and
should be high for a reduced path deviation.

Disturbance and tracking properties: When being excited by disturbances (e.g. by impacts), the gripper performs oscillations, the amplitude
and damping of which depend strongly on the robot's position and joint controller. On the other hand, a desired force or motion must be transmitted
to the end effector as directly as possible. All this must be performed with

as little expenditure of energy as possible. To meet these requirements, the equations of motion (5) are evaluated by time simulations for certain test inputs and three integral criteria are formulated:

(1) Damping of force induced gripper oscillations:

$$G_4 = \int_0^\infty \boldsymbol{x}^T(t)\boldsymbol{g}_S\boldsymbol{x}(t)\,\mathrm{dt}\,, \tag{12}$$

where $\boldsymbol{x}(t) = \begin{bmatrix} \boldsymbol{J}_{TG}(\boldsymbol{q}_0) \\ \boldsymbol{J}_{RG}(\boldsymbol{q}_0) \end{bmatrix} \boldsymbol{q}(t)$ and $\boldsymbol{\lambda}(t) = [0\,,\,0\,,\,\delta(t)\,,\,0\,,\,0\,,\,0]^T$ represents a unit impulse input to (5) in the direction of insertion. $\boldsymbol{g}_S$ in (12) gives the trade-off between end-effector oscillations in the different cartesian directions.

(2) Transmission of a desired force to the end-effector:

$$G_5 = \int_0^\infty (\boldsymbol{\lambda} - \boldsymbol{\lambda}_d)^T \boldsymbol{g}_F (\boldsymbol{\lambda} - \boldsymbol{\lambda}_d)\,\mathrm{dt}\,, \tag{13}$$

where $\boldsymbol{\tau}_{C,d}(t) = -\boldsymbol{I}_G^{-1}\boldsymbol{W}(\boldsymbol{q}_0)^T\boldsymbol{\lambda}_d(t)$ contains the motor torques needed to exert the desired end effector forces. $\boldsymbol{W}$ is the projection from working space into configuration space and $\boldsymbol{I}_G$ denotes the matrix of gear ratios. As a test signal $\boldsymbol{\lambda}_d(t) = [0,0,\sigma(t),0,0,0]^T$ is used, where $\sigma(t)$ is the unit step function.

(3) Joint torques: A perfect damping of gripper oscillations and perfect tracking properties would require infinite joint torques. Thus, as soon as control coefficients are being optimized, the necessary torques must be considered. The performance criterion to be minimized is

$$G_6 = \int_0^\infty \boldsymbol{\tau}_C^T \boldsymbol{g}_\tau \boldsymbol{\tau}_C\,\mathrm{dt} \tag{14}$$

with the same disturbance as in (12) and $\boldsymbol{\tau}_C$ being the joint torques from eq. (4).

The above list of optimization critria is of course not a complete list of possible objectives for robotic optimization. However, for a large class of manipulation tasks, a combination can be found suitable for the specific properties of the process. For any process the cartesian weighting factors g_M, g_Q, g_S, g_F and g_τ can be chosen from physical evidence: e. g. the impact intensities during insertion of a rigid peg into a hole will be worst in the cartesian z_G-direction. Thus, a large weight must be imposed on it in g_M.

3.2   Constraints

In order to obtain sensible results, which can be utilized in practice, certain constraints have to be imposed on the optimization problem. The highly nonlinear programming problem defined by (7) to (14) shows good convergence only when it is "properly" constrained, i. e. that the parameters are restricted to an area, where a minimum of the criteria can be reliably found. The constraints are in detail:

- **The linearized equations of motion** (5),

- **Joint angle limitations:**

$$q_{min} \leq q_0 \leq q_{max},\tag{15}$$

- **Joint torque limitations:**

$$\tau_{C,min} \leq \tau_C \leq \tau_{C,max}.\tag{16}$$

   Joint angle and joint torque limitations for our example are chosen according to those of the PUMA 562 robot.

- **Stability of the controller used:** As soon as control coefficients are being optimized, stability of the resulting system must be assured by suitable constraints. Since the robot dynamics has a linear time-invariant characteristics, the eigenvalues of the dynamic system matrix derived from eq. (5) are calculated and their real parts are restricted to be negative.

$$\mathrm{Re}\left\{ \mathrm{eig}\begin{bmatrix} 0 & I \\ -M^{-1}\tilde{Q} & -M^{-1}\tilde{P} \end{bmatrix} \right\} < 0.\tag{17}$$

- The **proximity to singularities** must be avoided. In such positions, the robot would not be able to move in the desired manner and the obtained results would be without any practical relevance. Furthermore, some of the optimization criteria tend to infinity at singular positions. Thus, punching out finite regions around them would improve the condition of the optimization problem. As a measure the condition number κ of the end effector's Jacobian is used, which is defined by $\kappa(A) = \|A\|\|A^{-1}\|$ and tends to infinity as the Jacobian becomes singular.

$$\kappa\left(\left[\begin{array}{c} J_{TG}(q_0) \\ J_{RG}(q_0) \end{array}\right]\right) < \varepsilon. \tag{18}$$

In the example ε is chosen $\varepsilon = 20$.

- In any industrial application, the **position and orientation** of the gripper are restricted by external constraints, such as obstacles within the working space, or the requirement that the parts should be assembled on a workbench with a given height. Position and orientation are calculated using the robot's forward kinematics, so that geometrical constraints can be stated in Cartesian space. For simplicity in our example, we restrict the robot's position to a cube, the edges of which are parallel to the base coordinate frame B of the robot, see Fig. 4:

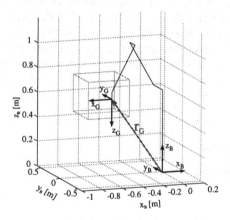

Fig. 4: Working space restrictions

$$_B r_{min} \leq_B r_G(q_0) \leq_B r_{max}. \tag{19}$$

Orientation restrictions are expressed using the rotational gripper transform A_{GB}. In the example, we choose the orientation to be restricted such that the z_G direction should have a negative component in each the x_B- and the z_B-direction, which means that the mating direction points downwards away from the robot's base.

3.3 Example: Rectangular Peg-in-Hole Insertion

We illustrate the method with the position controlled insertion of a rigid rectangular peg into a hole using a PUMA 562 manipulator, starting from the reference configuration $q_{0,ref}$, $K_{p,ref}$, $K_{d,ref}$ defined in eq. (20). This configuration is characterized by short effective lever arms that disturbances can work on, and small control coefficients, which increase gripper compliance for improved mating tolerance:

$$
\begin{aligned}
q_{0,ref} &= \begin{bmatrix} 2 & -152 & -4 & 0 & -19 & 179 \end{bmatrix}^{\circ} \\
K_{p,ref} &= \begin{bmatrix} 1.604 & 1.304 & 2.608 & 0.395 & 0.556 & 0.390 \end{bmatrix} \frac{Nm}{rad} \\
K_{d,ref} &= \begin{bmatrix} 0.055 & 0.013 & 0.019 & 0.00263 & 0.00280 & 0.00195 \end{bmatrix} \frac{Nm}{rad}
\end{aligned} \tag{20}
$$

Rigid peg-in-hole insertion is mainly characterized by rigid body contacts, the occurrence of which can not be predicted because of the limited positioning accuracy of the gripper. Thus, peg and hole will show lateral and angular offset between each other. This causes impacts between the peg and the chamfer, which result in gripper oscillations. On the other hand, compliance in the lateral directions is required in order to compensate for positioning errors. Therefore, for a rectangular peg-in-hole process, the criteria for impact sensitivity, mating tolerance and damping of gripper oscillations are the most relevant ones. For our case study we chose the vector of objective functions

$$
G = \begin{bmatrix} \dfrac{G_1}{G_{1,0}} \\ \dfrac{G_3}{G_{3,0}} \\ \dfrac{G_4 + G_6}{G_{4,0} + G_{6,0}} \end{bmatrix} \in \mathbb{R}^{n_G} \tag{21}
$$

normalized to the objective function values $G_{i,0}, i = 1, 3, 4, 6$ of the reference configuration $q_{0,ref}$, $K_{p,ref}$ and $K_{d,ref}$. Let us for an appropriate choice of cartesian weighting factors assume the lateral clearance in x-direction between the two parts to be smaller than the robot's positioning accuracy. In y-direction the clearance is assumed to be large enough to avoid contact with chamfers. Thus, impacts will occur mainly in x- and z-direction and optimizing for impact intensities means to find a position, where the effective end-effector masses in x and z are minimized. The mating tolerance for this process is determined by x-translational and φ_y-rotational cartesian stiffnesses. Also the vibration behavior is most critical in x- and z-direction and the weights for the motor torques are chosen according to the maximum motor torques of the PUMA 562 robot. Thus, the cartesian weights for our example problem write

$$
\begin{array}{llllllllll}
\boldsymbol{g}_M & = & \mathrm{diag}\big[\ 1 & 0 & 1 & 0 & 0 & 0\ \big], & \rightarrow & G_1 \\
\boldsymbol{g}_Q & = & \mathrm{diag}\big[\ 1 & 0 & 0 & 0 & 1 & 0\ \big], & \rightarrow & G_3 \\
\boldsymbol{g}_S & = & \mathrm{diag}\big[\ 1 & 0.1 & 1 & 0 & 0.1 & 0.1\ \big], & \rightarrow & G_4 \\
\boldsymbol{g}_\tau & = & \mathrm{diag}\big[\ 0.2 & 0.2 & 0.2 & 6.2 & 6.2 & 6.2\ \big]. & \rightarrow & G_6
\end{array} \tag{22}
$$

A sensitivity analysis was carried out, from which can conclude that the considered objectives are in accordance with physical evidence. Hence they reflect the physical behavior of the system in the sense that they become a minimum at the locations, where performance is best. On the other hand, they are highly nonlinear functions of the optimization parameters. This forms a nonlinear, nonconvex optimization problem. Moreover, some of the cost functions tend to infinity at singularities showing very high curvatures. Thus, for an efficient optimization, analytical derivatives of the objective functions, for which a calculation is possible at a reasonable cost may significantly improve convergence. This is done for G_1, G_2 and G_3 and for the singularity (18) and working space constraints (19) using analytical calculation software. The objective functions possess local minima, in which the optimization routine may converge, so that an optimization of the position makes sense only, if the problem is constrained to a certain region within the working space. However, in most cases practical considerations in a real environment yield working space restrictions anyway.

4 Vector Optimization Problem

The above mentioned criteria and constraints form a nonlinear vector problem for the position/controller optimization. Thereby the manipulation task to be carried out is charactarized by a specific combination of cost functions and weighting factors, which can be chosen by physical evidence, as shown before. Thus, the complete vector problem for our example writes

$$
\min_{\boldsymbol{q}_0, \boldsymbol{K}_p, \boldsymbol{K}_d} \ \{\boldsymbol{G} : \boldsymbol{f}_1 = 0; \boldsymbol{f}_2 \leq 0\} \tag{23}
$$

with $f_1(q_0, K_p, K_d)$ stemming from (5) and $f_2(q_0, K_p, K_d)$ being

$$
f_2(q_0, K_p, K_d) = \begin{bmatrix} q_{min} - q_0 \\ q_0 - q_{max} \\ \tau_{C,min} - \tau_C \\ \tau_C - \tau_{C,max} \\ -\mathrm{diag}(K_p) \\ -\mathrm{diag}(K_d) \\ \kappa\left(\begin{bmatrix} J_{TG}(q_0) \\ J_{RG}(q_0) \end{bmatrix}\right) - \varepsilon \\ {}_Br_{min} - {}_Br(q_0) \\ {}_Br(q_0) - {}_Br_{max} \\ A_{GB;1,3} \\ A_{GB;3,3} \end{bmatrix}
\begin{array}{l} \left.\vphantom{\begin{matrix}a\\a\end{matrix}}\right\} \text{ joint angles} \\ \left.\vphantom{\begin{matrix}a\\a\end{matrix}}\right\} \text{ joint torques} \\ \left.\vphantom{\begin{matrix}a\\a\end{matrix}}\right\} \text{ controller stability} \\ \left.\vphantom{\begin{matrix}a\\a\end{matrix}}\right\} \text{ avoid singularities} \\ \left.\vphantom{\begin{matrix}a\\a\end{matrix}}\right\} \text{ workspace} \\ \left.\vphantom{\begin{matrix}a\\a\end{matrix}}\right\} \text{ mating direction} \end{array}
\tag{24}
$$

It is known from the theory of vector optimization that (23) can not be uniquely solved, if any of the components of G are competing, [4]. Rather, the solution of (23) is a subspace of the parameter space of dimension $\mathbb{R}^{n_G-1}$ and denotes the Pareto-optimal set of all possible solutions of $f_1 = 0$, $f_2 \leq 0$. Pareto-optimality is reached if none of the objective functions can be improved without deteriorating at least one of the other criteria. Using the method of objective weighting [5], a substitute problem is stated with a scalar preference function P to be minimized. For this, a vector of weighting factors $w = [w_1 w_3 w_{46}] \in \mathbb{R}^{n_G}$ is introduced such that

$$
0 \leq w_i \leq 1; \qquad \sum_{i=1,3,46} w_i = 1;
$$
$$
P\left(G(q_0, K_p, K_d), w\right) = wG(q_0, K_p, K_d).
\tag{25}
$$

The Pareto-optimal set of solutions is then obtained by solving the scalar substitute problem

$$
\min_{q_0, K_p, K_d} \{P : f_1 = 0; f_2 \leq 0\}
\tag{26}
$$

for each vector w, which fulfills (25). Eq. (26) is solved for a systematic variation of w using a Sequential Quadratic Programming algorithm with the Hessian matrix of the Lagrangian function being updated at each iteration by a quasi-Newton approximation (BFGS), [10, 12].

5   Optimization Results

In the following considerations, the reference configuration from (20), which is already considered suitable for the regarded process, is used as starting point for the optimization. It is then compared to an 'optimal trade-off' configuration, which is chosen from the set of Pareto-optimal solutions. As shown in Fig. 5, the single cost-functions G_i can be considerably diminished with respect to the reference configuration if they are emphasized in the preference function P. The criterion for impact intensities can be reduced

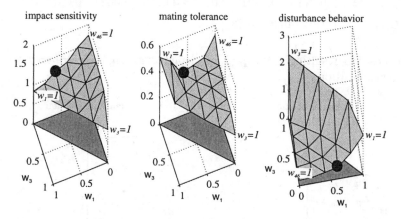

Fig. 5: Pareto-optimal set of solutions

by at most 17 %, that for mating tolerances by 78 % and the damping of end-effector oscillations and motor torques by up to 79 % with respect to the reference point. However, it is evident that such tremendous improvements cause deteriorations in other criteria. For example, optimizing for damping of oscillations only deteriorates the impact sensitivity by 72 % and optimization for mating tolerance only increases the oscillation criterion by 136 %. But there are also regions within the Pareto-optimal area, where all criteria are improved with respect to the reference configuration. Fig. 5 shows that over a wide range of possible weighting factors the criteria for impact sensitivity and for mating tolerance are not contradictory to each other: A simultaneous improvement of both cost functions can be observed for a large number of possible weights. Only if G_1 is strongly weighted in P, G_3 becomes worse. On the other hand, G_{46} is found to impose completely different demands on the position and on the controller. G_1 and G_3 have their largest value at the point, where G_{46} is minimized.

From these considerations it is clear that an 'optimal trade-off' must be found, which gives a satisfactory improvement in each of the criteria. The process of finding this optimal trade-off can hardly be formalized in a mathematical sense, because the trade-off between the cost-function weights is generally governed by criteria, which require human expertise. Thus, the Pareto-optimal region in Fig. 5 has to be judged in order to find an optimal solution. In our peg-in-hole example $w = \begin{bmatrix} 0.6 & 0 & 0.4 \end{bmatrix}$ is chosen, which yields the following configuration:

$$
\begin{aligned}
\boldsymbol{q}_{0,opt} &= \begin{bmatrix} 3.3 & -165.9 & 1.1 & 92.3 & -39.0 & 72.5 \end{bmatrix}^{o} \\
\boldsymbol{K}_{p,opt} &= \begin{bmatrix} 1.226 & 1.219 & 1.000 & 0.132 & 0.139 & 0.133 \end{bmatrix} \frac{Nm}{rad} \\
\boldsymbol{K}_{d,opt} &= \begin{bmatrix} 0.0715 & 0.0331 & 0.0145 & 0.0016 & 0.0024 & 0.0074 \end{bmatrix} \frac{Nm}{rad} \\
\boldsymbol{G}_{opt} &= \begin{bmatrix} 1.24 & 0.36 & 0.33 \end{bmatrix}^{T}
\end{aligned}
\tag{27}
$$

The position resulting from $\boldsymbol{q}_{0,opt}$ is depicted in Fig. 6, compared to the reference position $\boldsymbol{q}_{0,ref}$.

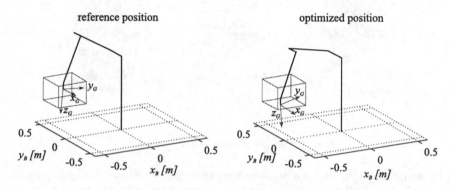

Fig. 6: Robot position for 'optimal trade-off' and reference configuration

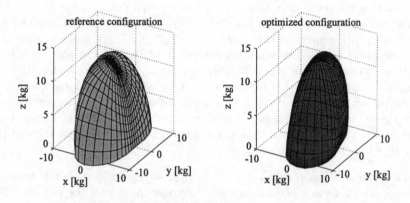

Fig. 7: Impact sensitivity ellipsoid for 'optimal trade-off' compared to reference configuration

It can be seen from $\boldsymbol{G}_{opt}$ in (27) that significant improvements in G_3 and G_{46} can be expected, which must be paid for by a slight deterioration in G_1.

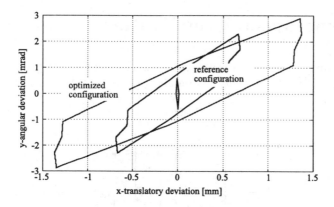

Fig. 8: Tolerance area for $\lambda_{max} = 10N$ for 'optimal trade-off' compared to reference configuration

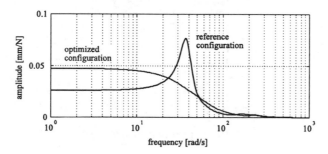

Fig. 9: Transfer function $f = \frac{G x_z}{G \lambda_z}$ for 'optimal trade-off' compared to reference configuration

This can be fully comprehended in Figs. 7 to 9. The volume of the impact ellipsoid has slightly increased and the main axis is rotated with respect to the y-axis by a small angle. The tolerance area for a given maximal mating force of $\lambda_{max} = 10N$, calculated from a quasistatic force equilibrium, [9], is significantly enlarged, Fig. 8. For the judgement of the disturbance behavior, the amplitude frequency response function for z_G gripper displacement related to z_G gripper force is depicted in Fig. 9. The resonance peak at the first natural frequency vanishes completely, which indicates gripper oscillations to be well damped. However, the starting amplitude is increased. This is due to the trade-off with the mating tolerance criterion, which reduces the cartesian end-effector stiffness.

6 Experimental Vertification

The optimized configuration is verified experimentally using the test-setup depicted in Fig. 10. A rigid rectangular peg is assembled into a rigid hole

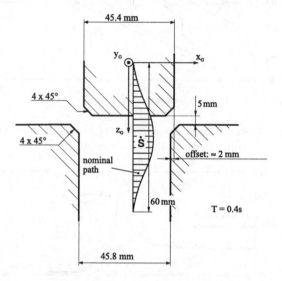

Fig. 10: Experimental test-setup

with a clearance of 0.3 mm in x_G-direction and an x_G offset of 2 mm using a PUMA 562 manipulator. The desired mating path is a straight line in z_G-direction with a length of 60 mm and an assembly time of 0.4 s. Forces are measured using a Schunk FTS 330/30 force-torque sensor installed at the robot's end-effector. The gripper position is reconstructed by measuring the joint encoder angles using the robot's forward kinematics.

For the judgement of impacts, the time histories of the z_G gripper force is considered, Fig. 11. Significantly, a force peak occurs at the time where the two parts are getting in contact for the first time, the height of which gives a measure for the impact intensity. Since the relative velocity, with which the parts meet, is almost equal in both cases (about 150 mm/s), the peak height gives a direct measure of the effective mass acting on the impacting bodies. Fig. 11 shows that in the regarded direction similar impact intensities can be expected and thus, the 'optimal trade-off' yields no improvement with respect to the impact behavior, which expresses itself also in G_{opt} and in Fig. 7.

In contrast to this, according to the values in G_{opt} and the tolerance area of Fig. 8, the optimal trade-off must show significant advantages with regard to the mating tolerance. This is verified with time histories of the x_G lateral

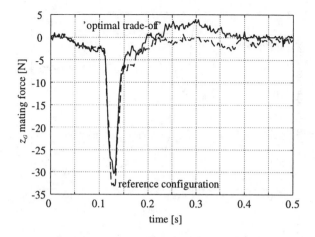

Fig. 11: Gripper force in mating direction during insertion
for 'optimal trade-off' compared to reference configuration

gripper force during insertion, Fig. 12. Although the lateral offset is approximately equal in both cases, the reduced lateral end-effector stiffness of the
optimized system allows compliant motion and thus reduces the strains on
the manipulator and on the mating parts significantly. Most of the improvement in this criterion is due to the change in control coefficients, since the
end-effector stiffness is essentially determined by K_p and K_d. The manipulator position q_0 defines the gripper's Jacobian and thus the lever arms the
compliant controllers can work on.

In Fig. 13 the z_G path deviation due to external forces is depicted for
both the reference and the optimized configurations. In fact, two different
sources exist, which excite the manipulator dynamics, external contact forces
and the desired movement. In order to separate those two effects in the
experiment, the trajectory is measured twice for each configuration. First,
the desired trajectory is performed without any external forces, particularly
with no contacts. The resulting path deviation is then subtracted from the
path deviation measured during manipulation, Fig. 13. This ensures that
only the path deviation resulting from contact events show up in Fig. 13. It
can be seen that the first amplitude peak resulting from the initial impact is
reduced by a factor of 3. Furthermore, the transient behavior is much better
damped in the optimized configuration and shows no oscillations. Thus, also
the vibrational performance is significantly improved by the optimization.

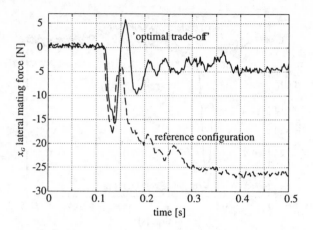

Fig. 12: Lateral gripper force during insertion for 'optimal trade-off' compared to reference configuration

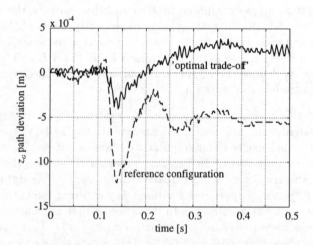

Fig. 13: z_G path deviation in mating direction due to mating forces during insertion for 'optimal trade-off' compared to reference configuration

7   Conclusions

We have shown the analysis and automatic synthesis of an automated assembly cell using numerical programming methods. The robot's response behavior is optimized by choosing appropriate position and controller gains. It was shown that the criteria give a good performance measure with respect to the physical effects to be optimized. Constraints have been introduced, which make the solution of the problem applicable in practice. The cost functions were found to be nonlinear and nonconvex functions of the optimization parameters and show local minima. Thus, for good convergence analytic function derivatives are necessary. The resulting vector problem is solved and the Pareto-optimal set of solutions is discussed for a rigid peg-in-hole insertion carried out by a PUMA 562 manipulator. Significant improvements can be gained in each of the criteria. However, they must be paid for with sometimes large deteriorations in other objectives. Impact sensitivity and mating tolerance are to a wide extent not contradictory to each other, whereas optimization for vibrational behavior imposes opposite demands on the natural dynamics of the manipulator, which makes a trade-off necessary. The performance of a compromise configuration is discussed. Although this configuration shows a slight deterioration for the impact sensitivity criterion with respect to a refernece, mating tolerance and vibratory behavior can be to a large extent improved. This is tested experimentally and the results show that the improvements gained by the optimizaiton can be reproduced in practice.

Acknowledgements

The authors are pleased to acknowledge the support of the German Research Association (DFG) for funding the research presented in the paper.

References

[1] Asada H., Ogawa K.: On the Dynamic Analysis of a Manipulator and its End Effector Interacting with the Environment; Proc. of the 1987 Int. Conf. on Robotics and Automation, Raleigh, NC, March 31-April 3, 1987, pp. 751-756.

[2] Barcio B. T., Walker I. D.: Impact Ellipsoids and Measures for Robot Manipulators; Proc. of the 1994 Int. Conf. on Robotics and Automation, San Diego, CA, May 8-13, 1994, pp. 1588-1594.

[3] Caine M. E., Lozano-Pérez T., Seering W. P.: Assembly Strategies for Chamferless Parts; - In: Proc. of IEEE Int. Conf. on Robotics and Automation, Scottsdale, Arizona, May 14-19, 1989, pp. 472-477.

[4] Da Cunha N. O., Polak E.: Constrained Minimization Under Vector-valued Criteria in Finite Dimensional Spaces; - In: J. Math. Anal. Appl., Vol. 19, pp. 103-124, 1967.

[5] Eschenauer H. A., Koski J., Osyczka A.: Multicriteria Optimization - Fundamentals and Motivation; - In: Multicriteria Design Optimization, Eds. H. Eschenauer, J. Koski, A. Osyczka, Springer, Berlin, 1990, pp. 1-32.

[6] Kiguchi K., Fukuda T.: Fuzzy Neural Friction Compensation Method of Robot Manipulation During Position/Force Control; - In: Proc. IEEE Int. Conf. on Robotics and Automation, Minneapolis, MS, April 22-28, 1996, pp. 372-377.

[7] Laval L., M'Sirdi N. K., Cadiou J.-Ch.: H^∞-Force Control of a Hydraulic Servo-Actuator with Environmental Uncertainties; - In: Proc. IEEE Intl. Conf. on Robotics and Automation, Minneapolis, MS, April 22-28, 1996, pp. 1566-1571.

[8] Meitinger Th., Pfeiffer F.: Dynamic Simulation of Assembly Processes; - In: Proc. IEEE/RSJ Intl. Conf. on Intelligent Robots and Systems, Pittsburgh, PA, August, 5-9, 1995, pp. 298-304.

[9] Pfeiffer F., Wapenhans H., Seyfferth W.: Dynamics and Control of Automated Assembly Processes; - In: Smart Structures, Nonlinear Dynamics and Control, Eds. A. Guran, D. J. Inman, Prentice Hall PTR, New Jersey, 1994, pp. 190-225.

[10] Powell, M. J. D.: Variable Metric Methods for Constrained Optimization; - In: Mathematical Programming: The State of the Art, Eds. A. Bachem, M. Grotschel, B. Korte, Berlin: Springer, 1983, pp. 288-311.

[11] Prokop G., Pfeiffer F.: Improved Robotic Assembly by Position and Controller Optimization; - In: Proc. IEEE Int. Conf. on Robotics and Automation, Minneapolis, MS, April 22-28, 1996, pp. 2182-2187.

[12] Schittkowski K.: NLQPL: A Fortran-subroutine Solving Constrained Nonlinear Programming Problems; - In: Operations Research, Vol. 5, 1985, pp. 485-500.

[13] Seyfferth W., Pfeiffer F.: Dynamics of Assembly Processes with a Manipulator; - In: Proc. of the 1992 IEEE/RSJ Int. Conf. on Intelligent Robots and Systems (IROS), July 7-10, 1992, Raleigh, NC, USA, pp. 1303-1310.

[14] Steinle J.: Entwicklung einer prozeßangepaßten Roboterregelung für Montagevorgänge; Fortschrittberichte VDI, Reihe 8, Nr. 548, Düsseldorf, 1996.

[15] Sturges R. H., Laowattana S.: Virtual Wedging in Three Dimensional Peg Insertion Tasks; - In: Proc. IEEE/RSJ Intl. Conf. on Intelligent Robots and Systems, Raleigh, NC, July 7-10, 1992, pp. 1295-1302.

[16] Tarokh M., Bailey S.: Force Tracking with Unknown Environmental Parameters using Adaptive Fuzzy Controllers; - In: Proc. IEEE Int. Conf. on Robotics and Automation, Minneapolis, MS, April 22-28, 1996, pp. 270-275.

[17] Wapenhans H.: Optimierung von Roboterbewegungen bei Manipulationsvorgängen; Fortschrittberichte VDI, Reihe 2, Nr. 304, Düsseldorf, 1994.

[18] Whitney D. E., Gustavson R. E., Hennessey M. P.: Designing Chamfers; - In: Intl. J. of Robotics Research, Vol. 2, No. 4, 1983, pp. 3-18.

[19] de Wit C., Noël P., Aubin A., Brobliato B.: Adaptive Friction Compensation in Robot Manipulation at Low Velocities; - In: Int. J. of Robotics Research, Vol. 10-3, 1991, pp. 189-199.

[20] Yoshikawa T.: Dynamic Manipulability of Robot Manipulators; Proc. of the 1985 IEEE Int. Conf. on Robotics and Automation, 1985, pp. 1033-1038.

Some Basic Principles of Reliability-Based Optimization (RBO) of Structures and Mechanical Components

M. Gasser[1] and G.I. Schuëller[2]

[1] Leitner AG, Sterzing (BZ), Italy; formerly Institute of Engineering Mechanics, Leopold-Franzens-Universität, A-6020 Innsbruck, Austria
[2] Institute of Engineering Mechanics, Leopold-Franzens-Universität, A-6020 Innsbruck, Austria

Key Words. Reliability Based Optimization, structural reliabilty, cost function, nonlinear programming, Monte Carlo Simulation

Abstract. Particularly in view of the introduction of product liability, reliability-based design procedures, and for that matter optimization (RBO) receive increasing attention. The analysis deals with statistical uncertainties inherent in structural, material, damage parameters, etc., which are modeled by random variables. The mechanical representation of the structure and the component respectively is generally modeled by Finite Elements (FE). In this paper basic mathematical formulations of design objectives and restrictions including reliability measures are discussed. Based on these models structural reliability analyses provide information for design modification and selection of an optimal design solution. As generally minimization of the expected total cost of the structure including initial costs and costs due to failure, minimization of the overall probability of failure and weight minimization with respect to reliability constraints are considered. Design problems commonly denoted as multiobjective or multicriteria optimization problems are treated. For the reliability analysis numerical methods are utilized to estimate the reliability measures. These procedures are already cast in an easy-to-use-software, denoted as COSSAN (Computational Stochastic Structural Analysis). In this context a concept is discussed, which is based on the *separation* of the tasks of reliability analysis and nonlinear mathematical programming techniques for which pertinent applicable software is already available. The RBO procedure utilizes approximation techniques for estimating the reliability measures. In particular, the reliability analysis makes use of the well known Response Surface Method (RSM) in context with Advanced Monte Carlo Simulation techniques, while the reliability based optimization procedure itself is controlled by the well known NLPQL-algorithm. Finally a number of numerical applications are shown in order to exemplify the approach.

1 Introduction

In the fields of structural optimization and reliability theory considerable progress has been made particularly in recent years. Deterministic optimization in structural and mechanical design as well as in other engineering disciplines has been used as a decision tool in order to obtain the *best design*. The requirement of identifying a "best design" includes that the risk of failure of mechanical components, systems and structures is considered in the decision making. Thus, the *quantification* of safety and reliability of structural systems gained considerable importance. Hence, a deterministic analysis of usually complex structures neglecting the random nature (uncertainties in loading, strength, etc.) is not satisfactory. As consequence the structural design process has to be improved by investigation of efficient reliability-based optimization methods. The combination of reliability analysis and structural optimization becomes very important for the rational and quantitative comparison between economy and safety. This fact was first recognized by *A.M. Freudenthal* (e.g. [3, 4]).

The performance of structural analysis is generally based on mechanical and mathematical models. In most cases the mechanical model of the structure is idealized by means of the Finite Elements (FE). Uncertainties of structural systems are modeled by random variables and/or stochastic processes whose characteristics must be estimated from measurements. Design objectives and restrictions, including reliability measures, have to be formulated mathematically. Based on these models structural and reliability analyses provide information for design modification and selection of an optimal design solution.

The complexity of the reliability-based optimal design problem depends strongly on the types of the considered design variables (sizing, shape, material and *global* topological variables) and their combinations. Most studies performed in the fields of sizing and shape optimization utilize standard optimization algorithms. Comparatively little work is available with respect to the consideration of material parameters in overall structural optimization. If parameters of the objective and constraint functions are considered as random variables these functions, of course, become also random, which, in turn, however, makes the design optimization problem extremely complicated. As to the present, no literature on *realistic* engineering examples of such stochastic structural design problems appears to be available. So far only statistical parameters of the random variables and reliability measures are included in the design problem formulation. It is important to note that design variables are always deterministic variables, because otherwise design modifications could not be carried out in practice.

The RBO approach viewed as a mathematical programming problem raises the question with respect to the meaning of an optimum solution. In the literature various alternatives are suggested: e.g. minimization of the expected total costs of the structure including initial costs and costs due to failure, minimization of the overall probability of failure, weight minimization with respect to reliability constraints etc.. Very often more than one objective is considered, e.g. the minimization of the expected total cost and the maximization of the overall system reliability. Such design problems are known as *multiobjective* or *multicriteria* optimization problems. Aside from the individual interests it has to be kept in mind that the information resulting from the optimization procedure is generally used for decision support only.

With respect to the definition of failure and the related costs different damage levels may also be considered for the assessment of the structural reliability. Total collapse, i.e. total failure of a structure under extreme loads as well as partial damage have to be taken into account in a realistic concept of design optimization. The latter damage criterion may influence the design decisions significantly. In most problem formulations time dependent behaviour of structural systems is not taken into account. Due to the fact that system parameters usually change during lifetime, e.g. due to change of load conditions, fatigue, repairs, etc., which, as a consequence, causes a change in reliability, it is necessary to extend the design objectives. When the structural reliability decreases with time it is often necessary to develop maintenance programs. Optimal inspection and repair strategies may be integrated in the reliability-based optimal design problem. Hence, efficient methods and strategies are needed to solve the extensive reliability-based optimal design problem. This applies to algorithms of structural, stochastic as well as optimization analyses, respectively.

As RBO problems are strongly nonlinear this applies particularly to nonlinear programming methods which are utilized to solve the design problems. For this purpose in the literature a series of algorithms has been applied. Usually, algorithms are utilized as a *black-box* and are directly connected to the routines for reliability analysis. This means that the subprogram for the reliability analysis is called by the optimization program through an interface. This is certainly the most general approach because the optimization problem is separated from the reliability problem. A proper data transfer through an interface is necessary. In RBO the NLPQL algorithm [12, 13] has proven to be one of the most powerful tools for nonlinear programming problems. As RBO is an interdisciplinary engineering task it requires the combination of different interacting analysis and synthesis tools. These disciplines are structural modeling and analysis, reliability modeling and analysis, decision modeling, mathematical programming, all within an efficient software environment.

In this context the NLPQL algorithm is implemented into an interactive, modular, and flexible software environment such that it can be applied to RBO or deterministic optimization problems in a most general form. *Single* as well as *multiple* optimization problems can be handled. The problem function values and their gradients can be provided on COSSAN-User input file level by *object oriented* programming [2].

2 Reliability-Based Optimization Problems and Procedures

2.1 Problem Formulations

One of the major goals of structural reliability theory is the optimization of structural design based on reliability concepts. For this purpose the quite different disciplines *reliability analysis* and *mathematical programming* are interrelated by an optimum design problem formulation.

2.2 Deterministic Problem Formulation

The general aim of optimization is to find extrema - that is, minima or maxima of commonly complicated nonconvex real-valued functions. These functions are used to compare various design solutions of the investigated system. In most design problems specific constraints must be satisfied. Depending on the respective optimization problem inequality constraints and/or equality constraints have to be considered in the formulation. Thus, a general mathematical model for deterministic design optimization can be defined.

Identify a vector $\mathbf{x} \in R^n$ of design variables to minimize an objective function defined within the domain $S \subset R^n$

$$f(\mathbf{x}) = f(x_1, x_2, \ldots, x_n) \to \min \tag{2.1}$$

subject to equality constraints and inequality constraints respectively

$$h_j(\mathbf{x}) \equiv h_j(x_1, x_2, \ldots, x_n) = 0 \qquad j = 1 \, to \, p \tag{2.2}$$

$$g_i(\mathbf{x}) \equiv g_i(x_1, x_2, \ldots, x_n) \leq 0 \qquad i = 1 \, to \, m \tag{2.3}$$

Lower and upper bounds of the design variables, the so-called side constraints, are included.

Depending on the types of the objective and constraint functions a series of different mathematical programming algorithms are available. As structural optimization problems are usually formulated as nonlinear programming problems, respective mathematical programming procedures are mentioned only, e.g. feasible direction method, gradient projection method, generalized reduced gradient method, linearization method, cost function bounding method, method of moving asymptotes, sequential quadratic programming methods, potential constraint strategy, etc. (see e.g. [1, 12, 13]). Other optimization algorithms based on Monte Carlo simulation (see e.g. [11]) and evolutionary procedures are known to solve very complex problems especially when discontinuous functions are considered.

The RBO problem can be defined in a similar form. The important difference lies in the properties of the design variables and the parameters of the problem function. Deterministic variables, stochastic variables, stochastic parameters of the random design variables and reliability measures may be introduced in the optimum design formulation.

2.3 Stochastic Problem Formulations

In RBO of structural systems it is assumed that both the loads and the member strengths respectively are random system parameters. If some of the function parameters are random, the optimum design formulations - objective function and constraints - become random functions. The general stochastic optimization problem writes:

$$f(\mathbf{x}, \mathbf{Y}, \mathbf{r}(\mathbf{x}, \mathbf{Y}, \mathbf{y}')) \to \min \tag{2.4}$$

subject to

$$h_j(\mathbf{x}, \mathbf{Y}, \mathbf{r}(\mathbf{x}, \mathbf{Y}, \mathbf{y}')) = 0 \qquad j = 1 \text{ to } p \tag{2.5}$$

$$g_i\left(\mathbf{x}, \mathbf{Y}, \mathbf{r}\left(\mathbf{x}, \mathbf{Y}, \mathbf{y'}\right)\right) \le 0 \qquad\qquad i = 1 \text{ to } m \qquad\qquad (2.6)$$

where

> $\mathbf{x}$ is the vector of deterministic design variables
> $\mathbf{Y}$ is the vector of random variables
> $\mathbf{r}$ is the vector of the considered probability functions
> $\mathbf{y'}$ are the statistical parameters of the random variables.

The statistical parameters $\mathbf{y'}$ may also be included in the deterministic design vector $\mathbf{x}$.

A somewhat involved situation would result from introducing random variables as design parameters into the design process. For the theoretical background see e.g. [7, 8, 9].

A RBO problem is a stochastic structural optimization problem. This is due the fact that probability functions with respect to the considered reliability measures have to be determined.

2.4 Reliability-Based Problem Formulations

Already in 1956 *Freudenthal* discussed the problem of specification of an acceptable risk "...*on the basis of economic balance between the cost of increasing the safety and the cost of failure*" [4]. Thereby the optimal *economic* probability of failure should make the sum of all anticipated costs a minimum.

2.4.1 Single Objective Problem

In probabilistic context several reliability-based optimization formulations have been suggested. Basically they can be written in the form

$$f\left(\mathbf{x}, \mathbf{y'}, \mathbf{r}\left(\mathbf{x}, \mathbf{Y}, \mathbf{y'}\right)\right) \to \min \qquad\qquad (2.7)$$

subject to

$$h_j \equiv h_j\left(\mathbf{x}, \mathbf{y'}, \mathbf{r}\left(\mathbf{x}, \mathbf{Y}, \mathbf{y'}\right)\right) = 0 \qquad\qquad j = 1 \text{ to } p \qquad\qquad (2.8)$$

$$g_i \equiv g_i\left(\mathbf{x}, \mathbf{y'}, \mathbf{r}\left(\mathbf{x}, \mathbf{Y}, \mathbf{y'}\right)\right) \le 0 \qquad\qquad i = 1 \text{ to } m \qquad\qquad (2.9)$$

It should be pointed out that in the problem formulation as shown in the above equations no random functions are considered. Parameters for the description of the joint distribution function of the basic random variables are denoted by $\mathbf{y'}$, and $\mathbf{r}$ represents the vector of probability functions or reliability measures e.g. failure probabilities with respect to element failure modes.

2.4.2 Multiobjective Problem

In a number of structural optimization problems more objectives rather than one are pursued. There are basically two very alternative ways to approach such problems:

a) the collection of objectives are rewritten by one objective function, and
b) the multiobjective problem formulation is treated by application of vector optimization methods. From the current works related to RBO it can be seen that in most cases the first strategy is pursued.

The collection of subobjectives may follow the approach

$$F = \sum_{i=1}^{m} w_i f_i\left(\mathbf{x}, \mathbf{y'}, \mathbf{r}(\mathbf{x}, \mathbf{Y}, \mathbf{y'})\right) \to \min \tag{2.10}$$

$$\mathbf{f} = \left(f_1, \dots, f_m\right)^T \tag{2.11}$$

where w_i are weights representing the importance of each single objective. The multiobjective problem then writes

$$\mathbf{f} \to \min \tag{2.12}$$

which means that the single objective functions should be minimized

$$f_i\left(\mathbf{x}, \mathbf{y'}, \mathbf{r}(\mathbf{x}, \mathbf{Y}, \mathbf{y'})\right) \to \min \tag{2.13}$$

where the design vectors $\mathbf{x}$ and $\mathbf{y'}$ are elements of the feasible set $S \subset R^n$ defined by the constraints. There is generally no unique point at which all the objectives (2.13) reach their minima simultaneously. Therefore, the approach (2.10) might be preferred for practical purposes

2.4.3 Standard Reliability-Based Formulations

The most common objectives include:
a) minimization of the expected total cost of the structure or
b) maximization of the expected overall utility of the structure
c) minimization of the probability of failure for a fixed structural cost, or
d) the minimization of the expected cost of the structure for a specified level of failure probability.

Cases a) and b) can be formulated as unconstrained optimization problems, while problems c) and d) include constraints either as equality or inequality functions. Various extensions and combinations of these approaches are also possible. Mathematical formulations of cases c) and d) are shown in the following.

Find the design vector $\mathbf{x} \in R^n$ such that it is a solution to

case c:

$$\beta_{system}(\mathbf{x}) \to \max \qquad \text{or} \qquad P_{f,system}(\mathbf{x}) \to \min \tag{2.14}$$

subject to

$$f(\mathbf{x}) \le f^{\max} \tag{2.15}$$
$$x_{il} \le x_i \le x_{iu} \qquad i = 1 \ \text{to} \ n$$

where $\beta_{system}(\mathbf{x})$ is the structural system reliability index, $p_{f,system}(\mathbf{x})$ is the system failure probability, $f^{\max}$ is the maximum acceptable initial cost or weight of the structure;

case d1:
$$f(\mathbf{x}) \to \min \tag{2.16}$$

subject to

$$\beta_i(\mathbf{x}) \ge \beta_i^{accept} \qquad i = 1 \ \text{to} \ m \tag{2.17}$$
$$x_{il} \le x_i \le x_{iu} \qquad i = 1 \ \text{to} \ n$$

where $\beta_i(\mathbf{x})$ is the reliability index related to failure mode i, and β_i^{accept} represents the acceptable lower limit;

case d2:
$$f(\mathbf{x}) \to \min$$

subject to

$$\beta_{system}(\mathbf{x}) \ge \beta_{system}^{accept} \qquad or \qquad p_{f,system}(\mathbf{x}) \le p_{f,system}^{accept} \tag{2.18}$$
$$x_{il} \le x_i \le x_{iu} \qquad i = 1 \ \text{to} \ n$$

where β_{system}^{accept} and $p_{f,system}^{accept}$ are the acceptable limits of the system reliability measures, and $f(\mathbf{x})$ is the cost of the structure.

It is quite obvious that with an increasing number of constraints the conditioning of the optimization problem becomes more difficult. The chance to find optimal solutions decreases. Thus, it is important to formulate the optimization problem in an appropriate manner.

2.5 Decision Models for Global Optimization

2.5.1 Physical Objectives

A most frequently used approach is to consider the weight of a mechanical structure as objective function which is to be minimized. In addition, with respect to RBO, failure probability or safety index constraints are introduced into the design problem. Design variables may be cross-sectional dimensions or global topological parameters. The idea to consider the structural weight as design objective is simply derived from pure deterministic design aims in civil engineering. In the simplest form the problem writes, e.g.

$$W = \sum_{j=1}^{m} A_j(\mathbf{x}) l_j \rho_j \rightarrow \min \tag{2.19}$$

subject to

$$p_{fj} \leq p_j^{accept} \qquad j = 1 \text{ to } p \tag{2.20}$$

where A_j represents the cross-sectional area of element j of the length l_j, and ρ_j is the unit weight. The design vector $\mathbf{x}$ contains cross-sectional dimensions whereas the global topology remains unchanged. In the above equation failure probability constraints related to the p considered failure criteria are formulated.

2.5.2 Expected Utility Objectives

In the next step of improved decision making *expected utilities* or expected monetary values are used as objective criterion. Cost expectations derived from weighted cost components due to product manufacturing and failure consequences are considered. For example, the following cost function is used to assess the design alternatives

$$C = c_I(\mathbf{x}) + I_S(\mathbf{x}, \mathbf{y}) c_S(\mathbf{x}) + I_C(\mathbf{x}, \mathbf{y}) c_C(\mathbf{x}) \rightarrow \min \tag{2.21}$$

where c_I is the initial cost, c_S are costs due to e.g. partial failure, and c_C is the cost due to total collapse failure. The factors I_S and I_C are indicator terms which are set to one if partial failure or collapse failure occurs, and they are set to zero if no failure occurs. The total cost C depends on random quantities and is therefore a random function. In conventional decision making the judgement is based on the expected value of C which can be estimated as follows

$$E[C] \approx c_I + p_S c_S + p_C c_C \rightarrow \min \tag{2.22}$$

where

$$p_S = P[g_S(\mathbf{x}, \mathbf{Y}) \leq 0] \tag{2.23}$$
$$p_C = P[g_C(\mathbf{x}, \mathbf{Y}) \leq 0] \tag{2.24}$$

with the occurrence probabilities p_S and p_C which are dependent on the design vector $\mathbf{x}$ and the limit state functions $g_s(\cdot)$ and $g_c(\cdot)$. In order to reflect realistic design it is assumed that p_C is small compared to p_S. The random vector $\mathbf{Y}$ follows the joint distribution function $F_\mathbf{Y}(\mathbf{y})$.

The cost representation is a continuous function of $\mathbf{x}$, but it is discrete due to the discrete failure indicators I_S and I_C. The cost as consequence to failure events either occur or do not occur depending on the state of $\mathbf{Y}$, respectively the j-th realization $\mathbf{y}_j$. The cost function becomes continuous in $\mathbf{y}$ too, if instead of failure indicators continuous damage indicators are used so that an infinite number of damage states are possible. Hence, the distribution function becomes continuous in $\mathbf{x}$ and $\mathbf{y}$. The objective of meeting this requirement may then be expressed as:

$$C = c_I(\mathbf{x}) + \sum_{k=1}^{m} C_{D,k}(\mathbf{x}, D_k(\mathbf{x}, \mathbf{Y})) \quad D_k \in \Omega \to [0,1] \tag{2.25}$$

with damage indicator D_k concerning the k-th damage criterion, e.g. displacements, plastic deformation, and the respective damage consequence cost C_{Dk} of m failure or damage elements. The reliability-based problem then writes

$$E[C] = c_I(\mathbf{x}) + \sum_{k=1}^{m} E\left[C_{Dk}(\mathbf{x}, D_k(\mathbf{x}, \mathbf{Y}))\right] \tag{2.26}$$

subject to

$$P\left[D_k(\mathbf{x}, \mathbf{Y}) > d_k\right] \le p_k^{accept} \tag{2.27}$$

where p_k^{accept} is an upper bound of the failure probability related to criterion k and fragility value D_k. Following such a strategy, a probabilistic overall assessment of design alternatives becomes possible and hence, a design process based on the probability structure of the objective can be performed in a most rational way.

2.5.3 Cost Probability Criteria

The aim of this type of decision strategy is to determine the distributions of cost functions and to use the expected pattern of outcomes as information for decision making. Higher statistical moments and cost probabilities can be utilized to make decisions of high quality. They may be considered as objectives or as constraints in the problem formulation.

If the distribution function of the total cost can be determined, i.e. in the following form:

$$F_C(c) = P[C \le c] \tag{2.28}$$

the decision problem respectively the design problem can be formulated e.g.

$$P[C > c_{max}] \to \min \quad or \quad P[C \le c_{max}] \to \max \tag{2.29}$$

or alternatively

$$P[c_{min} \leq C \leq c_{max}] \rightarrow max \qquad (2.30)$$

where a specified probability guides the design process. In the literature examples of such decision strategies are rarely applied, especially if the investigations are of more theoretical nature. Quite naturally much more attention to cost probability-based decision criteria comes from practitioners of economic decision making.

For the probabilities usually no analytical functions are available. Therefore, MCS techniques serve as convenient and appropriate tools for estimation of such probabilities. By simulation of the random quantities the cumulative cost frequencies are obtained and the probabilities of failure or other event occurrence probabilities are estimated in parallel. The results are available in each design iteration step of the design process. Hence, an iteration history of estimates of distribution functions is provided. The distribution function is estimated by

$$\overline{F}_{Ck}(c, \mathbf{x}_k) \approx F_C(c) \qquad (2.31)$$

which is due to the simulated cost function

$$c = c_{kj}(\mathbf{x}_k, \mathbf{y}_j) \qquad (2.32)$$

for a design vector $\mathbf{x}_k$ corresponding to the k-th design iteration, and $\mathbf{y}_j$ are realizations according to the j-th simulation of the random variables. Estimates of failure probabilities write

$$\overline{P}_{Sk}(g_{Sk}(\mathbf{x}_k, \mathbf{Y})) \approx p_S \qquad (2.33)$$
$$\overline{P}_{Ck}(g_{Ck}(\mathbf{x}_k, \mathbf{Y})) \approx p_C \qquad (2.34)$$

with p_S and p_C as occurrence probabilities due to partial and total structural failure respectively. Certainly, in most cases direct simulation of structural systems is practically impossible. Therefore, approximation techniques are required to reduce the numerical effort significantly. An approximate damage representation can be considered as one possibility to overcome the numerical problems and to estimate cost function probabilities utilized for structural optimization.

2.6 Decision Models for Local Optimization

The decision models may also be concerned with optimal manufacturing and maintenance of welding constructions taking into account fatigue crack growth within welding joints. Unstable fatigue crack growth is considered as failure criterion in the local RBO approach. Different optimization strategies are possibly dependent on product lifetime periods and the pursued quality assurance philosophies. All these approaches are based on expected cost formulations and probability constraints. However, other problem formulations are generally also possible. Needless to say that this issue will not be discussed here. For further details it is referred e.g. to [5] and the references mentioned therein.

2.7 Procedures

So far when dealing with problem formulations it was already mentioned that the main problem of RBO of structures is that the limit state function is usually not known in explicit form and furthermore it depends on the structural design. It has to be kept in mind that when a reliability analysis is performed the results are related to a specific structural design, and that the limit state function is defined as function of the random variables conditioned on that particular design status:

$$g = g(\mathbf{x}, \mathbf{y}) \qquad \mathbf{x} \in S \subset R^n \tag{2.35}$$

where a failure event $\mathbf{y}_j$ is identified by

$$g(\mathbf{y}_j) \le 0 \rightarrow \text{failure} \tag{2.36}$$

Hence, the estimated reliability measure which may be used in the optimization problem formulation depends also on the design status, e.g. the probability of failure writes by definition

$$p_f = p_f(\mathbf{x}) = P[\mathbf{Y}|g(\mathbf{x}, \mathbf{Y}) \le 0] \tag{2.37}$$

The difficulties in solving the RBO problem reduces considerably if one assumes that the limit state function is given as function of the design and random variables, and the requirements of the availability of an efficient and accurate reliability analysis procedure is fulfilled. For these cases the limit state function may have the form

$$
\begin{aligned}
g(\mathbf{x}, \mathbf{y}) &= \sum_{k=1}^{n_p} a_k M_k - S = \\
&= a_1 x_1 x_2^2 R_e + \ldots - S = \\
&= a_1 x_1 x_2^2 y_1 + \ldots - y_m
\end{aligned}
\tag{2.38}
$$

which corresponds to the limit state function of a frame structure with rectangular cross sections with plastic hinge collapse modes. The reliability measures - safety index or failure probability - can be determined quite simply for each design vector $\mathbf{x}$. But the function is generally an unknown function in real life of complex structures. Due to this fact it is at least necessary to approximate the limit state function $g(\mathbf{x}, \mathbf{y})$ by a design dependent so-called response surface $\bar{g}(\mathbf{x}, \mathbf{y})$, i.e.:

$$\bar{g}(\mathbf{x}, \mathbf{y}) \approx g(\mathbf{x}, \mathbf{y}). \tag{2.39}$$

In addition it is necessary to approximate reliability measures as functions of the design variables, primarily the failure probabilities.

3 Numerical Examples

3.1 Direct Approach of NLP

3.1.1 General Considerations

The aim of RBO is to develop procedures and software tools such that one is in the position to apply a mathematical programming method to any general nonlinear optimization problem with respect to RBO. For this purpose the various analysis and synthesis tools have to be preferably available in black-box form within a powerful software environment providing flexible interfaces. In other words, the reliability-based structural optimization approach is simply based on the *direct connection* of structural analysis (FE), reliability analysis (use of RSM for problems of engineering practice) and optimization methods (e.g. NLPQL for general problem formulations).

In the first step the decision model has to be developed and the problem functions have to be defined, e.g. the objective function determines the total expected cost including failure consequence costs subject to reliability constraints. In addition to that the reliability model has to take into account all uncertainties of the decision system so that the reliability measures can be estimated utilizing an *efficient* reliability analysis method. At this point the main problem occurs, for real complex structures the limit state function is usually not known in explicit form. Hence, single limit state points can be calculated only by application of Finite Element analysis methods which increases the numerical effort tremendously. For reasons of efficiency direct MCS to calculate response are not applicable. Therefore, approximate methods, such as the RSM, have to be utilized.

Following definitions and having available the calculation tools, the problem functions can be determined. This is the point now where the optimization algorithm can be executed. During the iterative procedure the problem functions and their gradients are evaluated according to the rules of the mathematical programming algorithm. The optimization procedure is executed in the socalled *reverse communication mode* (see [5] for details) so that external or third-party software can be utilized as well to calculate the problem functionals (e.g. FE-software). According to this the actual design will be modified. This causes, in turn, a modification of the mechanical system and/or the reliability model. Subsequently, the structural response quantities are calculated (use of FEM) to determine the points at the limit state surface and the *response surface* as well as other functionals as required. With the calculation of safety indices or failure probabilities (use of RSM and MCS) the problem function values are determined. Iteration and/or other parameters may be controlled interactively if the optimization procedure does not show convergence.

The numerical efforts result from repeated response and reliability calculations due to following tasks:
- iterations of mathematical programming algorithm
- line search calculations (if NLPQL is utilized)
- gradient calculations (all Newton Methods)
- limit state point search to determine the coefficients of the response surface, especially when nonlinear structural behaviour is taken into account
- additional effort for different classes of failure criteria (e.g. plastic hinge mechanisms and stability criteria considered simultaneously).

The major numerical efforts are due to limit state calculations, because FE analyses have to be performed several times. However, the situation changes completely if the limit state can be formulated in *explicit* form. In this case the *direct* optimization procedure does not change, but no response calculations are necessary. The solution of this type of problem does not cause severe difficulties as the failure probability can be estimated by simulation of the random variables only. The efficiency of the developed procedure and software is demonstrated by simple examples in the following subsections.

The first example shows the general logic of RBO when FEM and RSM as well as Importance Sampling simulation procedures are used. The intention of the subsequent comparative example is to show the efficiency and the quality of the numerical methods.

3.1.2 Simple Frame Structure Using the RSM

In this example a simple frame structure (see Fig. 3.1) is analyzed to demonstrate the basic principles of the *direct optimization procedure*. The NLPQL algorithm is utilized directly in connection with the RSM and Importance Sampling. The decision problem is formulated as an *unconstrained reliability based* minimization problem of the expected total costs. The initial or production costs are assumed to be proportional to the structural mass. The failure consequence costs are due to loss of serviceability and occurrence of collapse failure. First yielding in any component represents the loss of serviceability and collapse is modelled by complete plastification of the structure (see eq. (2.38)). The optimal results are calculated for various combinations of partial (first yielding) and total (collapse) failure costs.

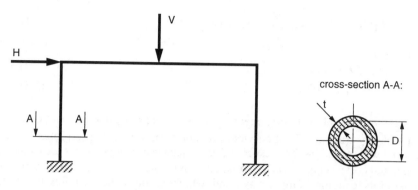

Fig. 3.1: Simple Frame Structure.

The simple one bay-one story frame structure is modelled by eight pipe elements and nine nodes and an ideal-elastic ideal-plastic material behaviour is assumed. The structure is exposed to both horizontal (H) and vertical (V) load components. The horizontal load component is assumed to be normally distributed with a zero mean value and a standard deviation of 20KN, the vertical loading follows a lognormal distribution LN(50KN, 20KN). For simplicity the uncertainties of the material properties are neglected. All elements have the same dimensions and properties. In the next step after the structural modeling and the definition of the sample space the actual reliability analysis can be performed. As two random variables are considered only, this task seems to be not very difficult to solve. For such a simple structure this is the case if the component limit states (plastic hinge mechanisms) can be formulated explicitly. However, FE methods and the RSM are used for general type problems. Then the reliability analysis procedure contains the following steps:

- calculation of a specific number of limit state points for each considered failure or serviceability criterion
- evaluation of the coefficients of the response surface, a 2nd-order polynomial is utilized
- determination of the design point utilized for importance sampling (suboptimization problem)
- weighted simulation of the random variables around the design point and estimation of the system failure probability (6048 simulations).

In this example five interpolation points of the response surface have to be determined according to the first yielding condition and the collapse criterion. In practice first yielding is represented by more than one iteration in a Newton-Raphson procedure. Structural collapse is modelled by exceedance of a "maximum" number of iterations to reach equilibrium.

The RBO problem for the structural system shown in Fig. 3.1 is to minimize the total expected costs including structural costs and failure consequence costs. As already mentioned, two load components are considered as basic variables. Design variables are the pipe diameter D and the wall thickness t of the pipes. The unconstrained RBO problem writes (according to eq. (2.22))

$$E[C] \approx c_0(\mathbf{x}) + c_s p_S(\mathbf{x}) + c_c p_C(\mathbf{x}) \to \min \qquad (3.1)$$

and

$$0.2 \le D \le 0.4[m] \qquad (3.2)$$
$$0.01 \le t \le 0.02[m] \qquad (3.3)$$

where the material cost component is given by $c_0(\mathbf{x}) = 50 \cdot m(\mathbf{x})$, m[kg] is the mass of the structure. The first yield probability and the collapse probability are denoted with p_S and p_C. The respective consequence cost components are c_s and c_C. The initial costs can be directly compared with failure consequence costs. Complex constraints can therefore be avoided which reduces the complexity of the optimization problem.

combination:	c_S [cost units]	c_C [cost units]
1	1.0e4	1.0e6
2	1.0e5	1.0e6
3	1.0e6	1.0e6
4	1.0e4	1.0e7
5	1.0e5	1.0e7
6	1.0e6	1.0e7
7	1.0e4	1.0e8
8	1.0e5	1.0e8
9	1.0e6	1.0e8

Tab. 3.1: Failure consequence cost factors used for sensitivity study.

A sensitivity study with respect to the variation of failure consequence costs is performed to show the importance of taking into account serviceability or partial failure criteria in addition to collapse criteria. The considered failure cost factors are listed in Table 3.1. The cost combinations where the "first yielding" costs are higher than the collapse costs are not of practical interest. However, from the mathematical point of view the general dependency of the optimal solution on the cost factors can be investigated.

The following two Figures 3.2 and 3.3 show the optimal results of the objective function and the pipe radius dependent on the failure consequence costs. An equivalent figure for the wall thickness is not shown as for all cost combinations the results are identical, in fact the value of this design variable is set to its lower bound.

It can be seen that in case of very high partial failure costs (first yielding, c_S=1.0e6) the objective and the design variable value show very little sensitivity to variations of collapse cost variations. In this RBO problem formulation the serviceability respectively the partial failure criterion is dominant. In contrast to this the optima are always sensitive to variations of the partial failure costs. From these results it may be concluded that it is important to introduce partial or serviceability criteria into the decision problem given that the expected consequence cost component $c_s p_S(\mathbf{x})$ is high enough. For example, the optimal results do not change very much if the cost c_S is changed from 1.0e4 to 1.0e5 cost units, but from 1.0e5 to 1.0e6 cost units the optima vary considerably. It is not surprising that the objective and the pipe radius show the same sensitivity due to cost variations. With an increasing pipe radius the failure probabilities decrease rapidly as the global structural stiffness increases and vice versa. Therefore, a more resistant structural system is required if higher failure consequence costs are expected.

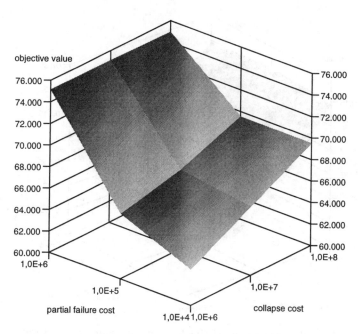

Fig. 3.2: Optimal *expected total costs* of simple frame structure due to various failure consequence cost combinations.

From the experience in carrying out this simple example it can be concluded that the proposed procedure is straight forward and generally applicable. However, from a practical point of view the numerical effort is still not satisfactory, although the RSM and importance sampling is applied. This fact becomes more obvious if a parameter study is performed. Convergence problems occur when the search directions of limit state points are inappropriately chosen. To the same class of problems belongs the problem of inaccurate determination of the limit state points. In both cases the limit state functions are not suitably approximated which causes convergence problems due to the fact that the failure probabilities are not accurately estimated. Hence, interactive control of the reliability analysis and the optimization procedure is necessary, at least for some trial calculations.

3.1.3 Simple Truss Structure Using Explicit Limit State Function

This second example deals, first, also with the general procedure of the *direct optimization approach* to RBO problems. The aim is to reproduce the presented results in [10] utilizing the COSSAN software package [2] and the algorithm NLPQL [12, 13] and to show the efficiency of the developed software tools for this purpose. Subsequently, this example is considered as a reference example for demonstration of the applicability of the *posterior approximation strategy* discussed subsequently.

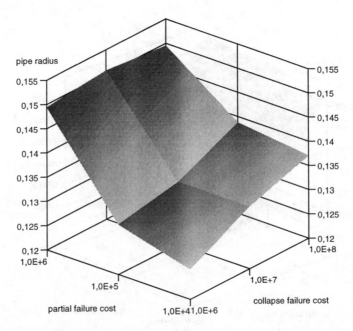

Fig. 3.3: Optimal *pipe radius* of simple frame structure due to various failure consequence cost combinations.

A schematic bridge structure modelled by 13 truss elements should be optimized such that the structural weight will be minimized satisfying a given system reliability constraint. The structural system, its dimensions, the element numbering and the load components are shown in Fig. 3.4. Seven design variables are introduced which correspond to the cross-sections of the symmetrical grouped truss elements. The structure is loaded by three vertical load components. These load components and the 13 resistance properties of the truss elements are considered as basic variables in this RBO problem. Due to the fact that the system is statically determined, system failure can be modelled by exceedance of the yield stress in any truss element (chain mechanism).

The element limit states can be formulated in explicit form as functions of the design variables and the random variables respectively. All element limit state functions are linearly dependent on the element resistances and the mentioned three load components. Hence, FE analyses are not necessary for structural response calculations to determine failure states, neither for single realizations of the basic variables nor for response surface calculations. The element and the system failure probabilities can simply be estimated by MCS of the random variables.

The yield stresses of the elements are considered as random resistance variables which have all the same mean value of 248.2 N/mm² and a coefficient of variation of 12% (9.76 N/mm²). The loads have as well same mean values of 66723 N and a COV of 16% (10675.7 N/mm²). All basic variables are assumed to be normally distributed and statistically independent.

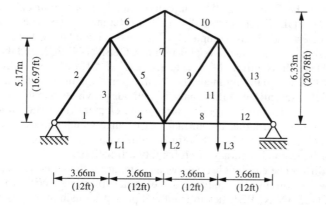

Fig. 3.4: Truss structure with 13 elements (data from [10])

The RBO problem as mentioned above is defined as a weight minimization problem where the objective function writes (eq. (2.19))

$$W(\mathbf{A}) = \sum_{i=1}^{13} A_i l_i \rho_i \rightarrow \min \tag{3.4}$$

with the cross sections A_i, the element lengths l_i, and the unit weight ρ_i. The design space is restricted due to the system reliability constraint (eq. (2.18)):

$$p_{fc}(\mathbf{A}) \leq 10^{-5}, \quad \beta_c(\mathbf{A}) \geq 4.256 \tag{3.5}$$

The design variables $\mathbf{A} = \mathbf{x} \in S \subseteq R^7$ correspond to seven groups of element cross-sections. The limit state function is given in explicit form. For each combination of the design variables the failure probability has to be estimated. Provided that the probability of failure can be estimated with high accuracy, the optimization procedure can then be executed in form of a "black-box" utilizing any available optimization algorithm. Nakib and Frangopol [10] applied the feasible direction method and a penalty function method (reference solution).

For this type of problem formulation and due to the assumptions made it can generally be expected that the optimum design is near the limit of the feasible domain, in other words, the constraint defined in eq. (3.5) becomes active at the optimum. This circumstance can be explained that there exist a direct relation between the objective functional (weight) and the structural stiffness. In this example the structural weight decreases with an increasing failure probability. Due to this fact the constraint function (eq. (3.5)) becomes active at the optimal solution (compare Table 3.2, values of β).

The optimum solutions calculated using different optimization algorithms are compared in the Table 3.2. The objective value at optimum calculated with COSSAN-NLPQL differs only 5% from the reference value. The constraints at optimum correspond almost completely. There exist a good agreement between the optimal design obtained due to application of COSSAN-NLPQL and the reference solution. Deviations might be due to the use of different reliability analysis methods (MCS vs. bounding methods). At this point it is indicated that, despite of some special cases, all reliability analysis methods provide estimates of the failure probability only. However, in this example the element limit state functions are linear and all basic variables are normally distributed which allows to consider the reference solution as the "true" optimal solution, since for this case bounding methods provide good accuracy of probability estimations.

A large difference can be observed when comparing the numerical effort taking the number of reliability analyses as reference measure. With 75 reliability analyses COSSAN-NLPQL lies in front and provides much more efficiency. The effort for calculating the design point used as center point for the weighted simulation and the number of simulations (4096) might put the numerical advantage of COSSAN into perspective. However, in case of explicitly formulated limit state functions the underlying measure is not of real significance since no response quantities have to be calculated during the iterative procedure. The situation changes completely if the limit state functions are not known apriori .

Assuming that the limit state functions are - due to nonlinear structural behavior - much more complex and, furthermore, not known explicitly one would still like to perform a RBO taking into account the same number of basic variables. As it was already mentioned before the RSM has to be applied for this purpose. There remains only the problem of calculation the limit state points. Using a second order polynomial approximation it would be necessary to determine 152 interpolation points. In context with RBO taking into account seven design variables and keeping in mind the gradient calculations this strategy is by no means practicable anymore. Hence, the application of additional approximate strategies is required. The *posterior approximation* approach of representing the failure probabilities as explicit functions of the design variables appears to be an efficient way to overcome this practical engineering problem.

variable number	reference solution				COSSAN using NLPQL	
A_i	feasible direction		penalty			
	[mm²]	[in²]	[mm²]	[in²]	[mm²]	[in²]
1	759.4	1.177	745.8	1.156	934.2	1.448
2	1190.3	1.845	1218.7	1.889	1156.8	1.793
3	734.8	1.139	752.9	1.167	670.3	1.039
4	759.4	1.177	745.2	1.155	941.9	1.460
5	231.6	0.359	227.1	0.352	193.5	0.300
6	843.2	1.307	840.0	1.302	733.5	1.137
7	514.2	0.797	521.9	0.809	877.4	1.360
objective fct.		804.2		809.2		847.2
ß		4.256		4.256		4.237
iterations		379		636		5
number of RA		379		636		75

Table 3.2: Comparison of optimal solutions obtained by COSSAN - NLPQL with reference solution (see [10])

3.2 Posterior Approximation of Reliability Measures

3.2.1 General Remarks

The basic aim of this approximate method is to reduce the numerical efforts of reliability analyses required for RBO and sensitivity analysis with respect to the decision parameters. For completeness the general RBO problem formulation is considered, which may be written as follows

$$f(\mathbf{x}, \mathbf{r}(\mathbf{x})) \to \min \tag{3.6}$$

defined on the feasible domain S

$$\mathbf{x} \in S \subset R^n, \qquad \mathbf{r} \in D \subset R^k \tag{3.7}$$

In each iteration the functionals and the actual gradients have to be evaluated which would require repeated calculations of the reliability measures $\mathbf{r}$. As already mentioned before the numerical efforts become impracticable, particularly for RBO's of larger complex systems. Therefore, having in mind the response surface methodology, one could try to represent the considered reliability measures using an approximate function, dependent on the design variables. The advantage is that during the optimization procedure the reliability measures $\mathbf{r}$ can be approximated by calculating the explicit functions $\bar{\mathbf{r}}$. In the following subsection it is discussed which type of reliability measure can be approximated with sufficient accuracy covering the complete design space. The optimization procedure is then demonstrated considering again the truss structure as treated above.

3.2.2 Approximation of Probability of Failure

Generally one could approximate the first order reliability index as well as the failure probability as functions of the design variables. However, using the RSM there is basically no need to introduce approximate first order reliability indices as there is no real additional numerical effort necessary to simulate the random variables to estimate the failure probabilities. Moreover, from practical experience it can be claimed that estimates of failure probabilities ensure at least to some extent numerical continuity during the iterative procedures which is not the case for first order reliability indices. It was suggested that for small values the failure

probability can be approximated with sufficient accuracy by simple exponential functions [6]. Provided that the failure probability is small over the complete design space the following approximation is proposed

$$\bar{\bar{p}}_f \approx p_f(\mathbf{x}) \tag{3.8}$$

where

$$\bar{\bar{p}}_f = \exp\left(a_0 + \mathbf{b}^T\mathbf{x} + \mathbf{x}^T\mathbf{C}\mathbf{x}\right) \tag{3.9}$$

The parameters a, $\mathbf{b}$, $\mathbf{C}$ are polynomial coefficients which have to be determined by solving a linear equation system. The equation system is assembled by estimating failure probabilities for specific design variable combinations and interpolating the polynomial function in the exponent of eq. (3.9) where the supporting points for the approximate probability function are given by

$$\ln \bar{\bar{p}}_f = \ln \bar{p}_f(\mathbf{x}_j) \qquad \forall \qquad g(\mathbf{x}_j, \mathbf{y}) = 0, \quad j = 1,\dots,k \tag{3.10}$$

The probabilities $\bar{p}_f(\mathbf{x}_j)$ are the estimated failure probabilities using the RSM and importance sampling. The number of interpolation points k taken into account depends on the number of considered design variables and can be determined according to the following relation:

$$k = 2n + \frac{n(n-1)}{2} \tag{3.11}$$

The number of interpolation points proposed in eq. (3.11) provides that the equation system for evaluation of the coefficients in eq. (3.9) is determinate. Certainly, the goodness of approximation of the RS may depend on the number of interpolation points used. However, this can not be claimed generally. A general rule for choosing the interpolation points is not available. The only criterion which can be taken into account is that the supporting points should cover almost the complete feasible domain. In most cases where the feasibility of design variable combinations can not be assessed in advance the interpolation points will be distributed within the range of the side constraints.

3.2.3 Simple Truss Structure

The posterior approximation strategy is illustrated by performance of RBO of the truss structure already investigated above. The same problem definition is considered where a weight minimization subject to a single system reliability constraint is carried out. The approximation of the system probability of failure respectively the constraint function takes place based on 70 interpolation points. For this purpose 70 reliability analysis calculations have been performed. The component limit state functions are given in explicit form. Following this calculation step a linear equation system is formed and solved to evaluate the probability function coefficients. At this point the actual optimization procedure can be executed *without* any response calculations required for the reliability analyses. The numerical effort is only due to the calculation of the interpolation points for the approximation of the probability function. Hence, the main advantage of this process is that the reliability analyses are reduced to almost simple function calculations.

variable number	reference solution				COSSAN using NLPQL	
A_i	feasible direction		penalty		70 supports	
	[mm²]	[in²]	[mm²]	[in²]	[mm²]	[in²]
1	759.4	1.177	745.8	1.156	645.2	1.000
2	1190.3	1.845	1218.7	1.889	1077.4	1.670
3	734.8	1.139	752.9	1.167	645.2	1.000
4	759.4	1.177	745.1	1.155	645.2	1.000
5	231.6	0.359	227.1	0.352	193.5	0.300
6	843.2	1.307	840.0	1.302	645.2	1.000
7	514.2	0.797	521.9	0.809	781.3	1.211
objective fct.		804.2		809.2		733.2
ß		4.256		4.256		4.256
iterations		379		636		6
number of RA		379		636		90

Table 3.3: Comparison of optimal results due to posterior approximation with reference solution, 70 interpolation points (see [10])

A further advantage is that when probability functions are available in explicit form the optimization problem can be individually modified without any severe additional numerical effort. For example, the failure consequence cost factors can be changed and also the complete problem definition can be rearranged. As long as no additional or alternative failure criteria or distribution parameters have to be introduced, the numerical effort is confined to the pure nonlinear optimization procedure carrying out explicit function calls. Sensitivity studies of parameters of the problem functions can be carried out without any problem.

The results determined applying the posterior approximation method are compared with the reference solution in Table 3.3. All seven design variables show differences and the optimal solution does not fit so well as the solution proposed by using the direct approach. However, the principal tendency of the design variables to the optimal solution is maintained. Hence, the results are sufficiently accurate for decision making within the scope of practical applications.

4 Summary

Some basic principles of RBO of structures and mechanical components have been presented. Two different approaches are discussed in some detail.

Direct Optimization Approach. It was shown that by the use of COSSAN-NLPQL and pursuing the direct optimization approach an almost exact agreement with the reference solutions, such as the feasible direction method and the penalty function method, are obtained. The required numerical effort is considerably less than when applying these reference optimization methods. Moreover it is possible with COSSAN-NLPQL to apply the RSM which is certainly a big advantage with respect to applications of practical significance. There is generally no restriction in using any reliability analysis method, e.g. the directional sampling method could be applied as well. Consequently, RBO of complex structures can be performed where other methods based on explicit and often simplified limit state formulations fail. This is primarily the case when nonlinear structural behaviour and non-Gaussian distributions have to be taken into account.

Posterior Approximation Approach. Based on the proposed approximation this strategy allows the utilization of reliability-based structural optimization in engineering practice. The numerical example showed that the approximate results are not of the same accuracy as those obtained by the direct approach. This fact is quite clear as, in addition to the limit state, the failure probability function is also approximated. Nevertheless, all the design variables tend to the optimal reference solution. From this follows that for problems where the limit state function can be formulated explicitly the posterior approximation method does not show any advantage. In this case it is possible to obtain better and more confidence in the results without considerable increase of the numerical efforts by the direct RBO approach. With regard to sensitivity studies this argument is still valid (recall: no additional structural analyses are required). The advantage of this strategy becomes particularly evident for those RBO problems where the number of design variables is low and the efforts within the scope of the reliability analysis are high.

Approximations are required in any case and hence, the use of the posterior approximation approach is most advantageous. Especially when sensitivity analyses become necessary for high quality decision making this method is preferable, although inaccuracies as those discussed here are to be expected.

Acknowledgement

This work has been partially supported by the Austrian Research Promotion Fund (FFF) under contract 6/636 which is gratefully acknowledged by the authors.

References

1   Arora, J.S., Introduction to optimum design, McGraw-Hill, New York, 1989

2   COSSAN "Computational Stochastic Structural Analysis", User's Manual, Institute of Engineering Mechanics, Leopold-Franzens University, Innsbruck, Austria, 1996.

3   Freudenthal, A.M., Gaither, W.S., Probabilistic approach to economic design of maritime structures, *Proceedings* of 22nd International Navigation Congress, Permanent International Association of Navigation Congress, Sec. 11/5, pp. 119-133 (1969)

4   Freudenthal, A.M., Safety and the probability of structural failure, *Transactions* ASCE, Vol. 121, pp. 1337-1397 (1956)

5   Gasser, M., Structural optimization based reliability criteria utilizing approximate methods, Dissertation, Leopold-Franzens University, Innsbruck, Austria, September 1996

6   Lind, N.C., Approximate analysis and economics of structures, *Journal of the Structural Division*, Vol. 102, No. ST6, pp. 1177-1196 (1976)

7   Marti, K., Stochastic optimization in structural design, *ZAMM - Z. angew. Math. Mech.*, 72, 6, T452-T464 (1992a)

8   Marti K., Semi-stochastic approximations by the response surface methodology (RSM)" *Optimization*, Vol. 25, pp. 209-230 (1992b)

9   Marti, K., Approximations and derivatives of probabilities in structural design" *ZAMM - Z. angew. Math. Mech.*, 72, 6, T575-T578 (1992c)

10   Nakib, R., Frangopol, D.M., RBSA and RBSA-OPT: Two computer programs for structural system reliability analysis and optimization" *Computers & Structures*, Vol. 36, No. 1, pp. 13-27 (1990)

11   Rubinstein, R.Y., Simulation and the Monte Carlo method, John Wiley & Sons, New York, 1981

12   Schittkowski, K., Theory, implementation, and test of a nonlinear programming algorithm" *Proceedings* of Euromech-Colloquium 164 on "Optimization Methods in Structural Design", Universität Siegen, 1982, Bibliographisches Institut, Mannheim, pp. 122-132 (1983a)

13   Schittkowski, K., On the convergence of a sequential quadratic programming method with an augmented Lagrangian line search function" *Math. Operationsforschung, Ser. Optimization*, Vol. 14, No. 2, pp. 197-216 (1983)

Stochastic Optimization Approach to Dynamic Problems with Jump Changing Structure

Vadim I. Arkin

Central Economics and Mathematics Institute,
Russian Academy of Sciences, Krasikova 32, Moscow, 117418, Russia

We consider dynamic optimization models the structure of which (functional, equations, constrains) can be changed at time depending on control strategy.

The main problem is a choice of optimal structure and a strategy which provides an optimal transition from one structure to another.

The problem under discussion is connected with modelling global changes in economic mechanism (for example, a transition from Centralization to Market), radical technological innovations and so on.

In deterministic case these problems are nonconvex and nonsmooth. We propose general approach to such models based on their stochastic approximations and obtain a stochastic programming problem with controlled measure.

This approach is illustrated by economic dynamic model with endogenous innovations.

1   General Economic Dynamic Model

We study the general multi-sector economic dynamic model with discretely expanding technologies. Emergence moments of new technological modes (new technological structure) are defined by given levels of expenditure on R&D (research and development) and by the strategy of investments into R&D. The model may be written in the following form:

$$\sum_{k=0}^{N} \sum_{t=\theta_k}^{\theta_{k+1}} \varphi_k(a_t, b_{t+1}) \Rightarrow \max,$$

$$(a_t, b_{t+1}) \in T_k, \quad \theta_k \le t < \theta_{k+1}, \quad (c_t, d_{t+1}) \in Q_k, \quad b_t \ge c_t + a_t$$

$$\theta_k = \min\{t : \sum_{j=\theta_{k-1}}^{t} d_j \ge \xi_k\}, \quad d_{\theta_k} = 0, k = 1, \ldots, N, \quad \theta_{N+1} = \tau - 1,$$

where convex technological set T_k (i.e. a set of 'input–output' vectors (a, b)) and concave utility function φ_k form structure of economic system; vector ξ_k is the level of expenditures necessary for the transition from the structure

(φ_k, T_k) to (φ_{k+1}, T_{k+1}); θ_k is the corresponding transition moment, and convex set Q_k specifies dynamics of assets in R&D for unit interval.

This model is nonconvex in general. We use a stochastic version of the initial model, which takes into account an uncertainty of expenditure levels on R&D and incompleteness of information on parameters of future technologies. The stochastic model is already locally convex. This fact allows us to formulate the stochastic maximum principle for the model and to find the optimal structure as well as optimal strategy of transition.

For simple case ($N = 1$) this problem was study in [1]. Model of structural transition from centralized economy to market economy was considered in [2]. Analisys of the above model essentially use the following generalization of stochastic maximum principle from [3].

2 Stochastic Maximum Principle with Controlled Measure

Let $\{\eta_t, t = 0, 1, \ldots, \tau\}$ be stochastic process with values in measurable space (S, E) with transition functions $\Pi^{t+1}(\eta^t, x_t, u_t, d\eta_{t+1})$, depending measurably on the process $x_t \in R^n$, and on control $u_t \in U$, where U is Polish space, $\eta^\tau = (\eta_0, \ldots, \eta_\tau)$, $\tau < \infty$. The process x_t is described by the system of difference equations:

$$x_{t+1} = f^{t+1}(\eta^{t+1}, x_t, u_t), \quad x_0 = x_0(\eta_0) \tag{1}$$

Suppose that measures Π^{t+1} are absolutely continuous with respect to some (fixed) transition measure $P^{t+1}(\eta^t, d\eta_{t+1})$; $\pi^{t+1}(\eta^t, x_t, u_t, \eta_{t+1})$ is density with respect to η_{t+1}. Each control $u_t = u_t(\eta^t)$ generates a measure on the space of sequences $\{\eta^\tau\}$. It is required to maximize the functional:

$$E^u \sum_0^{\tau-1} \tilde{\varphi}^t(\eta^t, x_t, u_t) \tag{2}$$

subject to restrictions

$$E^u \sum_0^{\tau-1} \bar{\varphi}(\eta^t, x_t, u_t) \geq 0, \quad g(\eta^t, x_t, u_t) \leq 0 \quad (u_t \in U), \quad (P - a.s.), \tag{3}$$

where P is a measure generated by the initial distribution $P_0(d\eta_0)$ and the transition function $P^{t+1}(\eta^t, d\eta_{t+1})$, $\tilde{\varphi}^t(\cdot) \in R^1$, $\bar{\varphi}^t(\cdot) \in R^m$, $g^t(\cdot) \in R^k$. Let denote

$$H^{t+1}(\eta^{t+1}, x, u) = \tilde{\alpha}\tilde{\varphi}^t(\eta^t, x, u) + \bar{\alpha}\bar{\varphi}^t(\eta^t, x, u) + \psi_{t+1}f^{t+1}(\eta^{t+1}, x, u)$$
$$- \lambda_t g^t(\eta^t, x, u).$$

Next we formulate assumptions on the functions $\varphi^t = (\tilde{\varphi}^t, \bar{\varphi}^t)$, $F^t = (f^{t+1}, \pi^{t+1})$, g^t required to obtain necessary optimality conditions for the above problem.

General conditions.

(A1)   The functions given by $\varphi^t(\eta^t, x, u)$, $t = 0, \ldots, \tau - 1$, are jointly measurable and the composition $\varphi^t(\cdot, x_t(\cdot), u_t(\cdot)) \in L_1(S^t)$ for all admissible pairs $\{(x_t, u_t)\}$. At each point (η^t, x, u) the functions given by $\varphi^t(\eta^t, x, u)$ are differentiable with respect to x and their derivatives $\varphi_x^t(\eta^t, x, u)$ (gradients with respect to x) are continuous in x and satisfy the following condition:

for any bounded set $C \subseteq R^n$ there exists a function $\rho_C^t \in L_1(S^t)$ such that for all $x \in C$

$$|\varphi_x^t(\eta^t, x, u)| \le \rho_C^t(\eta^t).$$

(A2)   The vector-valued functions given by $F^{t+1}(\eta^{t+1}, x, u), g^t(\eta^t, x, u)$, $t = 0, \ldots, \theta - 1$, are jointly measurable. The set $U_t(\eta^t)$ depends measurably on the parameter η^t, $t = 0, \ldots, \tau - 1$.

(A3)   For any bounded set $C \subseteq R^n$ there exists a constant K_C such that

$$|F^{t+1}(\eta^{t+1}, x, u(\eta^t))| + |g^t(\eta^t, x, u(\eta^t))| \le K_C \qquad a.s.$$

for all $x \in C$ and $t = 0, 1, \ldots, \tau - 1$. The derivatives of the constraint functions $F_x^{t+1}(\eta^{t+1}, x, u)$ and $g_x^t(\eta^t, x, u)$ with respect to x exist and the constant K_C which corresponds to every bounded set $C \subseteq R^n$ has the further properties that

$$|F_x^{t+1}(\eta^{t+1}, x, u(\eta^t))| + |g_x^t(\eta^t, x, u(\eta^t))| \le K_C \qquad a.s.$$

for all $x \in C$ and $t = 0, 1, \ldots, \tau - 1$, and

$$|F_x^{t+1}(\eta^{t+1}, x_1, u(\eta^t)) - F_x^{t+1}(\eta^{t+1}, x_2, u(\eta^t))|$$

$$+ |g_x^t(\eta^t, x_1, u(\eta^t)) - g_x^t(\eta^t, x_2, u(\eta^t))| \le K_C|x_1 - x_2| \qquad a.s.$$

for all $x_1, x_2 \in C$ and $t = 0, 1, \ldots, \tau - 1$.

Convexity condition.

(B)   For any set of parameters $\sigma := \{\eta^t, x, u', u'', \alpha\}$, where $u', u'' \in U_t(\eta^t)$, $x \in R^n$, $0 \le \alpha \le 1$ and $0 \le t \le \tau$, an element $u_\sigma \in U_t(\eta^t)$ may be found such that the following relations hold:

$$\varphi^t(\eta^t, x, u_\sigma) \ge \alpha\varphi^t(\eta^t, x, u') + (1 - \alpha)\varphi^t(\eta^t, x, u''),$$

$$F^{t+1}(\eta^{t+1}, x, u_\sigma) = \alpha F^{t+1}(\eta^{t+1}, x, u') + (1-\alpha)F^{t+1}(\eta^{t+1}, x, u'')$$
$$P^{t+1}(\eta^t, \cdot) - a.s.,$$

$$g^t(\eta^t, x, u_\sigma) \le \alpha g^t(\eta^t, x, u') + (1-\alpha)g^t(\eta^t, x, u'').$$

Regularity condition.

(C) *For each* t, $0 \le t \le \tau - 1$, *one can find a control* $\tilde{u}_t \in U_t(\eta^t)$ *a.s. for which* $E|\varphi^t(\eta^t, x_t^0, \tilde{u}_t)| < \infty$ *and*

$$g^t(\eta^t, x_t^0, \tilde{u}_t) < -\gamma e \qquad a.s.,$$

for some positive γ *and 'unit' vector* $e := (1, 1, \ldots, 1)$.

Theorem 1. (Maximum Principle) *Let* $\{x_t^*, u_t^*\}$ *be a solution of the problem (1)–(3), and* Π^* *be the corresponding measure on the space of histories* $\{\eta^\tau\}$. *Then there exist non-trivial vector* $\alpha = (\tilde{\alpha}, \bar{\alpha}) \in R^{m+1}$, $\alpha \ge 0$, *and functions* $\psi_t \in L_1^n(S^t, E^t, \Pi^*)$, $\lambda_t \in L_1^k(S^t, E^t, \Pi^*)$, $\lambda_t \ge 0$, $h_t \in L_1^1(S^t, E^t, P)$ *such that:*

1) $u_t^* = \arg\max_{u \in U} \{E^*[H^{t+1}(\eta^{t+1}, x_t^*, u)|\eta^t] + \int h_{t+1}\Pi_x^{t+1}(\eta^t, x_t^*, u, d\eta_{t+1})\}$
   $(\Pi^* - a.s.);$

2) $\psi_t = E^*[H_x^{t+1}(\eta^{t+1}, x_t^*, u_t^*)|\eta^t] + \int h_{t+1}\Pi_x^{t+1}(\eta^t, x_t^*, u, d\eta_{t+1}), \quad \psi_\tau = 0$
   $(\Pi^* - a.s.);$

3) $h_t = \alpha\varphi^t(\eta^t, x_t^*, u_t^*) + \int h_{t+1}\Pi_x^{t+1}(\eta^t, x_t^*, u, d\eta_{t+1}), \quad h_\tau = 0$
   $(P - a.s.);$

4) $\bar{\alpha}E^* \sum \bar{\varphi}^t(\eta^t, x_t^*, u_t^*) = 0;$

5) $\lambda_t g^t(\eta^t, x_t^*, u_t^*) = 0 \quad (\Pi^* - a.s.).$

This result extends similar ones in [3,4] and can be proved following the arguments used in [3].

3 The Model with Jump Changing Structure. Supporting Prices

Let us return to the model described in Section 1 and let's assume that level ξ_k which is necessary for the transition to the structure (φ_k, T_k) be a random vector with a given distribution function $\pi_k(y)$ $0 \le k \le N$. Moreover, we assume that before the moment θ_k when the structure (φ_k, T_k) emerges, characteristics of this structure are known incompletely. Formally it is expressed by a dependence of the set T_k and function φ_k on some parameter $s \in S_k$. Before the moment θ_k this parameter is assumed random, and its particular value s_k becomes known after emergence of the structure (φ_k, T_k) only, i.e. after the moment θ_k. We assume that $(S_k, \mathcal{E}_k, P_k)$ is a probability space, and distributions $P_k(ds)$ $(k = 1, \ldots, N)$ are mutually independent. Everywhere below we shall assume that all functions depending on $s_1, s_2, \ldots$ are measurable (with respect to the product $\mathcal{E}_1 \times \ldots \times \mathcal{E}_N$), and all relations between these functions hold almost sure with respect to product-measure $P_1 \times \ldots \times P_N$.

A *transition program* is defined as follows

$$Z = \{z_t^k(\theta_1, \ldots, \theta_k), \ k = 0, \ldots, N, \ t \geq \theta_k, \ 1 \leq \theta_1 < \theta_2 < \ldots < \theta_N < \tau\}$$
(4)

where functions

$$z_t^k(\theta_1, \ldots, \theta_k) = \{(a_t^k, b_{t+1}^k) := (a_t^k(\theta_1, s_1, \ldots, \theta_k, s_k), b_{t+1}^k(\theta_1, s_1, \ldots, \theta_k, s_k),$$
(5)

$$(c_t^k, d_{t+1}^k) := (c_t^k(\theta_1, s_1, \ldots, \theta_k, s_k), d_{t+1}^k(\theta_1, s_1, \ldots, \theta_k, s_k))\}, \quad t \geq \theta_k, \quad (6)$$

satisfy the constraints:

$$(a_t^k, b_{t+1}^k) \in T_k(s_k), \ (c_t^k, d_{t+1}^k) \in Q_k, \ b_t^k \geq c_t^k + a_t^k, \quad \theta_k \leq t < \theta_{k+1},$$

$$\theta_k = \min\{t : \sum_{j=\theta_{k-1}}^{t} d_j^{k-1} \geq \xi_k\}, \quad b_{\theta_{k-1}}^k = b_{\theta_{k-1}}^{k-1}, \quad d_{\theta_k}^k = 0.$$

The initial resource vector $\hat{b}_0$ is assumed given.

It is required to find a transition program Z (as in (4)–(6)) which maximize the functional:

$$E\left[\sum_{k=0}^{N} \sum_{t=\theta_k}^{\theta_{k+1}} \varphi^k(a_t^k, b_{t+1}^k, s_k)\right] \Rightarrow \max.$$
(7)

In further we shall assume that each function $\pi_k(y)$ is continuously differentiable in all the arguments, and $\pi_k(0) = 0$.

For almost all $(s_1, s_2, \ldots s_N) \in S_1 \times \ldots \times S_N$ a sequence of the corresponding technologies is assumed to be non-decreasing:

$$T_0 \subset T_1(s_1) \subset T_2(s_2) \subset \ldots \subset T_N(s_N).$$

The sets Q_k, T_0, $T_k(s)$, $k = 1, \ldots, N$ for each s are convex, and the functions $\varphi^k(a, b, s)$ are concave in $(a, b) \in T_k(s)$ for every $s \in S_k$, $k = $

$0, 1, \ldots, N$, in addition $\varphi^{k+1}(a, b, s_{k+1}) \geq \varphi^k(a, b, s_k)$ for each pair $(a, b) \in T_k(s_k)$. The sets Q_k are assumed bounded and contains zero point $(0, 0)$.

Definition. We say that the sequences of nonnegative vector functions

$$\psi_t^k = \psi_t^k(\theta_1, s_1, \ldots, \theta_k, s_k), \quad \alpha_t^k = \alpha_t^k(\theta_1, s_1, \ldots, \theta_k, s_k), \quad \theta_k \leq t < \theta_{k+1},$$

with values in R^n and R^{m_k} respectively, the sequence of nonnegative scalar functions $h_t^k = h_t^k(\theta_1, s_1, \ldots \theta_k, s_k)$ support the transition program Z (of the form (4)–(6)), if the following conditions are satisfied:
A. for each $\quad 0 \leq k \leq N, \theta_k \leq t < \theta_{k+1}$

$$(a_t^k, b_{t+1}^k) = \arg\max_{(a,b)\in T_k(s_k)} [\varphi^k(a, b, s_k) + \bar{\psi}_{t+1}^k b - \psi_t^k a],$$

where

$$\bar{\psi}_{t+1}^k = \frac{1 - \pi_{k+1}(y_{t+1}^k)}{1 - \pi_{k+1}(y_t^k)} \psi_{t+1}^k + \frac{\pi_{k+1}(y_{t+1}^k) - \pi_{k+1}(y_t^k)}{1 - \pi_{k+1}(y_t^k)} \int_{S_{k+1}} \psi_{t+1}^{k+1} P(ds_{k+1}),$$

$$y_t^k = \sum_{j=t_k}^t d_j^k, \qquad \psi_t^k b_t^k = \psi_t^k(a_t^k + c_t^k)$$

B. $(c_t^k, d_{t+1}^k) = \arg\max_{(c,d)\in Q_k} [\alpha_{t+1}^k d - \psi_t^k c], \quad \theta_k \leq t < \theta_{k+1}$,
C. The prices α_t^k satisfy the relationship

$$\alpha_t^k = \alpha_{t+1}^k \frac{1 - \pi_{k+1}(y_t^k)}{1 - \pi_{k+1}(y_{t_1}^k)} + h_t^k \frac{\pi'_{k+1}(y_t^k)}{1 - \pi_{k+1}(y_{t-1}^k)}, \quad [1]$$

$$h_t^k = W_t^{k+1} - \{\varphi^k(a_t^k, b_{t+1}^k, s_k) + W_{t+1}^{k+1}\},$$

$$\theta_k \leq t < \theta_{k+1},$$

where W_t^{k+1} is the value of functional on the program Z under the condition that the structure (φ_{k+1}, T_{k+1}) emerges at the moment t, i.e.

$$W_t^{k+1} = E[\sum_{j=k+1}^N \sum_{i=\theta_j}^{\theta_{j+1}} \varphi(a_i^j, b_{i+1}^j, s_i) | \theta_1 = t_1, s_{t_1} = s_1, \ldots, \theta_k = t_k,$$

$$s_{t_k} = s_k, \ \theta_{k+1} = t].$$

Theorem 2. *Let the transition program Z of the form (4)–(6) be optimal in problem (7) and let the following additional condition holds:*

---

[1] $\pi'_{k+1}(y)$ denotes a vector of partial derivatives (gradient) of the function $\pi_{k+1}(y)$

there are technological processes $(\hat{a}^k(s_k), \hat{b}^k(s_k)) \in T_k(s_k)$, *such that* $b_t^k > \hat{a}^k(s_k)$, $\theta_k \leq t < \theta_{k+1}$, $0 \leq k \leq N$.

Then there exist prices supporting the transition program Z.

Proof of this Theorem is based on the version of Stochastic Maximum Principle (Theorem 1) stated in Section 2.

Acknowledgement. This work is partially supported by Russian Foundation of Basic Researches (grant 97–01–00684) and INTAS–RFBR (project 95–0061).

References

1. V.I.Arkin. Economic dynamics with discrete changes in technology: stochastic approach. – *Matekon*, v. 26, 3 (1990)

2. V.I.Arkin, A.D.Slastnikov. Change in economic mechanism: model of transition from budgets regulation to competitive market. – *Economic Theory*, v.7, 2 (1996), p. 307–321.

3. V.I.Arkin, I.V.Evstigneev. Stochastic models of control and economic dynamics. London: Academic Press, 1987.

4. V.I.Arkin. On the maximum principle for a stochastic control problem, in *Proceedings of the Bernoulli World Congress*, VNU Sci. Press, Utrecht, 1987, p. 777–782.

Reflections on Robust Optimization

Jitka Dupačová

Department of Probability and Mathematical Statistics, Charles University, 186 00 Prague, Czech Republic

Abstract. Various aspect of the robust optimization approach [11] are discussed in the context of scenario based stochastic linear programs. The main items are the choice of the model parameter, which can be related to properties of nonlinearly perturbed linear programs [10] or of parametric quadratic programs [1], and an extension of the first results on the robustness of the optimal value with respect to probabilities of the selected scenarios and with respect to out-of-sample scenarios, cf. [5].

Keywords. Robust optimization, tracking model, scenario based stochastic programs, postoptimality, contamination technique, out-of-sample scenarios

1   Introduction

Let us consider various approaches to mathematical formulation of decisions problems under stochastic uncertainty about the future values of the system parameters. We assume that the initial available information consists of a finite number of possible batches of these parameters, called *scenarios* and that they are complemented by probabilities of their outcome. A frequent requirement is to decide before realization of one of these scenarios is known; there is an option to update this decision and/or to recompute the cost of the total decision procedure after the information about which scenario occures is revealed. Given the initial information, the goal is to get the best possible scenario-independent initial decision. The decision criterion can be, for instance, the lowest expected cost, the lowest variability of costs under individual scenarios, etc. Its choice depends on the problem to be solved.

Example 1. Scenario-based two-stage stochastic linear program (SLP) with *random relatively complete recourse* can be written in the form:
  Minimize

$$(1) \qquad \mathbf{c}^\mathsf{T}\mathbf{x} + \sum_{s=1}^{S} p_s \mathbf{q}_s^\mathsf{T} \mathbf{y}_s$$

Supported by the Grant Agency of the Czech Republic under grants No. 201/96/0230 and 402/96/0420

Typeset by $\mathcal{A}\mathcal{M}\mathcal{S}$-TEX

subject to

$$
\begin{aligned}
(2) \qquad \mathbf{A}\mathbf{x} \phantom{+\mathbf{W}_1\mathbf{y}_1} &= \mathbf{b} \\
\mathbf{T}_1\mathbf{x}+\mathbf{W}_1\mathbf{y}_1 \phantom{\mathbf{W}_2\mathbf{y}_2} &= \mathbf{h}_1 \\
\mathbf{T}_2\mathbf{x}+ \qquad \mathbf{W}_2\mathbf{y}_2 &= \mathbf{h}_2 \\
\vdots \qquad\qquad \ddots \qquad \vdots \\
\mathbf{T}_S\mathbf{x}+ \qquad \cdots \qquad + \mathbf{W}_S\mathbf{y}_S &= \mathbf{h}_S
\end{aligned}
$$

$$
\mathbf{x} \geq 0, \mathbf{y}_s \geq 0, s = 1, \ldots, S
$$

where $\omega_s = [\mathbf{q}_s, \mathbf{T}_s, \mathbf{W}_s, \mathbf{h}_s], s = 1, \ldots, S$ are scenarios and $p_s \geq 0, s = 1, \ldots, S$ are their probabilities, $\sum_s p_s = 1$. The first-stage decision $\mathbf{x}$ is scenario independent and, for each of considered scenarios, second-stage decisions $\mathbf{y}_s$ are introduced to maintain the constraints for the minimal additional cost. The assumption of the relatively complete recourse means that the set of feasible solutions (2) is nonempty.

The problem (1) – (2) can be rewritten as
minimize

$$
(3) \qquad \mathbf{c}^\mathsf{T}\mathbf{x} + \sum_{s=1}^{S} p_s q(\mathbf{x}, \omega_s)
$$

on the set

$$
(4) \qquad \mathcal{X} = \{\mathbf{x}|\mathbf{A}\mathbf{x} = \mathbf{b}, \mathbf{x} \geq 0\}
$$

where

$$
(5) \qquad q(\mathbf{x}, \omega_s) = \min_{\mathbf{y}_s}\{\mathbf{q}_s^\mathsf{T}\mathbf{y}_s|\mathbf{W}_s\mathbf{y}_s = \mathbf{h}_s - \mathbf{T}_s\mathbf{x}, \quad \mathbf{y}_s \geq 0\}
$$

For this type of problem, the criterion is the minimal expected cost of the two-stage decision process.

Example 2. The objective function (3) from Example 1 can be modified to minimize

$$
(6) \qquad -\sum_{s=1}^{S} p_s u(\mathbf{c}^\mathsf{T}\mathbf{x} + q(\mathbf{x}, \omega_s))
$$

on the set (4) and with notation (5). This criterion follows the principle of maximal expected utility and u is assumed to be a concave nondecreasing utility function.

Example 3. A robust optimization model (RO) (cf. [11]) which corresponds to (1), (2) is:
Minimize

(7)
$$\sum_{s=1}^{S} p_s \xi_s + \lambda \sum_{s=1}^{S} p_s \left[\xi_s - \sum_{j=1}^{S} p_j \xi_j \right]^2$$

subject to

(8)
$$\begin{aligned}
\mathbf{A}\mathbf{x} &= \mathbf{b} \\
\mathbf{T}_s\mathbf{x} + \mathbf{W}_s\mathbf{y}_s &= \mathbf{h}_s \\
\mathbf{c}^\mathsf{T}\mathbf{x} + \mathbf{q}_s^\mathsf{T}\mathbf{y}_s - \xi_s &= 0
\end{aligned}$$

$$\mathbf{x} \geq 0, \mathbf{y}_s \geq 0, s = 1, \dots, S$$

The newly introduced variables ξ_s are equal to the cost of the decision $\mathbf{x}$ plus the corresponding cost of its compensation or of the recourse activity $\mathbf{y}_s$ if the scenario ω_s occurs. The additional term in the objective function equals the variance of the random costs ξ and its weight in the objective function is expressed via a scalar parameter $\lambda \geq 0$. The objective function (7) can be related to a bicriteria optimization problem where the first criterion coincides with minimization of the objective function used in Example 1. Hence, the efficient solutions which are obtained by solving RO for different values of λ can be also computed as follows:
Minimize

(9)
$$\sum_{s=1}^{S} p_s \left[\xi_s - \sum_{j=1}^{S} p_j \xi_j \right]^2$$

subject to (8) and

$$\sum_{s=1}^{S} p_s \xi_s \leq \gamma$$

with an appropriately chosen parameter value γ. Moreover, similarly as in [9], the optimal solution of this program can be regarded as an approximation of the optimal solution for the (minus) expected utility criterion applied to the costs ξ_s. The parameter γ identifies the point about which the approximation of the utility function by its Taylor expansion is used.

Variance of the random costs ξ is only one of possible choices of additional criteria suggested in [11]; see also Example 7.

Example 4. The mean - variance model (M-V) for the minimal recourse costs $q(\mathbf{x}, \omega_s)$ can be written as minimize

$$(10) \quad \mathbf{c}^\mathsf{T}\mathbf{x} + \sum_{s=1}^{S} p_s q(\mathbf{x},\omega_s) + \lambda \sum_{s=1}^{S} p_s [q(\mathbf{x},\omega_s) - \sum_{s=1}^{S} p_s q(\mathbf{x},\omega_s)]^2$$

on the set (4) and with the minimal recourse costs $q(\mathbf{x},\omega_s)$ defined by (5). In contrast with the robust optimization problem, this criterion works with the *minimal recourse costs* $q(\mathbf{x},\omega_s)$ and it deals with two conflicting criteria - the minimal total expected costs and the minimal variability of these costs.

Example 5. The tracking model (see [2]) related to (1) – (2) can be formulated as follows: Let $v_s, s = 1,\ldots,S$ be the optimal values of the *individual* scenario problems

minimize

$$(11) \qquad\qquad\qquad \mathbf{c}^\mathsf{T}\mathbf{x} + \mathbf{q}_s^\mathsf{T}\mathbf{y}_s$$

subject to

$$(12) \qquad\qquad\qquad\begin{aligned} \mathbf{A}\mathbf{x} \qquad\quad &= \mathbf{b} \\ \mathbf{T}_s\mathbf{x} + \mathbf{W}_s\mathbf{y}_s &= \mathbf{h}_s \\ \mathbf{x} \geq 0, \mathbf{y}_s &\geq 0. \end{aligned}$$

Then the basic compromising or tracking model is

minimize

$$(13) \qquad \sum_{s=1}^{S} p_s \left(\|\mathbf{c}^\mathsf{T}\mathbf{x} + \mathbf{q}_s^\mathsf{T}\mathbf{y}_s - v_s\| + \|\mathbf{T}_s\mathbf{x} + \mathbf{W}_s\mathbf{y}_s - \mathbf{h}_s\| \right)$$

subject to

$$\mathbf{A}\mathbf{x} = \mathbf{b}$$
$$\mathbf{x} \geq 0, \mathbf{y}_s \geq 0, s = 1,\ldots,S.$$

The first and second stage solutions obtained by solving this problem track the optimal solutions of the individual scenario problems (11), (12) as closely as possible. The norm in (13) can be in principle chosen in an arbitrary way; its choice influences the solution procedure.

Example 6. SLP with restricted recourse [12] aims at limitation of the dispersion of the recourse *decisions*. According to the principles of multicriteria optimization, the objective function (1) should be extended to

$$\mathbf{c}^\mathsf{T}\mathbf{x} + \sum_{s=1}^{S} p_s \mathbf{q}_s^\mathsf{T}\mathbf{y}_s + \lambda \sum_{s=1}^{S} p_s \|\mathbf{y}_s - \sum_{j=1}^{S} p_j \mathbf{y}_j\|$$

with a nonnegative parameter λ or the constraints (2) extended for an additional constraint

$$(14) \qquad\qquad \sum_{s=1}^{S} p_s \|\mathbf{y}_s - \sum_{j=1}^{S} p_j \mathbf{y}_j\| \leq \epsilon$$

where ϵ is a chosen tolerance level. Models of this type capture features of both robust optimization and tracking model.

Example 7. Another possibility is to replace the objective function in the RO problem (7) – (8) by a general performance function of the decision variables $\mathbf{x}, \mathbf{y}_s, s = 1, \ldots, S$ and in addition, to relax the constraints; cf. [11]. The resulting model can be formulated as follows:

Minimize

$$\sigma(\mathbf{x}, \mathbf{y}_1, \ldots, \mathbf{y}_S) + \mu\rho(\mathbf{u}_1, \ldots \mathbf{u}_S)$$

subject to

(2')
$$\begin{aligned}
\mathbf{A}\mathbf{x} &= \mathbf{b} \\
\mathbf{T}_s\mathbf{x} + \mathbf{W}_s\mathbf{y}_s + \mathbf{u}_s &= \mathbf{h}_s
\end{aligned}$$

$$\mathbf{x} \geq 0, \mathbf{y}_s \geq 0, s = 1, \ldots, S$$

The first term in the objective function corresponds to (7), the second one with a parameter $\mu \geq 0$ penalizes possible violations of the inital second-stage constraints $\mathbf{T}_s\mathbf{x} + \mathbf{W}_s\mathbf{y}_s = \mathbf{h}_s$.

From the modeling point of view the choice among the introduced models depends on the nature of the real life problem to be solved, on the numerical tractebility of the resulting optimization problem and also on properties such as the sensitivity of the solution on the input data, i. e., on the choice of scenarios and their probabilities, on the errors in scenarios, etc. In this paper, we shall concentrate on properties of the optimal solution of the robust optimization problem (7), (8) in comparison with those for stochastic linear program (1) – (2) and for the corresponding mean-variance model; in terminology of [10] it means that our focus is exclusively on *solution robustness* but not on model robustness. In the next Section we shall give a comparison of optimal solutions of the three mentioned models on a simple example; we shall see that the differences between optimal solutions appear only for sufficiently large values of parameter λ. A detailed analysis of this phenomena in the general case is the main subject of Section 3. Section 4 is devoted to the resistance of the optimal value of the RO problem with respect to inclusion of additional scenarios; using the contamination technique we shall derive global bounds for the optimal value of RO under assumptions comparable with those for SLP. The numerical illustration of the obtained results comes from [3].

2 An Illustrative Example: The Newsboy Problem

The well known *newsboy problem* can be stated as follows:

A newsboy sells newspapers for the cost c each. Before he starts selling, he has to buy the daily supply at the cost b a paper, $c > b > 0$. The demand is random and the unsold newspapers are returned without refund at the end of the day. How many newspapers should he buy?

In the framework of scenario based stochastic decision models introduced in Section 1, one assumes that the demand is random with a known discrete distribution concentrated at $0 < \omega_1 < \omega_2 < \cdots < \omega_S$ with probabilities $p_s > 0, s = 1, \ldots, S, \sum p_s = 1$ and we get:

2.1 The SLP formulation

$$(15) \qquad \min_{x \geq 0}[(b-c)x + c\sum_{s=1}^{S} p_s(x - \omega_s)^+]$$

which can be also written as

$$(16) \qquad \min_{x \geq 0, y_s \geq 0 \forall s}[(b-c)x + c\sum_{s=1}^{S} p_s y_s]$$

subject to

$$(17) \qquad x - y_s \leq \omega_s, s = 1, \ldots, S$$

2.2 The M-V formulation
minimize
$$(18)$$
$$(b-c)x + c\sum_{s=1}^{S} p_s(x - \omega_s)^+ + \lambda c^2 \sum_{s=1}^{S} p_s[(x - \omega_s)^+ - \sum_{j=1}^{S} p_j(x - \omega_j)^+]^2$$

subject to $x \geq 0$.

2.3 The RO formulation
$$(19)$$
$$\min_{x \geq 0, y_s \geq 0 \forall s}\left\{(b-c)x + c\sum_{s=1}^{S} p_s y_s + \lambda c^2 \sum_{s=1}^{S} p_s[y_s - \sum_{j=1}^{S} p_j y_j]^2\right\}$$

subject to (17).

For to be able to understand the differences, we shall solve these three problems assuming that there are only two extremal scenarios of demand, $0 < \omega_1 < \omega_2$ with probabilities $p_1, p_2 = 1 - p_1$.

For the *SLP formulation 2.1* it is enough to evaluate and to compare the values of the objective function (15) at the points $x = \omega_1, x = \omega_2$ what gives $(b-c)\omega_1 < 0$, $(b-c)\omega_2 + p_1 c(\omega_2 - \omega_1)$, respectively. The optimal decision is

$$x_{SLP} = \omega_1 \quad \text{if} \quad b - p_2 c > 0$$

$$x_{SLP} = \omega_2 \quad \text{if} \quad b - p_2 c < 0$$

$$x_{SLP} \in [\omega_1, \omega_2] \quad \text{if} \quad b = p_2 c$$

Similarly for the *M-V formulation 2.2*, the objective function (18) attains its minimum in the interval $[\omega_1, \omega_2]$; this gives three possibilities

$$(20) \qquad x_{MV} = \omega_1 \qquad\qquad\qquad\qquad \text{if} \quad b - p_2 c \geq 0$$

$$x_{MV} = \omega_1 + \frac{p_2 c - b}{2\lambda c^2 p_1 p_2} \qquad\qquad \text{if} \quad b - p_2 c < 0$$

$$\text{and} \quad \lambda > \lambda_0 := \frac{p_2 c - b}{2c^2 p_1 p_2 (\omega_2 - \omega_1)}$$

$$x_{MV} = \omega_2 \qquad\qquad\qquad \text{if} \quad b - p_2 c < 0 \quad \text{and} \quad \lambda \leq \lambda_0$$

Hence, for small values of λ, the optimal decision x_{MV} is optimal for SLP, too.

The analysis of the *RO formulation 2.3* is more complicated. Consider first the response of the model on an *a priori chosen* first-stage decision x which is given by minimization of

$$(21) \quad c(p_1 y_1 + p_2 y_2) + \lambda c^2 \left\{ p_1 [y_1 - (p_1 y_1 + p_2 y_2)]^2 + p_2 [y_2 - (p_1 y_1 + p_2 y_2)]^2 \right\}$$

subject to

$$(22) \qquad\qquad y_s \geq (x - \omega_s)^+ := a_s(x), s = 1, 2$$

The objective function (21) can be further rearranged to

$$(23) \qquad\qquad \min c(p_1 y_1 + p_2 y_2) + \lambda c^2 p_1 p_2 [y_1 - y_2]^2$$

The optimal solution is attained at the boundary of the set (22). It equals $a_s(x), s = 1, 2$ for λ small enough,

$$(24) \qquad\qquad \lambda \leq \lambda^*(x) := [2cp_1(a_1(x) - a_2(x))]^{-1}$$

or for $x \leq \omega_1$ (i. e., for $a_1(x) = (x - \omega_1)^+ = 0$). Moreover, for all $x > \omega_1$, $\lambda^*(x) > \lambda_0$. It means that *for $\lambda \leq \lambda_0$ the RO model reduces to the M-V formulation so that their optimal decision x is identical and is optimal also for the SLP model.* (Compare with results of Section 3.2.)

For $x \leq \omega_1$, the optimal value in (23) equals zero, so that the overall objective function is $(b - c)x$ and its minimum is attained at $x = \omega_1$.

For $x > \omega_1$ and for $\lambda > \lambda^*(x)$, $y_1 = a_1(x) = (x - \omega_1)^+$ whereas $y_2 = a_1(x) - \frac{1}{2\lambda p_1 c} > a_2(x)$. This means that for an already selected $x > \omega_1$ and for large values of λ, the penalty for purchasing x can be calculated higher than necessary. Moreover, for $\lambda \to +\infty$ we have $y_2 \to a_1(x)$, i. e., $y_1 = y_2 = (x - \omega_1)^+$.

For a fixed sufficiently large value of λ and for $x > \omega_1$, the RO optimization problem is

$$\min_{x > \omega_1} \left\{ (b - c)x + c(x - \omega_1) - \frac{p_2}{4\lambda p_1} \right\}$$

The objective function increases in x and the optimal decision is not attained. However, for $x \to \omega_1$, "large" value of λ means $\lambda \geq \lambda^*(x)$ where $\lambda^*(x) \to +\infty$, so that $\frac{p_2}{4\lambda p_1} \to 0$ and the infimum equals $(b-c)\omega_1$.

Hence, the partial results valid for $x \leq \omega_1$ and for large λ with $x > \omega_1$ imply that *for $\lambda \to \infty$, the optimal decisions $x_{RO} \to \omega_1$, $x_{MV} \to \omega_1$ independently of probabilities p_1, p_2.* (Compare with optimal solution for SLP and with results of Section 3.3.)

3 Reflections on Robust Optimization

3.1 The basic properties

The model introduced in Example 3
   minimize

$$(7) \qquad \sum_{s=1}^{S} p_s \xi_s + \lambda \sum_{s=1}^{S} p_s \left[\xi_s - \sum_{j=1}^{S} p_j \xi_j \right]^2$$

   subject to

$$(8) \qquad \begin{aligned} \mathbf{Ax} &= \mathbf{b} \\ \mathbf{T}_s\mathbf{x} + \mathbf{W}_s\mathbf{y}_s &= \mathbf{h}_s \\ \mathbf{c}^\mathsf{T}\mathbf{x} + \mathbf{q}_s^\mathsf{T}\mathbf{y}_s - \xi_s &= 0 \\ \mathbf{x} \geq 0, \mathbf{y}_s \geq 0, s &= 1, \ldots, S \end{aligned}$$

is a quadratic convex program in variables $\mathbf{z} = [\mathbf{x}, \mathbf{y}_s, \xi_s, s = 1, \ldots, S]$ and with a nonnegative scalar parameter $\lambda \geq 0$. Let $\mathcal{Z} \neq 0$ be the set of feasible solutions defined by (8). The objective function (7) can be written as

$$(25) \qquad \mathbf{p}^\mathsf{T}\xi + \lambda \xi^\mathsf{T}\mathbf{P}\xi$$

where $\mathbf{p}$ is the S-vector of components p_s, ξ the S-vector of components ξ_s and

$$(26) \qquad \mathbf{P} = \mathrm{diag}\{\mathbf{p}\} - \mathbf{pp}^\mathsf{T}$$

is a singular matrix. Given scenarios $\omega_s, s = 1, \ldots, S$, the optimal value depends on λ and $\mathbf{p}$; to emphasize this fact, it will be denoted by $\varphi(\lambda, \mathbf{p})$. Similarly, the set of optimal solutions of (7) – (8) will be denoted by $\mathcal{Z}^*(\lambda, \mathbf{p})$.

The two-stage SLP (1) – (2) is equivalent to the robust optimization problem (7) – (8) with the parameter value $\lambda = 0$: It can be written as

$$(27) \qquad \text{minimize} \quad \sum_{s=1}^{S} p_s \xi_s \quad \text{on the set} \quad (8)$$

We assume that the set $\mathcal{Z}^* := \mathcal{Z}^*(0, \mathbf{p})$ of its optimal solutions is nonempty. Its optimal value $\varphi(0, \mathbf{p})$ is a lower bound for the value of the objective function (26) for all $\mathbf{z} \in \mathcal{Z}$ and $\lambda \geq 0$. Hence, it is easy to prove the following properties:

Proposition 3.1. Provided that there is an optimal solution of the RO problem for $\lambda = 0$, the RO problem has an optimal solution for all $\lambda \geq 0$.

Proposition 3.2. Assume that $\mathcal{Z}^* := \mathcal{Z}^*(0, \mathbf{p})$ is nonempty. Then the optimal value function $\varphi(\lambda, \mathbf{p})$ is a concave function of $\lambda \in [0, \infty)$.

Proposition 3.3. Assume that $\mathcal{Z}^* := \mathcal{Z}^*(0, \mathbf{p})$ is nonempty and bounded. Then there exists derivative of $\varphi(\lambda, \mathbf{p})$ at $\lambda = 0^+$

(28)
$$\varphi'(0^+, \mathbf{p}) = \min_{z \in \mathcal{Z}^*} \boldsymbol{\xi}^T \mathbf{P} \boldsymbol{\xi} \geq 0$$

(an application of [8, Theorem 17]).

Notice that the differentiability property 3.3 can be extended to all λ, for which $\mathcal{Z}^*(\lambda, \mathbf{p})$ is nonempty and bounded.

Proposition 3.4. Let z^*, z^{**} be two different optimal solutions of (7) – (8). Then for their corresponding parts $\boldsymbol{\xi}^*, \boldsymbol{\xi}^{**}$

(29)
$$\mathbf{p}^T \boldsymbol{\xi}^* = \mathbf{p}^T \boldsymbol{\xi}^{**} \quad \text{and} \quad \boldsymbol{\xi}^{*T} \mathbf{P} \boldsymbol{\xi}^* = \boldsymbol{\xi}^{**T} \mathbf{P} \boldsymbol{\xi}^{**}$$

(see [1, Theor. A.4.3]).

3.2 The case of small λ

The robust optimization problem (7) – (8) with small $\lambda > 0$ can be viewed as a perturbation of the SLP problem (1) – (2) or (27). At this moment we can exploit general results of [9]:

Proposition 3.5. Let $\mathcal{Z}^*(0, \mathbf{p})$ be nonempty and bounded. Then

(i) there exists $\hat{\lambda} > 0$ and $\left[\hat{\mathbf{x}}, \hat{\mathbf{y}}_s, \hat{\boldsymbol{\xi}}_s, s = 1, \ldots, S\right]$ which is an optimal solution of the two-stage SLP (27) and of the robust optimization problem (7), (8) for all $\lambda \in \left[0, \hat{\lambda}\right]$.

(ii) For λ small enough, $\mathcal{Z}^*(\lambda, \mathbf{p}) \subset \mathcal{Z}^*(0, \mathbf{p})$.

(iii) If $\bar{z} = [\bar{\mathbf{x}}, \bar{\mathbf{y}}_s, s = 1, \ldots, S, \bar{\boldsymbol{\xi}}]$ is an optimal solution of the convex quadratic program

(30)
$$\text{minimize} \quad \boldsymbol{\xi}^T \mathbf{P} \boldsymbol{\xi} \quad \text{on the set} \quad \mathcal{Z}^*(0, \mathbf{p})$$

then $\bar{z}$ is an optimal solution of the two-stage SLP (27) and of the robust optimization problem (7) – (8) for $\lambda \geq 0$ small enough.

Proof is a straightforward application of Theorem 1 of [10] in the case of statement (i) and of Theorems 3 and 4 ibid in the case of the remaining two statements.

In the sequel we shall assume that the set of optimal solutions of the two-stage SLP (1) – (2) or the set $\mathcal{Z}^*(0, \mathbf{p})$ is nonempty. Under this assumption and for small values of λ, the robust optimization problem (7) – (8) can be thus viewed as a *regulatization of the two-stage* SLP (27). Quite similar results hold true also for the M-V problem (10).

The above statements imply that differences between SLP and RO or M-V can come into force only for sufficiently large values of the parameter λ.

3.3 The case of large λ

For large values of λ, it is convenient to study instead of (25), (8) the following parametric convex quadratic program

$$(31) \qquad \min_{z \in Z} \{\alpha \mathbf{p}^{\mathsf{T}} \xi + 1/2 \xi^{\mathsf{T}} \mathbf{P} \xi\}$$

with a nonnegative parameter $\alpha = (2\lambda)^{-1}$ in the linear part of the objective function. Stability with respect to changes of parameter α follows from more general results by [1, Chapter 5.4 and 5.5] according to which the solubility set $\mathcal{A}$ of (31) can be decomposed into finitely many convex stability sets - open intervals and isolated points. Proposition 3.1 implies that the considered set of parameters $(0, \infty)$ belongs into the solubility set $\mathcal{A}$ of program (31); it means that there is $\alpha_1 > 0$ such that all parameter values $\alpha \in (0, \alpha_1)$ belong into the same stability set characterized by fixed indices of positive components of optimal solutions $\mathbf{x}(\alpha), \mathbf{y}_s(\alpha)$, $s = 1, \ldots, S$ and that these optimal soultions are linear functions of α on $(0, \alpha_1)$.

The interpretation of this result for the RO problem (7) – (8) explains the robustness property valid for large values λ:

Proposition 3. 6. There exists $\lambda_1 > 0$ such that the optimal first-stage solutions $\mathbf{x}(\lambda, \mathbf{p})$ and the optimal compensations $\mathbf{y}_s(\lambda, \mathbf{p})$ $\quad \forall s$ in the RO problem (7) – (8) keep fixed indices of zero components for all $\lambda > \lambda_1$.

This was the case of $y_1 = y_2 = 0$ for all large values of λ in the RO problem in Section 2. Notice that except for evaluation of λ_1, the result holds true for arbitrary values of probabilities $p_s, s = 1, \ldots, S$.

4 Resistance with respect to Selected Scenarios and their Probabilities

Up to now we have studied properties of the robust optimization problems in dependence on the parameter λ, assuming that the scenarios $\omega_s, s = 1, \ldots, S$ and their probabilities p_s have been fixed in advance. Now we shall turn our attention to the question of resistance of the results with respect to the mentioned input. This problem was opened in [5]. We shall continue in developing further the approach of [5] which is based on the contamination technique [4]. We refer to [6] for a similar treatment of the tracking model (Example 5) and for numerical experience related to an expected utility model (Example 2).

We shall substitute $\mathbf{c}^{\mathsf{T}} \mathbf{x} + \mathbf{q}_s^{\mathsf{T}} \mathbf{y}_s$ for ξ_s into the objective function (7) and we shall apply the contamination technique to the problem

minimize

$$(32) \qquad \sum_{s=1}^{S} p_s \left[\mathbf{c}^{\mathsf{T}} \mathbf{x} + \mathbf{q}_s^{\mathsf{T}} \mathbf{y}_s \right] + \lambda \sum_{s=1}^{S} p_s \left[\mathbf{q}_s^{\mathsf{T}} \mathbf{y}_s - \sum_{j=1}^{S} p_j \mathbf{q}_j^{\mathsf{T}} \mathbf{y}_j \right]^2$$

subject to

(33)
$$\mathbf{A}\mathbf{x} = \mathbf{b}$$
$$\mathbf{T}_s\mathbf{x}+\mathbf{W}_s\mathbf{y}_s = \mathbf{h}_s$$

$$\mathbf{x} \geq 0, \mathbf{y}_s \geq 0, s = 1,\dots,S$$

with fixed scenarios $\omega_s = [\mathbf{T}_s, \mathbf{W}_s, \mathbf{h}_s, \mathbf{q}_s], s = 1,\dots,S]$ and with a fixed parameter value λ. We denote by $\tilde{\mathcal{Z}}$ the corresponding canonical projection of $\mathcal{Z}$ – the fixed set of feasible solutions $\tilde{z} := [\mathbf{x}, \mathbf{y}_s, \forall s]$ described by (33), we assume that the set $\tilde{\mathcal{Z}}^*(\lambda, \mathbf{p})$ of optimal solutions of (32), (33) is nonempty and bounded and we denote by $\varphi(\lambda, \mathbf{p})$ the optimal value of (32), (33). The objective function $f(\tilde{z}; \lambda, \mathbf{p})$ defined by (32) is *concave* in probabilities $\mathbf{p}$ and convex with respect to all considered variables $\tilde{z} = [\mathbf{x}, \mathbf{y}_s, \forall s]$. It can be again written in the form (25) with $\xi_s = \mathbf{c}^\top\mathbf{x} + \mathbf{q}_s^\top\mathbf{y}_s$ or rewritten as

(34)
$$f(\tilde{z}; \lambda, \mathbf{p}) = \mathbf{c}^\top\mathbf{x} + \sum_s p_s\mathbf{q}_s^\top\mathbf{y}_s + \lambda\sum_s p_s[\mathbf{q}_s^\top\mathbf{y}_s - \sum_j p_j\mathbf{q}_j^\top\mathbf{y}_j]^2$$
$$= \mathbf{c}^\top\mathbf{x} + E_p\mathbf{q}^\top\mathbf{y} + \lambda\mathrm{var}_p\mathbf{q}^\top\mathbf{y}$$

Concerning the structure of the set of feasible solutions $\tilde{\mathcal{Z}}$ of (33), we shall assume that the sets of feasible compensations

(35)
$$\mathcal{Y}_s(\mathbf{x}) = \{\mathbf{y}_s | \mathbf{W}_s\mathbf{y}_s = \mathbf{h}_s - \mathbf{T}_s\mathbf{x}, \mathbf{y}_s \geq 0\}$$

are nonempty for all s and for any $\mathbf{x} \in \mathcal{X}$,

(36)
$$\mathcal{X} = \{\mathbf{x} | \mathbf{A}\mathbf{x} = \mathbf{b}, \mathbf{x} \geq 0\}$$

i.e., assumption of the relatively complete recourse.

For a fixed set of scenarios $\omega_s, s = 1,\dots,S$, the contamination by another distribution carried by these scenarios means simply a change of the original probabilities $p_s, s = 1,\dots,S$ to $p_s(t) = (1 - t)p_s + t\pi_s, s = 1,\dots,S$ with $0 \leq t \leq 1; \pi_s \geq 0, \sum\pi_s = 1$ denote probabilities of the given scenarios under the alternative distribution.

For *fixed* probabilities $\mathbf{p}, \boldsymbol{\pi}$, the objective function $f(\tilde{z}; \lambda, (1-t)\mathbf{p}+\boldsymbol{\pi})$ that corresponds to contamination of $\mathbf{p}$ by $\boldsymbol{\pi}$ results from (34) by replacing p_s by $(1-t)p_s + t\pi_s$. It depends on a scalar parameter $t \in [0, 1]$ and we shall denote it briefly $f_\lambda(\tilde{z}, t)$. It is evidently a concave quadratic function of t. We shall assume that the optimal value $\varphi(\lambda, 1)$ is finite, i.e., that the problem has an optimal solution when probabilities $p_s, s = 1,\dots,S$ are replaced in (34) by the alternative probabilities $\pi_s, s = 1,\dots,S$.

At any feasible point $\tilde{z} \in \tilde{\mathcal{Z}}$, the derivative

(37) $$\frac{d}{dt}f_\lambda(\tilde{z}, t) = f_\lambda(\tilde{z}, 1) - f_\lambda(\tilde{z}, 0) + \lambda(1 - 2t)\left(\sum_s(\pi_s - p_s)\mathbf{q}_s^\top\mathbf{y}_s\right)^2$$

and the derivative of the optimal value function $\varphi(\lambda, (1-t)\mathbf{p} + t\boldsymbol{\pi})$ of the contaminated problem with respect to t at the point $t = 0^+$ follows from a standard result of parametric programming; see e.g. [8, Theor. 17] or discussions in [3]–[5]:

$$(38) \quad \frac{\mathrm{d}}{\mathrm{d}t}\varphi(\lambda, 0^+) = \min_{\tilde{\mathbf{z}} \in \tilde{Z}^*(\lambda, \mathbf{p})} \left[f_\lambda(\tilde{\mathbf{z}}, 1) + \lambda \left(\sum_s (\pi_s - p_s)\mathbf{q}_s^\mathsf{T}\mathbf{y}_s \right)^2 \right] - \varphi(\lambda, \mathbf{p})$$

This derivative provides an information about the response of the optimal value function on small changes in probabilities and it can be used in construction of *global bounds* on the optimal value $\varphi(\lambda, (1-t)\mathbf{p} + t\boldsymbol{\pi})$ of the contaminated problem for all $t \in [0, 1]$:

$$(39) \quad (1-t)\varphi(\lambda, \mathbf{p}) + t\varphi(\lambda, \boldsymbol{\pi}) \leq \varphi(\lambda, (1-t)\mathbf{p} + t\boldsymbol{\pi}) \leq \varphi(\lambda, \mathbf{p}) + t\frac{\mathrm{d}}{\mathrm{d}t}\varphi(\lambda, 0^+)$$

whose analysis implies

Proposition 4.1. Let $\lambda \geq 0$ be fixed, the set $\tilde{Z}^*(\lambda, \mathbf{p})$ of optimal solutions of (32) – (33) be nonempty and bounded and the set $\tilde{Z}^*(\lambda, \boldsymbol{\pi}) \neq \emptyset$. Then for an arbitrary $t \in [0, 1]$

$$|\varphi(\lambda, (1-t)\mathbf{p} + t\boldsymbol{\pi}) - \varphi(\lambda, \mathbf{p})| \leq t \max \left\{ |\varphi(\lambda, \boldsymbol{\pi}) - \varphi(\lambda, \mathbf{p})|, |\frac{\mathrm{d}}{\mathrm{d}t}\varphi(\lambda, 0^+)| \right\}$$

Moreover,

$$\varphi(\lambda, \boldsymbol{\pi}) \geq \varphi(\lambda, \mathbf{p}) \Longrightarrow \varphi(\lambda, (1-t)\mathbf{p} + t\boldsymbol{\pi}) \geq \varphi(\lambda, \mathbf{p}) \quad \forall t$$

$$\frac{\mathrm{d}}{\mathrm{d}t}\varphi(\lambda, 0^+) \leq 0 \Longrightarrow \varphi(\lambda, (1-t)\mathbf{p} + t\boldsymbol{\pi}) \leq \varphi(\lambda, \mathbf{p}) \quad \forall t$$

If the set of optimal solutions of the initial problem is a singleton, $\tilde{Z}^*(0, \mathbf{p}) = \{\tilde{\mathbf{z}}^*(0, \mathbf{p})\}$, evaluation of the derivative (38) does not require any minimization – an essential simplification that was exploited in [5]. However, some simplification can be obtained under less stringent assumptions:

Proposition 4.2. Let the assumptions of Proposition 4.1 hold true and assume in addition that all probabilities p_s in the original problem (32) – (33) are positive and that for all optimal solutions of (32) – (33), the $\mathbf{x}$-part is fixed, denoted by $\mathbf{x}^*(\mathbf{p})$. Then the derivative

$$(40) \quad \frac{\mathrm{d}}{\mathrm{d}t}\varphi(\lambda, 0^+) = \mathbf{c}^\mathsf{T}\mathbf{x}^*(\mathbf{p}) - \varphi(\lambda, \mathbf{p}) + \sum_s \pi_s \mathbf{q}_s^\mathsf{T}\mathbf{y}_s^*(\mathbf{x}^*(\mathbf{p}))$$

$$+ \lambda \sum_s \pi_s [\mathbf{q}_s^\mathsf{T}\mathbf{y}_s^*(\mathbf{x}^*(\mathbf{p})) - \sum_j \pi_j \mathbf{q}_j^\mathsf{T}\mathbf{y}_j^*(\mathbf{x}^*(\mathbf{p}))]^2$$

$$+ \lambda \left(\sum_s (\pi_s - p_s)\mathbf{q}_s^\mathsf{T}\mathbf{y}_s^*(\mathbf{x}^*(\mathbf{p})) \right)^2$$

where $\mathbf{y}_s^*(\mathbf{x}^*(\mathbf{p}))$ is an arbitrary optimal compensation of $\mathbf{x}^*(\mathbf{p})$ under scenario s.

Proof follows from Proposition 3.4: for a unique $\mathbf{x}$-part of the optimal solution of (32) – (33), both the mean value of costs and the variance of these costs are fixed for all optimal compensations $\mathbf{y}_s^*(\mathbf{x}^*(\mathbf{p}))$ of $\mathbf{x}^*(\mathbf{p})$). This implies that for all scenarios that enter with a positive probability p_s the costs $\mathbf{q}_s^\top \mathbf{y}_s^*(\mathbf{x}^*(\mathbf{p}))$ must be fixed for all optimal compensations. Hence no minimization in (38) is needed and any of optimal compensations can be used to evaluate the derivative.□

Proposition 4.3. Under assumptions of Proposition 4.2 (or in the case when $\tilde{Z}^*(\lambda, \mathbf{p})$ is a singleton), the derivative $\frac{d}{dt}\varphi(\lambda, 0^+)$ is *linear* in π.
Proof follows by rearranging (40):

$$\frac{d}{dt}\varphi(\lambda, 0^+) = \mathbf{c}^\top \mathbf{x}^*(\mathbf{p}) - \varphi(\lambda, \mathbf{p}) + \sum_s \pi_s \mathbf{q}_s^\top \mathbf{y}_s^*(\mathbf{x}^*(\mathbf{p}))$$
$$+ \lambda \sum_s \pi_s (\mathbf{q}_s^\top \mathbf{y}_s^*(\mathbf{x}^*(\mathbf{p})))^2 + \lambda \left(\sum_s p_s \mathbf{q}_s^\top \mathbf{y}_s^*(\mathbf{x}^*(\mathbf{p})) \right)^2$$
$$- 2\lambda \sum_s p_s \mathbf{q}_s^\top \mathbf{y}_s^*(\mathbf{x}^*(\mathbf{p})) \sum_s \pi_s \mathbf{q}_s^\top \mathbf{y}_s^*(\mathbf{x}^*(\mathbf{p})) \quad \square$$

Also *sensitivity results with respect to additionally included scenarios* can be treated as sensitivity with respect to probabilities: We assume again that the problem (32) – (33) has been solved for scenarios $\omega_s, s = 1, \ldots, S$ and probabilities $p_s > 0 \forall s$. Inclusion of other scenarios, say, $\omega_s, s = S + 1, \ldots, S + S'$ means an essential extension of the solved problem. To quantify the local response on these changes of the input data and to get global bounds on the optimal value of the extended problem we shall start with the problem based on the pooled sample $\omega_1, \ldots, \omega_S, \omega_{S+1}, \ldots, \omega_{S+S'}$. The initial problem is identified by probabilities $p_s > 0, s = 1, \ldots, S, p_s = 0, s = S + 1, \ldots, S + S', \sum p_s = 1$, the contaminating problem by probabilities $\pi_s = 0, s = 1, \ldots, S, \pi_s \geq 0, s = S + 1, \ldots, S + S', \sum \pi_s = 1$. The formally extended initial problem (32) – (33) is

minimize

$$(32) \qquad \sum_{s=1}^S p_s \left[\mathbf{c}^\top \mathbf{x} + \mathbf{q}_s^\top \mathbf{y}_s \right] + \lambda \sum_{s=1}^S p_s \left[\mathbf{q}_s^\top \mathbf{y}_s - \sum_{j=1}^S p_j \mathbf{q}_j^\top \mathbf{y}_j \right]^2$$

subject to

$$(41) \qquad \begin{aligned} \mathbf{Ax} &= \mathbf{b} \\ \mathbf{T}_s\mathbf{x} + \mathbf{W}_s\mathbf{y}_s &= \mathbf{h}_s \end{aligned}$$

$$\mathbf{x} \geq 0, \mathbf{y}_s \geq 0, s = 1, \ldots, S + S'$$

and the formula (38) applies under standard assumptions (the set of optimal solutions of (32) – (33) nonempty and bounded and the existence of an optimal solution for the problem based on the additional scenarios $\omega_{S+1}, \ldots, \omega_{S+S'}$ and their probabilities $\pi_{S+1}, \ldots, \pi_{S+S'}$). To simplify (38), assume again that the **x**-part of the optimal solution of (32) – (33), resp. (41) is unique and equals $\mathbf{x}^*(\mathbf{p})$. Similarly as in Proposition 4.2, the costs of optimal compensations are constant for all original scenarios $s = 1, \ldots, S$ and the minimization in (38) will be carried only over optimal compensations of $\mathbf{x}^*(\mathbf{p})$ for scenarios $\omega_{S+1}, \ldots, \omega_{S+S'}$ that enter (32), (41) with zero probabilities. The sets of these optimal compensations coincide with the sets of all possible compensations $\mathcal{Y}_s(\mathbf{x}^*(\mathbf{p})), s = S + 1, \ldots, S + S'$. The derivative

$$(42) \quad \frac{d}{dt}\varphi(\lambda, 0^+) = \mathbf{c}^\top \mathbf{x}^*(\mathbf{p}) - \varphi(\lambda, \mathbf{p}) + \lambda \left(\sum_{s=1}^{S} p_s \mathbf{q}_s^\top \mathbf{y}_s^*(\mathbf{x}^*(\mathbf{p})) \right)^2$$

$$+ \min_{\mathbf{y}_{S+s} \in \mathcal{Y}_{S+s}(\mathbf{x}^*(\mathbf{p})), s=1,\ldots,S'} \sum_{s=1}^{S'} \pi_{S+s} \left(\mathbf{q}_{S+s}^\top \mathbf{y}_{S+s} + \lambda (\mathbf{q}_{S+s}^\top \mathbf{y}_{S+s})^2 \right.$$

$$\left. -2\lambda \mathbf{q}_{S+s}^\top \mathbf{y}_{S+s} \sum_{s=1}^{S} p_s \mathbf{q}_s^\top \mathbf{y}_s^*(\mathbf{x}^*(\mathbf{p})) \right)$$

$$= \mathbf{c}^\top \mathbf{x}^*(\mathbf{p}) + \lambda \left(\sum_{s=1}^{S} p_s \mathbf{q}_s^\top \mathbf{y}_s^*(\mathbf{x}^*(\mathbf{p})) \right)^2 + \sum_{s=1}^{S'} \pi_{S+s} \min_{\mathbf{y}_{S+s} \in \mathcal{Y}_{S+s}(\mathbf{x}^*(\mathbf{p}))}$$

$$\left[\mathbf{q}_{S+s}^\top \mathbf{y}_{S+s} + \lambda (\mathbf{q}_{S+s}^\top \mathbf{y}_{S+s})^2 - 2\lambda \mathbf{q}_{S+s}^\top \mathbf{y}_{S+s} \sum_{s=1}^{S} p_s \mathbf{q}_s^\top \mathbf{y}_s^*(\mathbf{x}^*(\mathbf{p})) \right]$$

$$- \varphi(\lambda, \mathbf{p})$$

If the contaminating distribution is a degenerated one that assigns probability 1 to one new scenario ω_* for which the corresponding single scenario problem is solvable, (42) reduces to the optimal value of the simple quadratic program introduced in [5]:

$$(43) \quad \frac{d}{dt}\varphi(\lambda, 0^+) = \mathbf{c}^\top \mathbf{x}^*(\mathbf{p}) - \varphi(\lambda, \mathbf{p})$$

$$+ \min_{\mathbf{y}_* \in \mathcal{Y}_*(\mathbf{x}^*(\mathbf{p}))} \left\{ \mathbf{q}_*^\top \mathbf{y}_* + \lambda \left[\mathbf{q}_*^\top \mathbf{y}_* - \sum_{s=1}^{S} p_s \mathbf{q}_s^\top \mathbf{y}_s^*(\mathbf{x}^*(\mathbf{p})) \right]^2 \right\}$$

On the other hand, (42) can be written as an average (with weights $\pi_{S+s}, s = 1, \ldots, S'$) of expressions (43) computed for each of included out-of-sample scenarios $\omega_{S+s}, s = 1, \ldots, S'$ separately.

For separability of the derivative (42) with respect to additional scenarios, assumption of a unique first stage solution $\mathbf{x}^*(\mathbf{p})$ of the original problem is essential. For multiple first stage solutions, (42) computed at any of these solutions is an upper bound for the derivative $\frac{d}{dt}\varphi(\lambda, 0^+)$, so that (39) with (42) at the place of $\frac{d}{dt}\varphi(\lambda, 0^+)$ are still valid bounds but they are less tight.

5 Conclusions

Our analysis implies that a genuine difference between optimal solutions of the two-stage SLP and of the robust optimization problem appears only for λ sufficiently large, say, for $\lambda \geq \lambda_0$, $\lambda_0 > 0$. In the limit case of $\lambda \to \infty$, say, for $\lambda \geq \lambda_1$, the indices of zero components of the optimal first- and second-stage solutions of the considered RO problem do not change any more. The numerical values of λ_0, λ_1 depend on the input data.

Postoptimality and sensitivity results for robust optimization problems can be obtained by an extension of the contamination technique as suggested in [5]. The first numerical studies [3] indicate that the upper bound for the optimal value based on the pooled sample is quite precise; see Table. The results concern a modified metal melting problem from [7] with 64 equiprobable scenarios of the input composition (elements of the technology matrices T_s) and with one additional scenario. The value of t has been fixed to $1/65$.

The original results of [5] were based on the assumption of unique first- and *second-stage* optimal solutions of the initial RO problem. This paper demonstrates that, similarly as for the two-stage SLP, the assumption can be relaxed to *uniqueness of the first-stage solutions*. In this case, the desired additional information about resistance with respect to additional scenarios or with respect to changes of probabilities follows by evaluation (or estimation) of the optimal value based on the alternative/contaminating distribution and on solution of simple quadratic programs (43) for all new scenarios.

Acknowledgement. This paper has benefited from discussions with Stavros Zenios (University of Cyprus).

References

[1] B. Bank et al.: Non-Linear Parametric Optimization. Akademie-Verlag, Berlin 1982.

[2] R. S. Dembo: Scenario optimization. Annals of Oper. Res. 30 (1991) 63–80.

[3] P. Dobiáš: Robust optimization. Diploma thesis, Dept. of Statistics, Charles University, Prague, 1997.

[4] J. Dupačová: Stability and sensitivity analysis for stochastic programming. Annals of Oper. Res. 27 (1990) 115–142.

[5] J. Dupačová: Scenario based stochastic programs: Resistance with respect to sample. Annals of Oper. Res. 64 (1996) 21–38.

[6] J. Dupačová, M. Bertocchi and V. Moriggia: Postoptimality for scenario based financial planning models with an application to bond portfolio management. To appear in: World Wide Asset and Liability Modeling (W. T. Ziemba and J. Mulvey, eds.), Cambridge University Press 1997.

[7] W. H. Evers: A new model for stochastic linear programming. Management Science 13 (1967) 680–693.

[8] E. G. Gol'shtein: Theory of Convex Programming, Translations of Mathe-
    matical Monographs 36 (American Mathematical Society, Providence RI,
    1972).

[9] A. J. King: Asymmetric risk measures and tracking models for portfolio
    optimization under uncertainty, Annals of Oper. Res. 45 (1993) 165–178.

[10] O. L. Mangasarian and R. R. Meyer: Nonlinear perturbation of linear
    programs, SIAM J. on Control and Optimization 17 (1979) 745–752.

[11] J. M. Mulvey, R. J. Vanderbei and S. A. Zenios: Robust optimization of
    large scale systems, Oper. Res. 43 (1995) 264–281.

[12] H. Vladimirou and S. A. Zenios: Stochastic linear programs with re-
    stricted recourse. To appear in EJOR.

Table

λ	64 scenarios			bounds			65 scenarios		
	cost	variance	objective	lower b.	upper b.	objective	objective	cost	variance
0,000	66,2186	51,3593	66,2186	66,1225	66,1439	66,1439	66,1439	66,1439	50,9259
0,001	66,2186	51,3593	66,2699	66,1730	66,1948	66,1948	66,1948	66,1439	50,9259
0,010	66,2849	39,3461	66,6784	66,5752	66,5996	66,5995	66,5995	66,2088	39,0745
0,030	66,3192	37,2115	67,4356	67,3207	67,3528	67,3526	67,3526	66,2418	37,0265
0,050	66,3400	36,6822	68,1741	68,0479	68,0880	68,0877	68,0877	66,2620	36,5131
0,070	66,3511	36,4908	68,9055	68,7680	68,8161	68,8157	68,8157	66,2735	36,3172
0,080	66,3546	36,4441	69,2701	69,1271	69,1791	69,1786	69,1786	66,2770	36,2694
0,090	66,3573	36,4120	69,6344	69,4857	69,5417	69,5411	69,5411	66,2798	36,2366
0,100	66,4321	35,6401	69,9961	69,8419	69,9028	69,9019	69,9019	66,3229	35,7902
0,120	66,8805	31,5095	70,6616	70,4972	70,5714	70,5706	70,5706	66,7752	31,6282
0,140	67,3270	28,0844	71,2588	71,0851	71,1708	71,1699	71,1699	67,2066	28,3090
0,170	68,2136	22,3373	72,0109	71,8257	71,9299	71,9290	71,9290	68,0987	22,5312
0,200	68,9305	18,4542	72,6213	72,4267	72,5453	72,5444	72,5444	68,8149	18,6475
0,250	69,9096	14,0482	73,4217	73,2147	73,3522	73,3515	73,3515	69,8171	14,1375
0,300	70,8592	10,6180	74,0446	73,8281	73,9828	73,9818	73,9818	70,7336	10,8273
0,400	72,4519	5,9726	74,8410	74,6122	74,7946	74,7939	74,7939	72,3577	6,0903
0,500	73,4075	3,8225	75,3188	75,0826	75,2817	75,2811	75,2811	73,3322	3,8978
0,800	74,8410	1,4932	76,0355	75,7883	76,0123	76,0119	76,0119	74,7939	1,5226
1,000	75,3188	0,9556	76,2744	76,0236	76,2558	76,2555	76,2555	75,2811	0,9745
1,500	75,9558	0,4247	76,5929	76,3372	76,5806	76,5804	76,5804	75,9307	0,4331

On Constrained Discontinuous Optimization

Yuri Ermoliev[1] and Vladimir Norkin[2]

[1] International Institute for Applied System Analysis, A-2361 Laxenburg, Austria
[2] Glushkov Institute of Cybernetics, 252207 Kiev, Ukraine

Abstract. In this paper we extend the results of Ermoliev, Norkin and Wets [8] and Ermoliev and Norkin [7] to the case of constrained discontinuous optimization problems. General optimality conditions for problems with nonconvex feasible sets are obtained. Easily implementable random search technique is proposed.

Key words. Discontinuous systems, necessary optimality conditions, averaged functions, mollifier subgradients, stochastic optimization

1   Introduction

In this paper we elaborate further results of Ermoliev, Norkin and Wets [8] and Ermoliev and Norkin [7] to a general constrained discontinuous optimization problem:

$$\text{minimize } F(x) \tag{1}$$

$$\text{subject to } x \in K \subseteq R^n, \tag{2}$$

where $F(x)$ is a (strongly) lower semicontinuous function, K is a compact set.

As we showed in [7] the class of strongly lower semicontinuous functions is appropriate for modeling and optimization of abruptly changing systems without instantaneous jumps and returns. In particular, we analyzed risk control problems, optimization of stochastic networks and discrete event systems, screening irreversible changes and stochastic pollution control. Another important application may be stochastic jumping processes describing risk reserves of interdependent insurance and reinsurance companies. In a rather general form the risk reserves can be understood as "reservoirs", where risk premiums are continuously flowing in and random claims at random time moments abruptly draining them out. A sample path of such process is a strongly lower semicontinuous function with random jumps at claim occurrence times.

In a sense the main aim of this article is to provide proofs of optimality conditions for general discontinuous constrained optimization problems discussed in [7]. In Section 2 we analyze situations when the expectation function belongs to the class of strongly lower semicontinuous functions. General idea of discontinuous optimization is presented in Section 3. Optimality conditions for discontinuous functions and general constraints are analysed in Section 4. Section 5 outlines possible computational procedures.

2   Some classes of discontinuous functions

In nonsmooth analysis different classes of continuous functions are introduced and studied. The same is necessary for discontinuous functions. We basically restrict possible discontinuity to the case of strongly lower semicontinuous functions, which seem to be most important for applications.

Definition 2.1 *A function* $F : R^n \longrightarrow R^1$ *is called strongly lower semicontinuous at* x, *if it is lower semicontinuous at* x *and there exists a sequence* $x^k \longrightarrow x$ *with* F *continuous at* x^k *(for all* k*) such that* $F(x^k) \longrightarrow F(x)$. *The function* F *is called strongly lower semicontinuous (strongly lsc) on* $X \subseteq R^n$ *if this holds for all* $x \in X$.

Definition 2.2 *Lower semicontinuous function* $F : R^n \longrightarrow R^1$ *is called directionally continuous at* x *if there exists an open (direction) set* $D(x)$ *containing sequences* $x^k \in D(x)$, $x^k \longrightarrow x$ *such that* $F(x^k) \longrightarrow F(x)$. *Function* $F(x)$ *is called directionally continuous if this holds for any* $x \in R^n$.

Definition 2.3 *Function* $F(x)$ *is called piecewise continuous if for any open set* $A \subset R^n$ *there is another open set* $B \subset A$ *on which* $F(x)$ *is continuous.*

Proposition 2.1 *If function* $F(x)$ *is piecewise continuous and directionally continuous then it is strongly lower semicontinuous.*

Proof. By definition of piecewise continuity for any open vicinity $V(x)$ of x we can find an open set $B \subset D(x) \cap V(x)$ on which function F is continuous. Hence there exists sequence $x^k \in D(x)$, $x^k \longrightarrow x$ with F continuous at x^k. By definition of directional continuity $F(x^k) \longrightarrow F(x)$.□

Properties of directional continuity, peicewise continuity and strong lower semicontinuity can be easily verified for one dimensional functions. For instance, if one dimensional function $F(x)$, $x \in R$, is (i) lower semicontinuous, (ii) continuous almost everywhere in R and (iii) at each point of discontinuity $x \in R$ function $F(x)$ is continuous either from the left or from the right, then $F(x)$ is strongly lsc. Next proposition clarifies the structure of multidimensional discontinuous functions of interest.

Proposition 2.2 *If $F(x) = F_0(F_1(x_1), \ldots, F_m(x_m))$, where $x = (x_1, \ldots, x_m)$, $x_i \in R^{n_i}$, function $F_0(\cdot)$ is continuous and functions $F_i(x_i)$, $i = 1, \ldots, n$ are strongly lsc (directionally continuous), then the composite function $F(x)$ is also strongly lsc (directionally continuous). If $F(x) = F_0(F_1(x), \ldots, F_m(x))$, $x \in R^n$, where $F_0(\cdot)$ is continuous and $F_i(x)$, $i = 1, \ldots, m$, are piecewise continuous, then $F(x)$ is also piecewise continuous.*

In particular, strong lsc, directional continuity and piecewise continuity are preserved under continuous transformations.

Proof is evident.

The next proposition gives a sufficient condition for a mathematical expectation function $F(x) = \mathbf{E}f(x, \omega)$ to be strongly lower semicontinuous.

Proposition 2.3 *Assume function $f(\cdot, \omega)$ is locally bounded around x by an integrable (in ω) function, piecewise continuous around x and a.s. directionally continuous at x with direction set $D(x, \omega) = D(x)$ (independent of ω). Suppose ω takes only a finite or countable number of values. Then $F(x) = \mathbf{E}f(x, \omega)$ is strongly lsc at x.*

Proof. Lower semicontinuity of F follows from Fatu lemma. The convergence of $F(x^k)$ to $F(x)$ for $x^k \longrightarrow x$, $x^k \in D(x)$ follows from Lebesgue's dominant convergence theorem. Hence F is directionally continuous at x in $D(x)$. It remains to show that in any open set $A \subset R^n$ which is close to x there are points of continuity of F. For the case when

ω takes finite number of values $\omega_1, \ldots, \omega_m$ with probabilities $p_1, \ldots, p_m$ the function $F(\cdot) = \sum_{i=1}^{m} p_i f(\cdot, \omega_i)$ is clearly piece-wise continuous. For the case when ω takes a countable number of values there is a sequence of closed balls $B_i \subseteq B_{i-1} \subset A$ convergent to some point $y \in A$ with $f(\cdot, \omega_i)$ continuous on B_i. We shall show that $F(\cdot) = \sum_{i=1}^{\infty} p_i f(\cdot, \omega_i)$ is continuous at y. By assumption $|f(x, \omega_i)| \leq C_i$ for $x \in A$ and $\sum_{i=1}^{\infty} p_i C_i < +\infty$. Then

$$
\begin{aligned}
F(x) - F(y) &= \sum_{i=1}^{\infty} p_i (f(x, \omega_i) - f(y, \omega_i)) \\
&= \sum_{i=1}^{m} p_i (f(x, \omega_i) - f(y, \omega_i)) + \delta_m(x, y),
\end{aligned}
$$

$$
|\delta_m(x, y)| \leq \sum_{i=m+1}^{\infty} 2 p_i C_i \quad x, y \in A.
$$

Thus for any $x^k \longrightarrow y$

$$
\limsup_k F(x^k) \leq F(y) + \sum_{i=m+1}^{\infty} 2 p_i C_i;
$$

$$
\liminf_k F(x^k) \geq F(y) - \sum_{i=m+1}^{\infty} 2 p_i C_i.
$$

Since $\sum_{i=m+1}^{\infty} 2 p_i C_i \longrightarrow 0$ as $m \longrightarrow \infty$ then $\lim_k F(x^k) = F(y)$. $\square$

Let us remark that functions of the form $f(x, \omega) = f(x - \omega)$, $x, \omega \in R^n$, with $f(\cdot)$ piecewise and directionally continuous have $D(x)$ independent of ω.

Propositions 2.1-2.3 provide a certain calculous for strongly lsc functions.

3 Averaged functions and mollifier subgradients

In order to optimize discontinuous functions we approximate them by so-called averaged functions which are often considered in optimization theory (see Antonov and Katkovnik [1], Katkovnik and Kulchitsky [13], Archetti and Betrò [2], Warga [21], Katkovnik [12], Gupal [9], [10], Gupal and Norkin [11], Rubinstein [20], Batuhtin and Maiboroda [4], Mayne and Polak [15], Mikhalevich, Gupal and Norkin [16], Ermoliev and Gaivoronski [6], Kreimer and Rubinstein [14], Batuhtin [3], Ermoliev, Norkin and Wets [8]). The convolution of a discontinuous function with appropriate mollifier (probability density function) improves continuity and differentiability, but on the other hand increases computational complexity of resulting problems since it transfers a deterministic function $F(x)$ into an expectation function defined as multiple integral. Therefore, this operation is meaningful only in combination with appropriate stochastic optimization techniques. Our purpose is to introduce such technique and to develop a certain subdifferential calculous for discontinuous functions. Let us introduce necessary notions and facts which are generalized in the next section to the case of constrained problems.

Definition 3.1 *A set (family) of bounded integrable functions $\{\psi_\theta : R^n \longrightarrow R_+, \ \theta \in R_+\}$ satisfying conditions*

$$\int_{R^n} \psi_\theta(z)dz = 1, \quad supp\psi_\theta := \{z \in R^n \mid \psi_\theta(z) > 0\} \subseteq \rho_\theta \mathbf{B}$$

with a unit ball $\mathbf{B}$, $\rho_\theta \downarrow 0$ as $\theta \downarrow 0$, is called a family of mollifiers. Mollifiers $\{\psi_\theta\}$ are called smooth if functions $\psi_\theta(\cdot)$ are continuously differentiable.

Given a locally integrable (discontinuous) function $F : R^n \longrightarrow R^1$ and a family of mollifiers $\{\psi_\theta\}$ the associated family $\{F^\theta, \ \theta \in R_+\}$ of averaged functions is defined by

$$F^\theta(x) := \int_{R^n} F(x - z)\psi_\theta(z)dz = \int_{R^n} F(z)\psi_\theta(x - z)dz. \tag{3}$$

Mollifiers may also have unbounded support (see [8]).

Example 3.1 *Assume $F(x) = \mathbf{E}_\omega f(x,\omega)$. If $f(x,\omega)$ is such that $\mathbf{E}_\omega|f(x,\omega)|$ exists and grows in the infinity not faster than some polynom of x and random vector η has standard normal distribution, then for $\xi_\theta(x,\eta,\omega) = \frac{1}{\theta}[f(x + \theta\eta, \omega) - f(x, \omega)]\eta$ or $\xi_\theta(x,\eta,\omega) = \frac{1}{2\theta}[f(x + \theta\eta, \omega) - f(x - \theta\eta, \omega)]\eta$, $\theta > 0$, we have $\nabla F^\theta(x) = \mathbf{E}_{\eta\omega}\xi_\theta(x,\eta,\omega)$. The finite difference approximations $\xi_\theta(x,\eta,\omega)$ are unbiased estimates of $\nabla F^\theta(x)$. As in [7], we can call them stochastic mollifier gradient of $F(x)$.*

Definition 3.2 *(See, for example, Rockafellar and Wets [17]). A sequence of functions $\{F^k : R^n \longrightarrow \overline{R}\}$ epi-converges to $F : R^n \longrightarrow \overline{R}$ relative to $X \subseteq R^n$ if for any $x \in X$*

(i) $\liminf_{k\to\infty} F^k(x^k) \geq F(x)$ for all $x^k \longrightarrow x$, $x^k \in X$;

(ii) $\lim_{k\to\infty} F^k(x^k) = F(x)$ for some sequence $x^k \longrightarrow x$, $x^k \in X$.

The sequence $\{F^k\}$ epi-converges to F if this holds relative to $X = R^n$.

For example, if $g : R^n \times R^m \longrightarrow \overline{R}$ is (jointly) lsc at $(\overline{x}, \overline{y})$ and is continuous in y at $\overline{y}$, then for any sequence $y^k \longrightarrow \overline{y}$, the corresponding sequence of functions $F^k(\cdot) = g(\cdot, y^k)$ epi-converges to $F(\cdot) = g(\cdot, y)$.

The following important property of epi-convergent functions shows that constrained optimization of a discontinuous function $F(x)$ can be in principle carried out through optimization of approximating epi-convergent functions $F^k(x)$.

Theorem 3.1 *If sequence of functions $\{F^k : R^n \longrightarrow \overline{R}\}$ epi-converges to $F : R^n \longrightarrow \overline{R}$ then for any compact $K \subset R^n$*

$$\lim_{\epsilon \downarrow 0}(\liminf_k(\inf_{K_\epsilon} F^k)) = \lim_{\epsilon \downarrow 0}(\limsup_k(\inf_{K_\epsilon} F^k)) = \inf_K F, \qquad (4)$$

where $K_\epsilon = K + \epsilon \mathbf{B}$, $\mathbf{B} = \{x \in R^n \,|\, \|x\| \leq 1\}$. If $F^k(x_\epsilon^k) \leq \inf_{K_\epsilon} F^k + \delta_k$, $x_\epsilon^k \in K_\epsilon$, $\delta_k \downarrow 0$ as $k \longrightarrow \infty$, then

$$\limsup_{\epsilon \downarrow 0}(\limsup_k x_\epsilon^k) \subseteq argmin_K F, \qquad (5)$$

where $(\limsup_k x_\epsilon^k)$ denotes the set X_ϵ of cluster points of the sequence $\{x_\epsilon^k\}$ and $(\limsup_{\epsilon \downarrow 0} X_\epsilon)$ denotes the set of cluster points of the family $\{X_\epsilon, \epsilon \in R_+\}$ as $\epsilon \downarrow 0$.

Proof. Note that $(\inf_{K_\epsilon} F^k)$ monotonously increases (non decreases) as $\epsilon \downarrow 0$, hence the same holds for $\liminf_{k \to \infty} \inf_{K_\epsilon} F^k$ and $\limsup_{k \to \infty} \inf_{K_\epsilon} F^k$. Thus limits over $\epsilon \downarrow 0$ in (4) exist.

Let us take arbitrary sequence $\epsilon_m \downarrow 0$, indices k_m^s and points x_m^s such that under fixed m

$$\liminf_k(\inf_{K_{\epsilon_m}} F^k) = \lim_{s \to \infty}(\inf_{K_{\epsilon_m}} F^{k_m^s}) \lim_{s \to \infty} F^{k_m^s}(x_m^s).$$

Thus

$$\begin{aligned}
\lim_{\epsilon \downarrow 0}(\limsup_k(\inf_{K_\epsilon} F^k)) &\geq \lim_{\epsilon \downarrow 0}(\liminf_k(\inf_{K_\epsilon} F^k)) \\
&= \lim_{m \to \infty} \lim_{s \to \infty} F^{k_m^s}(x_m^s) \\
&= \lim_{m \to \infty} F^{k_m^{s_m}}(x_m^{s_m})
\end{aligned}$$

for some indices s_m. By property (i) of epi-convergence $\lim_{m \to \infty} F^{k_m^{s_m}}(x_m^{s_m}) \geq \inf_K F$. Hence

$$\lim_{\epsilon \downarrow 0}(\limsup_k(\inf_{K_\epsilon} F^k)) \geq \lim_{\epsilon \downarrow 0}(\liminf_k(\inf_{K_\epsilon} F^k)) \geq \inf_K F.$$

Let us proof the opposite inequality. Since F is lower semicontinuous, then $F(x) = \inf_K F$ for some $x \in K$. By condition (ii) of epi-convergence there exists sequence $x^k \longrightarrow x$ such that $F^k(x^k) \longrightarrow F(x)$. For k sufficiently large $x^k \in K_\epsilon$, hence $\inf_{K_\epsilon} F^k \leq F^k(x^k)$ and

$$\lim_{\epsilon \downarrow 0}(\liminf_k(\inf_{K_\epsilon} F^k)) \leq \lim_{\epsilon \downarrow 0}(\limsup_k(\inf_{K_\epsilon} F^k)) \leq F(x) = \inf_K F.$$

The proof of (4) is completed.

Now prove (5). Let $x^k_\epsilon \in K_\epsilon$ and $F^k(x^k_\epsilon) \leq \inf_{K_\epsilon} F^k + \delta_k$, $\delta_k \downarrow 0$. Denote $X_\epsilon = \limsup_k x^k_\epsilon \subseteq K_\epsilon$. Let $\epsilon_m \downarrow 0$, $x_{\epsilon m} \in X_{\epsilon m}$ and $x_{\epsilon m} \longrightarrow x \in K$ as $m \longrightarrow \infty$. By construction of X_ϵ for each fixed m there exist sequences $x^{k^s_m}_m \longrightarrow x_{\epsilon m}$ satisfying $F^{k^s_m}(x^{k^s_m}_m) \leq \inf_{K_{\epsilon m}} F^{k^s_m} + \delta_{k^s_m}$, $\delta_{k^s_m} \downarrow 0$ as $s \longrightarrow \infty$. By property (i)

$$F(x_{\epsilon m}) \leq \liminf_s F^{k^s_m}(x^{k^s_m}_m) \leq \liminf_s(\inf_{K_{\epsilon m}} F^{k^s_m}) \leq \limsup_k(\inf_{K_{\epsilon m}} F^k).$$

Due to lower semicontinuity of F and (4) we obtain

$$F(x) \leq \liminf_{m \longrightarrow \infty} F(x_{\epsilon m}) \leq \liminf_{\epsilon m \downarrow 0}(\limsup_k(\inf_{K_{\epsilon m}} F^k)) = \inf_K F,$$

hence $x \in \operatorname{argmin}_K F$, that proves (5).$\square$

Remark that in Theorem 3.1 we could relax constraint set K in different ways, for instance, if $K = \{x \in R^n \mid G(x) \leq 0\}$ with some lower semicontinuous function $G(x)$, then we could define $K_\epsilon = \{x \in R^n \mid G(x) \leq \epsilon\}$, $\epsilon \geq 0$.

Let us illustrate the result of Theorem 3.1 by the following example.

Example 3.2 *Consider a discontinuous optimization problem*

$$\min_{x \geq 0}\left[F(x) = \begin{cases} 0, & x \leq 0, \\ 1, & x > 0 \end{cases}\right] \tag{6}$$

Let $F^\theta(x)$ be a family of averaged functions for F associated with a family of mollifiers $\psi_\theta(y) = \psi(y/\theta)$, $\theta > 0$, where function $\psi(\cdot)$ is symmetric with respect to point $y = 0$. Obviously, functions $F^\theta(x)$ epi-convege to F and $\min_{x \geq 0} F^\theta(x) = F^\theta(0) = 1/2$. If we don't relax constraint set $\{x \mid x \geq 0\}$ then optimization of approximate functions $F^\theta(x)$ over set $\{x \mid x \geq 0\}$ leads to a wrong result

$$\lim_{\theta \to 0} \min_{x \geq 0} F^\theta(x) = \frac{1}{2}.$$

The relaxation according to Theorem 3.1 leads to the true optimal value of the problem:

$$\lim_{\theta \to 0} \min_{x \geq -\epsilon} F^{\theta}(x) = 0$$

and thus

$$\lim_{\epsilon \to 0} \left(\lim_{\theta \to 0} \min_{x \geq -\epsilon} F^{\theta}(x) \right) = 0 = \min_{x \geq 0} F(x).$$

The following statement jointly with Theorem 3.1 shows that the averaged functions can be used for optimization of discontinuous functions.

Theorem 3.2 *(Ermoliev et al. [8]). For any strongly lower semicontinuous, locally integrable function* $F : R^n \longrightarrow R$, *any associated sequence of averaged functions* $\{F^{\theta_k}, \theta_k \downarrow 0\}$ *epi-converges to* F.

Jointly with Propositions 2.1, 2.3 Theorem 3.2 gives sufficient conditions for average functions to epi-converge to original discontinuous expectation function.

A subdifferential calculous for nonsmooth and discontinuous functions can be developed on the basis of their mollifier approximations.

Definition 3.3 *Let function* $F : R^n \longrightarrow R$ *be locally integrable and* $\{F^k := F^{\theta_k}\}$ *be a sequence of averaged functions generated from* F *by means of the sequence of mollifiers* $\{\psi^k := \psi_{\theta_k} : R^n \longrightarrow R\}$ *where* $\theta_k \downarrow 0$ *as* $k \longrightarrow \infty$. *Assume that the mollifiers are such that the averaged functions* F^k *are smooth (of class* C^1). *The set of* ψ-*mollifier subgradients (subdifferential) of* F *at* x *is by definition*

$$\partial_{\psi} F(x) := \lim_{k} \sup \{\nabla F^k(x^k) | x^k \longrightarrow x\},$$

i.e. $\partial_{\psi} F(x)$ *consists of the cluster points of all possible sequences* $\{\nabla F^k(x^k)\}$ *such that* $x^k \longrightarrow x$.

For example, for function (6) $\partial_{\psi} F(0) = \{g \in R^1 | g \geq 0\}$.

The subdifferential $\partial_{\psi} F(x)$ has the following properties (see Ermoliev, Norkin and Wets [8]):

$\partial_{\psi} F(x) = \partial F(x)$ for a convex functions $F(x)$;

convex hull of $\partial_{\psi} F(x)$ coincides with Clarke subdifferential $\partial F(x)$ for a locally Lipschitzian function $F(x)$;

$\partial_{\psi} F(x)$ coincides with Warga subdifferential [21] $\partial_W F(x)$ for a continuous function $F(x)$.

Theorem 3.3 *(Ermoliev et al. [8]). Suppose that $F : R^n \longrightarrow R$ is strongly lower semicontinuous and locally integrable. Then for any sequence $\{\psi_{\theta_k}\}$ of smooth mollifiers, we have $0 \in \partial_\psi F(x)$ whenever x is a local minimizer of F.*

4 Optimality conditions

Theorem 3.3 can be used for constrained optimization problems if exact penalties are applicable. Unfortunately, this operation can practically remove some important minimums of the original problem. Consider problem (6). Here function $F(x)$ is strongly lsc and point $x = 0$ is a reasonable solution of the problem. We could replace this problem, for example, by the following one:

$$\min_{x \geq 0} \left[\overline{F}(x) = \left\{ \begin{array}{ll} F(x), & x \geq 0, \\ -x + 1, & x < 0. \end{array} \right. \right].$$

The penalty function $\overline{F}(x)$ has single discontinuity point $x = 0$, where $\overline{F}$ achieves its global minimum $\overline{F}(0) = 0$. Thus penalty functions may lead to isolated minimums, which are difficult to discover.

Besides, we also encounter the following difficulties. Consider

$$\min\{\sqrt[3]{x} \mid x \geq 0\}. \tag{7}$$

In any reasonable definition of gradients the gradient of the function $\sqrt[3]{x}$ at point $x = 0$ equals to $+\infty$. Hence to formulate necessary optimality conditions for such problems and possibly involving discontinuities we need a special notion which incorporates infinite quantities. An appropriate notion is a cosmic vector space $\overline{R^n}$ introduced by Rockafellar and Wets [18]. Denote $R_+ = \{x \in R \mid x \geq 0\}$ and $\overline{R_+} = R_+ \cup \{+\infty\}$.

Definition 4.1 *Define a (cosmic) space $\overline{R^n}$ as a set of pairs $\overline{x} = (x, a)$, where $x \in R^n$, $\|x\| = 1$ and $a \in \overline{R_+}$. All pairs of the form $(x, 0)$ are considered identical and are denoted as $\overline{0}$.*

A topology in the space $\overline{R^n}$ is defined by means of cosmically convergent sequences.

Definition 4.2 *Sequence $(x_k, a_k) \in \overline{R^n}$ is called (cosmically) convergent to an element $(x, a) \in \overline{R^n}$ (denoted c-$\lim_{k \to \infty}(x_k, a_k)$) if either*

$\lim_k a_k = a = 0$ *or there exist both limits* $\lim_k x_k \in R^n$, $\lim_k a_k \in \overline{R^n}$ *and* $x = \lim_k x_k$, $a = \lim_k a_k \neq 0$, *i.e.*

$$c\text{-}lim_k(x_k, a_k) = \begin{cases} (\lim_k x_k, \lim_k a_k) & if \quad (\lim_k a_k) < +\infty, \\ (\lim_k x_k, +\infty) & if \quad a_k \longrightarrow +\infty, \\ (\lim_k x_k, +\infty) & if \quad a_k = +\infty. \end{cases}$$

Denote

$$c\text{-}\mathrm{Limsup}_k(x_k, a_k) = \{(x,a) \in \overline{R^n}| \exists\{k_m\} : (x,a) = c\text{-}\lim_{k\to\infty}(x_{k_m}, a_{k_m})\}.$$

For closed set $K \subset R^n$ denote a tangent cone

$$T_K(x) = \limsup_{\tau} \frac{K - x}{\tau},$$

to the set K at point x, normal cones

$$\hat{N}_K(x) = \{v \in R^n| < v, \omega > \le 0 \text{ for all } \omega \in T_K(x)\},$$

$$N_K(x) = \limsup_{\overline{x} \to x} \hat{N}_K \overline{x},$$

and extended normal cone

$$\overline{N}_K(x) = \{(y, b) \in \overline{R^n}| y \in N_K(x), \|y\| = 1, b \in \overline{R_+}\}.$$

For what follows we need the following closeness property of normal cone mapping $(x, \epsilon) \longrightarrow N_{K_\epsilon}(x)$.

Lemma 4.1 *Let* $K_\epsilon = K + \epsilon \times B$, $B = \{x \in R^n| \|x\| \le 1\}$. *Then for any sequences* $x \longrightarrow \overline{x} \in K$ *and* $\epsilon \longrightarrow 0$,

$$\limsup_{x \to \overline{x}, \epsilon \to 0} N_{K_\epsilon}(x) \subseteq N_K(\overline{x}).$$

Proof. For $x \in R^n$ define $y(x) \in K$ such that

$$\|y(x) - x\| = \inf_{y' \in K} \|x - y\|.$$

Let us show that $T_K(y(x)) \subseteq T_{K_\epsilon}(x)$. Let $w \in T_K(y(x))$, i.e.

$$w = \lim_{\nu \to \infty} \frac{y^\nu - y(x)}{\tau_\nu}, \quad \text{where } y^\nu \in K, \ y^\nu \longrightarrow y(x), \ \tau_\nu \longrightarrow 0.$$

Denote $x^\nu = y^\nu + (x - y(x)) \in K_\epsilon$. Then by definition

$$w = \lim_{\nu \to \infty} \frac{x^\nu - x}{\tau_\nu} \in T_{K_\epsilon}(x)$$

and thus $T_K(y(x)) \subseteq T_{K_\epsilon}(x)$. This inclusion implies $\hat{N}_{K_\epsilon}(x) \subseteq \hat{N}_K(y(x))$ and $N_{K_\epsilon}(x) \subseteq N_K(y(x))$. Hence

$$\limsup_{x \to \bar{x}, \epsilon \to 0} N_{K_\epsilon}(x) \subseteq \limsup_{x \to \bar{x}} N_K(y(x)) \subseteq N_K(\bar{x}). \square$$

Corollary 4.1 *For extended normal cones we have the same closeness property,*

$$\limsup_{x \to \bar{x}, \epsilon \to 0} \overline{N}_{K_\epsilon}(x) \subseteq \overline{N}_K(\bar{x}).$$

Remark. We could use another sort of relaxation for set K. Suppose K is convex and is given by an inequality constraint:

$$K = \{x \in R^n \mid G(x) \leq 0\}$$

with some convex function $G(x)$. Consider a relaxed set

$$K_\epsilon = \{x \in R^n \mid G(x) \leq \epsilon\}.$$

Normal cones to K_ϵ and $K = K_0$ are formed by subdifferentials $\partial G(x)$, $x \in K_\epsilon$, of function G,

$$N_{K_\epsilon}(x) = \begin{cases} \{\lambda \partial G(x) \mid \lambda \geq 0\} & \text{if } G(x) = \epsilon, \\ 0, & \text{if } G(x) < \epsilon, \end{cases} \quad \epsilon \geq 0.$$

Now closeness property of mapping $(x, \epsilon) \longrightarrow N_{K_\epsilon}$, stated in Lemma 4.1 follows from closeness of subdifferential mapping $x \longrightarrow \partial G(x)$.

Definition 4.3 *Let function $F : R^n \longrightarrow R$ be locally integrable and $\{F^k := F^{\theta_k}\}$ be a sequence of averaged functions generated from F by convolution with mollifiers $\{\psi^k := \psi_{\theta_k} : R^n \longrightarrow R\}$ where $\theta_k \downarrow 0$ as $k \longrightarrow \infty$. Assume that the mollifiers are such that the averaged functions F^k are smooth (of class C^1). The set of the extended ψ-mollifier subgradients of F at x is by definition*

$$\overline{\partial}_\psi F(x) := c\text{-}Limsup_k \left\{ \left(\frac{\nabla F^k(x^k)}{\|\nabla F^k(x^k)\|}, \|\nabla F^k(x^k)\| \right) \mid x^k \longrightarrow x \right\},$$

where expression $\frac{\nabla F^k(x^k)}{\|\nabla F^k(x^k)\|}$ is replaced by any unit vector if $\nabla F^k(x^k) = 0$, i.e. $\overline{\partial}_\psi F(x)$ consists of the cluster points (in cosmic space $\overline{R^n}$) of all possible sequences $\{(\frac{\nabla F^k(x^k)}{\|\nabla F^k(x^k)\|}, \|\nabla F^k(x^k)\|)\}$ such that $x^k \longrightarrow x$. The full (extended) Ψ-mollifier subgradient set is $\partial_\Psi F(x) := \cup_\psi \partial_\psi F(x)$ where ψ ranges over all possible sequences of mollifiers that generate smooth averaged functions.

The extended mollifier subdifferential $\overline{\partial}_\psi F(x)$ is always a non-empty closed set in $\overline{R^n}$.

Now we can formulate optimality conditions (8) for constrained discontinuous optimization problem: $\min\{F(x)\mid x \in K\}$, where $F(x)$ may have the form of the expectation. Theorems 4.1, 4.2 and Corollaries 4.2, 4.3 clarify the structure of the set of points satisfying optimality condition (8).

Theorem 4.1 Let K be a closed set in R^n. Assume that a locally integrable function F has a local minimum relative to K at some point $x \in K$ and there is a sequence $x^k \in K$, $x^k \longrightarrow x$ with F continuous at x^k and $F(x^k) \longrightarrow F(x)$. Then, for any sequence $\{\psi^k\}$ of smooth mollifiers, one has

$$- \overline{\partial}_\psi F(x) \cap \overline{N}_K(x) \neq \emptyset, \tag{8}$$

where $-\overline{\partial}_\psi F(x) = \{(-g, a) \in \overline{R^n}\mid (g, a) \in \overline{\partial}_\psi F(x)\}$.

Proof. Let x be a local minimizer of F on K. For a sufficiently small compact neighborhood V of x, define $\phi := F(z) + \|z - x\|^2$. The function ϕ achieves its global minimum on $(K \cap V)$ at x. Consider also the averaged functions

$$\phi^k(z) = \int_{R^n} \phi(y - z)\psi^k(y)dy = F^k(z) + \beta^k(x, z),$$

where

$$F^k(z) = \int_{R^n} F(y - z)\psi^k(y)dy, \quad \beta^k(x, z) = \int |y - z - x|^2\psi^k(y)dy.$$

In [8] it is shown that (i) functions ϕ^k are continuously differentiable, (ii) they epi-converge to ϕ relative to $K \cap V$ and (iii) their global minimums z^k on $K \cap V$ converge to x as $k \longrightarrow \infty$. For sufficiently large k the following necessary optimality condition is satisfied:

$$-\nabla F^k(z^k) = n(z^k) \in N_K(z^k), \quad z^k \in K.$$

If $\nabla F^{km}(z^{km}) = 0$ for some $\{z^{km} \longrightarrow x\}$ then also $\bar{0} \in \bar{\partial}_\psi F(x)$ and $\bar{0} \in \overline{N}_K(x)$. If $\nabla F^{km}(z^{km}) \longrightarrow g \neq 0$ for some $\{z^{km} \longrightarrow x\}$ then

$$-\frac{\nabla F^{km}(z^{km})}{\|\nabla F^{km}(z^{km})\|} \longrightarrow -\frac{g}{\|g\|} \in N_K(x),$$

and $(\frac{g}{\|g\|}, \|g\|) \in \bar{\partial}_\psi F(x)$, $(-\frac{g}{\|g\|}, \|g\|) \in \overline{N}_K(x)$. If $\limsup_k \|\nabla F^k(z^k)\| = +\infty$ then for some $\{z^{km} \longrightarrow x\}$

$$-\frac{\nabla F^{km}(z^{km})}{\|\nabla F^{km}(z^{km})\|} \longrightarrow -g \in N_K(x),$$

and $(g, +\infty) \in \bar{\partial}_\psi F(x)$, $(-g, +\infty) \in \overline{N}_K(x)$. $\square$

Corollary 4.2 *For a continuous function F condition (8) is necessary, i.e. (8) is satisfied for all local minimizers of F on K.*

Next proposition shows that optimality conditions are also satisfied for limits X' of some local minimizers x_ϵ of relaxed problems $\min\{F(x)| \ x \in K_\epsilon = K + \epsilon\mathbf{B}\}$. It follows that, although the global minimum $x = 0$ of problem (6) does not satisfy conditions of Theorem 4.1, it falls in the scope of the next statement and thus satisfy optimality condition (8).

Corollary 4.3 *Let x_ϵ be a local minimizer such that there exists sequence $x_\epsilon^k \longrightarrow x_\epsilon$, $x_\epsilon^k \in K_\epsilon$, with F continuous at x_ϵ^k and $F(x_{\epsilon^k}) \longrightarrow F(x_\epsilon)$ as $k \longrightarrow \infty$. Assume $x_{\epsilon m} \longrightarrow x$ for some $\epsilon_m \downarrow 0$ as $m \longrightarrow \infty$. Then (8) is satisfied at x.*

Proof follows from Theorem 4.1 and closeness of (extended) mollifier subdifferential mapping $x \longrightarrow \bar{\partial}_\psi F(x)$ and (extended) normal cone mapping $(x, \epsilon) \longrightarrow \overline{N}_{K_\epsilon}(x)$ (Corollary 4.1).

Theorem 4.2 *If F is strongly lsc and the constraint set K is compact then the set X^* of points, satisfying optimality condition (8), is nonempty and contains at least one global minimizer of F in K.*

Proof. Construct a sequence of differentiable averaged functions F^k epi-converging to F (what is possible by Theorem 3.2). Relax constraint set K, i.e. define $K_\epsilon = K + \epsilon \times B$, where $B = \{x| \ \|x\| \leq 1\}$, $\epsilon \geq 0$.

Find a global minimizer x_ϵ^k of F^k over K_ϵ. For x_ϵ^k we have necessary optimality condition (see Rockafellar and Wets [19]):

$$- \nabla F^k(x_\epsilon^k) \in N_{K_\epsilon}(x_\epsilon^k). \tag{9}$$

We can assume that $x_\epsilon^k \longrightarrow y_\epsilon \in K_\epsilon$. From here it follows

$$-\overline{\partial} F(y_\epsilon) \cap \overline{N}_{K_\epsilon}(y_\epsilon) \neq \emptyset.$$

Now let $y_\epsilon \longrightarrow y \in K$, $\epsilon \longrightarrow 0$. By Theorem 3.1 y is a global minimizer of F on K. Then by closeness of mappings $\overline{\partial} F(\cdot)$ and $\overline{N}_{K_\epsilon}(\cdot)$ we finally obtain

$$- \overline{\partial} F(y) \cap \overline{N}_K(y) \neq \emptyset, \tag{10}$$

i.e. $y \in X^*.\square$

The proof of Theorem 4.2 also clarifies the structure of the set X^*: (9) is satisfied for local minimums of F^k on K_ϵ, hence (10) is satisfied for their limit points.

Now let us come back to problem (7) and show how the developed theory resolves the exposed difficulties.

Example 4.1 *Consider again optimization problem (7). Then we have*

$$\overline{\partial}_\psi \sqrt[3]{x}|_{x=0} = (+1, +\infty), \quad \overline{N}_{x \geq 0}(0) = \cup_{a \in \overline{R_+}}(-1, a)$$

and thus

$$-\overline{\partial}_\psi \sqrt[3]{x}|_{x=0} \cap \overline{N}_{x \geq 0}(0) = (-1, +\infty) \neq \emptyset.$$

5 On numerical optimization procedures

Theorems 3.1,4.2 immediately give at least the following idea for the approximate solution of problem (1), (2). Let us fix a small smoothing parameter θ and a small constraint relaxation parameter ϵ, choose a mollifier $\psi_\theta(\cdot) = \psi(\cdot/\theta)$ and instead of original discontinuous optimization problem consider a relaxed smoothed optimization problem:

$$\min[F^\theta(x)| \ x \in K_\epsilon], \tag{11}$$

where $F^\theta(x)$ is defined by (3). Then stochastic gradient method to solve (11) has the form:

x^0 is an arbitrary starting point;

$$x^{k+1} = \Pi_{K_\epsilon}(x^k - \rho_k \xi_\theta(x^k)), \quad k = 0, 1, \ldots; \tag{12}$$

where $\mathbf{E}\{\xi_\theta(x^k)|x^k\} = \nabla F^\theta(x^k)$, Π_{K_ϵ} denotes the orthogonal projection operator on the set K_ϵ, positive step multipliers ρ_k satisfy conditions

$$\sum_{k=0}^{\infty} \rho_k = +\infty, \quad \sum_{k=0}^{\infty} \rho_k^2 < +\infty, \tag{13}$$

Vectors $\xi_\theta(x^k)$ can be called stochastic mollifiers gradients.

The convergence of such kind stochastic gradient method to a stationary set

$$X_\epsilon^\theta = \{x \in K_\epsilon | -\nabla F^\theta(x) \in N_{K_\epsilon}(x)\}$$

(containing local and global minimums of $F^\theta(x)$ on K_ϵ) follows from results of [5]. Now coming to the limit in $\theta \longrightarrow 0$ and then in $\epsilon \longrightarrow 0$ we see that limit points $[\limsup_\epsilon(\limsup_\theta X_\epsilon^\theta)]$ satisfy optimality condition (8).

6 Conclusions

In the paper we formulated optimality conditions (8) based on extended mollifier subdifferentials $\overline{\partial}_\psi F(x)$ for a constrained ($x \in K$) optimization problems with strongly lower semicontinuous objective function $F(x)$. In unconstrained cases these conditions are reduced to a familiar form $0 \in \partial_\psi F(x)$ (Theorem 3.3) and are necessary, i.e., they are satisfied for all local minimums of the problem. Optimality conditions (8) are necessary also in the case of continuous objective function $F(x)$. In discontinuous cases the situation is more complicated: we cannot guarantee that all local minimizers and even all global minimizers satisfy (8). The reason for this is that not all local and global minimizers of F on K are achievable through the minimization of averaged functions. Nevertheless the set X^* of points satisfying (8) is nonempty, contains at least one global minimizer and its structure is clarified by Theorems 4.1, 4.2 and Corollaries 4.2, 4.3. Optimality condition (8) is constructive: it leads to numerical procedure (12) to find elements of X^*. Limits (as $\theta \longrightarrow 0$ and $\epsilon \longrightarrow 0$) of local and global minimizers of problems (11) belong to X^*.

References

[1] Antonov G.E. and Katkovnik V.Ya. (1970), Filtration and smoothing in extremum search problems for multivariable functions, Avtomatika i vychislitelnaya tehnika, N.4, Riga. (In Russian).

[2] Archetti F. and Betrò B. (1975), Convex programming via stochastic regularization, *Quaderni del Dipartimento di Ricerca Operativa e Scienze Statistiche*, N 17, Università di Pisa.

[3] Batuhtin,B.D. (1994), On one approach to solving discontinuous extremal problems, Izvestia AN Rossii. Tehnicheskaia kibernetika (Communications of Russian Academy of Sciences. Technical Cybernetics), No. 3, pp.37-46. (In Russian).

[4] Batuhtin,B.D. and Maiboroda L.A. (1984), *Optimization of discontinuous functions*, Moscow, Nauka. (In Russian).

[5] Dorofeev P.A. (1986), A scheme of iterative minimization methods, U.S.S.R. Comput. Math. Math. Phys., Vol. 26, No. 2, pp.131-136. (In Russian).

[6] Ermoliev Yu. and Gaivoronski A. (1992), Stochastic programming techniques for optimization of discrete event systems, *Annals of Operations Research*, Vol. 39, pp.120-135.

[7] Ermoliev Yu.M. and Norkin V.I. (1995), On Nonsmooth Problems of Stochastic Systems Optimization, Working Paper WP-95-096, Int. Inst. for Appl. Syst. Anal., Laxenburg, Austria.

[8] Ermoliev Yu.M, Norkin V.I. and Wets R.J-B. (1995), The minimization of semi-continuous functions: Mollifier subgradients, SIAM J. Contr. and Opt., No.1, pp.149-167.

[9] Gupal A.M. (1977), On a method for the minimization of almost differentiable functions, Kibernetika, No. 1, pp.114-116. (In Russian, English translation in: Cybernetics, Vol. 13, N. 1, pp.115-117).

[10] Gupal A.M. (1979), *Stochastic methods for solving nonsmooth extremal problems*, Naukova dumka, Kiev. (In Russian).

[11] Gupal A.M. and Norkin V.I. (1977), An algorithm for minimization of discontinuous functions, Kibernetika, No. 2, 73-75. (In Russian, English translation in: Cybernetics, Vol. 13, N. 2, 220-222).

[12] Katkovnik V.Ya. (1976), *Linear Estimates and Stochastic Optimization Problems*, Nauka, Moscow. (In Russian).

[13] Katkovnik V.Ya. and Kulchitsky Yu. (1972), Convergence of a class of random search algorithms, *Automat. Remote Control*, No. 8, pp. 1321-1326. (In Russian).

[14] Kreimer J. and Rubinstein R.Y. (1992), Nondifferentiable optimization via smooth approximation: general analytical approach, Annals of Oper. Res., Vol. 39, pp.97-119.

[15] Mayne D.Q. and Polak E. (1984), Nondifferentiable optimization via adaptive smoothing, J. of Opt. Theory and Appl., Vol. 43, pp.601-613.

[16] Mikhalevich V.S., Gupal A.M. and Norkin V.I. (1987), *Methods of nonconvex optimization*, Nauka, Moscow. (In Russian).

[17] Rockafellar R.T. and Wets R.J-B. (1984), Variational systems, an introduction, in: *Multifunctions and Integrands*, G.Salinetti, ed., Lecture Notes in Mathematics 1091, Springer-Verlag, Berlin, pp.1-54.

[18] Rockafellar R.T. and Wets R.J-B. (1991), Cosmic convergence, in: *Optimization and Nonlinear Analysis*, eds. A.Ioffe, M.Marcus and S.Reich, Pitman Research Notes in Mathematics Series 244, Longman Scientific & Technical, Essex, U.K., pp. 249-272.

[19] Rockafellar R.T. and Wets R.J-B. (1995), *Variational Analysis*, a monograph to be published in Springer-Verlag.

[20] Rubinstein R.Y. (1983), Smoothed functionals in stochastic optimization, Math. Oper. Res, Vol. 8, pp.26-33.

[21] Warga J. (1975), Necessary conditions without differentiability assumptions in optimal control, J. Diff. Equations, Vol. 15, pp.41-61.

On the Equivalence in Stochastic Programming with Probability and Quantile Objectives

Yu.S.Kan and A.A.Mistryukov

Department of Probability Theory, Moscow Aviation Institute
4, Volokolamskoe Shosse, Moscow, 125871 RUSSIA

Abstract. The equivalence between the quantile functional minimization and the probability functional maximization under the assumption that the probability measure may depend on the optimized strategy is discussed. The equivalence ensures an opportunity to obtain a solution of each of these problems by solving another one. The weakened sufficient conditions for the equivalence are presented. These conditions are more general and can be verified easier than the known ones. They are applied to prove the optimality of a satellite orbital correction with respect to a quantile performance index.

1 Introduction

The subject of this article is an interconnection between solutions for two stochastic programming models. The first one contains an objective which is a probability functional. The second model contains the quantile functional. Such models have a lot of applications in Engineering and Economics. The probability functional can be used when we intend either to increase the reliability of a technical system under some technical constraints or to reduce the risk of obtaining the undesirable outcomes when making a financial decision. As a rule, the probability functional has to be maximized with respect to a strategy. The quantile objective is an inverse function to the probability functional. It appears when we should optimize another performance (e.g. the cost) of either a system or a decision under the given level of either the reliability or the risk. We set

$$\text{reliability} = 1 - \text{risk}.$$

In this relation the quantile objectives were first introduced in [1] where they were called confidence limits. We assume that the quantile objective has to be minimized. We see that these optimization problems are similar to one another, so an interconnection between their solutions should exist. This interconnection is called an equivalence. The equivalence allows us to get a solution for each problem by solving another one. Its rigorous definition will be presented in sec. 2.

The motivation of this research is caused by the following three reasons.

Firstly, the state of the art in this field is such that there are methods [2] to solve each optimization problem in question. These methods have different conditions of applicability. For this reason we can be able to solve one of the problems meanwhile another problem can be hard. In such a situation the above-mentioned interconnection can turn out to be very useful.

Secondly, there are sufficient conditions [2] for the equivalence between the problems in question. These conditions will be discussed in sec. 2, where we shall note that they can be very restrictive and hard to be verified.

Thirdly, the known results on the subject concern of problems where the probability measure does not depend on the optimized strategy. This condition is not typical for the infinite-dimensional models which arise in Multi-Stage Stochastic Programming and Stochastic Optimal Control.

Taking into account these reasons we formulate our goal as generalization of the known equivalence conditions in order to overcome the obstacles mentioned above. The basic results will be given in sec. 3 where we shall prove two theorems. We shall also obtain two-sided bounds for the optimal values of the objectives in question.

In sec. 4 we shall apply a presented theorem to verify the optimality of a satellite orbital correction. We shall deal there with a model containing a controlled probability measure.

2 Problem statement and some remarks

Let $(\Omega, \mathcal{B})$ be a measurable space, U be a set of strategies u, $\mathcal{P}_u$ be a family of probability measures defined on $\mathcal{B}$, $\Phi(u, \omega)$ and $Q(u, \omega)$ be $\mathcal{B}$-measurable functionals defined on $U \times \mathcal{B}$. The probability functional is defined as follows:

$$P_\varphi(u) \overset{def}{=} \mathcal{P}_u\{\omega : \Phi(u, \omega) \leq \varphi, \; Q(u, \omega) \leq 0\}, \tag{1}$$

where φ is a scalar parameter. If u is fixed then the probability functional as function of φ is a distribution function for the improper random variable

$$\xi_u(\omega) \overset{def}{=} \begin{cases} \Phi(u, \omega), & \text{if } Q(u, \omega) \leq 0 \\ +\infty, & \text{if } Q(u, \omega) > 0. \end{cases}$$

Therefore $P_\varphi(u)$ is a right-continuous non-decreasing function of φ.

Let us define the quantile functional in the following way

$$\Phi_\alpha(u) \overset{def}{=} \min\{\varphi : P_\varphi(u) \geq \alpha\}, \tag{2}$$

where $\alpha \in (0, 1)$ is a given probability. Since $\xi_u(\omega)$ is improper, we set by definition $\Phi_\alpha(u) = +\infty$ if there is no φ satisfying the probabilistic constraint $P_\varphi(u) > 0$. We surely deal with such a situation when $\alpha > P^* \overset{def}{=} \sup_{u \in U} \mathcal{P}_u\{\omega : Q(u, \omega) \leq 0\}$ and can meet it when $\alpha = P^*$.

We should note that if we fix u and investigate the behaviour of $P_\varphi(u)$ and $\Phi_\alpha(u)$ with respect to parameters φ and α there are no new mathematical problems caused by the controlled measure, i.e. by the fact that the probability measure depends on the strategy. The detailed survey can be found in [2], where the basic properties of the functions $\Phi_\alpha(u)$ and $P_\varphi(u)$ are also described for the constant measure $\mathcal{P}_u = \mathcal{P}$ and $U \subset R^n$. In particular, it is known that the quantile functional considered as function of α is left-continuous and non-decreasing.

The subject of this paper is the interconnection between the solutions for two optimization problems. The first problem is the probability functional maximization

$$\text{maximize } P_\varphi(u) \text{ subject to } u \in U. \tag{3}$$

The second one is the quantile functional minimization

$$\text{minimize } \Phi_\alpha(u) \text{ subject to } u \in U. \tag{4}$$

Most of the known methods [2] for solving these problems are hardly applicable here due to the controlled measure, since the basic properties of the functionals in question are not clear in this case. Moreover, the problem of establishing these properties seems to be hard, since we need a tool in order to describe the controlled measure $\mathcal{P}_u$.

There is an idea [3] for transforming problems (3) and (4) into one another. It is easy to see that the functionals (1) and (2) considered as functions of parameters φ and α are inverse to one another in the generalized sense. We emphasize that they are inverse if they are strictly increasing in their domains.

Let us denote by U_φ and U^α the sets of optimal strategies u_φ and u^α for problems (3) and (4), respectively, $M^* \stackrel{def}{=} (0, P^*)$, N^* be the range of $\Phi(u, x)$ for $Q(u, x) \leq 0$, $u \in U$ and $x \in \Omega$. The following two definitions and a theorem (Theorem 4.4) are taken from [2].

Definition 1. Let $\alpha \in M^*$, $u^\alpha \in U^\alpha \neq \emptyset$ and $\varphi \stackrel{def}{=} \Phi_\alpha(u^\alpha)$. If $U^\alpha = U_\varphi$ then problem (4) is equivalent to (3) with the parameter φ.

Definition 2. Let $\varphi \in N^*$, $u_\varphi \in U_\varphi \neq \emptyset$ and $\alpha \stackrel{def}{=} P_\varphi(u_\varphi)$. If $U^\alpha = U_\varphi$ then problem (3) is equivalent to (4) with the parameter α.

Theorem 1. *Let $\mathcal{P}_u$ do not depend on u and the following conditions hold:*

(i) *for every $\alpha \in A \subset M^*$ the set U^α is non-empty, and $\Phi_\alpha(u^\alpha) \in B \subset N^*$;*

(ii) *for every $\varphi \in B_1 \subset B \subset N^*$ the set U_φ is non-empty, and $P_\varphi(u_\varphi) \in A$;*

(iii) *$\text{int}(A) \neq \emptyset$ and $\text{int}(B_1) \neq \emptyset$;*

(iv) $P\{\omega : \Phi(u,\omega) = \varphi, \ Q(u,\omega) \le 0\} = 0$ *and*
$P\{\omega : |\Phi(u,\omega) - \varphi| < \varepsilon, \ Q(u,\omega) \le 0\} > 0$ *for all* $\varepsilon > 0$, $u \in U$ *and*
$\varphi \in B$.

Then

$$U^\alpha = \bigcup_{\varphi \in B_1} \{u_\varphi : P_\varphi(u_\varphi) = \alpha\} \qquad \text{for all } \alpha \in A$$

and

$$U_\varphi = \bigcup_{\alpha \in A} \{u^\alpha : \Phi_\alpha(u^\alpha) = \varphi\} \qquad \text{for all } \varphi \in B.$$

We see from this theorem that if its conditions hold we can obtain a solution of each problem in question by solving another one. For example, suppose we deal with problem (4) under the conditions of Theorem 1 and a solution u_φ for (3) can easily be obtained for every φ. Then solving the equation $P_\varphi(u_\varphi) = \alpha$ with respect to φ we can assert that the corresponding strategy u_φ is optimal for (4).

Note that the conditions of Theorem 1 are very restrictive, since they must hold, in particular, for every $u \in U$. This fact often leads to hard problems when the strategy u belongs to some functional space. Moreover, this theorem requires that the probability measure does not depend on u.

3 Equivalence conditions

We first obtain two-sided bounds on the optimal values of the objectives $P_\varphi(u)$ and $\Phi_\alpha(u)$. The bounds are stated on the basis of the monotonicity of the functions $P_\varphi(u)$, $\Phi_\alpha(u)$ and their optimal values

$$F(\varphi) \overset{def}{=} \sup_{u \in U} P_\varphi(u), \qquad G(\alpha) \overset{def}{=} \inf_{u \in U} \Phi_\alpha(u). \tag{5}$$

Lemma 1. *Let a and b be real numbers such that $b > a$ and $F(b) > F(a)$. Then $a \le G(\alpha) \le b$ for every $\alpha \in (F(a), F(b))$.*

Proof. Let $\alpha \in (F(a), F(b))$. Suppose that $G(\alpha) < a$. Then according to the definition of the infimum function there exists a strategy u, such that $h \overset{def}{=} \Phi_\alpha(u) < a$. From (2) it follows that $P_h(u) \ge \alpha$, hence $F(h) \ge \alpha > F(a)$. But this inequality is impossible, since the function $F(\cdot)$ is not decreasing. From the obtained contradiction we conclude that $G(\alpha) \ge a$. Further on, from $F(b) > \alpha$ it follows according to the definition of the supremum function that there exists a strategy u for which $P_b(u) \ge \alpha$. By (2) we obtain $\Phi_\alpha(u) \le b$. It follows that $G(\alpha) \le b$.

The lemma is proved.

Lemma 2. *Let γ and β be some probabilities such that $\gamma > \beta$ and $G(\gamma) > G(\beta)$. Then $\beta \le F(\varphi) \le \gamma$ for every $\varphi \in (G(\beta), G(\gamma))$.*

Proof. Let $\varphi > G(\beta)$. Then there exists u such that $\varphi > \Phi_\beta(u)$. Hence $P_\varphi(u) \geq \beta$. It follows that $F(\varphi) \geq \beta$.

Assume that $F(\varphi) > \gamma$ for some $\varphi \in (G(\beta), G(\gamma))$. Then there exists u such that $P_\varphi(u) > \gamma$. It follows that $\Phi_\gamma(u) \leq \varphi$, hence $G(\gamma) \leq \varphi$. The last inequality contradicts the above condition $\varphi \in (G(\beta), G(\gamma))$. Thus $F(\varphi) \leq \gamma$. The lemma is proved.

Following to the equivalence idea we need to deal with the equations $F(\varphi) = \alpha$ and $G(\alpha) = \varphi$. We know that the functions $F(\varphi)$ and $G(\alpha)$ are always non-decreasing meanwhile the problem of their continuity seems to be hard. We emphasize that the monotonicity and continuity of these functions are necessary to use the conventional notion of the equation root. To overcome the continuity problem we slightly modify this notion.

Definition 3. Let $f(x)$ be a non-decreasing function of a scalar argument. A point x_0 such that

$$f(x_0 - \varepsilon) \leq 0 \leq f(x_0 + \varepsilon)$$

for every $\varepsilon > 0$ is called *a root* of the equation $f(x) = 0$. This root is *single* if

$$f(x_0 - \varepsilon) < 0 < f(x_0 + \varepsilon).$$

Theorem 2. *Let φ_α be a single root of the equation $F(\varphi) = \alpha$. Then $G(\alpha) = \varphi_\alpha$. Moreover, if for $\varphi = \varphi_\alpha$ there exists a solution u_φ of problem (3) and the inequality $F(\varphi_\alpha) \geq \alpha$ holds then u_φ is also a solution of problem (4).*

Proof. The first assertion of the theorem is an easy consequence from Lemma 1. In fact, for every $\varepsilon > 0$ we have

$$F(\varphi_\alpha - \varepsilon) < \alpha < F(\varphi_\alpha + \varepsilon).$$

Hence by Lemma 1 we can write

$$\varphi_\alpha - \varepsilon \leq G(\alpha) \leq \varphi_\alpha + \varepsilon.$$

Since ε is an arbitrary positive number, we conclude that $G(\alpha) = \varphi_\alpha$.

Let us prove the second assertion. Let u_φ be a solution of (3) for $\varphi = \varphi_\alpha$. According to the definition of the infimum function we have

$$\varphi_\alpha \leq \Phi_\alpha(u_\varphi) \tag{6}$$

On the other hand, due to the theorem condition $F(\varphi_\alpha) \geq \alpha$ it follows that $P_{\varphi_\alpha}(u_\varphi) = F(\varphi_\alpha) \geq \alpha$. Hence from (2) we deduce

$$\varphi_\alpha \geq \Phi_\alpha(u_\varphi) \tag{7}$$

Inequalities (6) and (7) are compatible iff $\varphi_\alpha = \Phi_\alpha(u_\varphi)$. This means that u_φ is optimal for (4).

The theorem is proved.

Theorem 2 gives us a tool to get a solution of quantile minimization problem (4) by solving problem (3). We see that we need to verify that the root φ_α is single and $F(\varphi_\alpha) \geq \alpha$. The first condition can be verified, for example, by constructing the plot of $F(\varphi)$ in a neighbourhood of φ_α. The second one can be verified by the Monte Carlo simulation.

The next proposition allows us to carry out an inverse transformation, i.e. to reduce problem (3) to (4).

Theorem 3. *Let α_φ be a single root of the equation $G(\alpha) = \varphi$. Then $F(\varphi) = \alpha_\varphi$. Moreover, if for $\alpha = \alpha_\varphi$ there exists a solution u^α of problem (4) and $G(\alpha_\varphi) \leq \varphi$ then u^α is also a solution of (3).*

Proof. The first assertion follows directly from Lemma 2. In fact, since α_φ is a single root, for every $\varepsilon > 0$ we have

$$G(\alpha_\varphi - \varepsilon) < \varphi < G(\alpha_\varphi + \varepsilon).$$

Applying Lemma 2 we obtain

$$\alpha_\varphi - \varepsilon \leq F(\varphi) \leq \alpha_\varphi + \varepsilon.$$

Therefore $F(\varphi) = \alpha_\varphi$.

Let us prove the second assertion. Since $F(\varphi)$ is a supremum function, the following inequality

$$\alpha_\varphi \geq P_\varphi(u^\alpha) \tag{8}$$

holds. On the other hand, from the theorem condition $G(\alpha_\varphi) \leq \varphi$ it follows that $\Phi_{\alpha_\varphi}(u^\alpha) = G(\alpha_\varphi) \overset{def}{=} \psi \leq \varphi$. Hence

$$P_\varphi(u^\alpha) \geq P_\psi(u^\alpha) \geq \alpha_\varphi. \tag{9}$$

Inequalities (8) and (9) are satisfied simultaneously iff $\alpha_\varphi = P_\varphi(u^\alpha)$ which means that u^α is optimal for (3).

The theorem is proved.

Remark. In Theorems 2 and 3 it is supposed that there exist solutions u_φ and u^α of problems (3) and (4). Both theorems take into account cases where several such solutions exist, since it is easy to see that the proofs of these theorems are valid for all strategies u_φ and u^α which are optimal for problems (3) and (4), respectively.

Example. Let $\xi(\omega)$ be a random variable distributed uniformly over $[0,1]$, $Q(u,\omega) \equiv 0$, $U = \mathbb{R}^1$ and

$$\Phi(u,\omega) = \begin{cases} |u|, & u \neq 0 \\ 0, & \xi(\omega) \in [0,p] \text{ и } u = 0 \\ 1, & \xi(\omega) \in (p,1] \text{ и } u = 0. \end{cases}$$

It is easy to verify that in this case a solution u_φ of problem (3) exists and can be obtained analytically for every φ. The supremum function $F(\varphi)$ takes the following form

$$F(\varphi) = \begin{cases} 0, & \varphi < 0 \\ p, & \varphi = 0 \\ 1, & \varphi > 0. \end{cases}$$

Thus the equation $F(\varphi) = \alpha$ has the single root $\varphi_\alpha = 0$ for every $\alpha \in (0,1)$. According to Theorem 2 we can guarantee that $u_\varphi|_{\varphi=0}$ (it turns out to be equal to zero) is optimal for (4) if $\alpha \le p$. For $\alpha > p$ the condition $F(\varphi_\alpha) \ge \alpha$ of Theorem 2 does not hold. In this case it can be shown that for such α there are no optimal solutions for every φ. Thus the condition $F(\varphi_\alpha) \ge \alpha$ is quite essential to allow us to utilize the equivalence idea.

4 A satellite orbital correction

In this section we illustrate numerical techniques for the application of the above theoretical results by an example where we deal with optimization of the last orbital correction for a satellite which is desired to be geosynchronous. The following mathematical model for the correction is taken from [3]. Let d be a terminal drift speed which is an observed longitudinal bias per a satellite revolution after the correction, ξ_0 be an initial drift speed before the correction and $u = u(\xi_0)$ be a correcting impulse which is implemented with an error $u \cdot \xi_1$. Then

$$d = \xi_0 + u \cdot (1 + \xi_1).$$

Suppose that $\xi_0 = \xi_0(\omega)$ and $\xi_1 = \xi_1(\omega)$ be independent random variables with normal distributions such that $E[\xi_0] = m$, $\text{var}[\xi_0] = \sigma_0^2$, $E[\xi_1] = 0$ and $\text{var}[\xi_1] = \sigma_1^2$. Suppose also that every Borel-measurable function $u(\cdot)$ be a feasible correcting strategy. The probability measure $\mathcal{P}_u$ is induced on the Borel algebra $\mathcal{B}(\mathbb{R}^1)$ by a distribution of d. Set $\Phi(u(\cdot), \omega) = |d|$, $Q(u(\cdot), \omega) \equiv 0$ and consider problem (4) with the given reliability level $\alpha \in (0,1)$.

In [3] the following relations defining a strategy u_φ which is optimal for (3) for every $\varphi > 0$ have been obtained:

$$u_\varphi(\xi_0) = \begin{cases} -2\xi_0/(2 + a + b), & x_\varphi > 1 \\ 0, & x_\varphi \le 1, \end{cases} \quad x_\varphi = \frac{|\xi_0|}{\varphi},$$

$$b = \frac{2 + (x_\varphi + 1)a}{x_\varphi - 1}, \quad b^2 - a^2 = 2\sigma_1^2 \ln\left(\frac{x_\varphi + 1}{x_\varphi - 1}\right), \quad 2 + a + b > 0.$$

It is easy to verify that

$$u_\varphi(\xi_0) = \begin{cases} -2\xi_0/\left(1 + \sqrt{1 + 2\sigma_1^2 x_\varphi \ln\left(\frac{x_\varphi+1}{x_\varphi-1}\right)}\right), & x_\varphi > 1 \\ 0, & x_\varphi \le 1, \end{cases}$$

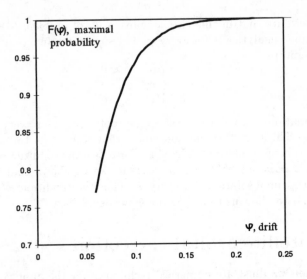

Figure 1: The supremum function $F(\varphi)$

There is a question: can we assert that this feedback control with the inert zone is optimal for (4) if we choose φ as a root of the equation $F(\varphi) = \alpha$, i.e. can we apply here the equivalence idea? Following to the above recommendations on the application of Theorem 2 we have constructed the plot of $F(\varphi) = P_\varphi(u_\varphi)$ (see figure) by using a dense net of points with the distance 0.001 between each two neighbouring points. The density of the net allows us to disregard errors of plotting. The calculation of $F(\varphi)$ has been carried out by the Monte Carlo method with 150 000 realizations for $m = 1°$, $\sigma_0 = 0.3°$, and $\sigma_1 = 0.05°$. We note also that the function $F(\varphi)$ is similar to the usual distribution function. It is known that by application of the Monte Carlo simulation we usually obtain a jumping curve which estimates the distribution function if the number of realizations is small. Increasing the number of realizations we decrease the jumps. In our opinion, the quite smooth behaviour of $F(\varphi)$ allows us to disregard calculation errors. Thus we conclude that the supremum function $F(\varphi)$ is most probably continuous and strictly increasing, so we can believe that all the conditions of Theorem 2 likely hold. Thus we can surely utilize the equivalence idea and assert that the strategy u_φ is optimal for (4) under an appropriate value $\varphi = \varphi_\alpha$ which is defined as a root of the equation $F(\varphi) = \alpha$ where $\alpha \in (0.77, 0.99)$. The limits for α can be widened by the more careful calculation and plotting.

5   Conclusion

The weakened sufficient conditions for transforming problems (3) and (4) into one another have been presented above by Theorems 2 and 3. They are verified as a rule after a strategy which is optimal for an alternative problem and seems to be optimal for the problem in question has been obtained. Their practical verification can be performed as shown in sec. 4.

References

[1] MOESEKE P.V. Stochastic linear programming. *Yale Economic Essays*, 1965, v.5, pp.197–253.

[2] KIBZUN A.I., KAN YU.S. *Stochastic programming problems with probability and quantile functions.* Wiley, Chichester, 1996, 310 p.

[3] MALYSHEV V.V., KIBZUN A.I. *Analysis and Synthesis of High-Precision Aircraft Control.* Mashinostroenie, Moscow, 1987, 304 p. [in Russian].

A Note on Multifunctions in Stochastic Programming

Vlasta Kaňková[1]

Abstract. Two–stage stochastic programming problems and chance constrained stochastic programming problems belong to deterministic optimization problems depending on a probability measure. Surely, the probability measure can be considered as a parameter of such problems and, moreover, it is reasonable to investigate stability with respect to it. However, to investigate the stability of the above mentioned problems it means, mostly, to investigate simultaneously behaviour of the objective functions and multifunctions corresponding to the constraints sets. The aim of the paper is to investigate stability of the multifunctions and summarize by it (from the constraints point of view) problems with the stable behaviour.

Key words: Stochastic programming problems, multifunctions, Hausdorff distance, Kolmogorov metric.

1 Introduction

It is generally known that to take a decision in practice, very often it is suitable (or even necessary) to deal with an optimization problem in the form:
Find

$$\inf\{g_\nu(x)|x \in \mathcal{K}(\nu)\} = \varphi(\nu), \tag{1}$$

where $g_\nu(x)$ is a real–valued function defined on E_n, $n \geq 1$ and $\mathcal{K}(\nu)$ is a multifunction mapping a parametric space into the space of the subsets of E_n; ν is a parameter of the problem. (E_n, $n \geq 1$ denotes an n–dimensional Euclidean space.)

If the "value" of the parameter ν is known and, moreover, the solution can be find out with respect to its actual "value", then (1) is a problem of the deterministic optimization. However, it happens very often that the "value" of ν has to be replaced by some approximate one. Consequently, it is reasonable (or even necessary) to investigate the stability of (1) with respect to a perturbation of the parameter ν. Evidently, the behaviour of $\varphi(\nu)$ depends, essentially, on the behaviour of the multifunction $\mathcal{K}(\nu)$.

Let (Ω, S, P) be a probability space; $\xi^1 = \xi^1(\omega) = (\xi_1^1(\omega), \ldots, \xi_{s_1}^1(\omega))$ and $\xi^2 = \xi^2(\omega) = (\xi_1^2(\omega), \ldots, \xi_{s_2}^2(\omega))$ be s_1 and s_2–dimensional random vectors defined on (Ω, S, P); $F^{\xi^1}(z^1)$, $F^{\xi^2|\xi^1}(z^2|z^1)$, $z^1 \in E_{s_1}$, $z^2 \in E_{s_2}$ be

---

[1] Institute of Information Theory and Automation, Academy of Sciences of the Czech Republic

the distribution function of $\xi^1(\omega)$ and the conditional distribution function respectively ($\xi^2(\omega)$ conditioned by $\xi^1(\omega)$).

Let, moreover, $f^2(\bar{x}^2, \bar{z}^2)$, $g_i^2(\bar{x}^2, z^1)$, $i = 1, 2, \ldots, l_2$ and $g^1(x^1, z^1)$ be (real–valued) functions defined on $E_{n_1+n_2} \times E_{s_1+s_2}$, $E_{n_1+n_2} \times E_{s_1}$ and $E_{n_1} \times E_{s_1}$; $\bar{x}^2 = [x^1, x^2]$, $\bar{\xi}^2 = \bar{\xi}^2(\omega) = [\xi^1(\omega), \xi^2(\omega)]$, $\bar{z}^2 = [z^1, z^2]$; $X^1 \subset E_{n_1}$, $X^2 \subset E_{n_2}$ be nonempty sets; $Z^1 = Z_{\xi^1}^1 \subset E_{s_1}$, $Z^2 = Z_{\xi^2}^2 \subset E_{s_2}$ denote the supports of the probability measures corresponding to the distribution functions $F^{\xi^1}(\cdot)$, $F^{\xi^2}(\cdot)$.

A rather general two–stage stochastic nonlinear programming problem can be introduced in the following form:

I. Find

$$\inf E_{F^{\xi^1}}\{Q(x^1, \xi^1(\omega))|x^1 \in X^1\} \quad (= \varphi^1(F^{\xi^1})), \tag{2}$$

where for $x^1 \in E_{n_1}$, $z^1 \in E_{s_1}$,

$$Q(x^1, z^1) = g^1(x^1, z^1) + \inf E_{\xi^2|\xi^1 = z^1}\{f^2(\bar{x}^2, \bar{\xi}^2(\omega))|x^2 \in \mathcal{K}^2(x^1, z^1)\}, \tag{3}$$

$$\mathcal{K}^2(x^1, z^1) = \{x^2 \in X^2 : g_i^2(\bar{x}^2, z^1) \le 0, i = 1, 2, \ldots, l_2\}, \tag{4}$$

$E_{F^{\xi^1}}$, $E_{F^{\xi^2}|\xi^1}$ denote the operators of mathematical expectation corresponding to $F^{\xi^1}(z^1)$, $F^{\xi^2}|\xi^1(z^2|z^1)$.

Evidently, (for given $F^{\xi^1}(\cdot)$ and x_1) $\nu := z^1$ can be considered as a parameter of the inner problem:

Find

$$\inf E_{F^{\xi^2}|\xi^1 = z^1}\{f^2(\bar{x}^2, \bar{\xi}^2(\omega))| x^2 \in \mathcal{K}^2(x^1, z^1)\}. \tag{5}$$

The corresponding multifunction $\mathcal{K}(\nu)$ and the function $\varphi(\nu)$ fulfil the relations

$$\mathcal{K}(\nu) := \mathcal{K}^2(x^1, z^1), \quad \varphi(\nu) := Q(x^1, z^1) \quad \text{by fixed } x^1 \in X^1. \tag{6}$$

Moreover, $F^{\xi^1}(\cdot)$ can be consider as a parameter of the outer problem (2).

Stochastic programming problems with the individual and joint probability constraints belong to the optimization problems in which the probability measure occurs in the constraints. If $g_i^1(x^1, z^1)$, $i = 1, 2, \ldots, l_1$ are real–valued functions defined on $E_{n_1} \times E_{s_1}$, then the stochastic programming problem with the individual probability constraints can be introduced as the problem:

II. Find

$$\inf E_{F^{\xi^1}}\{g^1(x^1, \xi^1(\omega))|x^1 \in \bar{X}_{F^{\xi^1}}(\bar{\alpha})\} \quad (= \bar{\varphi}(F^{\xi^1}, \bar{\alpha})), \tag{7}$$

$$\bar{X}_{F^{\xi^1}}(\bar{\alpha}) = \bigcap_{i=1}^{l_1} X_{F^{\xi^1}}^i(\alpha_i), \tag{8}$$

$$X_{F^{\xi^1}}^i(\alpha_i) = \{x^1 \in X^1 : P_{F^{\xi^1}}\{\omega : g_i^1(x^1, \xi^1(\omega)) \leq 0\} \geq \alpha_i\}, \; i = 1, \ldots, l_1,$$

where $\bar{\alpha} = (\alpha_1, \ldots, \alpha_{l_1})$, $\alpha_i \in \langle 0, 1 \rangle$, $i = 1, \ldots, l_1$ is a (well–known, fixed) parameter of the problem. $P_{F^{\xi^1}}(\cdot)$ denotes the probability measure corresponding to the distribution function $F^{\xi^1}(\cdot)$.

Evidently in this case we can consider $\nu := F^{\xi^1}(\cdot)$ and, consequently,

$$\mathcal{K}(\nu) := \bar{X}_{F^{\xi^1}}(\bar{\alpha}). \tag{9}$$

It is known from the literature that the stochastic programming problem with the joint probability constraints can be introduced as the problem:

III. Find

$$\inf E_{F^{\xi^1}}\{g^1(x^1, \xi^1(\omega)) \mid x^1 \in X_{F^{\xi^1}}(\alpha)\} \quad (= \hat{\varphi}(F^{\xi^1}, \alpha)), \tag{10}$$

$$X_{F^{\xi^1}}(\alpha) = \{x^1 \in X^1 : P_{F^{\xi^1}}\{\omega : g_i^1(x^1, \xi^1(\omega)) \leq 0, i = 1, 2, \ldots, l_1\} \geq \alpha\}, \tag{11}$$

where $\alpha \in \langle 0, 1 \rangle$ is a (well–known, fixed) parameter of the problem.

We can consider $\nu := F^{\xi^1}(\cdot)$ and, consequently,

$$\mathcal{K}(\nu) := X_{F^{\xi^1}}(\alpha). \tag{12}$$

The stability of the multifunctions (corresponding to the stochastic programming problems) has been already investigated in the literature, see e. g. [4], [13], [18], where the continuity of $\bar{X}_{F^{\xi^1}}(\bar{\alpha})$, $X_{F^{\xi^1}}(\alpha)$ with respect to the Kuratowski convergence was investigated. The continuity and the Lipschitz property of the multifunctions (in case of the multistage programming problems) with respect to the Hausdorff distance has been already investigated in [8], [9]. Moreover, the dependence on the continuity (and generally stability and estimates) of the optimal value and the optimal solution set on the corresponding multifunction behaviour appears in [2], [17].

In this paper, we focus our investigation mostly on the Lipschitz property (with respect to the Hausdorff distance) of $\mathcal{K}^2(x^1, z^1)$, $\bar{X}_{F^{\xi^1}}(\bar{\alpha})$, $X_{F^{\xi^1}}(\alpha)$. In particular, first, we shall deal with the Lipschitz property of the multifunction $\mathcal{K}^2(x^1, z^1)$. (We recall that deterministic parametric optimization problems were investigated in the literature many times, see e.g. [1].) Furthermore, we employ some of these assertions to investigate the problems with probability constraints.

2 Assumptions and Auxiliary Assertions

In this section we shall introduce some assumptions and auxiliary assertions. To this end, let first $f_i^*(x)$, $i = 1, 2, \ldots, l$ be real–valued functions defined on E_n $(n, l \geq 1)$ and $\mathcal{K}^*(y)$ be a multifunction defined by the relation

$$\mathcal{K}^*(y) = \{x \in X : f_i^*(x) \leq y_i, i = 1, \ldots, l, y = (y_1, \ldots, y_l)\}, y \in Y, \tag{13}$$

where $X \subset E_n$, $Y \subset E_l$ are nonempty sets.

Furthermore, let Y_0 denote the convex hull of Y; $Y_0(\varepsilon)$, $\varepsilon > 0$ the ε-neighbourhood of Y_0. We shall introduce the systems of the assumptions.

i.1   a. $f_i^*(x)$, $i = 1, 2, \ldots, l$ are linear functions, $X = E_n$; without loss of generality, we can consider in this case the constraints in (13) to be in the form of equalities,

      b. for every $y \in Y_0(\varepsilon)$, $\mathcal{K}^*(y)$ is a nonempty, compact set,

      c. the matrix A of the type $(l \times n)$, $l \leq n$ fulfils the relation

$$\mathcal{K}^*(y) = \{x \in X : Ax = y\}, \ y \in Y$$

and, moreover, all its submatrices of the types $(l \times l)$, $A(1)$, $\ldots$, $A(m)$ are nonsingular,

      d. $C = l \max_{i,r,s} |a_{i\,r}(s)|$, where $a_{i,r}(s)$ for $s \in \{1, 2, \ldots, m\}$ denote elements of the inverse matrix to $A(s)$,

i.2 There exist real–valued constants $d_1, \gamma_1, \varepsilon > 0$ such that:

      a. If for $x \in X, y \in Y_0(\varepsilon)$, $y = (y_1, \ldots, y_l)$, $f_i^*(x) \leq y_i$, $i = 1, \ldots, l$ and simultaneously $f_j^*(x) = y_j$ for at least one $j \in \{1, \ldots, l\}$, then there exists a vector $x(0) \in E_n$ (generally depending on x), $\|x(0)\| = 1$ such that

$$x + dx(0) \in X, \ f_i^*(x) - f_i^*(x + dx(0)) \geq \gamma_1 d$$

for every $d \in (0, d_1)$, $i = 1, 2, \ldots, l$,

      b. for every $y \in Y_0(\varepsilon)$, $\mathcal{K}^*(y)$ is a nonempty, compact set,

      c. $C = \frac{1}{\gamma_1}$,

$(\| \cdot \| = \| \cdot \|_n$, $n \geq 1$ denotes the Euclidean norm in E_n.)

i.3   a. X is a convex, compact set, $M_1 = \sup\limits_{x(1), x(2) \in X} \|x(1) - x(2)\|$,

      b. $f_i^*(x)$, $i = 1, \ldots, l$ are convex functions on X,

      c. for every $y \in Y_0(\varepsilon)$, $\mathcal{K}^*(y)$ is a nonempty set,

      d. $C = \frac{M_1}{\varepsilon_0}$ for an $\varepsilon_0 \in (0, \varepsilon)$.

Lemma 1. Let the relation (13) be satisfied. If at least one of the systems of the assumptions i.1, i.2, i.3 is fulfilled, then

$$\Delta[\mathcal{K}^*(y(1)), \mathcal{K}^*(y(2))] \leq C\|y(1) - y(2)\| \quad \text{for every } y(1), y(2) \in Y.$$

($\Delta[\cdot, \cdot] = \Delta_n[\cdot, \cdot]$, $n \geq 1$ denotes the Hausdorff distance of the subset E_n; for definition see e.g. [12].)

Proof. If the system of the assumptions i.1 or i.2 are fulfilled, then the assertions of Lemma 1 follows from the well–known results of linear programming and convex analysis [16]. The assertion of Lemma 1 under the assumption i.3 is proven in [8] (see also [9]).

3 Main Results

3.1 Two–Stage Stochastic Programming Problem

Investigating the multifunctions corresponding to the two–stage stochastic programming problems, first, we shall restrict to the case when

$$g_i^2(\bar{x}^2, z^1) = f_i^2(x^2) - h_i^2(x^1, z^1), \ i = 1, \ldots, l_2, \ x^1 \in X^1, \ x^2 \in X^2, \ z^1 \in Z^1,$$

where $f_i^2(x^2)$, $h_i^2(x^1, z^1)$, $i = 1, \ldots, l_2$ are real–valued functions defined on E_{n_2} and $E_{n_1} \times E_{s_1}$. Consequently, in this case

$$\mathcal{K}^2(x^1, z^1) = \{x^2 \in X^2 : f_i^2(x^2) \le h_i^2(x^1, z^1), \ i = 1, \ldots, l_2\},$$
$$x^1 \in E_{n_1}, z^1 \in E_{s_1}. \tag{14}$$

To investigate $\mathcal{K}^2(x^1, z^1)$, in this special case, we substitute (into (13))

$$l := l_2, \ n := n_2, \ X := X^2, \ f_i^*(x) := f_i^2(x^2), \ i = 1, 2, \ldots, l_2, \tag{15}$$

$$Y \quad := \quad \{y \in E_{l_2}, y = (y_1, \ldots, y_{l_2}) : \text{there exist } x^1 \in X^1,$$
$$z^1 \in Z^1 \text{ such that } y_i = h_i^2(x^1, z^1), \ i = 1, \ldots, l_2\},$$

$$\mathcal{K}^*(y) \quad := \quad \{x^2 \in X^2 : f_i^2(x^2) \le y_i, \ i = 1, \ldots, l_2, y = (y_1, \ldots, y_{l_2})\}, \ y \in E_{l_2}.$$

Evidently, in this case we can obtain

$$\mathcal{K}^*(y) = \mathcal{K}^2(x^1, z^1), \tag{16}$$

$$y = (y_1, \ldots, y_{l_2}), \ y_i = h_i^2(x^1, z^1), i = 1, 2, \ldots, l_2, \ x^1 \in E_{n_1}, z^1 \in E_{l_1}.$$

Proposition 1. Let the relations (14), (15) be satisfied. If

1. at least one of the systems of the assumptions i.1, i.2, i.3 is fulfilled,

2. for every $x^1 \in X^1$, $h_i^2(x^1, z^1), i = 1, \ldots, l_2$ are Lipschitz functions on Z^1 with the Lipschitz constants L_i,

then

$$\Delta[\mathcal{K}^2(x^1, z^1(1)), \mathcal{K}^2(x^1, z^1(2))] \le C\|z^1(1) - z^1(2)\| \sum_{i=1}^{l_2} L_i$$

for every $x^1 \in X^1$, $z^1(1), z^1(2) \in Z^1$.

Proof. The assertion of Proposition 1 follows immediately from the substitution (15), the relation (16), the assertion of Lemma 1 and the assumptions.

We have investigated the case when the variables corresponding to the first and to the second stage can be separated. Furthermore, we shall consider more general case. To this end we introduce the new systems of the assumptions.

i.4   a. X^2, Z^1 are convex sets, X^2 moreover compact,
$$M_1 = \sup_{x^2(1),\,x^2(2)\in X^2} \|x^2(1) - x^2(2)\|,$$

     b. for some $\varepsilon > 0$ and every $x^1 \in X^1$, $g_i^2(\bar{x}^2, z^1)$, $i = 1, 2, \ldots, l_2$ are convex functions on $X^2 \times Z^1(\varepsilon)$,

     c. for $x^1 \in X^1$, $z^1 \in Z^1(\varepsilon)$, $\mathcal{K}^2(x^1, z^1)$ is a nonempty, compact set,

     d. $C = \frac{M_1}{\varepsilon_0}$ for an $\varepsilon_0 \in (0, \varepsilon)$,

i.5 there exist real–valued constants $\underline{D} > 0$, $\bar{D} > 0$, $\varepsilon > 0$, $d_1 > 0$, $d_1 \geq \frac{\bar{D}}{\underline{D}}\varepsilon$ such that

     a. if $x^1 \in X^1$, $z^1(1)$, $\bar{z}^1(1) \in Z_0^1(\varepsilon)$, $x^2 \in X^2$,
$\|z^1(1) - \bar{z}^1(1)\| \leq \varepsilon$, $\bar{x}^2(1) = [x^1, x^2]$ fulfil the inequalities

$$g_i^2(\bar{x}^2(1), z^1(1)) \leq 0 \quad \text{for every} \qquad i \in \{1, \ldots, l_2\},$$
$$g_j^2(\bar{x}^2(1), \bar{z}^1(1)) > 0 \quad \text{for at least one} \quad j \in \{1, \ldots, l_2\},$$

then for $i \in \{1, \ldots, l_2\}$,

$$g_i^2(\bar{x}^2(1), \bar{z}^1(1)) - g_i^2(\bar{x}^2(1), z^1(1)) \leq \bar{D}\|z^1(1) - \bar{z}^1(1)\|,$$

and simultaneously there exists $x^2(0) = x^2(0, \bar{x}^2(1), \bar{z}^1(1))$,
$\|x^2(0)\| = 1$ such that $\bar{x}^2(2) = [x^1, x^2(2)]$, $x^2(2) = x + dx^2(0)$
$\in X^2$ for every $d \in (0, d_1)$ and, moreover, for $i \in \{1, \ldots, l_2\}$

$$g_i^2(\bar{x}^2(1), \bar{z}^1(1)) - g_i^2(\bar{x}^2(2), \bar{z}^1(1)) \geq \underline{D}\|x^2(1) - x^2(2)\|,$$

     b. for every $x^1 \in X^1$, $z^1 \in Z_0^1(\varepsilon)$, $\mathcal{K}^2(x^1, z^1)$ is a nonempty, compact set,

     c. $C = \frac{\bar{D}}{\underline{D}}$.

(Z_0^1 denotes the convex hull of Z^1, $Z_0^1(\varepsilon)$, $\varepsilon > 0$ the ε–neighbourhood of Z_0^1.)

Proposition 2. If the system of the assumptions i.4 or i.5 is fulfilled, then

$$\Delta[\mathcal{K}^2(x^1, z^1(1)), \mathcal{K}^2(x^1, z^1(2))] \leq C\|z^1(1) - z^1(2)\|$$

for every $x^1 \in X^1$, $z^1(1)$, $z^1(2) \in Z^1$.

Proof. If the assumptions i.4 are fulfilled, then the assertion of Proposition 2 is only a rather generalized assertion of [3] (Lemma 1). Since the proof of the original assertion and the genralized one are very similar each other it remains only to prove the assertion of Proposition 2 under the assumptions i.5. To this end, let $x^1 \in X^1$, $x^2(1) \in X^2$, $z^1(1)$, $z^1(2) \in Z^1$ be arbitrary such that $x^2(1) \in \mathcal{K}(x^1, z^1(1))$. According to the fact that $Z_0^1(\varepsilon)$ is a convex set there exist points $\hat{z}^1, \hat{z}^2, \ldots, \hat{z}^k \in Z_0^1(\varepsilon)$ such that

$$\hat{z}^1 = z^1(1), \quad \hat{z}^k = z^1(2), \quad \|\hat{z}^j - \hat{z}^{j+1}\| \leq \varepsilon,$$

$$\hat{z}^j = \lambda_j z(1) + (1 - \lambda_j)z(2) \text{ for some } \lambda_j \in \langle 0, 1 \rangle, \ j = 1, 2, \ldots, k. \tag{17}$$

Two cases can happen:

    a. $g_i^2(\bar{x}^2(1), \hat{z}^2) \leq 0$ for every $i \in \{1, 2, \ldots, l_2\}$, $\bar{x}^2(1) := [x^1, x^2(1)]$,

    b. $g_j^2(\bar{x}^2(1), \hat{z}^2) > 0$ for at least one $j \in \{1, 2, \ldots, l_2\}$.

Evidently, if a. happens, then $x^2(1) \in \mathcal{K}^2(x^1, \hat{z}^2)$. If the case b. happens, then according to i.5 a there exists $x^2(0) = x^2(0, \bar{x}^2(1), \hat{z}^2)$ such that

$$\|x^2(0)\| = 1, \quad x^2(2) = x^2(1) + dx^2(0) \in X^2 \quad \text{for } d \in (0, d_1)$$

and, moreover, the following inequalities are fulfilled for $\bar{x}^2(2) = [x^1, x^2(2)]$

$$g_i^2(\bar{x}^2(1), \hat{z}^2) - g_i^2(\bar{x}^2(1), z^1(1)) \leq \bar{D}\|z^1(1) - \hat{z}^2\|,$$

$$g_i^2(\bar{x}^2(1), \hat{z}^2) - g_i^2(\bar{x}^2(2), \hat{z}^2) \geq \underline{D}\|x^2(1) - x^2(2)\|, \ i = 1, 2, \ldots, l_2.$$

However, we can successively obtain from these inequalities that

$$g_i^2(\bar{x}^2(2), \hat{z}^2) \ \leq g_i^2(\bar{x}^2(1), \hat{z}^2) - \underline{D}\|x^2(2) - x^2(1)\| \leq$$

$$g_i^2(\bar{x}^2(1), z^1(1)) + \bar{D}\|z^1(1) - \hat{z}^2)\| - \underline{D}\|x^2(1) - x^2(2)\|,$$

$$i = 1, \ldots, l_2.$$

Consequently, if $\|x^2(2) - x^2(1)\| = \frac{\bar{D}}{\underline{D}}\|z^1(1) - \hat{z}^2\|$, then $x^2(2) \in \mathcal{K}^2(x^1, \hat{z}^2)$. Since $x^2(1) \in \mathcal{K}^2(x^1, z^1)$ was an arbitrary point we can see that

$$\Delta[\mathcal{K}^2(x^1, z^1(1)), \ \mathcal{K}^2(x^1, \hat{z}^2)] \leq C\|z^1(1) - \hat{z}^2\|.$$

Replacing successively, $z^1(1) := \hat{z}^j, \hat{z}^2 := \hat{z}^{j+1}, j = 2, \ldots, k - 1$ and employing the fact that the Hausdorff distance is a metric in the space of the closed subsets of E_{n_2} (see e.g. [12]) we obtain the assertion of Proposition 2.

    Proposition 1 and Proposition 2 introduce the assumptions under which for every $x^1 \in X^1$, $\mathcal{K}^2(x^1, z^1)$ is a Lipschitz multifunction on Z^1 (with respect to the Hausdorff distance). Employing, furthermore the results of [5], [6] we can see that this property together with some additional assumptions on the probability measure and the objective function garantee the validity of the inequality

$$|\varphi^1(F^{\xi^1}) - \varphi^1(G)| \leq C_0(\sup|F(z^1) - G(z^1)|)^{\frac{1}{t}}, \quad C_0 \geq 0 \quad \text{a constant}$$

for a ("near" to $F^{\xi^1}(\cdot)$) s_1–dimensional distribution function $G(\cdot)$. Furthermore, if we interchange the position of z^1 and x^1 in Propositions 1 and 2, then according to [7] we can employ these results (together with some additional assumptions on the objective functions) to garantee the "best" possible convergence rate of empirical estimates of optimal value (for independent and some weak dependent random samples) and "nearly best" rate of convergence for some others types of weak dependent random samples.

3.2 Stochastic Programming Problems with Individual Probability Constraints

In the case II we shall restrict to the case when $s_1 = l_1$

$$g_i^1(x^1, z^1) = f_i^1(x^1) - z_i^1, \quad i = 1, 2, \ldots, l_1, \; z^1 = (z_1^1, z_2^1, \ldots, z_{l_1}^1), \quad (18)$$

where $f_i^1(x^1)$, $i = 1, \ldots, l_1$ are functions defined on E_{n_1}. Evidently, then

$$\bar{X}_{F^{\xi^1}}(\bar{\alpha}) = \bigcap_{i=1}^{l_1} X_{F_i^{\xi^1}}^i(\alpha_i), \qquad (19)$$

$$X_{F_i^{\xi^1}}^i(\alpha_i) = \{x^1 \in X^1 : P_{F^{\xi^i}}\{\omega : f_i^1(x^1) \le \xi_i^1(\omega)\} \ge \alpha_i\}, \; i = 1, \ldots, l_1.$$

where $F_i^{\xi}(z_i^1)$, $z_i^1 \in E_1$, $P_{F_i^{\xi^i}}(\cdot)$, $i = 1, \ldots, l_1$ denote one–dimensional marginal distribution functions and the corresponding probability measure. If we define the new multifunctions $\bar{K}_i(z_i^1)$, $i = 1, \ldots, l_1$ and the quantils $k_{F_i^{\xi^1}}(\alpha_i)$, $\alpha_i \in (0, 1)$, $i = 1, \ldots, l_1$, by

$$\hat{K}_i^1(z_i^1) = \{x^1 \in X^1 : f_i^1(x^1) \le z_i^1\}, \; z_i^1 \in E_1, \; i = 1, \ldots, l_1,$$
$$k_{F_i^{\xi^1}}(\alpha_i) = \sup\{z_i^1 : P_{F_i^{\xi^1}}\{\omega : z_i^1 \le \xi_i^1(\omega)\} \ge \alpha_i\}, \; i = 1, \ldots, l_1,$$

then

$$X_{F_i^{\xi^1}}^i(\alpha_i) = \hat{K}_i^1(k_{F_i^{\xi^1}}(\alpha_i)). \qquad (20)$$

Furthermore, if $\bar{K}(z^1)$, $z^1 = (z_1^1, \ldots, z_{l_1}^1)$ is defined by the relation

$$\bar{K}^1(z^1) = \bigcap_{i=1}^{l_1} \hat{K}_i^1(z_i^1),$$

then also

$$\bar{X}_{F^{\xi^1}}(\bar{\alpha}) = \bar{K}^1(k_{F^{\xi^1}}(\bar{\alpha})), \quad \text{where } k_{F^{\xi^1}}(\bar{\alpha}) = (k_{F_1^{\xi^1}}(\alpha_1), \ldots, k_{F_i^{\xi^1}}(\alpha_{l_1})).$$

$$(21)$$

If $G(z^1)$, $z^1 \in E_{l_1}$ is an arbitrary l_1–dimensional distribution function, then to estimate $\Delta[\bar{X}_{F^{\xi^1}}(\bar{\alpha}) \; \bar{X}_G(\bar{\alpha})]$ let for $\delta_i > 0$, $i = 1, \ldots, l_1$, $\varepsilon > 0$

$$\underline{F}_{i, \delta_i}(z_i^1) = F_i^{\xi^1}(z_i^1 - \delta_i), \quad \bar{F}_{i, \delta_i}(z_i^1) = F_i^{\xi^1}(z_i^1 + \delta_i), \qquad (22)$$

$$\bar{Y} = \prod_{i=1}^{l_1} Y_i, \quad Y_i = (k_{F_i^{\xi^1}}(\alpha_i) - \delta_i - \varepsilon, \; k_{F_i^{\xi^1}}(\alpha_i) + \delta_i + \varepsilon). \qquad (23)$$

If we substitute into (13)

$$n := n_1,\, l := l_1,\, Y := \bar{Y},\, X := X^1,\, f_i^*(x) := f_i^1(x^1),\, i = 1, \ldots, l_1, \quad (24)$$

then the following assertion holds.

Proposition 3. Let $\delta_i > 0$, $i = 1, \ldots, l_1$, $\varepsilon > 0$ be arbitrary. Let, moreover, relation (24) is fulfilled. If

1. at least one of the systems of the assumptions i.1, i.2, i.3 is fulfilled,
2. for $i = 1, \ldots, l_1$, $F_i^{\varepsilon^1}(z_i^1)$ is an increasing function on Y_i,
3. $G(z^1)$ is an arbitrary l_1–dimensional distribution function such that

$$G_i(z_i^1) \in \langle \underline{F}_{i,\delta_i}(z_i^1),\, \bar{F}_{i,\delta_i}(z_i^1) \rangle \quad \text{for } z_i^1 \in Y_i,\, i = 1, 2, \ldots, l_1,$$

then

$$\Delta[\bar{X}_{F^{\varepsilon^1}}(\bar{\alpha}),\, \bar{X}_G(\bar{\alpha})] \le C \| k_{F^{\varepsilon^1}}(\bar{\alpha}) - k_G(\bar{\alpha}) \| \le C \sum_{i=1}^{l} \delta_i. \quad (25)$$

($G_i(z_i^1)$, $i = 1, 2, \ldots, l_1$ denotes the one–dimensional marginal distribution functions corresponding to $G(z^1)$.)

Proof. If we substitute the relation (24) into (13) then the assertion of Proposition 3 follows immediately from the assertion of Lemma 1, the relations (20), (21).

In the case when one–dimensional marginal distribution functions are absolutely continuous with respect to one–dimensional Lebesgue measure, then

Corollary. Let $\varepsilon > 0$, $\delta_i > 0$, $i = 1, 2, \ldots, l_1$. Let, moreover, the relation (24) be satisfied. If

1. at least one of the systems of the assumptions i.1, i.2, i.3, is fulfilled,

2. for $i = 1, \ldots, l_1$, $P_{F_i^{\varepsilon^1}}(\cdot)$ are absolutely continuous (with respect to the Lebesgue measure in E_1) probability measures such that the corresponding probability density $\bar{h}_i(z_i^1)$ fulfil the inequality

$$\bar{h}_i(z_i^1) \ge \vartheta_i \quad \text{for every } z_i^1 \in Y_i \quad \text{and some } \vartheta_i > 0,$$

then

$$\Delta[\bar{X}_{F^{\varepsilon^1}}(\bar{\alpha}),\, \bar{X}_G(\bar{\alpha})] \le C \sum_{i=1}^{l_1} \frac{2 \sup\limits_{Y_i} |F_i^{\varepsilon^1}(z_i^1) - G_i(z_i^1)|}{\vartheta_i},$$

whenever $G(\cdot)$ is an l_1–dimensional distribution function with one dimensional marginals $G_i(\cdot)$ such that

$$Z_{G_i}^1 \subset Z_{F_i^{\varepsilon^1}}^1 \left(\frac{\sup\limits_{Y_i} |F_i^{\varepsilon^1}(z^1) - G_i(z^1)|}{\vartheta_i} \right), \quad i = 1, \ldots, l_1.$$

$Z^1_{G_i}$, $Z^1_{F^{\xi^1}_i}(\varepsilon)$, $i = 1, \ldots, l_1$, $\varepsilon > 0$ denote the support and ε–neighbourhood (of the $Z^1_{F^{\xi^1}_i}$) corresponding to distribution functions $G_i(\cdot)$ and $F^{\xi^1}_i$.

Proof. The assertion of Corollary follows from the assertion of Proposition 3. Namely, if we set

$$\delta_i := \frac{2 \sup_{z_i \in Y_i} |F^{\xi^1}_i(z_i) - G_i(z_i)|}{\vartheta_i}, \quad i = 1, 2, \ldots, l_1,$$

then the assumptions of Proposition 3 are fulfilled.

Employing, the results of [10] we can see that (under some special assumptions on the objective functions and on the probability measure) we can obtain

$$|\bar{\varphi}(F^{\xi^1}, \bar{\alpha}) - \bar{\varphi}(G, \bar{\alpha})| \le \bar{C}_0 \sum_{i=1}^{s_1} \sup |F^{\xi^1}_i(z_i) - G_i(z_i)|, \quad \bar{C}_0 \ge 0 \quad \text{a constant,}$$

where $G(\cdot)$ is a ("near" to $F^{\xi^1}(\cdot)$) l_1–dimensional distribution function with the marginals $G_i(\cdot)$, $i = 1, \ldots, l_1$. Furthermore, under rather general asumptions on the objective function according to [7] we can obtain the "best" possible convergence rate for empirical estimates of the optimal value (in the case of independent and some types of weak dependent random samples) and "nearly best" rate of convergence for some others types of weak dependent random samples.

3.3 Stochastic Programming Problems with Joint Probability Constraints

To investigate the stability of the chance constrained stochastic programming problems we (again) restrict to the case $s_1 = l_1$ and, moreover, to the relation (18). Evidently, in this case

$$X_{F^{\xi^1}}(\alpha) = \{x^1 \in X^1 : P_{F^{\xi^1}}\{\omega : f^1_i(x^1) \le \xi^1_i(\omega), i = 1, 2, \ldots, l_1\} \ge \alpha\}.$$

To introduce the suitable system of the assumptions we first define for $\varepsilon > 0$, $\beta \in \langle \alpha - \varepsilon, \alpha \rangle$, $\alpha \in (0, 1)$, $z^1 \in E_{l_1}$ the following sets:

$$Z^0_{F^{\xi^1}}(\beta) = \{z^1 \in E_{l_1} : P_{F^{\xi^1}}\{\omega : z^1 \le \xi^1(\omega) \text{ componentwise}\} = \beta\},$$
$$Z_{F^{\xi^1}}(\beta) = \{z^1 \in E_{l_1} : P_{F^{\xi^1}}\{\omega : z^1 \le \xi^1(\omega) \text{ componentwise}\} \ge \beta\},$$

$$\hat{\mathcal{K}}^1(z^1) = \{x^1 \in X^1 : f^1_i(x^1) \le z^1_i, i = 1, \ldots, l_1, \ z^1 = (z^1_1, \ldots, z^1_{l_1})\}$$

and by the symbol $Z_{F^{\xi^1}}(\beta, \delta)$ we denote the δ–neighbourhood of $Z^0_{F^{\xi^1}}(\beta)$.

ii.1 $P_{F^{\xi^1}}(\cdot)$ is absolutely continuous with respect to the Lebesgue measure in E_{l_1}. We denote by $h(z^1)$ the corresponding probability density.

ii.2 there exists a constant $\vartheta_1 > 0$ and for every $\beta \in \langle \alpha - \varepsilon, \alpha \rangle$, $z^1 \in Z^0_{F^{\xi^1}}(\beta)$ a point $\bar{z}^1 \in Z^0_{F^{\xi^1}}(\beta)$, $\|z^1 - \bar{z}^1\| \le \sqrt{l_1}(\frac{2\varepsilon}{\vartheta_1})^{\frac{1}{l_1}}$ such that

$$\vartheta_1 \le h(z^1) \text{ for } z^1 \in Z^-_{F^{\xi^1}}(\beta, \bar{z}^1, \delta'), \ \delta' > 2\sqrt{l_1}(\frac{2\varepsilon}{\vartheta_1})^{\frac{1}{l_1}},$$

$$Z^-_{F^{\xi^1}}(\beta, \bar{z}^1, \delta') = \{z^1 \in E_{l_1} : z^1 = \bar{z}^1 + \hat{z}^1, \hat{z}^1 \in \mathcal{B}(\delta'),$$

$$z^1 \le \bar{z}^1 \text{ componentwise}\},$$

$(\mathcal{B}(\delta')$ denotes $\delta-$ neighbourhood of $0 \in E_{l_1})$,

ii.3 there exists a constant $L_{F^{\xi^1}} > 0$ such that for $\beta \in \langle \alpha - \varepsilon, \alpha + \varepsilon \rangle$

$$z^1, \bar{z}^1 \in Z_{F^{\xi^1}}(\beta, \delta'), \ \delta' > \frac{\varepsilon}{L_{F^{\xi^1}}}, \ z^1 \le \bar{z}^1 \text{ componentwise} \Longrightarrow$$

$$F^{\xi^1, c}(z^1) - F^{\xi^1, c}(\bar{z}^1) \ge L_{F^{\xi^1}} \|z^1 - \bar{z}^1\|.$$

$$F^{\xi^1, c}(z^1) = P_{F^{\xi^1}}\{\omega : z^1 \le \xi^1(\omega) \text{ componentwise}\}.$$

The next two assertions express a (mathematical underlying) relationship between the stability of the deterministic set (corresponding to the inner problem of two–stage stochastic programming problems) and the "stability" constraints in the chance constrained case.

Lemma 2. Let $\varepsilon > 0$, $\beta \in \langle \alpha - \varepsilon, \alpha, \rangle$, $\alpha \in (0, 1)$ be arbitrary, $\delta' > 0$. Let, moreover, C' be a real–valued constant such that

$$\Delta[\hat{\mathcal{K}}^1(z^1(1)), \hat{\mathcal{K}}^1(z^1(2))] \le C' \|z^1(1) - z^1(2)\| \text{ for } z^1(1), z^1(2) \in Z_{F^{\xi^1}}(\beta, \delta').$$

If
1. the assumptions ii.1 and ii.2 are fulfilled, $\delta' > 2\sqrt{l_1}(\frac{2\varepsilon}{\vartheta_1})^{\frac{1}{l_1}}$, then

$$\Delta[X_{F^{\xi^1}}(\alpha - \varepsilon), X_{F^{\xi^1}}(\alpha + \varepsilon)] \le 2C'(\frac{2\varepsilon}{\vartheta_1})^{\frac{1}{l_1}} \sqrt{l_1},$$

2. the assumptions ii.1 and ii.3 are fulfilled, $\delta' > \frac{\varepsilon}{L_{F^{\xi^1}}}$, then

$$\Delta[X_{F^{\xi^1}}(\alpha - \varepsilon), X_{F^{\xi^1}}(\alpha + \varepsilon)] \le 2C' \frac{\varepsilon}{L_{F^{\xi^1}}}.$$

Proof. First, we shall deal with the case 1. To this end, let $z^0 \in Z^0_{F^{\xi^1}}(\alpha - \varepsilon)$, $z^0 = (z^0_1, z^0_2, \ldots, z^0_{l_1})$ be arbitrary. According to the assumption ii. 2 there exist points z, z'; $z \in Z^0_{F^{\xi^1}}(\alpha - \varepsilon)$ such that $\|z - z^0\| \le \sqrt{l_1}(\frac{2\varepsilon}{\vartheta_1})^{\frac{1}{l_1}}$,

$z' = z'(z) = (z'_1, \ldots, z'_{l_1}), \ z'_i = z_i - (\frac{2\varepsilon}{\vartheta_1})^{\frac{1}{l_1}}, \ \|z - z'\| = \sqrt{l_1}(\frac{2\varepsilon}{\vartheta_1})^{\frac{1}{l_1}},$
$\|z' - z^0\| \leq 2\sqrt{l_1}(\frac{2\varepsilon}{\vartheta_1})^{\frac{1}{l_1}}$. And, moreover,

$$P_{F^{\xi^1}}\{\omega : z' \leq \xi^1(\omega) \text{ componentwise}\} \geq P_{F^{\xi^1}}\{\omega : z \leq \xi^1(\omega) \text{ componentwise}\}$$

$$+P_{F^{\xi^1}}\{\omega : z' \leq \xi^1(\omega) \leq z \text{ componentwise}\} \geq \alpha - \varepsilon + \vartheta_1(\frac{2\varepsilon}{\vartheta_1}).$$

Consequently, $z' \in Z_{F^{\xi^1}}(\alpha + \varepsilon)$. Since, evidently, $Z_{F^{\xi^1}}(\beta) \subset Z_{F^{\xi^1}}(\beta')$ for every $\beta' \leq \beta$, $\beta, \beta' \in (0, 1)$ we can see that

$$\Delta[Z_{F^{\xi^1}}(\alpha - \varepsilon), Z_{F^{\xi^1}}(\alpha + \varepsilon)] \leq 2\sqrt{l_1}(\frac{2\varepsilon}{\vartheta_1})^{\frac{1}{l_1}}.$$

However, according to the last inequality and to the fact that that for $\beta \in (0, 1)$

$$x^1 \in X_{F^{\xi^1}}(\alpha) \iff (f_1^1(x^1), \ldots, f_{l_1}^1(x^1)) \in Z_{F^{\xi^1}}(\alpha)$$

the following implication follows from the assumptions

$$x^1(1) \in X_{F^{\xi^1}}(\alpha - \varepsilon) \implies \quad \text{there exist } x^1(2) \in X_{F^{\xi^1}}(\alpha + \varepsilon) \text{ such that}$$

$$\|x^1(1) - x^1(2)\| \leq 2C'\sqrt{l_1}(\frac{2\varepsilon}{\theta_1})^{\frac{1}{l_1}}.$$

Since $X_{F^{\xi^1}}(\alpha + \varepsilon) \subset X_{F^{\xi^1}}(\alpha - \varepsilon)$ the first assertion of Lemma 2 is valid.

The proof of the second assertion of Lemma 2 is very similar and, consequently, we omit it.

To introduce the next assertion, let

$$U(F^{\xi^1}, \alpha) = \{z^1 \in E_{l_1} : F^{\xi^1, c}(z^1) \in \langle \alpha - \varepsilon, \alpha + \varepsilon \rangle\}.$$

Lemma 3. Let $\alpha, \varepsilon > 0$, $\alpha - \varepsilon, \alpha + \varepsilon \in (0, 1)$. Let, moreover, $X_{F^{\xi^1}}(\alpha - \varepsilon)$, $X_{F^{\xi^1}}(\alpha + \varepsilon)$ be nonempty, compact sets. If ii.1 is fulfilled and if

1. $\Delta[X_{F^{\xi^1}}(\alpha - \varepsilon), X_{F^{\xi^1}}(\alpha + \varepsilon)] \leq d \quad$ for some $d > 0$,

2. $G(z)$ is an arbitrary l_1–dimensional distribution function such that
$|F^{\xi^1, c}(z^1) - G^c(z^1)| \leq \frac{\varepsilon}{2}, \quad z^1 \in U(F^{\xi^1}, \alpha),$

then

$$\Delta[X_{F^{\xi^1}}(\alpha), X_G(\alpha)] \leq d.$$

$(G^c(z^1) = P_G\{\omega : z^1 \leq \xi^1 \text{ componentwise}\}.)$

Proof. First, we shall prove the validity of the relation

$$X_{F^{\xi^1}}(\alpha + \varepsilon) \subset X_G(\alpha) \subset X_{F^{\xi^1}}(\alpha - \varepsilon). \tag{26}$$

Evidently, to prove the last relation it is sufficient to prove

$$P_G\{\omega : z^1 \le \xi^1(\omega) \text{ componentwise}\} = \alpha, \ z^1 \in E_{l_1} \implies$$
$$P_{F^{\xi^1}}\{\omega : z^1 \le \xi^1(\omega) \text{ componentwise}\} \in \langle\alpha - \varepsilon, \ \alpha + \varepsilon\rangle. \tag{27}$$

We shall prove the relation (27) by the contradiction. To this end we shall assume that there exists $z^1 \in Z_{F^{\xi^1}}$ such that simultanously

$$P_G\{\omega : z^1 \le \xi^1(\omega) \text{ componentwise}\} = \alpha \text{ and}$$
$$P_{F^{\xi^1}}\{\omega : z^1 \le \xi^1(\omega) \text{ componentwise}\} \notin \langle\alpha - \varepsilon, \ \alpha + \varepsilon\rangle. \tag{28}$$

Of course then, just one of the following assertions must hold

a. $P_{F^{\xi^1}}\{\omega : z^1 \le \xi^1(\omega) \text{ componentwise}\} < \alpha - \varepsilon,$

b. $P_{F^{\xi^1}}\{\omega : z^1 \le \xi^1(\omega) \text{ componentwise}\} > \alpha + \varepsilon.$

Let us, first, to consider the case a. It means that

$$P_G\{\omega : z^1 \le \xi^1(\omega) \text{ componentwise}\} = \alpha \text{ and simultaneously}$$
$$P_{F^{\xi^1}}\{\omega : z^1 \le \xi^1(\omega) \text{ componentwise}\} < \alpha - \varepsilon. \tag{29}$$

It follows from the assumptions that then there exists $\bar{z}^1 \in E_{l_1}, \ \bar{z}^1 \le z^1$ componentwise $\bar{z}^1 \ne z^1$ such that

$$P_{F^{\xi^1}}\{\omega : \bar{z}^1 \le \xi^1(\omega) \text{ componentwise}\} = \alpha - \varepsilon.$$

However then, it follows from the assumptions that

$$P_G\{\omega : \bar{z}^1 \le \xi^1(\omega) \text{ componentwise}\} \ge \alpha - \tfrac{3}{2}\varepsilon \text{ and simultanously}$$
$$P_G\{\omega : \bar{z}^1 \le \xi^1(\omega) \text{ componentwise}\} \le \alpha - \tfrac{1}{2}\varepsilon.$$

Since $\bar{z}^1 \le z^1$ componentwise $z^1 \ne \bar{z}^1$ the last inequality is in the contradiction to the first relation in (28). It remains to consider the case b. Since the corresponding proof is very similar to the first one we can it omit. Consequently, we have proven the relation (26). Since, furthermore, evidently

$$X_{F^{\xi^1}}(\alpha + \varepsilon) \subset X_{F^{\xi^1}}(\alpha) \subset X_{F^{\xi^1}}(\alpha - \varepsilon),$$

the assertion of Lemma 3 follows already immediately from the properties of the Hausdorff distance.

If we substitute $n := n_1, \ l := l_1, \ Y := Z^1, \ K^*(y) := \bar{K}^1(z^1), f_i^*(x) := f_i^1(x^1) :=, \ i = 1, \ldots, l_1$ in i.1, i.2, i.3, then we can obtain:

Proposition 4. Let $\alpha \in (0, 1), \ \varepsilon > 0, \ \alpha - \varepsilon, \ \alpha + \varepsilon \in (0, 1)$. Let, moreover, relation (18) be fulfilled and $X_{F^{\xi^1}}(\alpha+\varepsilon), \ X_{F^{\xi^1}}(\alpha-\varepsilon)$ be nonempty, compact sets. If $G(z)$ is an arbitrary l_1–dimensional distribution function such that

$$\sup |F^{\xi^1}(z^1) - G(z^1)| \le \frac{\varepsilon}{2^{l_1+1}}$$

and, moreover, at least one of the assumptions i.1, i.2, i.3 is fulfilled, then

1. the assumptions ii.1 and ii.2 imply

$$\Delta[X_{F^{\xi^1}}(\alpha),\, X_G(\alpha)] \leq 4C \left(\frac{4\sup |F^{\xi^1}(z^1) - G(z^1)|}{\vartheta_1} \right)^{\frac{1}{l_1}} \sqrt{l_1},$$

2. the assumptions (ii.1) and (ii.2) imply

$$\Delta[X_{F^{\xi^1}}(\alpha),\, X_G(\alpha)] \leq 2C \frac{\sup |F^{\xi^1}(z) - G(z)|}{L_{F^{\xi^1}}}.$$

Proof. First, it follows from the relation (18) and Lemma 1 that

$$\Delta[\hat{\mathcal{K}}(z^1(1)),\, \hat{\mathcal{K}}(z^1(2))] \leq C\|z^1(1) - z^1(2)\| \quad \text{for every } z^1(1),\, z^1(2) \in Z^1.$$

The assertion of Proposition 4 follows then from the last relation, from Lemma 2 and Lemma 3 and the properties of probability measures.

We can see that (completing the problem by some assumptions on the objective function and the probability measure) we can obtain

$$|\hat{\varphi}(F^{\xi^1},\, \alpha) - \hat{\varphi}(G,\, \alpha)| \leq \hat{C}_0 (\sup |F^{\xi^1}(z^1) - G(z^1)|)^{\frac{1}{l}}, \quad \hat{C}_0 \geq 0 \quad \text{a constant.}$$

In the case when the assumptions ii.1 and ii.3 are fulfilled, then (under some special additional assumptions) we can also obtain

$$|\hat{\varphi}(F^{\xi^1},\, \alpha) - \hat{\varphi}(G,\, \alpha)| \leq \hat{C}_0'(\sup |F^{\xi^1}(z^1) - G(z^1)|), \quad \hat{C}_0' \geq 0 \quad \text{a constant.}$$

and, furthermorem according to [7] under rather general assumptions on the objective function we can obtain also the "best" possible convergence rate for empirical estimates of the optimal value (in the case of independent and some weak dependent random samples) and "nearly best" rate of convergence for some others types of weak dependent random samples.

Acknowledgement This work was supported by the Grant Agency of the Academy of Sciences of the Czech Republic under grant No. 107 55 02.

References

1. B. Bank, J. Guddat, D. Klatte, R. Kummer, R. Tammer: Non–Linear Parametric Optimization. Akademie–Verlag, Berlin, 1982, New York 1977.

2. R. Henrion and W. Römisch: *Metric regularity and quantitative stability in stochastic programs with probabilistic constraints.* Humb. Univ. Berlin, Sekt. Mathem. Preprint Nr. 96-2 (1996).

3. V. Kaňková: *Approximative solution of problems of two–stage stochastic programming problems (in Czech).* Ekonomicko–matematický obzor 16 (1980), 64–74.

4. V. Kaňková: *Estimates in stochastic programming-chance constrained case.* Problems Control. Inform. Theory *18* (1989), 251–260.

5. V. Kaňková: *On stability in two-stage stochastic nonlinear programming.* In: Proceedings of the Fifth Prague Symposium (P. Mandl and M. Hušková, eds.), Springer Verlag, Berlin 1994, 329–340.

6. V. Kaňková: *On Distribution Sensitivity in Stochastic Programming.* Research Report ÚTIA No. 1826, Prague 1994.

7. V. Kaňková: *A note on estimates in stochastic progrmming.* Journal of Computational and Applied Mathematics *56* (1994), 97–112.

8. V. Kaňková: *On Objective Functions in Multistage Stochastic Programming Problems.* Research Report ÚTIA No. 1839, Prague 1995.

9. V. Kaňková: *A note on objective functions in multistage stochastic nonlinear programming problems.* In: System Modelling and Optimization, Prague 1995 (J. Doležal and J. Fiedler, eds.), Chapman & Hall, London 1996, 582–589.

10. V. Kaňková: *On the stability in stochastic programming; the case of individual probability constraints.* to appear in Kybernetika.

11. A. Prékopa: Stochastic Programming. Akadémiai Kiadó, Budapest and Kluwer Publisher, Dordrecht 1995.

12. G. Salinetti and R. J.-B. Wets: *On the convergence of sequences of convex sets in finite dimension.* SIAM Review *21* (1979), 18–33.

13. G. Salinetti: *Approximation for chance constrained programming problems.* Stochastics *10* (1983), 157–179.

14 S. M. Robinson: *Regularity and stability for convex multivalueed functions.* Math. Oper. Research *1*, 1976, 130–145.

15 S. M. Robinson: *Generalized equations and their solution, part II. Applications to nonlinear programming.* Mathematical Programming Study, 1982, 19, 200–221.

16 R. T. Rockafellar: Convex Analysis. Princeton University Press, New Jersey 1970.

17. S. Vogel: *On stability in multiobjective programming – a stochastic approach.* Mathematical Programming *50* (1992), 197–226.

18. J. Wang: *Continuity of feasible solution sets probabilistic constrained programs.* Journal Optimization Theory and Applications *53* (1989), 79–89.

Approximation of Extremum Problems with Probability Cost Functionals

Riho Lepp [1]

Institute of Cybernetics, Akadeemia 21, EE–0026 Tallinn, Estonia

Abstract. The problem of maximization of the probability of a reliability level under integrable decision rules is approximated by a sequence of finite dimensional problems with discrete distributions. Convergence conditions of the approximation are presented.

Keywords. Probability functional, finite dimensional approximation, discrete convergence

1 Introduction and Problem Formulation

Usually the solution of a stochastic program is considered as a deterministic vector. This is the well known approach in one– and two–stage programming. Still, there exist models, in which the solution is determined as a function from the random parameter. In the latter case the class of solution functions (decision rules) should be described earlier (as measurable or summable or continuous or linear or constant functions, see, e.g., [5]). Even more, two–stage stochastic programs are equivalent to an extremum problem in a function space (see, e.g., [3] – the equivalence of linear programs in L^p, [10] – the equivalence of convex ones in L^∞).

To ensure a certain level of reliability for the solution it has become a spread approach to introduce probabilistic (chance) constraints into the model. The stability analysis of chance constraint problems is rather complicated due to uncomfortable properties of the probability function

$$u(x) = P\{s \mid f(x,s) \leq t\}. \tag{1}$$

Here $f(x,s)$ is a real valued function, defined on $R^r \times R^r$, t is a fixed level of reliability, s is a random vector and P denotes the probability. Note that the function $u(x)$ is never convex, only in some cases (e.g., $f(x,s)$ linear in s and distribution of s normal), it is quasiconvex, e.t.c. Varied examples and models with probability function $u(x)$ and its "inverse," the quantile function

$$\min_t \{t \mid P\{s \mid f(x,s) \leq t\} \geq \alpha\}, \quad 0 < \alpha < 1.$$

[1] also, Estonian National Defence and Public Service Academy, 61 Kase Street, EE–0020, Tallinn, Estonia

are presented in [7], Ch. I. In Ch. II the authors present some models with such a complicated structure that we are forced to look for a solution x from a certain class of strategies, that means, the solution x itself depends on the random parameter s, $x = x(s)$.

This class of probability functions was introduced to stochastic programming by Ernst Raik, and lower semicontinuity and continuity properties in Lebesgue L^p-spaces were studied in [9].

In this paper we will consider the maximization of the probability functional $v(x)$:

$$\max_{x \in C} v(x) = \max_{x \in C} P\{s \mid f(x(s), s) \leq t\}, \tag{2}$$

where $f : R^r \times R^r \to R^1$, t is a fixed level of reliability, s is a random parameter with bounded support S, $s \in S \subset R^r$ and with atomless distribution σ,

$$\sigma\{s \mid |s - s_0| = const\} = 0 \quad \forall s_0 \in R^r. \tag{3}$$

Due to technical reasons it will be assumed that the decision rule $x(s)$ will be a $\sigma-$ integrable function, $x \in L^1(S, \Sigma, \sigma) = L^1(\sigma)$, and C will be a compact constraint set in $L^1(\sigma)$.

Since the problem (3) is formulated in the function space $L^1(\sigma)$ of σ-integrable functions, the first step in its approximate solution is the approximation step – replacing the initial problem (2) by a sequence of finite dimensional optimization problems with increasing dimension.

We will approximate the initial measure σ by a sequence $\{(m_n, s_n)\}$ of discrete measures which converge weakly to σ. The usage of the weak convergence of discrete measures in stochastic programming has its disadvantages and advantages. An example in [11] shows that, in general, the stability of a probability function with respect to weak convergence cannot be expected without additional smoothness assumptions on the measure σ. This is one of the reasons, why we should use only continuous measures with the property (3). An advantage of the usage of the weak convergence is that it allows us to apply in the approximation instead of conditional means the more simple, grid point approximation scheme.

Since the functional $v(x)$ is not convex, we are not able to exploit in the discrete stability analysis of the problem (1) the more convenient, weak topology, but only the strong (norm) topology. As the first step we in this paper will approximate $v(x)$ so, that the discrete analogue of the continuous convergence of the sequence of approximate functionals will be guaranteed.

Schemes of stability analysis (e.g., finite dimensional approximations) of extremum problems in Banach spaces require from the sequence of solutions of "approximate" problems certain kind of compactness. Assuming that the constraint set C is compact in $L^1(\sigma)$, we, as the second step, will approximate the set C by a sequence of finite dimensional sets $\{C_n\}$ with increasing dimension so, that the sequence of solutions of approximate problems is compact in a certain (discrete convergence) sense in $L^1(\sigma)$. Then the approximation scheme for the discrete approximation of (2) will follow formed schemes of approximation of extremum problems in Banach spaces, see e.g. [2], [4], [13], [8].

Redefine the functional $v(x)$ by using the Heaviside zero - one function χ :

$$v(x) = \int_S \chi(t - f(x(s), s))\sigma(ds), \tag{4}$$

where

$$\chi(t - f(x(s), s)) = \begin{cases} 1, & \text{if } f(x(s), s) \leq t, \\ 0, & \text{if } f(x(s), s) > t. \end{cases}$$

In the next two sections we will assume that the function $f(x, s)$ is continuous in (x, s) and satisfies following growth and "platform" conditions:

$$| f(x, s) | \leq a(s) + \alpha | x |, \ a \in L^1(\sigma), \ \alpha > 0. \tag{5}$$

$$\sigma\{s \mid f(x, s) = const\} = 0 \quad \forall(x, s) \in R^r \times S. \tag{6}$$

The continuity assumption is technical in order to simplify the description of the approximation scheme below. The growth condition (5) is essential: without it the superposition operator $f(x) = f(x(s), s)$ will not map an element from L^1 to L^1 (is even not defined). Condition (6) means that the function $f(x, s)$ should not have horizontal platforms with positive measure.

Constraint set C is assumed to be a set of integrable functions $x(s)$, $x \in L^1(\sigma)$, with properties

$$\int_S | x(s) | \sigma(ds) \leq M < \infty \quad \forall \, x \in C \tag{7}$$

for some $M > 0$ (C is bounded in $L^1(\sigma)$);

$$\int_D | x(s) | \leq K\sigma(D) \quad \forall \, x \in C \quad \text{and} \quad \forall \, D \in \Sigma \tag{8}$$

for some $K > 0$;

$$(x(s) - x(t), s - t) \geq 0 \quad for \ almost \ all \quad (a.a.) \ s, t \in S \tag{9}$$

(functions $x \in C$ are monotone almost everywhere).

Conditions (7),(8) guarantee that the set C is weakly compact (i.e., compact in the (L^1, L^∞) – topology, see, e.g., [6], Ch. 9.1.2). Condition (9) guarantees now, following [1], Lemma 3, that together with conditions (7),(8) the set C is strongly compact in $L^1(\sigma)$. Then, following [9], we can conclude that assumptions (5) – (9) together with atomless assumption of the measure σ guarantee the existence of a solution of problem (2) in the Banach space $L^1(\sigma)$ of σ–integrable functions (the cost functional $v(x)$ is continuous and the constraint set C is compact in $L^1(\sigma)$).

In the last section we will approximate an unconstrained maximum of the probability functional (4), where decision rule $x(s)$ will be an integrable with p–th power function, $x \in L^p(\sigma)$, $1 < p < \infty$, and the probability functional $v(x)$ will vanish in the infinity.

2 Discrete Approximation in L^p-spaces

Let the initial probability measure σ be approximated by a sequence of discrete measures $\{(m_n, s_n)\}$ with weights m_{in} at points s_{in}, $i = 1, 2, ..., n$, $n \in N = \{1, 2, 3, ...\}$ (weak convergence of discrete measures):

$$\sum_{i=1}^{n} h(s_{in}) m_{in} \to \int_S h(s) \sigma(ds), \ \ n \in N, \tag{10}$$

for any continuous on S function $h(s)$.

Since approximate problems will be defined in R^{rn}, we in order to analyze stability of approximations should define a connection operator p_n between spaces $L^1(\sigma)$ and R^{rn}. In L^p-spaces, $1 \leq p \leq \infty$, systems of connection operators $\mathcal{P} = \{p_n\}$ should be defined in a piecewise integral form (as conditional means):

$$(p_n x)_{in} = \sigma(A_{in})^{-1} \int_{A_{in}} x(s) \sigma(ds), \ \ i = 1, ..., n. \tag{11}$$

Here sets A_{in}, $i = 1, ..., n$, $n \in N$, that define connection operators (11), satisfy following conditions A1) – A7):

A1) $\sigma(A_{in}) > 0$;

A2) $A_{in} \cap A_{jn} = \emptyset$, $i \neq j$;

A3) $\cup_{i=1}^{n} A_{in} = S$;

A4) $\sum_{i=1}^{n} | m_{in} - \sigma(A_{in}) | \to 0$, $n \in N$;

A5) $\max_i diam \ A_{in} \to 0$, $n \in N$;

A6) $s_{in} \in A_{in}$;

A7) $\sigma(int \ A_{in}) = \sigma(A_{in}) = \sigma(cl \ A_{in})$,

where "int" and "cl" denote topological interior and closure of a set, respectively.

Remark 1 *Weak convergence (10) is equivalent to the partition $\{\mathcal{A}_n\}$ of S, $\mathcal{A}_n = \{A_{1n}, ..., A_{nn}\}$, with properties A1) – A7), see [12].*

Remark 2 *Collection of sets $\{A_{in}\}$ with the property A7) constitutes an algebra $\Sigma_0 \subset \Sigma$, and if $S = [0, 1]$ and if σ is Lebesgue measure on $[0, 1]$, then integrability relative to $\sigma |_{\Sigma_0}$ means Riemann integrability.*

Define now the discrete convergence for the space $L^1(\sigma)$ of σ–integrable functions.

Definition 1 *Sequence of vectors $\{x_n\}$, $x_n \in R^{rn}$, $\mathcal{P}$-converges (or converges discretely) to integrable function $x(s)$, if*

$$\sum_{i=1}^{n} | x_{in} - (p_n x)_{in} | m_{in} \to 0, \ \ n \in N. \tag{12}$$

Remark 3 *Note that in the space $L^1(\sigma)$ of σ-integrable functions we are also able to use the projection methods approach, defining convergence of $\{x_n\}$ to $x(s)$ as follows:*

$$\int_S |\, x(s) - \sum_{i=1}^n x_{in} \chi_{A_{in}}(s)\,|\, \sigma(ds) \to 0, \ n \in N.$$

Remark 4 *Projection methods approach does not work in the space $L^\infty(\sigma)$ of essentially bounded measurable functions with vraisup-norm topology.*

Define also the discrete analogue of the weak convergence in $L^1(\sigma)$.

Definition 2 *Sequence of vectors $\{x_n\}$, $x_n \in R^{rn}$, $n \in N$, wP-converges (or converges weakly discretely) to integrable function $x(s)$, $x \in L^1(\sigma)$, if*

$$\sum_{i=1}^n (z_{in}, x_{in}) m_{in} \ \to \ \int_S (z(s), x(s)) \sigma(ds), \ n \in N, \tag{13}$$

for any sequence $\{z_n\}$ of vectors, $z_n \in R^{rn}$, $n \in N$, and function $z(s)$, $z \in L^\infty(\sigma)$, such that

$$\max_{1 \le i \le n} |\, z_{in} - (p_n z)_{in}\,| \to \ 0, \ n \in N. \tag{14}$$

Here the space $L^\infty(\sigma)$ of essentially bounded measurable functions is topological dual of $L^1(\sigma)$.

Remark 5 *For piecewise integral connection system (11) discrete weak convergence (13) is equivalent to the convergence*

$$\sum_{i=1}^n ((p_n z)_{in}, x_{in}) m_{in} \ \to \ \int_S (z(s), x(s)) \sigma(ds), \ n \in N \quad \forall z \in L^\infty(\sigma)$$

(see, [13]).

In order to formulate the discretized problem and to simplify the presentation, we will assume that in partition $\{A_n\}$ of S, where $A_n = \{A_{1n}, ..., A_{nn}\}$, with properties A1) – A7), in property A4) we will identify m_{in} and $\sigma(A_{in})$, i.e. $m_{in} = \sigma(A_{in})$ (e.g. squares with decreasing diagonal in R^2).

Discretize now the probability functional $v(x)$:

$$v_n(x_n) = \sum_{i=1}^n \chi(t - f(x_{in}, s_{in})) m_{in}, \tag{15}$$

and formulate the discretized problem:

$$\max_{x_n \in C_n} v_n(x_n) \ = \ \max_{x_n \in C_n} \sum_{i=1}^n \chi(t - f(x_{in}, s_{in})) m_{in}, \tag{16}$$

where constraint set C_n will satisfy discrete analogues of conditions (7) – (9), covered to the set C :

$$\sum_{i=1}^{n} | x_{in} | m_{in} \leq M \ \forall \ x_n \in C_n, \tag{17}$$

$$\sum_{i \in I_n} | x_{in} | m_{in} \leq K \sum_{i \in I_n} m_{in} \ \forall \ x_n \in C_n \ \forall \ I_n \subset \{1, 2, ..., n\}, \tag{18}$$

$$\sum_{k=1}^{r} (x_{i_k n}^k - x_{j_k n}^k)(i_k - j_k) \geq 0 \ \forall \ i_k, j_k, \ s.t. \ i_k < j_k, \tag{19}$$

and such that $0 \leq i_k, j_k \leq n \ \ \forall n \in N$.

Definition 3 *Sequence of sets* $\{C_n\}$, $C_n \subset R^{rn}$, $n \in N$, *converges to the set* $C \subset L^1(\sigma)$ *in the discrete Mosco sense, if*
1) for any subsequence $\{x_n\}$, $n \in N' \subset N$, *such that* $x_n \in C_n$, *from convergence* $wP - lim \ x_n = x$ *it follows that* $x \in C$;
2) for any $x \in C$ *there exists a sequence* $\{x_n\}$, $x_n \in C_n$, *which* P-*converges to* x, $P - lim \ x_n = x$, $n \in N$.

Remark 6 *If in Definition 3 also "for any" part 1) is defined for* P-*convergence of vectors, then it is said that sequence of sets* $\{C_n\}$ *converges to the set* C *in the discrete Painleve–Kuratowski sense.*

3 Conditions of the Discrete Approximation

Denote optimal values and optimal solutions of problems (2) and (16) by v^*, x^* and v_n^*, x_n^*, respectively.

Proposition 1 *Let function* $f(x, s)$ *be continuous in both variables* (x, s) *and satisfy growth and platform conditions (5) and (6). Then from convergence* $P - lim \ x_n = x$, $n \in N$ *for any monotone a. e.* $x(s)$, *it follows convergence* $v_n(x_n) \to v(x)$, $n \in N$.

Proof. Let $P - lim \ x_n = x$, $n \in N$, i.e., let $\sum_{i=1}^{n} | x_{in} - (p_n x)_{in} | m_{in} \to 0$, $n \in N$. Divide the difference

$$| v_n(x_n) - v(x) | = | \sum_{i=1}^{n} \chi(t - f(x_{in}, s_{in})) m_{in} - \int_S \chi(t - f(x(s), s)) \sigma(ds) | \tag{20}$$

into two parts:

$$| v_n(x_n) - v(x) | \leq | v_n(x_n) - v_n(p_n x) | + | v_n(p_n x) - v(x) | . \tag{21}$$

Estimate the first difference $| v_n(x_n) - v_n(p_n x) |$. Let us show that for any small $\varepsilon > 0$ there exists an index N_1, such that $| v_n(x_n) - v_n(p_n x) | \leq \varepsilon$ for all $n \geq N_1$. In the space $L^1(\sigma)$ of σ-integrable functions we can approximate an integrable function by a continuous function. Approximate first the discontinuous integrand $\chi(t - f(x, s))$ by continuous one. Define the continuous function χ_c in the following way:

$$\chi_c(t - f(x, s)) = \begin{cases} 1, & \text{if } f(x, s) \leq t, \\ 1 - \delta^{-1}[f(x, s) - t], & \text{if } t < f(x, s) \leq t + \delta, \\ 0, & \text{if } f(x, s) > t + \delta \end{cases}$$

where $\delta > 0$ is chosen in such a way, that

$$\int_{\{s | t < f(x(s), s) \leq t + \delta\}} \sigma(ds) \leq \varepsilon/6. \tag{22}$$

Then $| v_n(x_n) - v_n(p_n x) | =$

$$= | \sum_{i=1}^{n} \chi(t - f(x_{in}, s_{in})) m_{in} - \sum_{i=1}^{n} \chi(t - f((p_n x)_{in}, s_{in})) m_{in} | \leq$$

$$\leq | \sum_{i=1}^{n} \chi(t - f(x_{in}, s_{in})) m_{in} - \sum_{i=1}^{n} \chi_c(t - f(x_{in}, s_{in})) m_{in} | +$$

$$+ | \sum_{i=1}^{n} \chi_c(t - f(x_{in}, s_{in})) m_{in} - \sum_{i=1}^{n} \chi_c(t - f((p_n x)_{in}, s_{in})) m_{in} | +$$

$$+ | \sum_{i=1}^{n} \chi_c(t - f((p_n x)_{in}, s_{in})) m_{in} - \sum_{i=1}^{n} \chi(t - f((p_n x)_{in}, s_{in})) m_{in} | .$$

Due to choise of $\delta > 0$ in (22) we can find indices n_1 and n_2, such that

$$| \sum_{i=1}^{n} \chi(t - f(x_{in}, s_{in})) m_{in} - \sum_{i=1}^{n} \chi_c(t - f(x_{in}, s_{in})) m_{in} | \leq \varepsilon/3 \quad \text{as} \quad n \geq n_1$$

and

$$| \sum_{i=1}^{n} \chi_c(t - f((p_n x)_{in}, s_{in})) m_{in} - \sum_{i=1}^{n} \chi(t - f((p_n x)_{in}, s_{in})) m_{in} | \leq \varepsilon/3 \quad \text{as} \quad n \geq n_2.$$

Estimate the second difference:

$$| \sum_{i=1}^{n} \chi_c(t - f(x_{in}, s_{in})) m_{in} - \sum_{i=1}^{n} \chi_c(t - f((p_n x)_{in}, s_{in})) m_{in} | \leq$$

$$\leq \sum_{i=1}^{n} | \chi_c(t - f(x_{in}, s_{in})) - \chi_c(t - f((p_n x)_{in}, s_{in})) | m_{in}.$$

Function $\chi_c(t - f(x, s))$ is uniformly continuous on (X, S) (as a continuous function on a bounded set), so, taking for a small $\varepsilon > 0$ a $\nu > 0$, such that for all $n \geq n_3$ from $|x_{in} - (p_n x)_{in}| < \nu$ it follows

$$|\chi_c(t - f(x_{in}, s_{in})) - \chi_c(t - f((p_n x)_{in}, s_{in}))| < \varepsilon/3,$$

we get

$$\sum_{i=1}^{n} |\chi_c(t - f(x_{in}, s_{in})) - \chi_c(t - f((p_n x)_{in}, s_{in}))| \, m_{in} < \varepsilon/3.$$

Consequently, for all $n \geq N_1 = max\{n_1, n_2, n_3\}$ we have

$$|v_n(x_n) - v_n(p_n x)| \leq \varepsilon.$$

Estimate second difference $|v_n(p_n x) - v(x)|$ of the sum (21). Take a continuous monotone function $x_c(s)$, $x_c \in C(S)$, such that $|v(x) - v(x_c)| < \varepsilon/4$ and estimate difference $|v(x) - v_n(p_n x)|$ as follows:

$$|v(x) - v_n(p_n x)| \leq$$

$$\leq |v(x) - v(x_c)| + |v(x_c) - v_n(p_n x_c)| + |v_n(p_n x_c) - v_n(p_n x)|.$$

By assumptions ($f(x, s)$ continuous in (x, s), assumption (6) and σ atomless) we have $\sigma(S_t) = 0$, where $S_t = \{s \mid f(x_c(s), s) = t\}$. Then

$$|v(x_c) - v_n(p_n x_c)| \leq |v(x_c) - v_n(p'_n x_c)| + |v_n(p'_n x_c) - v_n(p_n x_c)|,$$

where

$$(p'_n x_c)_{in} = x_c(s_{in}). \tag{23}$$

Note that for the continuous function $x_c(s)$ we are able in parallel to the piecewise integral connection system $\mathcal{P} = \{p_n\}$ to use the simpliest connection system $\mathcal{P}' = \{p'_n\}$ of the form (23). Systems $\mathcal{P}$ and $\mathcal{P}'$ are equivalent in the following sense:

$$\sum_{i=1}^{n} |\sigma(A_{in})^{-1} \int_{A_{in}} x_c(s)\sigma(ds) - x_c(s_{in})| \, m_{in} \to 0, \ n \in N.$$

Now $|v_n(p'_n x_c) - v_n(p_n x_c)| =$

$$= |\sum_{i=1}^{n} \chi(t - f(x_c(s_{in}), s_{in})) m_{in} - \sum_{i=1}^{n} \chi(t - f((p_n x_c)_{in}, s_{in})) m_{in}| \leq \varepsilon/8$$

as $n \geq n_4$ for some n_4 (the function $F(s) = \chi(t - f(x_c(s), s))$ has only the first kind discontinuity and the σ–measure of its discontinuity points is zero).

Take now an index n_5 so large that for all $n \geq n_5$ we have

$$|v(x_c) - v_n(p'_n x_c)| =$$

$$= | \int_S \chi(t - f(x_c(s), s)) \sigma(ds) - \sum_{i=1}^{n} \chi(t - f(x_c(s_{in}), s_{in})) m_{in} | \leq \varepsilon/8$$

(approximation of a Riemann integrable function by Riemann sums). Then $| v(x_c) - v_n(p_n x_c) | \leq \varepsilon/4$, as $n \geq max\{n_4, n_5\}$. Since continuous function $x_c(s)$ was taken so that $| v(x) - v(x_c) | < \varepsilon/4$, we have

$$| v_n(p_n x_c) - v_n(p_n x) | \leq \varepsilon/2.$$

for $n \geq n_6$. Consequently, for $n \geq N_2 = max\{n_4, n_5, n_6\}$ we have

$$| v_n(p_n x) - v(x) | \leq \varepsilon$$

and hence, for $n \geq max\{N_1, N_2\}$

$$| v_n(x_n) - v(x) | \leq 2\varepsilon.$$

Proposition is proved.

Denote sets that satisfy only conditions (7),(8) and (17),(18) by C' and C'_n, respectively.

Proposition 2 *Let constraint sets C' and C'_n satisfy conditions (7),(8) and (17),(18), respectively. Let discrete measures $\{(m_n, s_n)\}$ converge weakly to the measure σ. Then sequence of sets $\{C'_n\}$ converges to the set C' in the discrete Mosco sense.*

Proof. Consider "for any" part 1) of the Definition 3. Let $wP - lim\ x_n = x$, $n \in N$, i.e., let

$$| \sum_{i=1}^{n} ((p_n z)_{in}, x_{in}) m_{in} - \int_S (z(s), x(s)) \sigma(ds) | \rightarrow 0, \ n \in N \ \forall z \in L^\infty, \quad (24)$$

and prove that $x \in C$. Test condition (7). Assume that

$$\sum_{i=1}^{n} | x_{in} | m_{in} \leq M$$

for all $n \in N$. Now, assuming in contrary that there exists a $\delta > 0$ such that $\int_S | x(s) | \sigma(ds) \geq M + \delta$, we get contradiction $M \geq M + \delta$ (the pair of functionals $(\| x \|, \{\| x_n \|_n\})$ is weakly discretely lower semicontinuous, i.e., if the convergence (24) holds, then

$$\lim_{n \in N} \inf \sum_{i=1}^{n} | x_{in} | m_{in} \geq \int_S | x(s) | \sigma(ds)$$

see, e.g., [13]).

Consider condition (8). Assume that

$$\sum_{i\in I_n} |\, x_{in}\,|\, m_{in} \le K \sum_{i\in I_n} m_{in} \quad \forall\, I_n \subset \{1,2,3,...,n\}\, \forall\, n \in N, \tag{25}$$

and that $wP - \lim x_n = x$, $n \in N$. Then from (24) and (25) we get

$$\lim_{n\in N} \sum_{i\in I_n} |\, x_{in}\,|\, \sigma(A_{in}) = \lim_{n\in N} \int_S \sum_{i\in I_n} |\, x_{in}\,|\, \chi_{A_{in}}(s)\sigma(ds) \le$$

$$\le K \lim_{i\in I_n} \int_S \chi_{A_{in}}(s)\sigma(ds).$$

Define the sequence of piecewise constant integrable functions $\{x_n\}$, $x_n \in L^1(\sigma)$, taking

$$x_n(s) = \sum_{i\in I_n} |\, x_{in}\,|\, \chi_{A_{in}}(s).$$

Now

$$\lim_{n\in N} \int_S \chi_D(s)x_n(s)\sigma(ds) = \int_S \chi_D(s)x(s)\sigma(ds)$$

(sequence $\{x_n\}$ as a sequence of piecewise constant σ–integrable functions converges weakly to the σ–integrable function $x(s)$ since $\chi_D \in L^\infty(\sigma)$). Consequently,

$$\lim_{n\in N} sup \int_S \sum_{i\in I_n} |\, x_{in}\,|\, \chi_{A_{in}}(s)\chi_D(s)\sigma(ds) = \int_S \chi_D(s)\,|\, x(s)\,|\, \sigma(ds) =$$

$$= \int_D |\, x(s)\,|\, \sigma(ds) \le K \lim_{n\in N} sup \int_S \sum_{i\in I_n} \chi_{A_{in}}(s)\chi_D(s)\sigma(ds) = K\sigma(D).$$

We can conclude now that weak discrete limit points of the sequence $\{x_n\}$ satisfy conditions (7) and (8).

Consider "there exists" part 2) of Definition 3. Take an $x \in C'$ that satisfies (7) and (8) and take x_{in}-s in the form: $x_{in} = (p_n x)_{in} = \sigma(A_{in})^{-1} \int_{A_{in}} x(s)\sigma(ds)$. Consider condition (17):

$$\sum_{i=1}^n |\, x_{in}\,|\, m_{in} = \sum_{i=1}^n |\, \sigma(A_{in})^{-1} \int_{A_{in}} x(s)\sigma(ds)\,|\, m_{in} =$$

$$= \sum_{i=1}^n |\int_{A_{in}} x(s)\sigma(ds)\,| \le \sum_{i=1}^n \int_{A_{in}} |\, x(s)\,|\, \sigma(ds) \le M.$$

Consider condition (18): fix an $I_n \subset \{1,2,...,n\}$. Then

$$\sum_{i\in I_n} |\, x_{in}\,|\, m_{in} = \sum_{i\in I_n} |\, (p_n x)_{in}\, m_{in} =$$

$$= \sum_{i \in I_n} | \sigma(A_{in})^{-1} \int_{A_{in}} x(s)\sigma(ds) | \sigma(A_{in}) \leq \sum_{i \in I_n} \int_{A_{in}} | x(s) | \sigma(ds) =$$

$$= \int_S \sum_{i \in I_n} \chi_{A_{in}}(s) | x(s) | \sigma(ds)$$

since $A_{in} \cap A_{jn} = \emptyset$, $i \neq j$. Now

$$\lim_{n \in N} \int_S \sum_{i \in I_n} \chi_{A_{in}}(s) | x(s) | \sigma(ds) = \int_D | x(s) | \sigma(ds) \leq K\sigma(D)$$

for some $D \in \Sigma$. But

$$K\sigma(D) = K \int_D \sigma(ds) = K \lim_{n \in N} \int_S \sum_{i \in I_n} \chi_{A_{in}}(s)\sigma(ds) = K \lim_{n \in N} \sum_{i \in I_n} m_{in}$$

since it was assumed that $\sigma(A_{in}) = m_{in}$. Proposition is proved.

Proposition 3 *Let constraint sets C and C_n satisfy conditions (7) – (9) and (17) – (19), respectively. Let discrete measures $\{(m_n, s_n)\}$ converge weakly to the measure σ. Then sequence of sets $\{C_n\}$ converges to the set C in the discrete Painleve–Kuratowski sense.*

Proof. Let $\mathcal{P} - lim\ x_n = x$, $n \in N$. Consider condition (9) with its discrete analogue (19). Let $x_n \in C_n$, i.e.,

$$\sum_{k=1}^r (x_{i_k n}^k - x_{j_k n}^k)(i_k - j_k) \geq 0 \ \forall\ i_k, j_k, \ such\ that\ i_k < j_k, \ k = 1, ..., r,$$

and

$$0 \leq i_k, \ j_k \ \leq n, \ n \in N$$

(remember that $x_n = (x_{1n}, x_{2n}, ..., x_{nn})$ and each $x_{in}, i = 1, ..., n$, is an r-dimensional vector, $x_{in} = (x_{in}^1, x_{in}^2, ..., x_{in}^r)$). Define a piecewise constant function $(r_n x_n)(s)$ in the following way:

$$(r_n x_{in})(s) = (x_{in}^1, x_{in}^2, ..., x_{in}^r), \ if\ s_k \in [(i_k-1)/n, i_k/n)\ \forall k \in \{1, 2, ..., r\}. \quad (26)$$

Then

$$((r_n x_n)(s_{in}) - (r_n x_n)(s_{jn}), s_{in} - s_{jn}) \geq 0$$

and therefore,

$$((r_n x_n)(s) - (r_n x_n)(t), s - t) \geq 0,$$

where $s = (s_1, ..., s_r)$ and $t = (t_1, ..., t_r)$ are defined by (26). Assume that the $\mathcal{P}$-limit of (discrete) monotone vectors x_n is not nondecreasing on a set $\Delta \in \Sigma$ with a positive measure, $\sigma(\Delta) > 0$, and construct a contradiction.

Assuming contrary, we get

$$(x(s) - x(t), s - t) \leq -\delta < 0 \quad \forall s, t \in \Delta, \tag{27}$$

but

$$((r_n x_n)(s) - (r_n x_n)(t), s - t) \geq 0 \quad for \ a.a. \ s, t \in \Delta. \tag{28}$$

From assumption $\mathcal{P} - \lim x_n = x, \ n \in N$, we get

$$\int_S | x(s) - \sum_{i=1}^n x_{in} \chi_{A_{in}}(s) | \, \sigma(ds) \to 0, \ n \in N.$$

We can extract from sequence $x_n(s) = \sum_{i=1}^n x_{in} \chi_{A_{in}}(s), n \in N$, a subsequence, which converges almost surely to zero, and consequently,

$$x_n(s) \to x(s), \ n \in N' \subset N, \quad almost \quad surely.$$

Now inequality (27) together with inequality (28) give us a contradiction. Consequently, the limit of the (discrete) monotone sequence of vectors $\{x_n\}$ is an almost everywhere monotone function $x(s)$.

Let now $x(s)$ be an almost everywhere monotone function,

$$(x(s) - x(t), s - t) \geq 0, \quad a.\,e.$$

Define x_{in} via projectors p_n :

$$x_{in} = (p_n x)_{in} = \sigma(A_{in})^{-1} \int_{A_{in}} x(s)\sigma(ds), \ i = 1, ..., n, \ n \in N,$$

where A_{in}-s are parallelepipeds with sides $[(i_k - 1)/n, i_k/n], \ k = 1, ..., r$. Then

$$\sum_{k=1}^r (x_{i_k n}^k - x_{j_k n}^k)(i_k/n - j_k/n) = \sum_{k=1}^r ((p_n x)_{i_k n}^k - (p_n x)_{j_k n}^k)(i_k/n - j_k/n).$$

Define, similarly to (26), piecewise constant functions $(r_n(p_n x)_n)(s) = = (x_{in}^1, x_{in}^2, ..., x_{in}^r)$, if $s_k = (i_k - 1)/n, \ k = 1, ..., r$. Then we get

$$((r_n(p_n x)_n)(s) - (r_n(p_n x)_n)(t), s - t) \geq 0 \quad \forall \ s, t \in S.$$

Consequently, the "there exists" part of discrete convergence $C_n \to C, \ n \in N$, is also verified. Now, together with Proposition 2 we can conclude that sequence of sets $\{C_n\}$ converges to the set C of admissible solutions of the initial problem (7) – (9) in the discrete Painleve–Kuratowski sense. Proposition is proved.

Relying on Propositions 1 – 3, we can now formulate and prove the main result of the paper.

Theorem 1 *Let function $f(x, s)$ be continuous in both variables (x, s) and satisfy growth and platform conditions (5) and (6), constraint set C satisfy conditions (7) – (9) and let discrete measures $\{(m_n, s_n)\}$ converge weakly to the initial measure σ. Then $v_n^* \to v^*$, $n \in N$, and sequence of solutions $\{x_n^*\}$ of approximate problems (16) has a subsequence, which converges discretely to a solution of the initial problem (2).*

Proof. By assumptions and by Proposition 3 sequence of solutions $\{x_n^*\}$ of approximate problems (16) is discretely compact, i.e., $\mathcal{P} - \lim x_n^* = x$, $n \in N' \subset N$ and by Proposition 1 $\lim v_n(x_n^*) = v(x)$, $n \in N'$. Then clearly

$$v^* \geq v(x) = \lim v_n(x_n^*) = \lim v_n^*, \ n \in N'$$

(by Proposition 3 limit point x is admissible but could not be optimal).

Prove the opposite inequality. By Proposition 3 for any $y \in C$ there exists a sequence $\{y_n\}$, $y_n \in C_n$, such that $\mathcal{P} - \lim y_n = y$. Consider difference $v_n(x_n^*) - v(x^*)$. By Proposition 1 $v_n(y_n) \to v(x^*)$ if $y_n \to x^*$ (continuous convergence of discrete approximations). Then

$$limsup \ v_n^* = limsup \ v_n(x_n^*) \geq limsup \ v_n(y_n) = v(x^*) = v^*, \ n \in N.$$

Consequently,

$$v^* = \lim v_n^* = \lim v(x_n^*), \ n \in N. \tag{29}$$

We can get convergence of a subsequence of solutions of approximate problems to a solution of the initial problem assuming in contrary: let $\mathcal{P} - \lim x_n^* = x$, $n \in N' \subset N$, and assume $v(x) \neq v^*$. Now we get the contradiction with (29) together with continuous convergence of a sequence of approximate functionals $\{v_n(x_n)\}$ to probability functional $v(x)$ (Proposition 1). Theorem is proved.

Remark 7 *The usage of the space $L^1(\sigma)$ of integrable functions is essential. In reflexive L^p-spaces, $1 < p < \infty$, serious difficulties arise with application of the strong (norm) compactness criterion for a maximizing sequence. We will see it in the next section.*

4 Approximation of Unconstrained Problems

Approximation of an unconstrained maximization of the probability functional $v(x)$ has its advantages and disadvantages. An advantage is that we do not need to approximate a constraint set. But now we should guarantee discrete compactness of a sequence of solutions of approximate problems itself. This disadvantage brings along some restrictions to functional $v(x)$ and to space L^p. We will assume that $v(x)$ is vanishing in infinity and that solution $x(s)$ belongs to a reflexive L^p-space, $1 < p < \infty$, since in reflexive spaces a bounded sequence is weakly compact.

Formulate the problem:

$$\max_{x \in L^p(\sigma)} v(x) \qquad (30)$$

where $1 < p < \infty$. In order to guarantee that $v : L^p \to R^1$, we instead of (5) should assume

$$\mid f(x, s) \mid \leq a(s) + \alpha \mid x \mid^p, \ a \in L^p(\sigma), \ \alpha > 0. \qquad (31)$$

Discrete convergence $\mathcal{P} - \lim x_n = x, n \in N$, is in L^p-spaces defined as follows

$$(\sum_{i=1}^{n} \mid x_{in} - (p_n x)_{in} \mid^p m_{in})^{1/p} \to 0, \ n \in N,$$

and weak discrete convergence, $w\mathcal{P} - \lim x_n = x, n \in N$, similarly:

$$\sum_{i=1}^{n} ((p_n z)_{in}, x_{in}) m_{in} \ \to \ \int_S (z(s), x(s)) \sigma(ds), \ n \in N \ \ \forall z \in L^q(\sigma),$$

where $1/p + 1/q = 1$.

Formulate the approximate problem:

$$\max_{x_n \in R^{rn}} v_n(x_n), \qquad (32)$$

where $x_n \in R^{rn}$.

Proposition 4 *Let continuous in both variables (x, s) function $f(x, s)$ satisfy growth condition (31) and platform condition (6). Then from convergence $\mathcal{P} - \lim x_n = x, n \in N$, it follows convergence $v_n(x_n) \to v(x), n \in N$.*

The proof proposition is quite sumilar to the proof of Proposition 1 and will be omitted here.

From now we will assume that the probability functional $v(x)$ will vanish in the infinity:

$$v(x_n) \to 0, \ as \ \parallel x_n \parallel \ \to \ \infty, \ n \in N.$$

Definition 4 *Sequence $\{v_n(x_n)\}$ of approximate functionals is uniformly vanishing in infinity, if from*

$$\lim_{n \in N} \inf \ \parallel x_n \parallel_n \ = \ \infty \qquad (33)$$

it follows

$$\lim_{n \in N} \sup v_n(x_n) \ = \ 0. \qquad (34)$$

Let x^* and x_n^* be solutions of problems (30) and (32), respectively.

Theorem 2 *Let*

1) function $f(x, s)$ be continuous in both variables (x, s) and satisfy growth and platform conditions (29) and (6);

2) functionals $v_n(x_n)$, $n \in N$, vanish uniformly in the infinity;

3) discrete measures $\{(m_n, s_n)\}$ converge weakly to the initial measure σ.

4) norms $\| x_n^* \|_n$ of solutions of approximate problems (32) be bounded with norm $\| x \|$ of their weak discrete limit point x,

$$limsup \ \| x_n^* \|_n \ \leq \ \| x \|, \ n \in N,$$

where $x = w\mathcal{P} - lim \ x_n^*, n \in N$. Then

$$v_n^* \ \to \ v^*, \ n \in N,$$

and sequence of solutions $\{x_n^*\}$ of approximate problems (32) has a subsequence, which converges discretely to a solution of the initial problem (30).

Proof. Let $\{x_n^*\}$ be a sequence of solutions of approximate problems (32). Due to uniform vanishing assumption 2) it is bounded and hence, weakly discretely compact, i.e., for some integrable with the p–th power function $x(s)$, $x \in L^p(\sigma)$, the convergence

$$\sum_{i=1}^{n} ((p_n z)_{in}, x_{in}^*) m_{in} \to \int_S (z(s), x(s)) \sigma(ds), n \in N \quad \forall z \in L^q(\sigma)$$

holds, where $1/p + 1/q = 1$. Since the L^p-norm is a convex functional, it is weakly lower semicontinuous. In our, discrete approximation case, it means that

$$\| x \| \ \leq \ lim inf \ \| x_n^* \|_n, \ n \in N' \subset N,$$

and together with assumption 4) we get

$$lim \ \| x_n^* \|_n \ = \ \| x \|, \ n \in N'.$$

In reflexive L^p-spaces the discrete analogue of Radon–Riesz property holds, i.e., if $w\mathcal{P} - lim \ x_n^* = x$ and $\| x_n^* \|_n \to \| x \|$, $n \in N'$, then

$$\| x_n^* \ - \ p_n x \|_n \to \ 0, \ n \in N'.$$

Consequently,

$$v^* \ \leq \ v(x) \ = \ lim \ v_n(x_n^*) \ = \ v_n^*, \ n \in N' \subset N$$

(a limit point x could not be optimal yet).

Prove the opposite inequality. Consider the difference $v_n(x_n^*) - v(x^*)$. By definition of the discrete convergence $\mathcal{P} - lim \ p_n x^* = x^*$, $n \in N$. Hence,

$$v_n(x_n^*) \ - \ v(x^*) \ \leq \ v_n(p_n x^*) \ - \ v(x^*) \ \leq \ \varepsilon$$

for a sufficiently small ε and for n sufficiently large. Consequently,

$$v^* \leq v(x) \leq liminf \; v_n(x_n^*) \leq limsup \; v_n(x_n^*) \leq v(x^*) = v^*.$$

Hence,

$$lim \; v_n^* = v^*, \; n \in N.$$

Discrete convergence of a subsequence $\{x_n^*\}$ of solutions of approximate problems (32) to a solution x^* of the initial problem (30) follows now from the observation that from sequence $\{x_n^*\}$, $n \in N'' = N/N'$, we can again separate converging (to some x) subsequence

$$v^* \leq v(x) \leq liminf \; v_n^* \leq limsup \; v_n^* \leq v^*, \; n \in N''' \subset N/N'.$$

Discrete convergence of a subsequence of solutions of approximate problems (32) to an optimal solution of the initial problem (30) follows now from the discrete analogue of the Radon –Riesz property. Theorem is proved.

The author expresses his gratitude to the referee for helpful remarks and suggestions.

References

1. J. Banaš, Integrable solutions of Hammerstein and Urysohn integral equations, J. Austral. Math. Soc. (Series A), 46(1989), 61 – 68.

2. J.W. Daniel, The Approximate Minimization of Functionals, Prentice-Hall, New Jersey, 1971.

3. M. Eisner and P. Olsen, Duality for stochastic programming interpreted as L.P. in L_p–spaces, SIAM J. Appl. Math., 28(1975), 779 – 792.

4. H. Esser, Zur Diskretisierung von Extremalproblemen, Lecture Notes in Math., 333(1973), 69 – 88.

5. S.J. Garstka and R.J.-B. Wets, On decision rules in stochastic programming, Math. Programming 7(1974), 117 – 143.

6. A.D. Ioffe and V.M. Tikhomirov, Theory of Extremal Problems, North-Holland, Amsterdam, 1979.

7. A. Kibzun and Y. Kan, Stochastic Programming Problems with Probability and Quantile Functions, Wiley, New York, 1995.

8. R. Lepp, Discrete approximation of extremal problems with operator inequality constraints, USSR Comput. Maths. Math. Phys., 30(1990), 127 – 132.

9. E. Raik, On stochastic programming problem with probability and quantile functionals, Proc. Acad. Sci. Estonian SSR. Phys. Math., 21(1972), 142 – 148.

10. R.T. Rockafellar and R. J.-B. Wets, Stochastic convex programming – Basic duality, Pacific J. Math., 62(1976), 173 – 195.

11. W. Römisch and R. Schultz, On distribution sensitivity in chance constrained programming, Advances in Mathematical Optimization, (J. Guddat et al., Eds.), Math. Res. 45, Akademie-Verlag, Berlin, 1988, 161 – 168.

12. G.M. Vainikko, On convergence of the method of mechanical curvatures for integral equations with discontinuous kernels, Sibirskii Math. Zh. 12(1971), 40 – 53.

13. V.V. Vasin, Discrete approximation and stability in extremal problems, USSR Comput. Maths. Math. Phys. 22(1982), 57 – 74.

Global Optimization of Probabilities by the Stochastic Branch and Bound Method

Vladimir Norkin

Glushkov Institute of Cybernetics, 252207 Kiev, Ukraine

Abstract. In this paper we extend the Stochastic Branch and Bound Method, developed in [7], [8] for stochastic integer and global optimization problems, to optimization problems with stochastic (expectation or chance) constraints. As examples we solve a problem of optimization of probabilities and a chance constrained programming problem with discrete decision variables.

Key words. Branch and bound method, stochastic global optimization, optimization of probabilities, chance constrained programming

1 Introduction

The paper deals with a stochastic global optimization problem. The problem functions have the form of mathematical expectations or probabilities and depend on continuous and discrete variables. These variables satisfy some, maybe nonlinear, constraints. To solve this problem we develop a certain version of the branch and bound method, which uses some specific stochastic bounds for optimal values of subproblems, generated by the method. As examples, we consider a problem of optimization of probabilities (from pollution control area) and a chance constrained programming problem.

Certainly, there is a broad literature on optimization of probabilities and chance constrained programming (see, for instance, [9], [2], [3], [5], [10], [6] and references therein), but most of these papers consider either conditions of (quasi)concavity of the probability functions or continuous local optimization problems. In the present paper we consider the problem of global optimization of probabilities.

In the paper we extend the Stochastic Branch and Bound method, developed in [7], [8], to problems with stochastic (in particular, chance) constraints.

Formally, the problem under consideration has the form of finding the global maximum

$$\max_x[F(x) = \mathbf{E}f(x,\theta)], \qquad (1)$$

subject to constraints

$$F_k(x) = \mathbf{E}f_k(x,\theta) \leq 0, \quad k = 1,\ldots,K; \qquad (2)$$

$$x \in X \cap D \subset R^n, \qquad (3)$$

where $f(x,\theta)$, $f_k(x,\theta)$ are some nonconvex (for instance, quasi-convex) functions, X is a finite set with a simple structure (for example, the intersection of some discrete lattice $\mathcal{L}$ and a parallelepiped $\mathcal{K}$ in R^n), the set $D = \{x \in R^n \mid G(x) \leq 0\}$ is given by some deterministic function $G : R^n \longrightarrow R^1$, $\mathbf{E}$ denotes the mathematical expectation with respect to a random variable θ, defined on some probability space $(\Theta, \Sigma, \mathbf{P})$.

If the original problem has continuous decision variable $x \in \mathcal{K}$, then we can make it discrete assuming that x belongs to some lattice $\mathcal{L} \in R^n$.

In this paper we are especially interested in the case where some of the functions $F_k(x)$ have the form of the probability:

$$F_k(x) = \mathbf{P}\{g_k(x,\theta) \in B\}, \qquad (4)$$

where $g_k : R^n \times \Theta \longrightarrow R^m$ is a random vector function, $\theta \in \Theta$ is a random variable, $B \subset R^m$. For instance, let

$$F_k(x) = \mathbf{P}\{g_k(x,\theta) \geq 0\} = \mathbf{E}\chi(g_k(x,\theta)),$$

where $g_k : R^n \times \Theta \longrightarrow R^1$, $\chi(t) = 1$ for $t \geq 0$ and $\chi(t) = 0$ for $t < 0$. Then for a given θ the function

$$f_k(\cdot,\theta) = \chi(g_k(\cdot,\theta))$$

is nonconvex and discontinuous, and thus $F_k(x)$ can be nonsmooth, noncvonvex or even discontinuous.

To solve problem (1) - (3) we develop a special version of the stochastic branch and bound method, which takes into account nonlinear stochastic constraints (2). Besides, special stochastic lower bounds of the optimal value of problem (1) - (3), based on the interchange of the minimization and the expectation (or the probability) operators, are constructed. Earlier, similar bounds were used in [7] and [8] for solving some stochastic discrete and continuous global optimization problems. In the present paper we deal with probability objective functions (4) and stochastic (chance) constraints (2).

2 Illustrative Example: Pollution Control

As a possible application we consider a pollution control problem (see, for example, [1], [4], [7]).

In the simplest pollution control problem there are emission sources $j = 1, \ldots, n$ and receptor points $i = 1, \ldots, m$. For every source j a set X_j of possible emission levels x_j is available. Each solution $x_j \in X_j$ has the cost $c_j(x_j)$. The emissions are transferred to receptors and produce depositions

$$y_i(x, \theta) = \sum_{i=1}^{m} t_{ij}(\theta) x_j,$$

where $x = (x_1, \ldots, x_m)$, $\{t_{ij}(\theta)\}$ are some random transfer coefficients, θ is a random variable (weather conditions). For simplicity, we can assume that θ takes on a finite number of values (scenarios). There are some target levels (ambient norms) q_i of depositions for the receptors $i = 1, \ldots, m$. We consider two decision-making problems. The first one is to minimize the probability of violating the ambient norms:

$$\max_{x} F(x) = \mathbf{P}\{\sum_{j=1}^{n} t_{ij}(\theta) x_j \leq q_i, \ i = 1, \ldots, m\}, \tag{5}$$

under the budget constraint

$$G(x) = \sum_{j=1}^{n} c_j(x_j) \leq r; \tag{6}$$

$$x_j \in X_j, \ j = 1, \ldots, n, \tag{7}$$

where r denotes the available resource.

Another problem is to minimize the cost function

$$G(x) = \sum_{j=1}^{n} c_j(x_j) \tag{8}$$

under the risk constraint

$$F(x) = \mathbf{P}\{\sum_{j=1}^{n} t_{ij}(\theta) x_j \leq q_i, \ i = 1, \ldots, m\} \geq \alpha; \tag{9}$$

$$x_j \in X_j, \ j = 1, \ldots, n, \tag{10}$$

where α is some reliability level, $0 < \alpha < 1$.

The discreteness of X_j in these examples is provided either by the artificial discretion of possible emission levels or by the fact that there is a finite number of possible emission reduction technologies available at source j.

Note that for these particular problems both the probability function $F(x)$ and the cost function $G(x)$ are monotonously decreasing in $x = (x_1, \ldots, x_n)$.

3 Assumptions

Consider a special case of problem (1)-(3):

$$\max_x \{F(x) \mid x \in X \cap D\}. \tag{11}$$

We assume that this problem has feasible solutions. Denote the set of optimal solutions X^* and the optimal value F^*.

In the branch and bound method the original problem (11) is subdivided into subproblems:

$$\max_x \{F(x) \mid x \in X' \cap D\}, \quad X' \subseteq X.$$

Denote $F^*(X' \cap D)$ the optimal value of this subproblem (by definition $F^*(\emptyset) = +\infty$).

We assume that there are some lower $L(X')$ and upper $U(X')$ bounds (set defined functions) for optimal values $F^*(X' \cap D)$, $X' \subseteq X$.

As a lower bound of $F^*(X' \cap D)$ we use the value

(i) $L(X') = F(s(X')) \leq F^*(X' \cap D)$ of the objective function at some point $s(X') \in X' \cap D$. By definition, $L(X') = +\infty$ if $X' \cap D = \emptyset$. Obviously,

(ii) if X' is a singleton then $L(X') = F(X')$.

Assume that the set defined function U has the following properties.

(iii) $U(X') \geq F^*(X' \cap D)$ if $X' \cap D \neq \emptyset$. By definition, $U(X') = +\infty$ if $X' \cap D = \emptyset$.

(iv) If X' is a singleton then $U(X') = F(X')$.

More over, we assume that bounds $L(X')$ and $U(X')$ are not known exactly and there are only some statistical estimates converging to them in the following limit sense.

(v) Assume that for each subset $X' \subseteq X$ there is a probability space $(\Omega_{X'}, \Sigma_{X'}, \mathbf{P}_{X'})$ and a collection of random variables (stochastic bounds) $\xi^k(X', \omega'), \eta^k(X', \omega'), k = 0, 1, \ldots$, defined on this space, such that

$$\lim_k \xi^k(X', \omega') = L(X') \quad \mathbf{P}_{X'} - \text{a.s.,} \tag{12}$$

$$\lim_k \eta^k(X', \omega') = U(X') \quad \mathbf{P}_{X'} - \text{a.s..} \tag{13}$$

For example, as stochastic lower bounds we can use the following empirical estimates:

$$\xi^k(X', \omega') = \frac{1}{k} \sum_{i=1}^{k} f(s(X'), \theta_i),$$

where θ_i are i.i.d. observations of θ.

Methods to construct stochastic upper bounds for different classes of stochastic optimization problems (with expectation objective functions) are developed in [7], [8]. In Section 5 we develop such bounds for probability functions.

4   The Stochastic Branch and Bound Method

In the next algorithm for solving (11) we assume that one can relatively easily find a feasible point $x' \in X' \cap D$ for any simple (partition) set $X' \subset X$ or find out that $X' \cap D = \emptyset$. For instance, such a feasible point x' can easily be found for parallelepipeds X', X and the monotonous constraint function $G(x)$.

First of all we define a new probability space $(\Omega, \Sigma, \mathbf{P})$ on which the algorithm works. Let us reserve some probability space $(\Omega_A, \Sigma_A, \mathbf{P}_A)$ for stochastic components of the algorithm itself. Define

$$(\Omega, \Sigma, \mathbf{P}) = (\Omega_A, \Sigma_A, \mathbf{P}_A) \times \prod_{X' \subseteq X} (\Omega_{X'}, \Sigma_{X'}, \mathbf{P})$$

as the product of the probability spaces $(\Omega_A, \Sigma_A, \mathbf{P}_A)$ and $(\Omega_{X'}, \Sigma_{X'}, \mathbf{P}_{X'})$, where X' runs over all subsets of X.

Now we consider that all quantities, appearing in the algorithm (random partitions $\mathcal{P}_k$, bounds ξ_k, η_k, sets X^k, Y^k, indices M_k and etc.), are defined on this common probability space $(\Omega, \Sigma, \mathbf{P})$, in particular, for ξ^k and η^k we have

$$\lim_k \xi^k(X', \omega) = L(X') \quad \mathbf{P} - \text{a.s.,} \tag{14}$$

$$\lim_k \eta^k(X',\omega) = U(X') \quad \mathbf{P} - \text{a.s.,} \tag{15}$$

for all $X' \subseteq X$.

For brevity we skip the argument ω from all random quantities generated by the following stochastic branch and bound algorithm.

Initialization. Form initial partition $\mathcal{P}_0 = \{X\}$. Calculate the bounds $\xi_0 = \xi^{M_0}(X)$ and $\eta_0 = \eta^{M_0}(X)$. Set iteration number $k = 1$.

Iterations $(k = 1, 2, \ldots)$. At the beginning of k-th iteration there is a partition $\mathcal{P}_{k-1}$ (a collection of subsets of X). For each set $X' \in \mathcal{P}_{k-1}$ there are estimates $\xi_k(X')$ and $\eta_k(X')$. Each iteration consists of the following steps.

Partitioning (branching). Select a record subset

$$X^k \in \text{Arg max}\{\eta_{k-1}(X') : X' \in \mathcal{P}_{k-1}\}$$

and an approximate solution

$$y^k \in Y^k \in \text{Arg max}\{\xi_{k-1}(X') : X' \in \mathcal{P}_{k-1}\}.$$

If the record subset X^k is a singleton, then put $\mathcal{P}'_k := \mathcal{P}_{k-1}$ and go to the *Bound Estimation* step. Otherwise, construct a partition $\mathcal{P}''_k(X^k) = \{X^k_i, \ i = 1, \ldots, n_k\}$ of the record set X^k such that $X^k = X^k_1 \cup \ldots \cup X^k_{n_k}$. Define the new full partition

$$\mathcal{P}'_k = (\mathcal{P}_{k-1} \setminus X^k) \cup \mathcal{P}''_k(X^k).$$

Bound Estimation. For all partition elements $X' \in \mathcal{P}'_k$ select stochastic estimates $\xi_k(X') = \xi^{M_k(X')}(X')$ of $L(X')$ and $\eta_k(X') = \eta^{M_k(X')}(X')$ of $U(X')$, where $M_k(X')$ is a random index.

Deletion. Clean partition $\mathcal{P}'_k$ of infeasible subsets, defining

$$\mathcal{P}_k = \mathcal{P}'_k \setminus \{X' \in \mathcal{P}'_k : X' \cap D = \emptyset\}.$$

The End of an Iteration. Set $k := k + 1$ and go to *Partitioning* step.

Remark to Deletion. If the estimates $\xi_k(X')$, $\eta_k(X')$ are a.s. exact, i.e. if besides (12), (13) we have $\xi_k(X') \leq L(X')$ and $\eta_k(X') \geq U(X')$ a.s., then at the *Deletion* step one can also delete the so-called nonperspective sets $X' \in \mathcal{P}'_k$ for which $\eta_k(X') < \max_{X' \in \mathcal{P}'_k} \xi_k(X')$. Other stochastic deletion rules are discussed in [8].

Theorem 4.1 *(Norkin et al. [7]) Suppose that $X \cap D \neq \emptyset$. Let indices $M_k(X', \omega)$ be chosen in such a way that*

$$\lim_k M_k(X', \omega) = +\infty \quad a.s.$$

for any fixed $X' \subseteq X$. Then with probability one there exists an iteration number $K(\omega)$ such that for all $k \geq K(\omega)$
the record sets $X^k(\omega)$ are singletons and $X^k(\omega) \subset X^*$;
the approximate solutions $y^k(\omega) \in X^*$;
$\lim_k \xi_{k-1}(Y^k(\omega)) = \lim_k \eta_{k-1}(X^k(\omega)) = F^*$.

5 Optimization of Probabilities

Consider the problems

$$\max_{x \in X' \cap D} [F(x) = \mathbf{P}\{f(x, \theta) \in B\}] = F^*(X' \cap D), \qquad (16)$$

where $X' \subseteq X$, X and D have the same meaning as in (3), $\theta \in \Theta$, $(\Theta, \Sigma, \mathbf{P})$ is some probability space, $f(x, \theta) = (f_1(x, \theta), \ldots, f_m(x, \theta))$ is a random vector function, B is a closed subset of R^m. In the next two subsections we give upper bounds (estimates) for $F^*(X' \cap D)$.

5.1 The Interchange Relaxation

Let us estimate $F^*(X' \cap D)$ from above by interchange of the maximization and the probability operators (*the interchange relaxation*). Obviously,

$$\begin{aligned}
F^*(X' \cap D) &\leq \mathbf{P}\{\exists x'(\theta) \in X' \cap D : f(x'(\theta), \theta) \in B\} = U(X') \\
&\leq \mathbf{P}\{\exists x'(\theta) \in \overline{X'} \cap D : f(x'(\theta), \theta) \in B\} = \overline{U}(X'),
\end{aligned}$$

where $\overline{X'}$ denotes the convex hull of X'.

Let us introduce indicator functions $\chi_{A(X')}(\theta)$, $\chi_{\overline{A}(X')}(\theta)$ of the events

$$A(X') = \{\theta \in \Theta | \exists x'(\theta) \in X' \cap D : f(x'(\theta), \theta) \in B\},$$

$$\overline{A}(X') = \{\theta \in \Theta | \exists x'(\theta) \in \overline{X'} \cap D : f(x'(\theta), \theta) \in B\},$$

i.e.

$$\chi_{A(X')}(\theta) = \begin{cases} 1, & \theta \in A(X'); \\ 0, & \text{otherwise.} \end{cases} \qquad \chi_{\overline{A}(X')}(\theta) = \begin{cases} 1, & \theta \in \overline{A}(X'); \\ 0, & \text{otherwise.} \end{cases}$$

Then

$$U(X') = \mathbf{E}\chi_{A(X')}(\theta), \quad \overline{U}(X') = \mathbf{E}\chi_{\overline{A}(X')}(\theta),$$

To calculate the stochastic estimate $\xi(X', \theta) = \chi_{A(X')}(\theta)$ of $U(X')$ one has to check for a given θ the feasibility of conditions $f(x', \theta) \in B$, $x' \in X' \cap D$.

To calculate the stochastic estimate $\overline{\xi}(X', \theta) = \chi_{\overline{A}(X')}(\theta)$ of $\overline{U}(X')$ one has to check for a given θ the feasibility of conditions $f(x', \theta) \in B$, $x' \in \overline{X'} \cap D$.

If functions $f_i(x, \theta)$, $i = 1, \ldots, m$, are linear in x, B is a polyhedral set in R^m, and D is given by linear constraints, then the problem of checking feasibility of conditions $f(x', \theta) \in B$, $x' \in X' \cap D$ is a linear integer programming problem (and linear problem for conditions $f(x', \theta) \in B$, $x' \in \overline{X'} \cap D$).

If the set Θ is finite and feasibility is checked easily and quickly, then $U(X')$ and $\overline{U}(X')$ can be calculated exactly. Otherwise we have to use some statistical (empirical) estimates for $U(X')$ and $\overline{U}(X')$, based on observations of θ and the calculation of $\chi_{A(X')}(\theta)$ or $\chi_{\overline{A}(X')}(\theta)$.

An important particular case of (16) is the problem of the optimization of the probability to exceed a given threshold:

$$\max_{x \in X' \cap D}[F(x) = \mathbf{P}\{f(x, \theta) \geq c\}],$$

where $f(x, \theta)$ is a random function of x, c is a given threshold. In this case

$$U(X') = \mathbf{E}\chi(\max_{x \in X' \cap D} f(x, \theta) - c) = \mathbf{P}\{\max_{x \in X' \cap D} f(x, \theta) \geq c\},$$

$$\overline{U}(X') = \mathbf{E}\chi(\max_{x \in \overline{X'} \cap D} f(x, \theta) - c) = \mathbf{P}\{\max_{x \in \overline{X'} \cap D} f(x, \theta) \geq c\},$$

where $\chi(t) = 0$ if $t < 0$ and $\chi(t) = 1$ otherwise, $\overline{X'}$ is a convex hull of X'. If the problem of maximizing $f(x, \theta)$ over $x \in X' \cap D$ (or over $x \in \overline{X'} \cap D$) is relatively simple (linear, convex or concave on a polyhedral set), then the above quantities $U(X')$, $\overline{U}(X')$ are practical estimates of $F^*(X' \cap D)$.

5.2   Multiple Sampling Bounds

One can make bounds of section 5.1 sharper by using multiple sampling of θ. Let $\theta^l = (\theta_1, \ldots, \theta_l)$ be l i.i.d. observations of θ. Then

$$[\max_{x \in X' \cap D} F(x)]^l = \max_{x \in X' \cap D} F^l(x)$$
$$= \max_{x \in X' \cap D} \mathbf{P}\{f(x, \theta_1) \in B\} \times \ldots \times \mathbf{P}\{f(x, \theta_l) \in B\}$$
$$\leq \mathbf{P}\{\exists x'(\theta^l) \in \overline{X'} \cap D : f(x'(\theta^l), \theta_1) \in B, \ldots, f(x'(\theta^l), \theta_l) \in B\},$$

where $\overline{X'}$ is a convex hull of X', $\mathbf{P}$ is the appropriate probability measure on Θ^l. Thus

$$\overline{U}_l(X') = \mathbf{P}^{1/l}\{\exists x'(\theta^l) \in \overline{X'} \cap D : f(x'(\theta^l), \theta_1) \in B, \ldots, f(x'(\theta^l), \theta_l) \in B\}$$

is an upper bound for the probabilities $F(x)$, $x \in X' \cap D$, satisfying condition (iv). Similar bounds for expectations and probabilities were used in [8].

The estimate $\overline{U}_l(X')$ is relatively easily calculated in the case of linear functions $f(x, \theta)$.

For the probability to exceed a given threshold, the above estimate has the form:

$$\overline{U}_l(X') = \mathbf{P}^{1/l}\{\max_{x \in \overline{X'} \cap D} \min_{1 \leq i \leq l} f(x, \theta_i) \geq c\}.$$

Advanced estimates $\overline{U}_l(X')$ are not worse then the simple estimate $\overline{U}_1(X')$. Indeed,

$$\overline{U}_l(X') \leq \mathbf{P}^{1/l}\{\exists x_1(\theta_1), \ldots, x_l(\theta_l) \in \overline{X'} \cap D :$$
$$f(x_1(\theta_1), \theta_1) \in B, \ldots, f(x_l(\theta_l), \theta_l) \in B\}$$
$$= \left(\mathbf{P}\{\exists x_1(\theta_1) \in \overline{X'} \cap D : f(x_1(\theta_1), \theta_1) \in B\} \times \ldots \right.$$
$$\left. \times \mathbf{P}\{\exists x_l(\theta_l) \in \overline{X'} \cap D : f(x_l(\theta_l), \theta_l) \in B\}\right)^{1/l}$$
$$= \mathbf{P}\{\exists x(\theta) \in \overline{X'} \cap D : f(x(\theta), \theta) \in B\} = \overline{U}_1(X').$$

6   The Constrained Stochastic Branch and Bound Method

Now we consider global optimization problem (1)-(3) with nonconvex stochastic constraints (2). Then we cannot easily point out feasible points for subproblems of this problem. Such points have to be elaborated by the algorithm. In this section we develop a version of the branch and bound method which explicitly treats nonconvex (stochastic) constraints of the problem.

6.1   Assumptions

Consider the problem

$$\max\{F(x)|\; x \in X \cap D \cap C\},$$

where X and D have the same meaning as in (3), the set C is given by the inequality constraint:

$$C = \{x \in R^n|\; H(x) \leq 0\}.$$

For instance, for constraints (2) we can put $H(x) = \max_{1 \leq k \leq K} F_k(x)$.

We assume that $X \cap D \cap C \neq \emptyset$. To check this one can apply the Stochastic Branch and Bound Method from Section 4 to the problem

$$\max_x \{-H(x)|\; x \in X \cap D\}.$$

As before, we assume that for any $X' \subseteq X$ one can easily find a point $x' \in X' \cap D$ or find out that $X' \cap D = \emptyset$. Function $H(x)$ is supposed to be essentially nonlinear and stochastic. Let X' be a (partition) subset of X. Consider auxiliary optimization problems

$$\max_{x \in X' \cap D} F(x), \quad \text{and} \quad \min_{x \in X'} H(x).$$

Denote $F^*(X' \cap D)$ and $H^*(X')$ optimal values of these problems, respectively. For some positive ϵ denote $C_\epsilon = \{x \in R^n|\; H(x) \leq \epsilon\}$.

The Branch and Bound method uses lower and upper bounds for $F^*(X' \cap D)$, $X' \subseteq X$. We assume that there are set defined functions U and L satisfying requirements **(i)-(v)** of Section 3.

The method also uses lower bounds for $H^*(X')$, $X' \subseteq X$. So, we assume that there exist a set defined function $l(X')$ such that:

(vi) $l(X') \leq H^*(X')$, for any $X' \subset X$;

(vii) for any singleton X' $l(X') = H(X')$;

(viii) for all subsets $X' \subseteq X$ there are sequences of random variables $\lambda^k(X', \omega)$, defined on some probability space $(\Omega_{X'}, \Sigma_{X'}, \mathbf{P}_{X'})$, such that

$$\lim_k \lambda^k(X', \omega) = l(X') \quad \text{a.s.} \tag{17}$$

6.2 The Algorithm

In the following algorithm we don't assume that for any partition subset $X' \subset X$ a feasible point $x' \in X' \cap D \cap C$ is known. ϵ-Feasible points $x' \in X' \cap D \cap C_\epsilon$ are elaborated by the algorithm. To do this, stochastic estimates of the lower bounds $l(X')$ of the optimal values of $H^*(X')$ are used. Thus this algorithm can be applied to problems with nonlinear and nonconvex constraints such as (2), (9). To provide convergence of the method we relax a little (by $\epsilon > 0$) the constraint $H(x) \leq 0$. As in Section 4 we consider that the method works on the common probability space $(\Omega, \Sigma, \mathbf{P})$. For brevity we skip the argument ω from all random quantities generated by the algorithm.

Initialization. Form initial partition $\mathcal{P}_0 = \{X\}$. Calculate the bounds $\xi_0(X) = \xi^{M_0}(X)$, $\eta_0(X) = \eta^{M_0}(X)$ and $\lambda_0(X) = \lambda^{M_0}(X)$, M_0 is some initial index. Set the iteration number $k = 1$.

Iterations $(k = 1, 2, \ldots)$. At the beginning of k-th iteration there is a partition $\mathcal{P}_{k-1}$ of the remaining part of X. For each set $X' \in \mathcal{P}_{k-1}$ there are estimates $\xi_{k-1}(X')$, $\eta_{k-1}(X')$ and $\lambda_{k-1}(X')$. Each iteration consists of the following steps.

Partitioning (branching). If $\mathcal{P}_{k-1}^\epsilon = \{X' \in \mathcal{P}_{k-1} : \lambda_{k-1}(X') \leq \epsilon\} \neq \emptyset$ then select a record subset

$$X^k \in \text{Arg max}\{\eta_{k-1}(X') : X' \in \mathcal{P}_{k-1}^\epsilon\},$$

and an approximate solution

$$y^k \in Y^k \in \text{Arg max}\{\xi_{k-1}(X') : X' \in \mathcal{P}_{k-1}^\epsilon\}.$$

If $\mathcal{P}_{k-1}^\epsilon = \emptyset$ or the record subset X^k is a singleton, then set $\mathcal{P}_k' := \mathcal{P}_{k-1}$ and go to the *Bound Estimation* step to improve bounds. Otherwise, construct a partition $\mathcal{P}_k''(X^k) = \{X_i^k, i = 1, \ldots, n_k\}$ of the record set X^k such that $X^k = X_1^k \cup \ldots \cup X_{n_k}^k$. Define a new full partition

$$\mathcal{P}_k' = (\mathcal{P}_{k-1} \setminus X^k) \cup \mathcal{P}_k''(X^k).$$

Bound Estimation. For all partition elements $X' \in \mathcal{P}_k'$ select stochastic estimates $\xi_k(X') = \xi^{M_k(X')}(X')$ of $L(X')$, $\eta_k(X') = \eta^{M_k(X')}(X')$ of $U(X')$ and $\lambda_k = \lambda^{M_k(X')}(X')$ of $l(X')$, where $M_k(X')$ is a random index.

Deletion. Clean partition $\mathcal{P}_k'$ of infeasible subsets, defining

$$\mathcal{P}_k = \mathcal{P}_k' \setminus \{X' \in \mathcal{P}_k' : X' \cap D = \emptyset\}.$$

The End of an Iteration. Set $k := k+1$ and go to *Partitioning* step.

Remark to Partitioning. The branching strategy to partition the record set X^k guarantees that X^k is a.s. a singleton for all sufficiently large k. Indeed, due to finiteness of X after a finite number of iterations the partition $\mathcal{P}_k$ becomes stable (unchanging) and contains a set X_0' such that $l(X_0') \leq 0$. Then by (17) for all sufficiently large k $\lambda_k(X_0') \leq \epsilon$ a.s. Hence record subsets X^k, Y^k are selected and are singletons in this stable partition. The other strategy is to partition the largest set from the collection

$$\{X' \in \mathcal{P}_{k-1} \mid \xi_{k-1}(X') \geq \xi_{k-1}(X^k), \lambda_{k-1}(X') \leq \epsilon\}.$$

Such a strategy guarantees that both record sets X^k and Y^k are singletons for all sufficiently large k.

Remark to Deletion. If estimates $\lambda_k(X')$ are a.s. exact, i.e. besides (17) we have $\lambda_k(X') \leq l(X')$ a.s., then one can safely delete from $\mathcal{P}_k'$ infeasible sets X' such that $\lambda_k(X') > \epsilon$, $\epsilon \geq 0$.

Next theorem states a convergence result for the method.

Theorem 6.1 *Suppose that $X \cap D \cap C \neq \emptyset$. Let indices $M_k(X', \omega)$ be chosen in such a way that for any fixed $X' \subseteq X$*

$$\lim_k M_k(X', \omega) = +\infty \quad a.s. \tag{18}$$

Then with probability one there exists an iteration number $K(\omega)$ such that for all $k \geq K(\omega)$
(a) the record sets $X^k(\omega)$ are singletons, $X^k(\omega) \subset X \cap D \cap C_\epsilon$ and

$$\max_{x \in X \cap D \cap C} F(x) \leq F(X^k(\omega)) \leq \max_{x \in X \cap D \cap C_\epsilon} F(x),$$

(b) if the partition strategy guarantees that Y^k becomes a singleton as $k \longrightarrow \infty$ (see Remark to Partitioning), then the approximate solutions $y^k \in X \cap D \cap C_\epsilon$ and

$$\max_{x \in X \cap D \cap C} F(x) \leq F(y^k(\omega)) \leq \max_{x \in X \cap D \cap C_\epsilon} F(x),$$

Proof. Due to the finiteness of X for each ω after some iteration $k = K_0(\omega)$ the current partition $\mathcal{P}_k(\omega)$ becomes stable (unchanging), i.e. $\mathcal{P}_k(\omega) = \mathcal{P}_\infty(\omega)$. Denote Ω' the set of those $\omega \in \Omega$ that (14), (15), (17) and (18) take place for all $X' \subseteq X$, $\mathbf{P}(\Omega') = 1$. For a fixed $\omega \in \Omega'$ and $X' \in \mathcal{P}_\infty$ we have $\lim_k M_k(X', \omega) = +\infty$ and hence $\lim_k \xi_k(X', \omega) = L(X')$, $\lim_k \eta_k(X', \omega) = U(X')$, $\lim_k \lambda_k(X', \omega) = l(X')$.

Now prove **(a)**. Fix some $\omega \in \Omega'$. Current partition always contains a set X_0^k such that $l(X_0^k) \leq 0$. So for $\omega \in \Omega'$ by (17) the set $\cup\{X' \in \mathcal{P}_{k-1}(\omega) : \lambda_k(X', \omega) \leq \epsilon\}$ is nonempty beginning from some $k = K_1(\omega) \geq K_0(\omega)$. Thus for $k \geq K_1(\omega)$ the record subset $X^k(\omega)$ is selected and this record subset is a singleton, $X^k(\omega) = x^k(\omega)$. Let $X^{k_m}(\omega) = x^\infty(\omega) \in \mathcal{P}_\infty(\omega)$ for some $k_m \geq K_1(\omega)$, $m = 1, 2, \ldots$. Obviously, $x^\infty(\omega) \in X \cap D$. Since $\lambda_k(X^k(\omega), \omega) \leq \epsilon$ for $k \geq K_1(\omega)$, then $\lambda_{k_m}(x^\infty(\omega), \omega) \leq \epsilon$ for $k_m \geq K_1(\omega)$ and $H(x^\infty(\omega)) \leq \epsilon$. Thus $x^\infty(\omega) \in X \cap D \cap C_\epsilon$ and

$$F(x^\infty(\omega)) \leq \max_{x \in X \cap D \cap C_\epsilon} F(x).$$

On the other hand, by construction, $\eta_{k-1}(X^k(\omega), \omega) \geq \eta_{k-1}(X', \omega)$ for all $X' \in \mathcal{P}_{k-1}^\epsilon(\omega)$, $k > K_1(\omega)$, then

$$\begin{aligned} F(x^\infty(\omega)) &= \lim_m \eta_{k_m}(X^{k_m}(\omega), \omega) \geq \lim_m \eta_{k_m}(X', \omega) \\ &= U(X') \geq \max_{x \in X' \cap D} F(x), \end{aligned}$$

for $X' \in \mathcal{P}_\infty(\omega)$ such that $l(X') \leq \epsilon$. Since

$$X \cap D \cap C \subseteq \cup\{X' \in \mathcal{P}_\infty(\omega) : l(X') \leq \epsilon\},$$

then

$$\begin{aligned} F(x^\infty(\omega)) &\geq \max_{X'}\{\max_{x \in X' \cap D} F(x) : X' \in \mathcal{P}_\infty(\omega), l(X') \leq \epsilon\} \\ &\geq \max_{x \in X \cap D \cap C} F(x). \end{aligned}$$

Now prove the statement **(b)** of the theorem. For $\omega \in \Omega'$ beginning from $k > K_1(\omega)$ the set $\mathcal{P}_{k-1}^\epsilon(\omega)$ is nonempty. So the approximate solutions $Y^k(\omega)$ are selected and

$$\lambda_k(Y^k(\omega), \omega) \leq \epsilon. \tag{19}$$

By construction of the algorithm

$$\xi_{k-1}(Y^k(\omega), \omega) \geq \xi_{k-1}(X^k(\omega), \omega), \tag{20}$$

and by **(a)**

$$\liminf_k \xi_k(X^k(\omega), \omega) \geq \max_{x \in X \cap D \cap C} F(x). \tag{21}$$

Let for some $k_m \geq K_1(\omega)$ $Y^{k_m}(\omega)$ be a singleton and $Y^{k_m}(\omega) = y^{\infty}(\omega) \in \mathcal{P}_{\infty}(\omega)$, $m = 1, 2, \ldots$. Obviously, $y^{\infty}(\omega) \in X \cap D$. Then by conditions (vii), (viii), (21) we obtain

$$H(y^{\infty}(\omega)) = l(y^{\infty}(\omega)) = \lim_m \lambda_{k_m}(Y^{k_m}(\omega), \omega) \leq \epsilon$$

and thus $F(y^{\infty}(\omega)) \leq \max_{x \in X \cap D \cap C_{\epsilon}} F(x)$. By (ii), (v), (20) and statement **(a)** we obtain

$$
\begin{aligned}
F(y^{\infty}(\omega)) = L(y^{\infty}(\omega)) &= \liminf_m \xi_{k_m}(Y^{k_m}(\omega), \omega) \\
&\geq \liminf_m \xi_{k_m}(X^{k_m}(\omega), \omega) \\
&\geq \max_{x \in X \cap D \cap C} F(x). \square
\end{aligned}
$$

7 Numerical Experiments

In this section we present the results of some numerical experiments with pollution control problems from Section 2.

7.1 Optimization of Probabilities Subject to the Budget Constraint

The following problem of dimensions $n = 5$, 10 was solved by the exhaustive enumeration and by the Stochastic Branch and Bound Method:

$$\max_x [F(x, q) = \mathbf{P}_{\theta}\{\sum_{j=1}^{n} t_j(\theta) x_j \leq q\}]$$

subject to

$$G(x) = \sum_{j=1}^{5} c_j(x_j) \leq r, \quad (r = 4.0, \ 8.0),$$

$$x_j \in [0, 1], \quad j = 1, \ldots, n,$$

where

$$c_j(x_j) = \max_{1 \leq k \leq 10} (b_{jk} - a_{jk} x_j), \quad j = 1, \ldots, n,$$

the random variable θ takes on integer values $1, 2, \ldots, 10$ with probabilities p_{θ}, probabilities p_{θ} firstly are randomly chosen from [0,1] then are normalized to satisfy $\sum_{\theta=1}^{10} p_{\theta} = 1$; coefficients $t_j(\theta)$, a_{jk} b_{jk} are randomly chosen from the interval $[0, 1]$, $q = 0.5$.

The Stochastic Branch and Bound Method of Section 4 was implemented by means of Microsoft (R) FORTRAN Optimizing Compiler

5.00. The computations were carried out on IBM PC AT 386 DX40 (with i387 co-processor).

Note that due to the finite number of scenarios θ one can calculate values of $F(x)$ and bounds L, U exactly. For the subset

$$X' = \{x \in R^n \mid c \leq x \leq d \text{ (componentwise)}\}, \quad c, d \in R^n,$$

we take

$$U(X') = \begin{cases} F(c, q), & \text{if } G(d) \leq r, \\ -\infty, & \text{otherwise;} \end{cases}$$

$$L(X') = \begin{cases} F((1 - t')c + t'd, q), & \text{if } G(d) \leq r, \\ -\infty, & \text{otherwise,} \end{cases}$$

where t' is the maximal t such that $G((1 - t)c + td) \leq r$.

Firstly, this problem was solved by the exhaustive examination of 0.1-step lattices (for $n = 5$, $r = 4.0$) and 0.2-step lattice (for $n = 10$, $r = 8.0$). It took 1 min 23.43 sec and 3 hr 58 min 35.89 sec, respectively.

Then the problem (with $n = 5$, $r = 4.0$) was solved by the Stochastic Branch and Bound Method over the 0.1-step and 0.01-step lattices. It took 5.49 sec (167 branching iterations, maximum 52 subproblems in the list) for the 0.1-lattice and 45.80 sec (1279 branching iterations, maximum 389 subproblems in the list) for the 0.01-lattice.

In one more experiment the problem (with $n = 10$, $r = 8.0$) was solved by the Branch and Bound Method over 0.2-step lattice. It took 3 min 27.23 sec (3641 branching iterations, maximum 633 subproblems in the list) that is approximately 68 times faster than the exhaustive enumeration.

7.2 Cost Minimization Subject to the Reliability Constraint

The following problem was solved over 0.1-step lattice by the Stochastic Branch and Bound Algorithm of section 6.2:

$$\min_x G(x),$$

subject to

$$F(x, 2.5) \geq 0.999,$$

$$x_j \in [0, 1], \quad j = 1, \ldots, 10,$$

where $F(x, q)$ and $G(x)$ are the same as in section 7.1.

To achieve accuracy 1% (in function) it took 53 sec (1144 branching iterations, 569 subproblems in the queue).

References

[1] Amann M., Klaassen G. and Schöpp W. (1993), Closing the gap between the 1990 deposition and the critical sulfur deposition values, Tech. report, Int. Inst. for Appl. Syst. Anal., 2361 Laxenburg, Austria.

[2] Kall P. and S.W.Wallace (1994), *Stochastic programming*, John Wiley & Sons.

[3] Kibzun A.I. and Yu.S.Kan (1996), *Stochastic Programming Problems with Probability and Quantile Functions*, J.Willey & Sons.

[4] Marti K. (1993), Satisficing Techniques in Stochastic Linear Programming, Working paper, Universität der Bundeswehr, Munchen/Neubiberg.

[5] Marti K. (1995), Differentiation of Probability Functions: The Transformation Method, *Computers Math. Applic.*, Vol. 30, No. 3-6, pp. 361-382.

[6] Norkin V.I. (1993), The analysis and optimization of probability functions, Working Paper WP-93-6, Int. Inst. for Appl. Syst. Anal., Laxenburg, Austria.

[7] Norkin V.I., Ermoliev Yu.M. and Ruszczyński A. (1994), On Optimal Allocation of Indivisibles under Uncertainty, Working Paper WP-94-021, Int. Inst. for Appl. Syst. Anal., Laxenburg, Austria (to appear in *Operations Research*).

[8] Norkin V.I., Pflug G.Ch. and Ruszczyński A. (1996), A Branch and Bound Method for Stochastic Global Optimization, Working Paper WP-96-065, Int. Inst. for Appl. Syst. Anal., Laxenburg, Austria (to appear in *Mathematical Programming*).

[9] Prékopa A. (1995), *Stochastic Programming*, Kluwer Academic Publishers.

[10] Uryas'ev St. (1989), A differentiation formula for integrals over set given by inclusion, *Numer. Funct. Anal. and Optimiz.*, Vol. 10, pp. 827-841.

Robust Stability of Interval Matrices: a Stochastic Approach

B. T. Polyak

Institute for Control Science, Moscow, Russia, e-mail: boris@ipu.rssi.ru

Abstract. The problem of checking robust stability of interval matrices has been proved to be NP-hard. However a closely related problem can be effectively solved in the framework of the stochastic approach [1]. Moreover the deterministic interval robust stability radius happens to be very conservative for large dimensions from the probabilistic point of view.

Keywords. Robustness, stability, interval matrices, random matrices, probabilistic approach.

1 Introduction

The first attempts to check robust stability of interval matrices (i.e. to check that all matrices of the form $\underline{A} \leq A \leq \overline{A}$ are Hurwitz; the inequalities are understood in component-wise sense) have been performed just after the publication of the famous Kharitonov theorem on robust stability of interval polynomials. These attempts happened to be unsuccessful – the equivalent of the Kharitonov theorem does not hold for interval matrices [2]. Moreover, checking robust stability of edges and other low dimensional faces of the box of entries can not guarantee stability of the entire family of matrices [3]. The final point in this series of "negative" results was the proof of NP-hardness of the problem [4]. Meanwhile there exist numerous sufficient conditions for robust stability of interval matrices, most of them being very conservative. The references as well as recent computational results for low-dimensional matrices can be found in [5].

Our aim is to demonstrate that passing from the deterministic to the stochastic point of view can change the situation drastically. The general idea of such approach to robustness can be found in [1]; here we demonstrate the tecniques of the approach applied to robust stability of interval matrices. It is based on two fundametal results. The first one relates to probability theory — it is Geman's theorem on the norm of random matrices [6]. The second one is the famous recent formula for computation of the real stability radius due to Qiu a.o. [7].

In Section 2 we provide Geman's result and its nonasymptotic extension. Section 3 contains the main formula for calculation of stochastic interval stability radius. The examples are given in Section 4, they demonstrate that the stochastic approach allows to extend significantly the stability margin if compared with the standard deterministic characteristics. We discuss the results and the directions for future research in Section 5.

2 Norms of random matrices

Let $a_{ij}, i = 1, \ldots, j = 1, \ldots$ be independent identically distributed (i.i.d.) real random variables with

$$\mathbf{E}a = 0, \mathbf{E}a^2 = \sigma^2, \mathbf{E}a^4 < \infty,$$

here a stands for any a_{ij}. Consider a random matrix A_n with entries $a_{ij}, i = 1, \ldots, n, j = 1, \ldots, n$. Denote

$$v_n = (1/\sigma\sqrt{n})\|A_n\|,$$

where the norm is understood as operator one: $\|A\| = \sup_{\|x\|=1}\|Ax\|$, $\|x\|^2 = \sum_{i=1}^n x_i^2$.

Fact 1. *For $n \to \infty$ the norms of random matrices converge to 2 with probability 1:*

$$v_n \to 2 \quad a.s.$$

This result is mainly due to Geman [6]; the superfluous condition on the existence of all moments of entries in [6] has been relaxed in [8].

The behavior of the eigenvalues of random matrices is also known.

Fact 2 [9], [10]. *The distribution of the eigenvalues of the matrix $B_n = (1/\sigma\sqrt{n})A_n$ tends to uniform on the unit disk as $n \to \infty$.*

Now suppose that the entries of A_n are uniformly distributed on $[-\gamma, \gamma]$ (denote it as $a \in R[-\gamma, \gamma]$). Then $\sigma = \gamma/\sqrt{3}$. For this case Monte-Carlo simulation provides the estimates for the rate of convergence in Geman's theorem:

$$\mathbf{E}v_n \approx 2 - 1.3n^{-0.7}, \quad \mathbf{Var}(v_n) \approx 0.22n^{-1.12}.$$

Moreover the following nonasymptotic result is valid.

Proposition 1. *The probability for the norm of a random matrix to exceed the given level is*

$$P(v_n > 2.1) < 0.01,$$

$$P(v_n > 2.2) < 0.002,$$

uniformly over all n.

3 Robust stability of interval matrices

Assume M is a real $n \times n$ stable matrix, that is all its eigenvalues have negative real parts. Its *real stability radius* is defined as

$$r_R(M) = \inf\{\gamma : M + A \text{ is unstable}, \|A\| \le \gamma\}.$$

The algorithm for computation of $r_R(M)$ has been proposed in [7]. *The interval stability radius* of M is

$$r_I(M) = \inf\{\gamma : M + A \text{ is unstable}, |a_{ij}| \leq \gamma, \forall i, j\}.$$

It is not hard to get the estimate $n r_I(M) \leq r_R(M)$, which is tight for some matrices. However in general this estimate is too conservative.

Now let us introduce *the stochastic stability radius* $r_S(M)$. Fix some confidence level $0 < \alpha < 1$, e.g. $\alpha = 0.99$. Then

$$r_S(M) = \inf\{\gamma : P\{M + A \text{ is unstable}\} \geq 1 - \alpha, a_{ij} \in R[-\gamma, \gamma]\}.$$

Here it is assumed that a_{ij} are i.i.d. entries of A. We choose the uniform distribution for them; there is a strong reasoning that such distribution is in a sense *the worst* one [11]. The main result of the paper is the immediate corollary of Proposition 1 and the above definition.

Proposition 2. *For confidence level 0.99 and all n*

$$r_S(M) \geq \rho_1 = (c_1/\sqrt{n}) r_R(M), \quad c_1 = \sqrt{3}/2.1 = 0.825\ldots,$$

and for confidence level 0.998 and all n

$$r_S(M) \geq \rho_2 = (c_2/\sqrt{n}) r_R(M), \quad c_2 = \sqrt{3}/2.2 = 0.787\ldots.$$

Thus to calculate the estimate for the stochastic stability radius it suffices to find the real stability radius of a matrix via the algorithm from [7]. After calculation of ρ_1, ρ_2 we can guarantee that if the perturbations of all entries of M do not exceed $\rho_1(\rho_2)$ then with probability not less than 0.99 (0.998) the stability is preserved.

4 Examples

Example 1. Consider the negative unit matrix as $M : M = -I$. Then both stability radii can be computed explicitly: $r_R(M) = 1, r_I(M) = 1/n$, where n is the dimension of M. For $n = 40$ the Proposition 2 yields $r_S(M) \geq \rho_1 = 0.13, r_I(M) = 0.025$. Moreover $r_S(M)$ can be estimated more precisely for this example. Fact 2 provides that the eigenvalues of $M + A = -I + A$ (where A is a random matrix with entries uniformly distributed on $[-\gamma, \gamma]$) are approximately uniformly distributed in the disk centered at -1 and having the radius $\gamma\sqrt{40/3}$. Hence real parts of the eigenvalues are negative with large probability iff $\gamma \leq \sqrt{3/40} = 0.274$ and $r_S(M) \approx 0.274$. The direct Monte-Carlo simulation provides $r_S(M) = 0.25$ with confidence level 0.997. Thus the true interval stability radius is approximately 10 times conservative versus its stochastic counterpart! In other words, if we are ready to tolerate instability with probability 0.003, then we can extend the margin for interval perturbations 10 times for this example.

Example 2. The case of Metzlerian matrices (i.e. the matrices M with entries $m_{ii} < 0, m_{ij} \geq 0, i \neq j$) gives the opportunity to calculate stability radii explicitly [12], [13],[14]. Consider the example:

$$n = 40, m_{ii} = -60, m_{ij} = 2, i < j, m_{ij} = 1, i > j, \quad i, j = 1, \ldots 40.$$

The eigenvalues of M can be calculated analytically:

$$\lambda_k = -60 + (\tau \epsilon_k - 2)/(1 - \tau \epsilon_k)$$

where $\tau = \sqrt[40]{2}, \epsilon_k$ are values of $\sqrt[40]{-1}$. Hence $\max \Re(\lambda_k) = -3.79$ and M is stable. The formula for real stability radius of Metzlerian matrices is equivalent to $r_R(M) = \sigma_1(M) = 3.535$, $\sigma_1(M)$ being the least singular value of M. For interval Metzlerian matrices the violation of stability is met first when all values of perturbations are positive and have the largest values, i.e. a matrix A with entries $|a_{ij}| \leq \gamma$ can be destabilizing if $a_{ij} \equiv \gamma$. This yields $r_I(M) = 0.0913$ for M as above. We get $\rho_1 = 0.484$, and Monte-Carlo calculation gives $r_S(M) = 2$ with confidence level 0.995. Again the ratio $r_S(M)/r_I(M)$ is large, demonstrating the conservativeness of the deterministic approach.

Example 3. A Hurwitz polynomial of degree n with stable zeros was randomly generated, then a corresponding companion form matrix B was constructed and it was taken $M = C^{-1}BC$, where C was a random nonsingular matrix. The calculations via the algorithm of [7] provided $r_R(M)$, and the formulae of Proposition 2 were used to calculate ρ_1, ρ_2. Then random matrices A with i.i.d. entries uniformly distributed on $[-\rho, \rho]$ were generated and the stability of the perturbed matrices $M + A$ was checked. The calculations confirmed that the probability of instability was generally less than 0.001 even for $\rho = \rho_1$. Incidentally it was found that serious computational problems arise for the algorithm of [7] for moderate dimensions ($n \approx 20$).

5 Discussion

The results of the computation confirm the fact that the stochastic approach allows to extend significantly the deterministic interval stability radius.

The proposed formula provides a lower bound for $r_S(M)$; the true value of $r_S(M)$ may be much greater (compare the results in the examples 1,2). It is explainable: we guarantee (with the given confidence level) that the norm of perturbation does not exceed the real stability radius, meanwhile if the norm is larger, it does not imply instability necesserely. Thus one of the directions for future research is to develop a new version of stochastic approach, not relying on the formula for the real stability radius, which provides tighter bounds for the stochastic stability radius.

Another limitation of the proposed tecnique is the assumption on equality of ranges of perturbations (the bounding γ is the same for all entries of A). It is an open problem to get rid of this assumption.

6   Acknowledgements

The discussions with Bob Barmish and Arkadii Nemirovskii were very helpful. Olga Panchenko performed some numerical computations. The work was supported in part by NSF Grant ECS – 941 8709 and by Russian Foundation for Fundamental Research, Grant 96 – 01 – 0093.

References

[1] B.R.Barmish, B.T.Polyak, A new approach to open robustness problems based on probabilistic prediction formulae, *Proceedings of the 13th IFAC World Congress*, San-Francisco,CA,July 1996, Vol. **H**, p.1-6.

[2] B.R.Barmish, C.V.Hollot, Counter-example to a recent result on the stability of interval matrices by S.Bialas, *Intern. Journ. Contr.*, **39**, p.1103-1104, 1984.

[3] J.D.Cobb, C.L.DeMarco, The minimal dimension of stable faces to guarantee stability of a matrix polytope, *IEEE Trans.Autom.Contr.*, **34**, p.1990-1992, 1989.

[4] A.S.Nemirovskii, Several NP-hard problems arising in robust stability analysis, *Math. Contr.,Sign.,Syst.*, **6**, 1, p.99-105, 1994.

[5] L.H.Keel, S.P.Bhattacharyya, Robust stability of interval matrices: a computational approach, *Intern. Journ. Contr.*, **62**, 6, p.1491-1506, 1995.

[6] A.Geman, A limit theorem for the norm of random matrices, *Ann. Prob.*, **8**, 2, p.252-261, 1980.

[7] L.Qiu, B.Bernhardsson, A.Rantzer, E.J.Davison, P.M.Young, J.C.Doyle, A formula for computation of the real stability radius, *Automatica*, **31**, 6, p.878-890, 1995.

[8] Y.Q.Yin, Z.D.Bai, P.R.Krishnaiah, On the limit of the largest eigenvalue of the large dimensional sample covariance matrix, *Prob. Theory and Related Fields*, **78**, 4, p.509-521, 1988.

[9] V.L.Girko, The circular law, *Prob. Theory and Appl.*, **29**, 4, p.669-679, 1984.

[10] V.L.Girko, The Circular Low: ten years later, *Random Oper. and Stoch. Equ.*, **2**, 3, p.235-276, 1994.

[11] B.R.Barmish, C.M.Lagoa, The uniform distribution: a rigorous justification for its use in robustness analysis, *Proceedings of the 35th CDC*, Kobe, Japan, December 1996.

[12] B.Shafai, M.Kothandaraman, J.Chen, Real and complex stability radii for nonnegative and Metzlerian systems, *Proceedings of the 32nd CDC*, San-Antonio, TX, p.3482-3484, 1993.

[13] S.P.Bhattacharyya, H.Chapellat, L.H.Keel, *Robust Control: the Parametric Approach*, Prentice Hall, 1995.

[14] N.K.Son, D.Hinrichsen, Robust stability of positive continuous time systems, *Numer. Funct. Anal. and Optimiz.*, **17**, 5/6, p.649-659, 1996.

A Note on Preprocessing via Fourier-Motzkin Elimination in Two-Stage Stochastic Programming

Rüdiger Schultz

Konrad-Zuse-Zentrum für Informationstechnik Berlin
Takustraße 7
D-14195 Berlin, Germany

Abstract: Preprocessing in two-stage stochastic programming is considered from the viewpoint of Fourier-Motzkin elimination. Although of exponential complexity in general, Fourier-Motzkin elimination is shown to provide valuable insights into specific topics such as solving integer recourse stochastic programs or verifying stability conditions. Test runs with the computer code PORTA [5] are reported.

Key Words: Stochastic programming, preprocessing, Fourier-Motzkin elimination

1 Introduction

Two-stage stochastic programs arise as deterministic equivalents to random optimization problems that are characterized by a two-stage scheme of alternate decision and observation. First, a here-and-now decision has to be taken without knowing the outcomes of random problem data. After realization of the random data, a second-stage (recourse) decision is possible, which is the solution of a subordinate optimization problem depending on the first-stage solution and the outcome of the random data. In this paper, we consider problems where the second-stage is a linear program with possibly integer requirements on the variables. In two-stage stochastic programming one encounters an interplay of algebraic and probabilistic difficulties. Preprocessing in stochastic programming is directed to analyzing the underlying algebraic structures. This may be helpful for supporting solution procedures or improving structural understanding of the problem.

In this note, we consider stochastic programs of the following form

$$\min\{c^T x + Q(x) \ : \ x \in C\} \tag{1}$$

where

$$Q(x) = \int_{\mathcal{R}^s} \Phi(z - Ax)\mu(dz) \tag{2}$$

and

$$\Phi(t) = \min\{q^T y \ : \ Wy \geq t, \ y \in Y\} \quad \text{with} \quad Y = \mathcal{R}_+^m \quad \text{or} \quad Y = \mathcal{Z}_+^m. \tag{3}$$

Here, $C \subset \mathcal{R}^n$ is a non-empty polyhedron and c, q, W are vectors and matrices of proper dimensions. The above mentioned scheme of alternate decision and observation is reflected by the first-stage variables x, the random vector z with underlying probability measure μ and the second-stage variables y. The two-stage stochastic program aims at finding a first-stage decision x such that the sum of direct costs $c^T x$ and expected recourse costs $Q(x)$ is minimal. For further reading on basics in two-stage stochastic programming we refer to [8, 9]. Let us asssume that $Y = \mathcal{R}_+^m$ and impose the following assumptions

the matrix W has full rank, $\tag{4}$

$$M_D = \{u \in \mathcal{R}_+^s \ : \ W^T u \leq q\} \neq \emptyset, \tag{5}$$

$$\int_{\mathcal{R}^s} \|z\| \mu(dz) < +\infty. \tag{6}$$

These assumptions imply that M_D has vertices, and according to the decomposition theorem for polyhedra it admits a representation

$$M_D = conv(V_0) + cone(V_1) \tag{7}$$

where V_0, V_1 are finite sets of vectors in $\mathcal{R}^s$ and $conv$ and $cone$ denote the convex and conical hulls, respectively. Preprocessing, as discussed in this note, concerns the algorithmic transformation of the representation in (5) into the one of (7). To see that explicit knowledge of V_0, V_1 can be beneficial, recall the following identities

$$cone(V_1) = \{u \in \mathcal{R}_+^s \ : \ W^T u \leq 0\}$$
$$\text{(since } cone(V_1) \text{ is the recession cone of } M_D\text{)}$$
$$= \{u \in \mathcal{R}^s \ : \ u^T w_i \leq 0, i = 1, \ldots, m, \ u^T(-e_j) \leq 0, j = 1, \ldots, s\}$$
$$(w_i \text{ are the columns of } W \text{ and } e_j \text{ the canonical unit vectors})$$
$$= \{u \in \mathcal{R}^s \ : \ u^T(\sum_i \lambda_i w_i + \sum_j \lambda_j(-e_j)) \leq 0 \ \forall \lambda_i \geq 0, \forall \lambda_j \geq 0\}$$
$$= \{u \in \mathcal{R}^s \ : \ u^T w \leq 0 \ \forall w \in pos(W, -I)\}$$
$$= (pos(W, -I))^*$$

where pos denotes the positive span and $*$ indicates the polar cone. Therefore, the elements of V_1 are the coefficients in an inequality description of $pos(W, -I)$. In computations, the probability measure μ is in general assumed to be discrete, i. e., with mass points $z_1, \ldots, z_L$ and probabilities $p_1, \ldots, p_L$. Then, $Q(x)$ is well defined if $z_l - Ax \in pos(W, -I)$ for all $l = 1, \ldots, L$. In the literature, the latter are called induced constraints. These are implicit conditions that are made explicit if V_1 is known. Moreover, for $t \in pos(W, -I)$ it holds that

$$\Phi(t) = \max_{v_i \in V_0} v_i^T t. \tag{8}$$

such that computation of Φ becomes easy if V_0 is known, and the domain of definition of Φ becomes explicit if V_1 is known.

Nevertheless, the above considerations have only limited impact on solution procedures for state-of-the-art linear recourse problems (i.e., if $Y = \mathcal{R}_+^m$), since far too many elements arise in V_0 and V_1. Methods like L-shaped, regularized or stochastic decomposition [17, 11, 7], for instance, generate only those elements of V_0 and V_1 that are relevant for the solution process. On the other hand, the above transformation may be useful to support solution procedures for smaller problems of more complicated nature (e.g., if $Y = \mathcal{Z}_+^m$) or for answering questions in the theory of stochastic programming (e.g., verifying stability conditions). The emphasis of our paper is on these two specific issues. A more general view on preprocessing including its impact on modeling is adopted in [18, 19] and Chapter 5 in [8].

In Section 2, we recall the role of Fourier-Motzkin elimination when transforming an inequality description of a polyhedron into the representation as Minkowski sum of a convex and a conical hull. In Section 3, we show how preprocessing via Fourier-Motzkin elimination enters into an algorithm for integer recourse stochastic programs. Section 4 deals with the verification of stability conditions for stochastic programs with linear recourse. Here Fourier-Motzkin elimination can be beneficial in generating information about the polyhedral complex of lineality regions of the second-stage value function Φ. Finally, we have a conclusions section.

2   Theoretical Background

In this section we recall the essence of Fourier-Motzkin elimination and put it into the context of computing extreme points and extreme rays of the polyhedron M_D (cf. (5)). Enumeration of extreme points and extreme rays of a polyhedron is a well studied problem in the literature (see, e.g., [21] and the references therein). In connection with stochastic programming, an excellent account is given in [18]. Our intention here is to show the relation to Fourier-Motzkin elimination, which is interesting from the practical computations point of view, since there exists a freely available implementation ([5]) of the transformation procedure described below. This supplements the implementations reported in [18].

For a polyhedron $\Pi = \{x \in \mathcal{R}^n : Ax \leq b\}$, Fourier - Motzkin elimination provides an algorithmic way for projection along the $k-$th coordinate ($1 \leq k \leq n$), i.e., for eliminating the variable x_k from the system $Ax \leq b$. Let

$a_i, b_i (i = 1, \ldots, l)$ be the rows of A and components of b, respectively, and let a_{ik} denote the k-th component of a_i. We introduce subsets $I_>, I_<, I_=$ of $\{1, \ldots, l\}$ such that

$$
\begin{aligned}
a_{ik} &> 0 &&\text{for all } i \in I_>, \\
a_{ik} &< 0 &&\text{for all } i \in I_<, \\
a_{ik} &= 0 &&\text{for all } i \in I_=.
\end{aligned}
$$

Simple manipulations then provide for any $x \in \Pi$

$$
x_k \leq \frac{1}{a_{ik}}(a_{ik}x_k - a_i^T x + b_i) \quad \text{for all } i \in I_>
$$

and

$$
x_k \geq \frac{1}{-a_{ik}}(-a_{ik}x_k + a_i^T x - b_i) \quad \text{for all } i \in I_<.
$$

The right-hand sides in these inequalities are independent of x_k such that we obtain for the projection $\Pi_{(k)}$ of Π along the k-th coordinate

$$
\Pi_{(k)} = \Big\{ \ x_{(k)} \in \mathcal{R}^{(n-1)} :
$$

$$
\max_{i \in I_>}\{\frac{1}{-a_{ik}}(-a_{ik}x_k + a_i^T x - b_i)\} \ \leq \ \min_{i \in I_<}\{\frac{1}{a_{ik}}(a_{ik}x_k - a_i^T x + b_i)\},
$$

$$
a_i^T x \leq b_i \text{ for all } i \in I_= \Big\}
$$

The first inequality in the above description of $\Pi_{(k)}$ can be equivalently expressed by $|I_>| \cdot |I_<|$ many linear inequalities, where $|.|$ denotes cardinality. When projecting further down to smaller dimensions, the above scheme has to be iterated, which in general produces inequality systems of enormous size. This prevents algorithmic use of the method for large-scale problems. Strategies going back to Tschernikow ([16], see also [6]) allow to produce a description of $\Pi_{(k)}$ without redundant inequalities such that in practical computations intermediate inequality systems can be kept as small as possible.

Proposition 1. *Let $M_D = \{u \in \mathcal{R}_+^s \ : \ W^T u \leq q\}$ be such that $0 \in M_D$, then a representation $M_D = conv(V_0) + cone(V_1)$ can be computed by using the above elimination procedure.*

Proof. We consider the polar polyhedron M_D^* of M_D, which is given as follows

$$
M_D^* = \{\xi \in \mathcal{R}^s : \xi^T u \leq 1 \ \forall u \in M_D\}.
$$

Then it holds (cf. [12], Theorem 9.1) that M_D^* is a polyhedron again and that $M_D^{**} = M_D$. Let M_D be written as

$$
M_D = \{u \in \mathcal{R}_+^s \ : \ (W, -I)^T u \leq \begin{pmatrix} q \\ 0 \end{pmatrix}\}
$$

where W is scaled in such a way that $q_i \in \{0, 1\}$ for all components q_i of

q, $(i = 1, \ldots, m)$. The latter is possible since $0 \in M_D$ implies $q \geq 0$. Another standard result on polarity (cf. again [12], Theorem 9.1) now yields that

$$M_D^* = conv\Big(\{0\} \cup \bigcup_{q_i=1} \{w_i\}\Big) + cone\Big(\bigcup_{q_i=0} \{w_i\} \cup \bigcup_{j=1}^{m} \{-e_j\}\Big)$$

where w_i and e_j are as in Section 1. Therefore

$$M_D^* = \Big\{u \in \mathcal{R}^s : \quad \exists \lambda \in \mathcal{R}_+^{|q_i=1|} \; \exists \mu^1 \in \mathcal{R}_+^{|q_i=0|} \; \exists \mu^2 \in \mathcal{R}_+^m$$

$$u = \sum_i \lambda_i w_i + \sum_k \mu_k^1 w_k + \sum_j \mu_j^2(-e_j),$$

$$\sum_i \lambda_i = 1\Big\}.$$

Hence

$$M_D^* = \mathcal{M}_{D(\lambda,\mu^1,\mu^2)}^*$$

where

$$\mathcal{M}_D^* = \Big\{(u,\lambda,\mu^1,\mu^2) : \quad u = \sum_i \lambda_i w_i + \sum_k \mu_k^1 w_k + \sum_j \mu_j^2(-e_j),$$

$$\sum_i \lambda_i = 1, \lambda \geq 0, \mu^1 \geq 0, \mu^2 \geq 0\Big\} \qquad (9)$$

and $\mathcal{M}_{D(\lambda,\mu^1,\mu^2)}^*$ denotes the projection of $\mathcal{M}_D^*$ along (λ,μ^1,μ^2). Eliminating the variables λ,μ^1,μ^2 from the above system by the Fourier-Motzkin procedure yields an inequality description

$$M_D^* = \{u \in \mathcal{R}^s : H_0 u \leq h, \; H_1 u = 0\}$$

which has no redundant rows if we apply the above mentioned Tschernikow rules. Since also $0 \in M_D^*$, we may assume that, after proper scaling, $h_i \in \{0,1\}$ for all components h_i of h. Now, $M_D = M_D^{**}$, and again Theorem 9.1 in [12] implies

$$M_D = (M_D^*)^* = conv\Big(\{0\} \cup \bigcup_{h_i=1} \{h_{0i}\}\Big)$$

$$+cone\Big(\bigcup_{h_i=0} \{h_{0i}\} \cup \bigcup_i \{h_{1i}\} \cup \bigcup_i \{-h_{1i}\}\Big)$$

where h_{0i}, h_{1i} are the rows of H_0, H_1. This is the desired representation and our proof is complete.

Recall that, due to our basic assumptions (4), (5), the polyhedron M_D has vertices. Therefore H_1 in the above proof has to be the zero matrix.

The main algorithmic step in the above proof, the elimination of the variables λ,μ^1,μ^2 from the system in (9), is implemented in the code PORTA [5]. As input, the user has to supply an inequality description of the relevant polyhedron, in our situation $(W, -I)^T u \leq \binom{q}{0}$, $u \geq 0$. The output of PORTA contains the list of vertices and extreme rays, in our situation the row vectors of H_0.

In [18] the authors report on numerical experience with the algorithm *support* ([20]) that, although different in appearance, follows similar principles as the Fourier-Motzkin procedure. Our intention with the above proposition is to point out the relation to Fourier-Motzkin elimination and to give an impression on the key procedure implemented in PORTA.

The bottleneck of Fourier-Motzkin elimination, however, is that, in general, the size of the iteration system of linear inequalities grows quadratically per elimination of one variable. Moreover, Fourier-Motzkin elimination is an all-or-nothing procedure. The complete list of vertices and extreme rays is generated only in the very last step. If the method breaks down because the iteration system of linear inequalities becomes too big, then no partial list of vertices and extreme rays is available.

3 Lower Bounds for Integer Recourse Problems

The present section deals with two-stage stochastic programs where the second stage problem is an integer linear program. The basic model is again given by (1) - (3) but now $Y = \mathcal{Z}_+^m$. We also assume that all entries in W are rational numbers, and (3) reads

$$\Phi(t) = \min\{q^T y \ : \ Wy \geq t, \ y \in \mathcal{Z}_+^m\}. \tag{10}$$

This value function Φ is in general non-convex and discontinuous, in fact, lower semicontinuous, and these properties of Φ, obviously, are transferred to Q. Therefore, algorithms for linear recourse problems with continuous variables, essentially resting on convexity of Q, break down for this class of problems. If the underlying probability measure μ is discrete, which we assume also in the present section, then (1) - (3) may be equivalently rewritten as a large-scale mixed-integer linear program with dual block angular structure. Tackling this problem by primal decomposition leads to master problems whose objectives are essentially governed by Φ, and, again, we are facing lower semicontinuous objectives ([2]).

In [15], an algorithm for the above integer recourse stochastic program is proposed that combines enumeration of Q with an efficient procedure for computing its function values. The latter employs Gröbner bases methods from computational algebra: Using Buchberger's algorithm, a Gröbner basis of a polynomial ideal related to the integer program in (10) is computed. This basis only depends on the objective and the coefficient matrix of the integer program. For the various right-hand sides, solution of the integer programs then is accomplished by a scheme of generalized division of multivariate polynomials. The latter is much faster than solving anew the integer program with conventional methods each time another right-hand side arises. Computing the Gröbner basis, however, is the bottleneck of the method such that second-stage problems with moderate size can be handled only. For details of Gröbner bases theory and its application to integer programming we refer to [1, 3, 4].

As to the enumeration of Q, bounds restricting the search are most welcome. Here, preprocessing does an important job which we will explain now. Consider the continuous relaxation

$$\Phi_R(t) = \min\{q^T y \; : \; Wy \geq t, \; y \in \mathcal{R}_+^m\}$$

and the corresponding relaxed expected recourse function

$$Q_R(x) = \int_{\mathcal{R}^s} \Phi_R(z - Ax)\mu(dz).$$

Of course, $Q_R(x) \leq Q(x)$, and therefore any optimal solution to the integer recourse stochastic program belongs to the level set

$$\{x \in C \; : \; c^T x + Q_R(x) \leq c^T \bar{x} + Q(\bar{x})\} \tag{11}$$

where $\bar{x} \in C$ is an arbitrary feasible point. The enumeration part of the algorithm in [15] rests on searching the above level sets. Each time a feasible point $\bar{x}$ with improved objective function value is found the level set (11) can be shrunk.

In view of the representation (8), the function Q_R is convex piecewise-linear if the measure μ is discrete. Therefore, all the level sets in (11) are non-empty polyhedra. If we assume that $pos(W, -I) = \mathcal{R}^s$ then $\Phi_R(t) = \max_{v_i \in V_0} v_i^T t$ where V_0 is the vertex set of M_D. This leads to the following lower bound for Q_R

$$
\begin{aligned}
Q_R(x) &= \sum_{l=1}^{L} p_l \Phi_R(z_l - Ax) \\
&\geq \Phi_R(\sum_{l=1}^{L} p_l z_l - Ax) \\
&= \max_{v_i \in V_0} v_i^T (\sum_{l=1}^{L} p_l z_l - Ax) =: Q_{RL}(x).
\end{aligned}
$$

Here, the second line is a consequence of Jensen's inequality. If we replace Q_R in (11) by Q_{RL} then again any optimal solution to the integer recourse stochastic program belongs to the respective level set. The advantage over the previous situation is that Q_{RL} is explicitly known via the vertices of M_D. If the latter are obtained by the procedure described in the previous section, then the enumeration part of the algorithm in [15] becomes algorithmically feasible. The disadvantage that Fourier-Motzkin elimination breaks down at large-scale instances is not of great significance in this case, since the above mentioned bottleneck in Gröbner basis computation restricts application of the algorithm in [15] to second-stage problems of moderate size.

As an example let us consider the following model that is used as a test example in [15]

$$\max\{c^T x + Q(x) \; : \; x \in C\} \tag{12}$$

where

$$Q(x) = \int_{\mathcal{R}^s} \Phi(z - Ax)\mu(dz) \tag{13}$$

and

$$\Phi(t) = \max\{q^T y \; : \; Wy \le t, \; y \in \{0,1\}^m\} \tag{14}$$

with non-negative components in c, q and A, W.

Although purely academical, the above model can be given the following interpretation as a two-stage knapsack problem with random budget where decision variables (boolean and continuous ones) correspond to investments. The first-stage investment decision x in (12) - (14) is selected from some feasible set C and yields an immediate revenue $c^T x$. Further revenue is gained from projects for which investment is done in the second stage after having observed the random vector $z \in \mathcal{R}^s$ leading to the budget $z - Ax$. Spending money in the first stage decreases possibilities in the second stage. However, negative entries in x may be permitted leading to the possibility to contract loans in the first stage to enlarge possibilities in the second stage. The objective in (12) - (14) is to find a first-stage investment decision x such that the sum of direct revenue from the first stage and expected revenue from the second stage is maximal.

Computational experience with solving (12) - (14) is reported in [15]. Here, we concentrate on the preprocessing part, i.e., on finding a representation (7) for the dual polyhedron

$$M_D = \{u \in \mathcal{R}_+^{s+m} : (W^T, I)u \ge q\}. \tag{15}$$

Although the assumption that $pos(W, I) = \mathcal{R}^s$ is not met here, such that Φ is not defined on the whole of $\mathcal{R}^s$, vertices of M_D, nevertheless, can be used in the above way to bound enumeration.

Our experience with PORTA on a SPARCstation 20 Model 61 with 160 MB of main memory indicates that, within seconds, vertex sets with up to several hundreds of elements can be enumerated. If there are several hundred thousands of vertices, then there is a pretty high chance that PORTA breaks down due to excessive size of the iteration system. Vertex sets with up to one hundred thousand elements have a good chance to be enumerable, although this might cost several hours of CPU time. To illustrate these statements, we tested PORTA on some instances of the polyhedron (15). For a matrix W with 2 rows and 29 columns the complete list of 241 vertices was found after 2 seconds, for a 2×50 matrix W we ended up with 545 vertices after 14 seconds and for a 4×50 matrix W it took 5550 seconds to find all the 37887 vertices.

4 Verification of Stability Conditions

The purpose of this section is to demonstrate how Fourier-Motzkin elimination can be used to verify assumptions in theoretical considerations on stochastic programs.

In recent years, studies on the stability of the stochastic program (1) - (3) with respect to perturbations of the underlying measure μ have attracted some interest. This is mainly motivated by the incomplete information on μ that one often faces in applications and by numerical problems that arise in computations of the integral in (2) if μ is multivariate continuous. A crucial issue in this analysis is that sufficient stability conditions are verifiable from the data in the unperturbed problem, i.e., in our situation, from (1) - (3) with some fixed measure μ. In the following, we illustrate at a result on the stability of optimal solution sets how Fourier-Motzkin elimination can be employed to extend verifiability of stability conditions.

Let $Y = \mathcal{R}_+^m$ and consider problem (1) - (3) as a parametric program in μ

$$P(\mu) \quad \min\{c^T x + Q_\mu(x) \ : \ x \in C\} \tag{16}$$

where

$$Q_\mu(x) = \int_{\mathcal{R}^s} \Phi(z - Ax)\mu(dz) \tag{17}$$

and

$$\Phi(t) = \min\{q^T y \ : \ Wy = t, \ y \in \mathcal{R}_+^m\}. \tag{18}$$

(The only reason for writing the second-stage linear program in equality form is consistency with the settings in [10], [13].)

The following proposition was established in [10]. It provides a Lipschitz estimate for the Hausdorff distance of optimal solution sets $\psi(\mu), \psi(\nu)$ to stochastic programs $P(\mu)$ and $P(\nu)$, respectively. The estimate is in terms of some distance of probability measures $d(\mu, \nu; U)$ that we will not explain here. For details we refer to [10] where it is also shown that $d(\mu, \nu; U)$ can be majorized by the uniform distance of distribution functions of probability measures closely related to μ and ν. The function $\tilde{Q}_\mu$ arising in the statement is defined by $\tilde{Q}_\mu(\chi) = \int_{\mathcal{R}^s} \Phi(z - \chi)\mu(dz)$.

Proposition 2. Let pos $W = \mathcal{R}^s, M_D \neq \emptyset$ and $\int_{\mathcal{R}^s} \|z\|\mu(dz) < +\infty$. Suppose further that $\psi(\mu)$ is non-empty and bounded. Assume that there exists a convex open subset V of $\mathcal{R}^s$ such that $A(\psi(\mu)) \subset V$ and the function $\tilde{Q}_\mu$ is strongly convex on V. Let $U = \text{cl}\, U_o$, where U_o is an open, convex, bounded

set such that $\psi(\mu) \subset U_o$ and $A(U) \subset V$.
Then there exist constants $L > 0$, $\delta > 0$ such that

$$d_H(\psi(\mu), \psi(\nu)) \leq L \cdot d(\mu, \nu; U)$$

for all probability measures ν such that $\int_{\mathcal{R}^s} \|z\| \nu(dz) < +\infty$ and $d(\mu, \nu; U) < \delta$.

As to verification of assumptions, the critical part of the above statement is the strong convexity of $\tilde{Q}_\mu$, which means that there exists some $\kappa > 0$ such that for all $\chi, \chi' \in V$ and all $\lambda \in [0, 1]$

$$\tilde{Q}_\mu(\lambda \chi + (1 - \lambda)\chi') \leq \lambda \tilde{Q}_\mu(\chi) + (1 - \lambda)\tilde{Q}_\mu(\chi') - \kappa\lambda(1 - \lambda)\|\chi - \chi'\|^2.$$

Strong convexity of $\tilde{Q}_\mu$ can be verified via the following result from [13].

Proposition 3. *Let pos $W = \mathcal{R}^s$, the interior of M_D be non-empty and $\int_{\mathcal{R}^s} \|z\| \mu(dz) < +\infty$. Suppose further that there exist a convex open set $V \subset \mathcal{R}^s$, constants $r > 0, \rho > 0$ as well as a density θ of μ such that $\theta(t') \geq r$ for all $t' \in V_\rho := \{t' \in \mathcal{R}^s : dist\,(t', V) \leq \rho\}$. Then $\tilde{Q}_\mu$ is strongly convex on V.*

The density assumption in the above proposition restricts application of the result to stochastic programs (16) - (18) where all components of z are random, i.e., not constant almost surely. This is quite restrictive in applications, and [14] analyses models where only a part of $z \in \mathcal{R}^s$, say $z_1 \in \mathcal{R}^{s_1}$, is random, and the remaining part $z_2 \in \mathcal{R}^{s_2}$ $(s_1 + s_2 = s)$ is fixed. Then Q_μ becomes

$$Q_\mu(x) = \int_{\mathcal{R}^{s_1}} \Phi^o(z_1 - A_1 x, z_2 - A_2 x)\mu(dz_1)$$

with

$$\Phi^o(t_1, t_2) = \min\{q^T y \ : \ W_1 y = t_1, W_2 y = t_2 \ y \in \mathcal{R}_+^m\}. \tag{19}$$

Consider

$$\tilde{Q}_\mu^o(\chi_1, \chi_2) = \int_{\mathcal{R}^{s_1}} \Phi^o(z_1 - \chi_1, z_2 - \chi_2)\mu(dz_1)$$

which, for fixed $\bar{\chi}_2$, is studied in [14] as a function of the first argument χ_1

$$\tilde{Q}_\mu(\chi_1) = \int_{\mathcal{R}^{s_1}} \Phi^o(z_1 - \chi_1, z_2 - \bar{\chi}_2)\mu(dz_1). \tag{20}$$

The key result in [14] is a sufficient condition for strong convexity of the above function $\tilde{Q}_\mu$. Again investigation of strong convexity is motivated by stability considerations, now for the more general situation of partially random right-hand side.

To be able to state the mentioned result from [14] some preparation is needed. The dual polyhedron M_D^o (cf.(5)) belonging to the linear program in (19) reads

$$M_D^o = \{(u_1, u_2) \in \mathcal{R}^{s_1+s_2} \ : \ W_1^T u_1 + W_2^T u_2 \leq q\}.$$

We will impose an assumption on $\pi_1 M_D^o$, where π_1 denotes the projection from $\mathcal{R}^s$ to $\mathcal{R}^{s_1}$. Furthermore, the function Φ^o from (19) induces a piecewise linear function $\Phi(t_1) = \Phi^o(t_1, z_2 - \bar\chi_2)$. The lineality regions of Φ form a polyhedral complex in $\mathcal{R}^{s_1}$. We will consider all members $\mathcal{F}_{ij}^*$ of this complex that arise as unbounded facets, i.e. unbounded members of dimension $\tilde s_1 - 1$.

The counterpart to Proposition 3 for the generalized function $\tilde Q_\mu$ from (20) then reads ([14], Theorem 3.3):

Proposition 4. *Assume that the matrix $\binom{W_1}{W_2}$ has full rank and that for each $t_1 \in \mathcal{R}^{s_1}$ there exists a $y \in \mathcal{R}_+^m$ such that $W_1 y = t_1$ and $W_2 y = z_2 - \bar\chi_2$. Let the interior of $\pi_1 M_D^o$ be non-empty and $\int_{\mathcal{R}^{s_1}} \|z_1\| \mu(dz_1) < +\infty$. Suppose that there exist a convex open set $V \subset \mathcal{R}^{s_1}$, constants $r > 0, \rho > 0$, points $e_{ij}^* \in \mathcal{F}_{ij}^*$ and a density θ of μ such that $\theta(t') \geq r$ for all $t' \in \cup_{(i,j)}\{e_{ij}^* + V_\rho\}$. Then $\tilde Q_\mu$ is strongly convex on V.*

Concerning verification the above generalized density assumption is non-trivial. It says that in each unbounded facet $\mathcal{F}_{ij}^*$ we have to find some point e_{ij}^* around which the density is uniformly bounded below by a positive constant. In Proposition 3 this is less involved due to the simplicity of the polyhedral complex of lineality regions. There, lineality regions of Φ coincide with the outer normal cones to the dual polyhedron M_D, which is compact in this case such that the polyhedral complex is a fan of cones with common vertex zero. Therefore, the points e_{ij}^* can all be selected as zero and the condition $t' \in \cup_{(i,j)}\{e_{ij}^* + V_\rho\}$ turns into $t' \in V_\rho$.

To see how the complex looks like in the more general situation we recall that, by duality,

$$\Phi(t_1) = \max\{t_1^T u_1 + \bar t_2^T u_2 \;:\; (u_1, u_2) \in M_D^o\}. \tag{21}$$

where $\bar t_2 = z_2 - \bar\chi_2$. Suppose that M_D^o has vertices, and let $\tilde d_1, \ldots, \tilde d_N$ be the vertices of M_D^o that arise as optimal ones in (21) when t_1 varies in $\mathcal{R}^{s_1}$. Denoting by $\tilde d_{i1}, \tilde d_{i2}$ $(i = 1, \ldots, N)$ the projections of $\tilde d_i$ on $\mathcal{R}^{s_1}$ and $\mathcal{R}^{s_2}$, respectively, we obtain

$$\Phi(t_1) = t_1^T \tilde d_{i1} + \bar t_2^T \tilde d_{i2}$$

for all $t_1 \in \mathcal{R}^{s_1}$ such that $(t_1, \bar t_2)$ belongs to the outer normal cone $\mathcal{K}_i$ to M_D^o at the vertex $\tilde d_i$. The lineality regions $\mathcal{K}_i^*$ of Φ then are given by

$$\mathcal{K}_i^* = \pi_1(\mathcal{K}_i \cap \{\mathcal{R}^{s_1} \times \{\bar t_2\}\}), \quad i \in \{1, \ldots, N\}. \tag{22}$$

They form a polyhedral complex that arises by intersecting the fan of outer normal cones to M_D^o with the affine subspace $\mathcal{R}^{s_1} \times \{\bar t_2\}$. The sets $\mathcal{F}_{ij}^*$ are the unbounded facets in this complex.

Some insight into the polyhedral complex in (22) can be gained by using PORTA [5]. Let us illustrate this at the following example treated in more detail in [14].

Example 1. In (19), (20) we put

$$q^T = (21, 21, 21, 21, 7, 7, 3, 3, 1, 0) \in \mathcal{R}^{10},$$

$$W_1 = \begin{pmatrix} -3 & -3 & -3 & 3 & -1 & 1 & 0 & 0 & 0 & 0 \\ -5 & 1 & 2 & -2 & -2 & 2 & -1 & 1 & 0 & 0 \end{pmatrix} \in L(\mathcal{R}^{10}, \mathcal{R}^2),$$

$$W_2 = \begin{pmatrix} 12 & 12 & 9 & 9 & 3 & 3 & 0 & 0 & 1 & -1 \end{pmatrix} \in L(\mathcal{R}^{10}, \mathcal{R}^1),$$

$$z_2 = 1, \quad \bar{\chi}_2 = 0.$$

Using PORTA [5], we computed the vertices of M_D^o together with a vertex-inequality incidence table displaying the binding inequalities for each vertex. Gradients of the binding (linear) inequalities then generate the respective outer normal cones:

$$\tilde{d}_1 = (-7, 0, 0),$$
$$K_1 = cone\ \{(0, 0, -1), (-1, -2, 3), (-3, 2, 9), (-3, 1, 12), (-3, -5, 12)\},$$

$$\tilde{d}_2 = (-1, -3, 0),$$
$$K_2 = cone\ \{(0, 0, -1), (0, -1, 0), (-1, -2, 3)\},$$

$$\tilde{d}_3 = (5, -3, 0),$$
$$K_3 = cone\ \{(0, 0, -1), (0, -1, 0), (3, -2, 9)\},$$

$$\tilde{d}_4 = (7, 0, 0),$$
$$K_4 = cone\ \{(0, 0, -1), (1, 2, 3), (3, -2, 9)\},$$

$$\tilde{d}_5 = (1, 3, 0),$$
$$K_5 = cone\ \{(0, 0, -1), (1, 2, 3), (0, 1, 0)\},$$

$$\tilde{d}_6 = (-5, 3, 0),$$
$$K_6 = cone\ \{(0, 0, -1), (0, 1, 0), (-3, 2, 9)\},$$

$$\tilde{d}_7 = (-3, 0, 1),$$
$$K_7 = cone\ \{(0, 0, 1), (-3, 1, 12), (-3, -5, 12)\},$$

$$\tilde{d}_8 = (2, -3, 1),$$
$$K_8 = cone\ \{(0, 0, 1), (0, -1, 0), (-1, -2, 3), (3, -2, 9), (-3, -5, 12)\},$$

$$\tilde{d}_9 = (4, 0, 1),$$
$$K_9 = cone\ \{(0, 0, 1), (1, 2, 3), (3, -2, 9)\},$$

$$\tilde{d}_{10} = (-2, 3, 1),$$
$$K_{10} = cone\ \{(0, 0, 1), (1, 2, 3), (0, 1, 0), (-3, 2, 9), (-3, 1, 12)\}.$$

One possibility to fulfill the density assumption from Proposition 4 is to select e_{ij}^* as vertices of unbounded facets $\mathcal{F}_{ij}^*$. To avoid further algorithmic effort for distinguishing between bounded and unbounded members of the lineality complex one could include *all* vertices of the complex into the list of points e_{ij}^*. Then the above generators can be helpful since they allow to compute all vertices of the complex (22). Indeed, each intersection of a positive multiple of some generator with the affine subspace $\mathcal{R}^2 \times \{1\}$ yields a vertex and each vertex has to arise as such an intersection. Hence, there is a one-to-one correspondence between the vertices and the generators with positive third component. In this way, we obtain a list of 7 vertices of which 3 do not belong to unbounded facets. For more complicated examples, the latter extraction may be non-trivial. In such cases the generalized density assumption can be fulfilled by using all vertices of (22) instead of the points e_{ij}^*.

Another possibility to fulfill the generalized density assumption is to claim that $\theta(t') \geq r$ for all $t' \in \mathcal{B} + V_\rho$ where $\mathcal{B}$ is a bounded set containing all the vertices from (22). To this end, we compute an upper bound for the norm of these vertices. Again preprocessing is helpful. The outer normal cone $\mathcal{K}_i$ to M_D^o at $\tilde{d}_i$ can be written as

$$\mathcal{K}_i = \{u \in \mathcal{R}^s : u = W(i)v, v \geq 0\}$$

where $W(i) \in L(\mathcal{R}^{m_i}, \mathcal{R}^s)$ is given by the generators of $\mathcal{K}_i$ computed above. Then it holds that

$$\begin{aligned} \mathcal{K}_i^* &= \{u_1 \in \mathcal{R}^{s_1} : u_1 = W(i)_1 v, \, \bar{t}_2 = W(i)_2 v, \, v \geq 0\} \\ &= W(i)_1(\{v \in \mathcal{R}^{m_i} : W(i)_2 v = \bar{t}_2, \, v \geq 0\}). \end{aligned}$$

hence, the vertices of $\mathcal{K}_i^*$ are among the $W(i)_1$−images of the vertices $v(i)_j$ $(j = 1, \ldots, J_i)$ of $\{v \in \mathcal{R}^{m_i} : W(i)_2 v = \bar{t}_2, \, v \geq 0\}$. The latter can be computed using PORTA [5]. For the desired upper bound all possible basis submatrices of all $W(i)_2$ have to be extracted which, together with $\bar{t}_2$ and submatrices of the $W(i)_1$, yields representations for the vertices of $\mathcal{K}_i^*$. These are bounded above by the usual estimates using matrix norms (see [14] for details).

It is evident that the procedures discussed in this section are not suitable for large-scale problems. However, they can serve well for problems with moderate size.

5   Conclusions

Fourier-Motzkin elimination provides an elegant way to look at preprocessing in stochastic programming. Moreover there is a well tested code, PORTA [5], that is based on Fourier-Motzkin elimination, such that, in addition to the codes reported in [18], [19], another convenient computer tool for preprocessing in stochastic programming is available. It is well known that Fourier-Motzkin elimination is of exponential complexity such that application of the method has to be restricted to problems of moderate size. In the present paper we discussed two applications with natural size limitation. For stochastic programs with integer recourse, preprocessing is helpful for restricting the search in enumeration algorithms. In the stability analysis of stochastic programs, Fourier-Motzkin elimination can be used to widen the class of problems for which sufficient stability conditions can be verified.

Acknowledgement: I wish to thank Thomas Christof (Universität Heidelberg) and Andreas Löbel (Konrad-Zuse-Zentrum für Informationstechnik Berlin) as well as Stein W. Wallace (Norwegian Institute of Technology Trondheim) and Roger J-B Wets (University of California, Davis) for beneficial discussions. Further thanks are due to an anonymous referee for constructive criticism.

References

[1] Adams, W.W., Loustaunau, P.: An Introduction to Gröbner Bases, American Mathematical Society, Graduate Studies in Mathematics, Volume 3, Providence, 1994.

[2] Carøe, C.C., Tind, J.: L-shaped decomposition of two-stage stochastic programs with integer recourse, Technical Report, Institute of Mathematics, University of Copenhagen, 1995.

[3] Conti, P., Traverso,C.: Buchberger algorithm and integer programming, in: Proceedings AAECC-9 (New Orleans), Lecture Notes in Computer Science 539, Springer-Verlag, Berlin, 1991, 130–139.

[4] Cox, D., Little, J., O'Shea, D.: Ideals, Varieties and Algorithms, Springer-Verlag, New York, 1992.

[5] Christof, T.: PORTA – A Polyhedron Representation Transformation Algorithm, version 1.2.1., 1994, written by T. Christof (Univ. Heidelberg), revised by A. Löbel and M. Stoer, available from the ZIB electronic library eLib by anonymous ftp from elib.zib.de, in the directory /pub/mathprog/polyth.

[6] Duffin, R.J.: On Fourier's analysis of linear inequality systems, Math. Programming Study 1, American Elsevier Publishing Company, New York, 1974.

[7] Higle, J.L., Sen, S.: Stochastic decomposition: an algorithm for two-stage linear programs with recourse, Mathematics of Operations Research 16 (1991) 650–669.

[8] Kall, P., Wallace, S.W.: Stochastic Programming, Wiley, Chichester, 1994.

[9] Prékopa, A.: Stochastic Programming, Kluwer Academic Publishers, Dordrecht, 1995.

[10] Römisch, W., Schultz, R.: Lipschitz stability for stochastic programs with complete recourse, SIAM Journal on Optimization 6 (1996) 531–547.

[11] Ruszczyński, A,: A regularized decomposition method for minimizing a sum of polyhedral functions, Mathematical Programming 35 (1986) 309–333.

[12] Schrijver, A.: Theory of Linear and Integer Programming, Wiley, Chichester, 1986.

[13] Schultz, R.: Strong convexity in stochastic programs with complete recourse, Journal of Computational and Applied Mathematics 56 (1994) 3–22.

[14] Schultz, R.: Strong convexity in stochastic programs with complete recourse II: partially random right-hand side, Konrad-Zuse-Zentrum für Informationstechnik Berlin, Preprint SC 95-21, 1995.

[15] Schultz, R., Stougie, L., van der Vlerk, M.H.: Solving stochastic programs with integer recourse by enumeration: a framework using Gröbner basis reductions, Mathematical Programming, (to appear).

[16] Tschernikow, S.N.: Lineare Ungleichungen, Deutscher Verlag der Wissenschaften, Berlin, 1971.

[17] van Slyke, R., Wets, R.J-B: L-shaped linear programs with applications to optimal control and stochastic programming, SIAM Journal of Applied Mathematics 17 (1969), 638–663.

[18] Wallace, S.W., Wets, R.J-B: Preprocessing in stochastic programming: the case of linear programs, ORSA Journal on Computing 4 (1992) 45–59.

[19] Wallace, S.W., Wets, R.J-B: Preprocessing in stochastic programming: the case of capacitated networks, ORSA Journal on Computing 7 (1995) 44–62.

[20] Wets, R.J-B: Elementary, constructive proofs of the Theorem of Farkas, Minkowski and Weyl, in: J. Gabszewicz, J.F. Richard, L.A. Wolsey (eds.) "Economic Decision Making: Games, Econometrics and Optimization: Contributions in Honour of Jacques Drèze, North Holland, Elsevier Science, Amsterdam, 1990, 427–432.

[21] Ziegler, G.M.: Lectures on Polytopes, Springer - Verlag, New York, 1995.

Bounds for the Reliability of k–out–of–connected–(r,s)–from–(m,n):F Lattice Systems [1]

T. Szántai

Technical University of Budapest, Department of Mathematics

Abstract. In the paper new bounds for the reliability of k–out–of–connected–(r,s)–from–(m,n):F lattice systems are given. These bounds are based on the Boole–Bonferroni bounding techniques and one of them on the Hunter–Worsley bound. All of these bounds need the calculation of the first few binomial moments of a random variable introduced according to the special reliability system. The main results of the paper are formulae for calculation of these binomial moments.

Keywords. Reliability system, Boole–Bonferroni bounds, Binomial moments.

1 Introduction

The k–out–of–connected–(r,s)–from–(m,n):F lattice systems are straightforward generalizations of the linear consecutive k–out–of–r–from–n:F failing systems. The later system was introduced by S. KOUNIAS and M. SFAKIANAKIS ([5],[6]) and was investigated by M.V. KOUTRAS and S.G. PAPASTAVRIDIS ([7]). Boole–Bonferroni type bounds were applied in this context first by M. SFAKIANAKIS, S. KOUNIAS and A. HILLARIS ([11]) and some sharper bounds of this type were given later by A. HABIB and T. SZÁNTAI ([3]). A conscientious survey paper of the topic was written by M.T. CHAO, J.C. FU and M.V. KOUTRAS ([2]).

The 2–dimensional generalizations of the above mentioned reliability systems were investigated first by A.A. SALVIA and W.C. LASHER ([10]) and then by T.K. BOEHME, A. KOSSOW and W. PREUSS ([1]). Lower and upper bounds for the 2–dimensional generalizations was published first in the paper by J. MALINOWSKI and W. PREUSS ([8]). They investigated the so called connected–(r,s)–out–of–(m,n):F lattice system. In this paper we are dealing with a more general reliability system which can be called k–out–of–connected–(r,s)–from–(m,n):F lattice system. If in this system one takes k to be equal to r×s gets back the former system. So this system can be regarded as a generalization of the former one.

[1] This work was partly supported by grants from the National Scientific Research Fund, Hungary, T014102

2 Definition of the investigated lattice system

Let us have n equal or unequal components (or elements) which are either operating or failing independently of each other in stochastic sense.

k–out–of–n:F failing system

The system of the n elements itself is failing if and only if there exists at least k failing elements in the system.

Consecutive–k–out–of–n:F failing system

The system of the n elements itself is failing if and only if there exists at least k consecutive failing elements in the system.

Consecutive–k–out–of–r–from–n:F failing system

The system of the n elements itself is failing if and only if there exists an r–element consecutive part of the elements with at least k failing elements in it.

In the above systems the elements can be arranged in a line or in a circle. If the elements of the system are arranged in a line then the system is called **linear** and if the elements of the system are arranged in a circle then the system is called **circular**.

In the two–dimensional case the system consists of $m \times n$ components ordered like the elements of an (m, n)–matrix.

Connected–(r,s)–out–of–(m,n):F lattice system

The system of the $m \times n$ elements itself is failing if and only if there exists at least one connected (r, s) submatrix of the system with all failing elements.

k–out–of–connected–(r,s)–out–of–(m,n):F lattice system

The system of the $m \times n$ elements itself is failing if and only if there exists at least one connected (r, s) submatrix of the system with at least k failing elements in it.

According to the (i, j) indices of the left–upper corner element in the (r, s)–submatrices we have

$$1 \leq i \leq m - r + 1; \quad 1 \leq j \leq n - s + 1.$$

Let us denote by

$$M = m - r + 1; \quad N = n - s + 1,$$

then there exist $M \times N$ events causing system failure:

$$A_{i_1 j_1} = \{\text{at least } k \text{ elements in } E_{i_1 j_1} \text{ are failing}\},$$

$$1 \leq i_1 \leq M; \quad 1 \leq j_1 \leq N,$$

where

$$E_{i_1 j_1} = \{e_{ij} : i = i_1, ..., i_1 + r - 1; j = j_1, ..., j_1 + s - 1\}.$$

According to the correspondence

$$A_l \leftrightarrow A_{i_1 j_1},$$

where

$$i_1 = \left\lfloor \frac{(l-1)}{N} \right\rfloor + 1,$$
$$j_1 = l - (i_1 - 1)N,$$

or in opposite way

$$l = (i_1 - 1)N + j_1,$$

one can order the $M \times N$ events causing system failure in a linear order and the probability of the system failure is

$$F = \Pr\left\{\sum_{l=1}^{MN} A_l\right\}.$$

The reliability of the system obviously is $R = 1 - F$.

3 Bounding techniques of the system failure

In this section we list some results according to the so called Boole–Bonferroni type and Hunter–Worsley inequalities. We follow the treatment of the book by A. Prékopa (see [9]) which is based on formulation of linear programming problems according to the best possible Boole–Bonferroni type inequalities.

Let us regard a set of events $A_1, ..., A_n$ on an arbitrary probability space. If we want to give lower and upper bounds on the probability that at least one of these events occur i.e. on the probability of the sum of them

$$P = \Pr\{A_1 + \cdots + A_n\},$$

then it is useful to introduce the random variable μ designating the number of those events $A_1, ..., A_n$ which occur.

For the binomial moments $E\{\binom{\mu}{k}\}$ of the random variable μ we have

$$E\left\{\binom{\mu}{k}\right\} = \sum_{1 \leq i_1 < \ldots < i_k \leq n} \Pr\{A_{i_1} \cdots A_{i_k}\} = S_k, \quad k = 1, \ldots, n. \quad (3.1)$$

We remark that the $S_k, k = 1, \ldots, n$ terms in the formula (3.1) are the same as they are in the well–known inclusion–exclusion formula:

$$\Pr\{A_1 + \ldots + A_n\} = \sum_{k=1}^{n} (-1)^{k-1} S_k.$$

The simplest proof of the equality (3.1) that we reproduce here was given by Takács (see [12]). Let $\mu_1, \ldots, \mu_n$ be the indicator random variables of the events $A_1, \ldots, A_n$, i.e. $\mu_i = 1$ if A_i occurs and $\mu_i = 0$ otherwise, $i = 1, \ldots, n$. Thus, $\mu = \mu_1 + \ldots + \mu_n$. By a well–known formula of Cauchy for binomial coefficients, we have

$$\binom{\mu}{k} = \sum_{\substack{j_1 + \ldots + j_n = k, \\ j_1 \geq 0, \ldots, j_n \geq 0}} \binom{\mu_1}{j_1} \cdots \binom{\mu_n}{j_n}.$$

Since $\mu_1, \ldots, \mu_n$ take only the values 0,1, it follows that if any of the numbers $j_1, \ldots, j_n$ is different from 0 or 1, then the term in the above sum is 0. Thus,

$$\binom{\mu}{k} = \sum_{1 \leq i_1 < \ldots < i_k \leq n} \mu_{i_1} \cdots \mu_{i_k}.$$

If we take expectations on both sides and use the equality

$$E(\mu_{i_1} \cdots \mu_{i_k}) = \Pr\{A_{i_1} \cdots A_{i_k}\},$$

(3.1) follows.

On the other hand if the discrete probability distribution of the random variable μ is denoted by $p_j, j = 0, 1, \ldots, n$, i.e.

$$p_j = \Pr\{\mu = j\}, \quad j = 0, 1, \ldots, n,$$

then taking into account that $\binom{j}{k}$ equals zero when $j < k$ by definition of the expected value we get the linear equation system

$$E\left\{\binom{\mu}{k}\right\} = \sum_{j=1}^{n} \binom{j}{k} p_j, \quad k = 1, \ldots, n. \quad (3.2)$$

From the equations (3.1) and (3.2) we get the following linear equation system

$$S_k = \sum_{j=1}^{n} \binom{j}{k} p_j, \quad k = 1, ..., n.$$

Now if we know all of the binomial moments $S_1, ..., S_n$ then the unique solution of the above linear equation system gives the probabilities $p_1, ..., p_n$ and the sum of them equals to the probability value P what we were looking for.

In the case when only the first few binomial moments $S_1, ..., S_m$ are known where $m \ll n$ we can regard the probabilities $p_1, ..., p_n$ also as unknown variables and formulate the linear programming problems

$$
\begin{array}{llllll}
\min & \{\, p_1 \;+\; & p_2 \;+\; ... \;+\; & p_m \;+\; ... \;+\; & p_n \,\} & \\
& p_1 \;+\; \binom{2}{1}p_2 \;+\; ... \;+\; & \binom{m}{1}p_m \;+\; ... \;+\; & \binom{n}{1}p_n & = S_1 \\
& p_2 \;+\; ... \;+\; & \binom{m}{2}p_m \;+\; ... \;+\; & \binom{n}{2}p_n & = S_2 \\
& \ddots \qquad \vdots & \ddots \qquad \vdots & \vdots \\
& & p_m \;+\; ... \;+\; & \binom{n}{m}p_n & = S_m \\
& p_1 \geq 0, \quad p_2 \geq 0, \quad ... \quad p_m \geq 0, \quad ... \quad p_n \geq 0 &
\end{array}
$$

and

$$
\begin{array}{llllll}
\max & \{\, p_1 \;+\; & p_2 \;+\; ... \;+\; & p_m \;+\; ... \;+\; & p_n \,\} & \\
& p_1 \;+\; \binom{2}{1}p_2 \;+\; ... \;+\; & \binom{m}{1}p_m \;+\; ... \;+\; & \binom{n}{1}p_n & = S_1 \\
& p_2 \;+\; ... \;+\; & \binom{m}{2}p_m \;+\; ... \;+\; & \binom{n}{2}p_n & = S_2 \\
& \ddots \qquad \vdots & \ddots \qquad \vdots & \vdots \\
& & p_m \;+\; ... \;+\; & \binom{n}{m}p_n & = S_m \\
& p_1 \geq 0, \quad p_2 \geq 0, \quad ... \quad p_m \geq 0, \quad ... \quad p_n \geq 0 &
\end{array}
$$

The optimal solutions of the above linear programming problems provide us with the best possible lower respectively upper bounds on the probability value

$$P = \Pr\{\mu \geq 1\} = \Pr\{A_1 + ... + A_n\}.$$

For small values of m there are known formulae for the optimal solutions of the linear programming problems.

Case $m = 2$:

$$\frac{2}{i^* + 1} S_1 - \frac{2}{i^*(i^* + 1)} S_2 \leq P \leq S_1 - \frac{2}{n} S_2, \tag{3.3}$$

where

$$i^* = \left\lfloor \frac{2S_2}{S_1} \right\rfloor.$$

Case $m = 3$:

$$\frac{j^* + 2n - 1}{(j^* + 1)n} S_1 - \frac{2(2j^* + n - 2)}{j^*(j^* + 1)n} S_2 + \frac{6}{j^*(j^* + 1)n} S_3 \leq$$

$$\leq P \leq \tag{3.4}$$

$$\leq S_1 - \frac{2(2k^* - 1)}{k^*(k^* + 1)} S_2 + \frac{6}{k^*(k^* + 1)} S_3,$$

where

$$j^* = \left\lfloor \frac{-6S_3 + 2(n - 2)S_2}{-2S_2 + (n - 1)S_1} \right\rfloor + 1,$$

and

$$k^* = \left\lfloor \frac{3S_3}{S_2} \right\rfloor + 2.$$

A further upper bound was given independently by D. HUNTER ([4]) and by K.J. WORSLEY ([13]). This bound is based only on S_1 and some individual probability values involved in S_2 but it proves to be sharper than the Boole–Bonferroni upper bound based on S_1, S_2 and sometimes even sharper than the Boole–Bonferroni bound based on S_1, S_2, S_3. This bound is given by

$$P \leq S_1 - \sum_{(i,j) \in T^*} \Pr\{A_i A_j\}, \tag{3.5}$$

where T^* is a maximal spanning tree in the non–oriented complete graph with n nodes for which the probabilities $\Pr\{A_i\}$ respectively $\Pr\{A_i A_j\}$ are assigned to the nodes respectively to the arcs as weights.

4   Binomial moment calculations

Let us introduce the notation $q = 1 - p$ then we have the following.

Calculation of S_1

$$S_1 = \sum_{l=1}^{MN} \Pr\{A_l\}, \tag{4.1}$$

where the index l with the following row resp. column index can be associated

$$i_1 = \left\lfloor \frac{(l-1)}{N} \right\rfloor + 1,$$
$$j_1 = l - (i_1 - 1)N.$$

The probability $\Pr\{A_{i_1 j_1}\}$ can be calculated as the sum of the appropriate binomial probabilities:

$$\Pr\{A_{i_1 j_1}\} = \sum_{x_1=l_1}^{u_1} \binom{u_1}{x_1} q^{x_1} p^{u_1 - x_1}, \tag{4.2}$$

where l_1 and u_1 are the lower resp. upper limits of the summation:

$$u_1 = rs,$$
$$l_1 = k.$$

Calculation of S_2

$$S_2 = \sum_{l_1=1}^{MN-1} \sum_{l_2=l_1+1}^{MN} \Pr\{A_{l_1} A_{l_2}\}, \tag{4.3}$$

where the indices l_1, l_2 with the following row resp. column indices can be associated

$$i_1 = \left\lfloor \frac{(l_1-1)}{N} \right\rfloor + 1,$$
$$j_1 = l_1 - (i_1 - 1)N,$$
$$i_2 = \left\lfloor \frac{(l_2-1)}{N} \right\rfloor + 1,$$
$$j_2 = l_2 - (i_2 - 1)N.$$

The calculation of the probability $\Pr\{A_{i_1 j_1} A_{i_2 j_2}\}$ now depends on the number of common elements in the sets $E_{i_1 j_1}, E_{i_2 j_2}$:

$$n_{12} = \max\{0, r - (i_2 - i_1)\}\max\{0, s - |j_2 - j_1|\}.$$

If $n_{12} = 0$ then the events $A_{i_1 j_1}$, $A_{i_2 j_2}$ are obviously independent and so

$$\Pr\{A_{i_1 j_1} A_{i_2 j_2}\} = \Pr\{A_{i_1 j_1}\}\Pr\{A_{i_2 j_2}\}. \tag{4.4}$$

If $n_{12} > 0$ then the theorem of total probability can be applied by making condition on the exact number of common elements in the sets $E_{i_1 j_1}$, $E_{i_2 j_2}$. After some algebra we get

$$\Pr\{A_{i_1 j_1} A_{i_2 j_2}\} = \sum_{x_1=l_1}^{u_1} \sum_{x_2=l_2}^{u_2} \sum_{x_3=l_3}^{u_3} \binom{u_1}{x_1}\binom{u_2}{x_2}\binom{u_3}{x_3} q^x p^{u-x}, \tag{4.5}$$

where

$$
\begin{aligned}
u_1 &= n_{12}, \\
u_2 &= rs - n_{12}, \\
u_3 &= u_2, \\
l_1 &= \max\{0, k - n_{12}\}, \\
l_2 &= \max\{0, k - x_1\}, \\
l_3 &= l_2, \\
x &= x_1 + x_2 + x_3, \\
u &= u_1 + u_2 + u_3.
\end{aligned}
$$

Calculation of S_3

$$S_3 = \sum_{l_1=1}^{MN-2} \sum_{l_2=l_1+1}^{MN-1} \sum_{l_3=l_2+1}^{MN} \Pr\{A_{l_1} A_{l_2} A_{l_3}\}, \tag{4.6}$$

where the indices l_1, l_2, l_3 with the following row resp. column indices can be associated

$$
\begin{aligned}
i_1 &= \left\lfloor \frac{(l_1-1)}{N} \right\rfloor + 1, \\
j_1 &= l_1 - (i_1 - 1)N, \\
i_2 &= \left\lfloor \frac{(l_2-1)}{N} \right\rfloor + 1, \\
j_2 &= l_2 - (i_2 - 1)N, \\
i_3 &= \left\lfloor \frac{(l_3-1)}{N} \right\rfloor + 1, \\
j_3 &= l_3 - (i_3 - 1)N.
\end{aligned}
$$

The calculation of the probability $\Pr\{A_{i_1 j_1} A_{i_2 j_2} A_{i_3 j_3}\}$ now depends on the number of common elements in the sets $E_{i_1 j_1}$, $E_{i_2 j_2}$; $E_{i_1 j_1}$, $E_{i_3 j_3}$ and $E_{i_2 j_2}$, $E_{i_3 j_3}$:

$$n_{12} = \max\{0, r - (i_2 - i_1)\}\max\{0, s - |j_2 - j_1|\},$$
$$n_{13} = \max\{0, r - (i_3 - i_1)\}\max\{0, s - |j_3 - j_1|\},$$
$$n_{23} = \max\{0, r - (i_3 - i_2)\}\max\{0, s - |j_3 - j_2|\}.$$

If at least two of n_{12}, n_{13}, n_{23} equals to zero the calculation of the probability $\Pr\{A_{i_1j_1}A_{i_2j_2}A_{i_3j_3}\}$ is relatively easy. Because of the independency it can be carried out by applying the formulae given earlier.

If $n_{12} = 0, n_{13} = 0, n_{23} = 0$ then

$$\Pr\{A_{i_1j_1}A_{i_2j_2}A_{i_3j_3}\} = \Pr\{A_{i_1j_1}\}\Pr\{A_{i_2j_2}\}\Pr\{A_{i_3j_3}\}. \qquad (4.7)$$

If $n_{12} > 0, n_{13} = 0, n_{23} = 0$ then

$$\Pr\{A_{i_1j_1}A_{i_2j_2}A_{i_3j_3}\} = \Pr\{A_{i_1j_1}A_{i_2j_2}\}\Pr\{A_{i_3j_3}\}. \qquad (4.8)$$

If $n_{12} = 0, n_{13} > 0, n_{23} = 0$ then

$$\Pr\{A_{i_1j_1}A_{i_2j_2}A_{i_3j_3}\} = \Pr\{A_{i_1j_1}A_{i_3j_3}\}\Pr\{A_{i_2j_2}\}. \qquad (4.9)$$

If $n_{12} = 0, n_{13} = 0, n_{23} > 0$ then

$$\Pr\{A_{i_1j_1}A_{i_2j_2}A_{i_3j_3}\} = \Pr\{A_{i_1j_1}\}\Pr\{A_{i_2j_2}A_{i_3j_3}\}. \qquad (4.10)$$

If more than one of n_{12}, n_{13}, n_{23} are positive then the theorem of total probability can be applied recursively by making condition on the exact number of common elements in the sets $E_{i_1j_1}, E_{i_2j_2}; E_{i_1j_1}, E_{i_3j_3}$ and $E_{i_2j_2}, E_{i_3j_3}$ after each other. Again after some algebra we get the following formulae.

If

$$n_{12} > 0, n_{13} > 0, n_{23} = 0,$$

or

$$n_{12} > 0, n_{13} = 0, n_{23} > 0,$$

or

$$n_{12} = 0, n_{13} > 0, n_{23} > 0,$$

then

$$\Pr\{A_{i_1j_1}A_{i_2j_2}A_{i_3j_3}\} = \sum_{x_1=l_1}^{u_1} \cdots \sum_{x_5=l_5}^{u_5} \binom{u_1}{x_1}\cdots\binom{u_5}{x_5}q^x p^{u-x}, \qquad (4.11)$$

where

$$
\begin{aligned}
u_1 &= rs - nc_1 - nc_2, \\
u_2 &= nc_1, \\
u_3 &= nc_2, \\
u_4 &= rs - nc_1, \\
u_5 &= rs - nc_2, \\
l_1 &= \max\{0, k - nc_1 - nc_2\}, \\
l_2 &= \max\{0, k - rs\}, \\
l_3 &= l_2, \\
l_4 &= \max\{0, k - x_2\}, \\
l_5 &= \max\{0, k - x_3\}, \\
x &= x_1 + x_2 + x_3 + x_4 + x_5, \\
u &= u_1 + u_2 + u_3 + u_4 + u_5,
\end{aligned}
$$

and

$$
\begin{aligned}
nc_1 &= n_{13}, & nc_2 &= n_{23}, & &\text{if } n_{12} = 0, \\
nc_1 &= n_{12}, & nc_2 &= n_{23}, & &\text{if } n_{13} = 0, \\
nc_1 &= n_{12}, & nc_2 &= n_{13}, & &\text{if } n_{23} = 0.
\end{aligned}
$$

Finally, if $n_{12} > 0, n_{13} > 0, n_{23} > 0$, then

$$
\Pr\{A_{i_1 j_1} A_{i_2 j_2} A_{i_3 j_3}\} = \sum_{x_1 = l_1}^{u_1} \cdots \sum_{x_7 = l_7}^{u_7} \binom{u_1}{x_1} \cdots \binom{u_7}{x_7} q^x p^{u-x}, \tag{4.12}
$$

where

$$
\begin{aligned}
u_1 &= \bigl(r - (i_3 - i_1)\bigr)\bigl(s - \max\{(j_2 - j_1), (j_3 - j_1), (j_3 - j_2)\}\bigr), \\
u_2 &= n_{12} - u_1, \\
u_3 &= n_{13} - u_1, \\
u_4 &= n_{23} - u_1, \\
u_5 &= rs - u_1 - u_2 - u_3, \\
u_6 &= rs - u_1 - u_2 - u_4, \\
u_7 &= rs - u_1 - u_3 - u_4, \\
l_1 &= \max\{0, k - (rs - u_1)\}, \\
l_2 &= \max\{0, k - x_1 - (rs - n_{12})\}, \\
l_3 &= \max\{0, k - x_1 - (rs - n_{13})\}, \\
l_4 &= \max\{0, k - x_1 - (rs - n_{23})\}, \\
l_5 &= \max\{0, k - x_1 - x_2 - x_3\}, \\
l_6 &= \max\{0, k - x_1 - x_2 - x_4\}, \\
l_7 &= \max\{0, k - x_1 - x_3 - x_4\}, \\
x &= x_1 + x_2 + x_3 + x_4 + x_5 + x_6 + x_7, \\
u &= u_1 + u_2 + u_3 + u_4 + u_5 + u_6 + u_7.
\end{aligned}
$$

Economic calculation of S_1 and S_2

It is obvious that the single event probability $\Pr\{A_{i_1 j_1}\}$ is independent of the indices i_1, j_1 so we can introduce the notation

$$P_1 = \Pr\{A_{i_1 j_1}\},$$

and

$$S_1 = MNP_1. \tag{4.13}$$

It is also obvious that the pair event probabilities $\Pr\{A_{i_1 j_1} A_{i_2 j_2}\}$ are depending on the indices $i_1, j_1; i_2, j_2$ only across the differencies

$$
\begin{aligned}
id_{12} &= i_2 - i_1, \\
jd_{12} &= j_2 - j_1,
\end{aligned}
$$

so we can introduce the notation

$$h(id_{12}, jd_{12}) = \Pr\{A_{i_1 j_1} A_{i_2 j_2}\}.$$

Then after some combinatorical considerations one can get the formula

$$S_2 = \left\{ \binom{N-s+1}{2}\left(M + 2Miu_{12} - iu_{12}(iu_{12}+1)\right) + N^2 \binom{M-r+1}{2} \right\} P_1^2$$

$$+ M \sum_{jd_{12}=1}^{ju_{12}} \binom{N - jd_{12}}{1} h(0, jd_{12}) \tag{4.14}$$

$$+ \sum_{id_{12}=1}^{iu12} \binom{M - id_{12}}{1} \left\{ Nh(id_{12}, 0) + 2 \sum_{jd_{12}=1}^{ju_{12}} \binom{N - jd_{12}}{1} h(id_{12}, jd_{12}) \right\},$$

where

$$
\begin{aligned}
iu_{12} &= \min\{r - 1, N - 1\}, \\
ju_{12} &= \min\{s - 1, M - 1\}.
\end{aligned}
$$

5 Examples

In the following tables there are given the results according to all of the test problems investigated by J. MALINOWSKI and W. PREUSS ([8]). In the heading of the tables the following abbreviations are used:

p   —   reliability of one element in the lattice system,
lo (M–P)   —   lower bound by Malinowski and Preuss,
up (M–P)   —   upper bound by Malinowski and Preuss,
exact   —   exact value for the reliability of the system,
$lo(S_2)$ —   Boole–Bonferroni lower bound based on S_1 and S_2,
$up(S_2)$ —   Boole–Bonferroni upper bound based on S_1 and S_2,
$lo(S_3)$ —   Boole–Bonferroni lower bound based on S_1, S_2 and S_3,
$up(S_3)$ —   Boole–Bonferroni upper bound based on S_1, S_2 and S_3,
lo(Hu)   —   Hunter–Worsley lower bound.

We remark that the bounds given in the paper are according to the system failing probability. When we turn to the bounds according to the reliability of the system from lower bounds we get upper and vice versa. So the Hunter–Worsley upper bound becomes lower bound according to the reliability of the system. This is the reason why in the tables the Hunter–Worsley bounds always appear as lower bound. In the tables the uniquely best lower resp. upper bounds are denoted by bold faced characters.

The test problems are fully defined by the values of m, n, r, s and p, as we always take $k = rs$ to get the special cases investigated by J. MALINOWSKI and W. PREUSS ([8]). The lo(M–P), up(M–P) and exact values were taken from the above paper by J. MALINOWSKI and W. PREUSS. The $lo(S_2)$, $up(S_2)$ and $lo(S_3)$, $up(S_3)$ values were calculated by the formulae (3.3) and (3.4), while for the calculation of the lo(Hu) values the formula (3.5) was used. For applying formulae (3.3)–(3.5) the binomial moments S_1, S_2 and S_3 are necessary only. The calculation procedures of these moments is fully defined in the Chapter 4. For completeness S_1 can be calculated by (4.1), using (4.2); S_2 can be calculated by (4.3), using (4.4)–(4.5) and S_3 can be calculated by (4.6), using (4.7)–(4.12). It is important to remark here that while the economic calculation form (4.13) for S_1 can be applied always, the economic calculation form (4.14) for S_2 can be applied only in the case when we don't want to calculate the lo(Hu) values as these latter need not only the knowledge of the value S_2 but also the individual probabilities $\Pr\{A_{i_1 j_1} A_{i_2 j_2}\}$.

Table 1
$m = 4, n = 4, r = 2, s = 2$

p	lo(M–P)	lo(S_3)	exact	up(S_3)	up(M–P)
0.5	0.601	**0.621**	0.644	**0.699**	0.712
0.7	0.933	**0.937**	0.937	**0.938**	0.955
0.9	0.999	0.999	0.999	0.999	0.999

Table 2
$m = 4, n = 3, r = 2, s = 2$

p	lo(M–P)	lo(S_3)	exact	up(S_3)	up(M–P)
0.5	0.712	**0.729**	0.740	**0.755**	0.793
0.7	0.955	**0.957**	0.957	**0.957**	0.969
0.9	0.999	0.999	0.999	0.999	0.999

Table 3
$m = 5, n = 5, r = 2, s = 2$

p	lo(M–P)	lo(Hu)	lo(S_3)	up(S_3)	up(M–P)
0.5	**0.409**	0.234	0.396	**0.583**	0.640
0.7	0.885	0.881	**0.891**	**0.896**	0.941
0.9	0.998	0.998	0.998	**0.998**	0.999

Table 4
$m = 10, n = 10, r = 2, s = 2$

p	lo(M–P)	lo(Hu)	lo(S_3)	up(S_3)	up(M–P)
0.5	**0.012**	0.000	0.000	0.245	**0.086**
0.7	0.541	0.402	**0.544**	**0.666**	0.711
0.9	0.992	0.992	0.992	**0.992**	0.996

Table 5
$m = 10, n = 10, r = 3, s = 3$

p	lo(M–P)	lo(Hu)	lo(S_3)	up(S_3)	up(M–P)
0.5	0.894	0.890	**0.903**	**0.919**	0.959
0.7	0.999	0.999	0.999	0.999	0.999
0.9	0.999	0.999	0.999	0.999	0.999

Table 6
$m = 20, n = 20, r = 3, s = 3$

p	lo(M–P)	lo(Hu)	lo(S_3)	up(S_3)	up(M–P)
0.6	0.923	0.920	**0.928**	**0.931**	0.974
0.7	0.994	0.994	0.994	**0.994**	0.998
0.8	0.999	0.999	0.999	0.999	0.999
0.9	0.999	0.999	0.999	0.999	0.999

Table 7
$m = 50, n = 50, r = 3, s = 3$

p	lo(M–P)	lo(Hu)	lo(S_3)	up(S_3)	up(M–P)
0.6	0.567	0.435	**0.570**	**0.673**	0.828
0.7	0.957	0.956	**0.958**	**0.958**	0.985
0.8	0.999	0.999	0.999	0.999	0.999
0.9	0.999	0.999	0.999	0.999	0.999

Table 8
$m = 100, n = 100, r = 3, s = 3$

p	lo(M–P)	lo(Hu)	lo(S_2)	up(S_2)	up(M–P)
0.6	**0.094**	0.000	0.000	**0.336**	0.451
0.7	**0.832**	0.816	0.811	**0.840**	0.940
0.8	0.995	0.995	0.995	**0.995**	0.998
0.9	0.999	0.999	0.999	0.999	0.999

6 Conclusions

The Boole–Bonferroni upper bounds are sharper than the upper bounds proposed by Malinowski and Preuss, except one case with extremely low system reliability. The upper bounds given in the paper by Malinowski and Preuss ([8]) are sometimes sharper than the Boole–Bonferroni upper bounds. This is the case especially when the system is large and the calculation of the more accurate Boole–Bonferroni bounds using the first three binomial moments becomes impossible. Finally we remark that while J. MALINOWSKI and W. PREUSS are dealing with the special connected–(r,s)–out–of–(m,n):F lattice systems only, the bounds given in this paper are according to the more general k–out–of–connected–(r,s)–from–(m,n):F lattice systems.

Acknowledgement. The author would like to express his thanks to the referee for the invaluable comments and advices.

References

[1] Boehme, T.K., Kossow, A. and Preuss, W. "A generalization of consecutive–k–out–of–n:F systems", *IEEE Trans. Reliability*, Vol. **41**, 1992, 451-457.

[2] Chao, M.T., Fu, J.C. and Koutras, M.V. "Survey of reliability studies of consecutive-k-out-of-n:F & related systems", *IEEE Trans. Reliability*, Vol. **44**, 1995, 120-127.

[3] Habib, A. and Szántai, T. "New bounds on the reliability of the consecutive k–out–of–r–from–n:F system", submitted to *Reliability Engineering and System Safety*.

[4] Hunter, D. "An upper bound for the probability of a union", *J. Appl. Prob.*, Vol. **13**, 1976, 597-603.

[5] Kounias, S. and Sfakianakis, M. "The reliability of a linear system and its connection with the generalized birthday problem", *Statistica Applicata*, Vol. **3**, 1991, 531-543.

[6] Kounias, S. and Sfakianakis, M. "A combinatorial problem associated with the reliability of a circular system", *Proc. HERMIS'92* (Lipitakis, *ed*), 1992, 187-196.

[7] Koutras, M.V. and Papastavridis, S.G. "Application of the Stein-Chen method for bounds and limit theorems in the reliability of coherent structures", *Nav. Research Logistics*, Vol. **40**, 1993, 617-631.

[8] Malinowski, J. and Preuss, W. "Lower and upper bounds for the relia-bility of connected–(r,s)–out–of–(m,n):F lattice systems", *IEEE Trans. Reliability*, Vol. **45**, 1996, 156-160.

[9] Prékopa, A. *Stochastic Programming*. Kluwer Academic Publishers, Dordrecht, 1995.

[10] Salvia, A.A. and Lasher, W.C. "2–Dimensional consecutive–k–out–of–n:F models", *IEEE Trans. Reliability*, Vol. **39**, 1990, 382-385.

[11] Sfakianakis, M., Kounias, S. and Hillaris, A. "Reliability of a consecutive-k-out-of-r-from-n:F system", *IEEE Trans. Reliability*, Vol. **41**, 1992, 442-447.

[12] Takács, L. "On the general probability theorem", *Communications of the Dept. of Math. and Physics of the Hungarian Acad. Sci.*, Vol. **5**, 1955, 467-476.

[13] Worsley, K.J. "An improved Bonferroni inequality and applications", *Biometrika*, Vol. **69**, 1982, 297-302.

On a Relation between Problems of Calculus of Variations and Mathematical Programming

Tamás RAPCSÁK[1] and Anna VÁSÁRHELYI[2]

[1] *Laboratory of Operations Research and Decision Systems, Computer and Automation Institute, Hung. Acad. of Sciences, P.O.Box 63, H-1518 Budapest, Hungary*

[2] *Technical Univ. of Budapest, Műegyetem rkp. 3. H-1111 Budapest, Hungary*

Abstract. Important practical problems of calculus of variations can be transformed into one-parametric optimization ones. Simple engineering examples are presented to show that the two kinds of solution coincide.

Keywords: Parametric optimization, calculus of variations

1. Introduction

Several problems in mechanics can be described by the help of systems of differential equations with initial and boundary conditions. In the case of boundary value problems, there is no difference between the solution of a system of differential equations and of a corresponding problem of calculus of variations. Problems of calculus of variations seem not to be well fitted to problems of initial value, or can be rendered by difficult approximations, only. The majority of numerical methods determine an approximating solution, but the question is how precise this approximation is. If both initial and boundary conditions belong to a system of differential equations related to problems in mechanics, the simultaneous handling of different conditions is difficult.

For solving numerical problems in the case of loading changing in time, generally two approaches are discussed in the literature related to problems of boundary values.

1. The problem is solved through fixing the values of the load function in several given times. In this way, an arbitrary number of function values defining the change of state variables depending on time can be determined. In this case, no influence of values calculated in different periods can be directly taken into account.

2. The problem is solved by supposing separability of the function in the independent variables related to place and time, which seems to be a strong restriction.

The elaboration of a solving method different from the previous ones is justified by the above. Paragraph 2 deals with the relation between problems of calculus of variations and mathematical programming. In Paragraph 3, two examples show how to transform variational problems obtained by Hu-Washizu principle into principle into mathematical programming ones in such a way that the solution provided by the two different methods be equal.

2. Relation between problems of calculus of variations and mathematical programming

Let $A \subseteq R^n$ be a set. Consider the parametric problems of calculus of variations

$$\min J(\mathbf{x}) = \int_{t_1}^{t_2} f(t, \mathbf{x}(t)) \, dt,$$

$$\mathbf{x}(t) = (x_1(t), x_2(t), \ldots, x_n(t)) \in \overline{C}[t_1, t], \qquad (2.1)$$

$$f(t, \mathbf{x}) \in C([t_1, t_2] \times A), \quad f : [t_1, t_2] \times A \to R,$$

where the aim is to determine the unknown vector function $\mathbf{x}(t)$. Here, $C([t_1, t_2] \times A)$ and $\overline{C}[t_1, t_2]$ mean the class of continuous functions on the set $[t_1, t_2] \times A$ and of piecewise continuous functions on interval $[t_1, t_2]$, respectively. Consider the optimization problem

$$\min f(t, \mathbf{x})$$

$$(t, \mathbf{x}) \in [t_1, t_2] \times A, \ f(t, \mathbf{x}) \in C([t_1, t_2] \times A), \quad f : [t_1, t_2] \times A \to R, \quad (2.2)$$

where the minimum must be determined for every fixed value $t \in [t_1, t_2]$.

Definition 2.1. Let $A \subseteq R^n$ be an arbitrary subset. It is said that the set $R_0 = [t_1, t_2] \times A \subseteq R^{n+1}$ is feasible if for every element $(t_0, \mathbf{x}_0) \in [t_1, t_2] \times A$, $\exists \mathbf{x}(t)$, $t \in [t_0 - \delta, t_0 + \delta]$, a continuous vector function for which $\mathbf{x}(t_0) = \mathbf{x}_0$ and $(t, \mathbf{x}(t)) \in [t_0 - \delta, t_0 + \delta] \times A$ hold.

A vector function $\mathbf{x}(t)$, $t \in [t_1, t_2]$, is feasible if $(t, \mathbf{x}(t)) \in R_0$, $\mathbf{x}(t) \in \overline{C}^n[t_1, t_2]$. In the Hestenes book, the equivalence of problems (2.1) and (2.2) is stated as follows:

Theorem 2.1. Functional $J(\mathbf{x})$ is minimized by a feasible vector function $\mathbf{x}_0(t)$, $t \in [t_1, t_2]$, over the feasible piecewise continuous functions defined on the feasible region $R_0 = [t_1, t_2] \times A$ if and only if the inequality

$$f(t, \mathbf{x}) \geq f(t, \mathbf{x}_0(t)), \quad t \in [t_1, t_2], \qquad (2.3)$$

fulfils for every element $(t,\mathbf{x}) \in R_0$.

If the vector function $\mathbf{x}_0(t)$, $t \in [t_1, t_2]$, is the minimum of functional $J(\mathbf{x})$, then $f(t, \mathbf{x}_0(t))$ is continuous on $t \in [t_1, t_2]$.

Let us examine the case if instead of (2.1), the problem is given in the form of

$$\min \ J_1(\mathbf{x}, \upsilon, \mu) = \int_{t_1}^{t_2} L(t, \mathbf{x}(t), \upsilon(t), \mu(t)) \, dt, \qquad (2.4)$$

$\mathbf{x}(t), \upsilon(t), \mu(t) \in \overline{C}^{\,n+q+m}[t_1, t_2]$, $L(t, \mathbf{x}, \upsilon, \mu) \in C([t_1, t_2] \times A \times B \times D)$,

where $A \subseteq R^n, B \subseteq R^q, D \subseteq R^m$.

Let us consider

$$L(t, \mathbf{x}, \upsilon, \mu) = f(t, \mathbf{x}) + \sum_{j=1}^{q} \mu_j h_j(t, \mathbf{x}) + \sum_{i=1}^{m} \upsilon_i g_i(t, \mathbf{x}), \qquad (2.5)$$

$f, h_j, g_i \in C([t_1, t_2] \times A)$, $j = 1, \ldots, q$, $i = 1, \ldots, m$,

$\mu(t) \in \overline{C}^{\,m}[t_1, t_2]$, $\upsilon(t) \in \overline{C}^{\,q}[t_1, t_2]$.

Mathematical programming problems of type (2.2) corresponding to problems of calculus of variations (2.4) can be stated in the form of

$$\min L(t, \mathbf{x}, \upsilon, \mu)$$
$$(t, \mathbf{x}, \upsilon, \mu) \in [t_1, t_2] \times A \times B \times D,$$
$$L(t, \mathbf{x}, \upsilon, \mu) \in C([t_1, t_2] \times A \times B \times D), \qquad (2.6)$$
$$L : [t_1, t_2] \times A \times B \times D \to R.$$

In this case, $R_0^1 = [t_1, t_2] \times A \times B \times D \subseteq R^{n+q+m+1}$

Corollary 2.1. Functional $J_1(\mathbf{x}, \upsilon, \mu)$ is minimized by a feasible vector function $(\mathbf{x}_0(t), \upsilon_0(t), \mu_0(t))$, $t \in [t_1, t_2]$, over the set of feasible piecewise continuous functions defined on the feasible region $R_0^1 = [t_1, t_2] \times A \times B \times D$ if and only if the inequality

$$L(t, \mathbf{x}, \upsilon, \mu) \geq L(t, \mathbf{x}_0(t), \upsilon_0(t), \mu_0(t)), \ t \in [t_1, t_2], \qquad (2.7)$$

fulfils for every element $(t, \mathbf{x}, \upsilon, \mu) \in R_0^1$.

If the vector function $(\mathbf{x}_0(t), \upsilon_0(t), \mu_0(t)), t \in [t_1, t_2]$, is the minimum of functional $J_1(\mathbf{x}, \upsilon, \mu)$, then the function $L(t, \mathbf{x}_0(t), \upsilon_0(t), \mu_0(t))$, is continuous on the interval $t \in [t_1, t_2]$,

Minimization problems (2.6) may correspond to a pair of constrained optimization problems:

$$\min f(t, \mathbf{x})$$

$$h_j(t, \mathbf{x}) = 0, \quad j = 1, \dots, q,$$

$$g_i(t, \mathbf{x}) \geq 0, \quad i = 1, \dots, m,$$

$$(t, x_1, x_2, \dots, x_n) \in [t_1, t_2] \times A, \tag{2.8}$$

$$f, h_j, g_i \in C([t_1, t_2]) \times A, \quad j = 1, \dots, q, \quad i = 1, \dots, m,$$

$$f, h_j, g_i : C([t_1, t_2]) \times A \to R, \quad j = 1, \dots, q, \quad i = 1, \dots, m,$$

and

$$\max \left(f(t, \mathbf{x}) + \sum_{j=1}^{q} \mu_j h_j(t, \mathbf{x}) + \sum_{i=1}^{m} \upsilon_i g_i(t, \mathbf{x}) \right)$$

$$\nabla_x f(t, \mathbf{x}) + \sum_{j=1}^{q} \mu_j \nabla_x h_j(t, \mathbf{x}) + \sum_{i=1}^{m} \upsilon_i \nabla_x g_i(t, \mathbf{x}) = 0$$

$$\upsilon_i g_i(t, \mathbf{x}) = 0, \ \upsilon_i \geq 0, \ i = 1, \dots, m, \ (t, \mathbf{x}, \upsilon, \mu) \in [t_1, t_2] \times A \times B \times D, \tag{2.9}$$

$$f, h_j, g_i \in C([t_1, t_2] \times A), \ j = 1, \dots, q, \ i = 1, \dots, m,$$

$$f, h_j, g_i : [t_1, t_2] \times A \to R \ j = 1, \dots, q, \ i = 1, \dots, m.$$

The relations between the optimality conditions of problems (2.6), (2.7) and (2.8) are given by the Lagrange theorem and duality theorem (see, e.g., Hestenes book). Thus, problems of calculus of variations can be transformed into unconstrained or constrained optimization problems.

3. Applications in elastic analysis of structures

Elastic problems can be formulated by the Hu-Washizu functional in the form of

$$\min \ \Pi_I = \iiint_V [A(\varepsilon(\mathbf{x})) - P(\mathbf{x})^* u(\mathbf{x})] dV - \iiint_V [(\varepsilon(\mathbf{x}) - \mathbf{B}u(\mathbf{x}))^* \sigma(\mathbf{x})] dV$$

$$- \iint_{S_I} \overline{P}(\mathbf{x})^* u(\mathbf{x}) dS - \iint_{S_{II}} \overline{u}(\mathbf{x}))^* p(\mathbf{x}) dS,$$

$$\tag{3.1}$$

where Π_I is the potential energy function, V the volume of the structure, **x** the vector of position, $\varepsilon(\mathbf{x})$ the vector of the strain functions, $A(\varepsilon(\mathbf{x}))$ the elastic energy function, $\mathbf{P}(\mathbf{x})$ the vector of the external load functions, $\mathbf{u}(\mathbf{x})$ the vector functions of displacement, **B** the transfer matrix which contains the corresponding derivatives of $\mathbf{u}(\mathbf{x})$, $\sigma(\mathbf{x})$ the vector function of stress, S_I force boundary condition, S_{II} displacements boundary, and the symbols $^-$ and $*$ a prescribed value and the transpose, respectively.

In some cases, problems (3.1) lead to problems of calculus of variations as it will be shown in the following two examples.

First, the deflection function of simple supported beam will be approximated with an orthogonal polynom system. Let us consider a simple supported beam with 2π length and unit rigidity. The external load is q(x)=2sin (x)+3sin(3x) and the problem is to determine the deflection function u(x). Under boundary conditions u(0)=0, u(2π)=0 and (u"(0)=0, u"(2π)=0), the problem of calculus of variations is

$$\min J(u) = \int_0^{2\pi} \left(\frac{1}{2}\left(\frac{d^2 u(x)}{dx^2} \right)^2 + q(x)u(x) \right) dx. \qquad (3.2)$$

In problem (3.2), choose a basis N_i, i=1,...,6, in the class of the feasible functions in the form of

$$N_1(x) = \sin(x), \quad N_2(x) = \sin(2x), \quad N_3(x) = \sin(3x),$$
$$N_4(x) = \cos(x), \quad N_5(x) = \cos(2x), \quad N_6(x) = \cos(3x). \qquad (3.3)$$

In order to satisfy the boundary conditions, the functions N_4, N_5 and N_6 will leave the basis, then the unknown vector function $u(x)$ and the given function q(x) will be expressed by the basis as follows:

$$u(x) = \sum_{i=1}^{3} a_i N_i(x), \quad q(x) = \sum_{i=1}^{3} b_i N_i(x), \quad N_i(x) \in L^2, \qquad (3.4)$$

where the multipliers a_1, a_2, a_3, are unknown.

The boundary conditions are satisfied by the basis:

$$u(0) = \sum_{i=1}^{3} a_i \sin_i(ix) = 0, \quad u(2\pi) = \sum_{i=1}^{3} a_i \sin_i(i2\pi) = 0,$$

$$u"(0) = \sum_{i=1}^{3} -i^2 a_i \sin_i(i0) = 0, \quad u"(2\pi) = \sum_{i=1}^{3} -i^2 a_i \sin_i(i2\pi) = 0,$$

Thus, we have that

$$\min_{a} \int_0^{2\pi} \frac{1}{2} \left(\frac{d^2 (\sum\limits_{i=1}^{3} a_i N_i(x))}{dx^2} \right)^2 + \sum_{i=1}^{3} b_i N_i(x) \sum_{i=1}^{3} a_i N_i(x) \Bigg| dx, \ \mathbf{a} \in R^3. \quad (3.5)$$

By differentiation in (3.5), we obtain that

$$\min_{a} \int_0^{2\pi} \left(\frac{1}{2} \left(\sum_{k=1}^{3} a_k(-k^2)\sin(kx) \sum_{j=1}^{3} a_j(-j^2)\sin(jx) \right) + \right.$$

$$\left. + \sum_{j=1}^{3} b_j \sin(jx) \sum_{k=1}^{3} a_k \sin(kx) \right) dx. \quad (3.6)$$

By using orthogonality, problems (3.7) and (3.8) are obtained:

$$\min_{a} \int_0^{2\pi} \left(\frac{1}{2} \sum_{k=1}^{3} k^4 a_k^2 \sin^2(kx) + \sum_{k=1}^{3} a_k b_k \sin^2(kx) \right) dx, \quad \mathbf{a} \in R^3 \quad (3.7)$$

$$\min_{a} \left(\frac{1}{2} \sum_{k=1}^{3} k^4 a_k^2 \int_0^{2\pi} \sin^2(kx)dx + \sum_{k=1}^{3} b_k a_k \int_0^{2\pi} \sin^2(kx)dx \right), \quad \mathbf{a} \in R^3. \quad (3.8)$$

By the Lagrange multiplier rule, the stationary function can be determined as follows:

$$k^4 a_k + b_k = 0, \quad k = 1,2,3,$$

from which

$$a_k = -\frac{b_k}{k^4}, \quad k = 1,2,3.$$

In the case of $q(x) = 2\sin(x) + 3\sin(3x)_1$, $b_1 = 2$, $b_2 = 0$, $b_3 = 3$, the approximation of the unknown function is

$$u(x) = -2\sin(x) - 1/27 \sin3(x). \quad (3.9)$$

Solution (3.9) is obtained by calculus of variations.

Now, let us consider problem (3.7) and the corresponding mathematical programming one based on the correspondence $t \Rightarrow x$, $x_k(t) \Rightarrow a_k \sin(kx)$, k=1,2,3, then

$$f(\mathbf{x}(t)) \Rightarrow \frac{1}{2} \sum_{k=1}^{3} k^4 a_k^2 \sin^2(kx) + \sum_{k=1}^{3} a_k b_k \sin^2(kx),$$

$$(a_1 N_1(x), a_2 N_2(x), a_3 N_3(x)) \in C^3 [0, 2\pi],$$

$$f(\mathbf{x}(t)) \Rightarrow f(\sum_{k=1}^{3} a_k N_k(x)) \in C([0, 2\pi] \times R), \quad f : R \to R.$$

Let $J_k(x) = a_k \sin(kx)$, $k=1,2,3$, then the parametric mathematical programming problem can be formulated as the following unconstrained optimization problem for every fixed value $x \in [0, 2\pi]$:

$$\min \left(\frac{1}{2} \sum_{k=1}^{3} k^4 y_k^2 + \sum_{k=1}^{3} b_k \sin(kx) y_k \right), \quad \mathbf{y} \in R^3. \qquad (3.10)$$

By the Lagrange multiplier rule,

$$k^4 y_k + b_k \sin(kx) = 0, \quad k = 1,2,3,$$

from which

$$y_k = -\frac{b_k \sin(kx)}{k^4}, \qquad k=1,2,3. \qquad (3.11)$$

So, $a_1 = -2$, $a_2 = 0$, $a_3 = -1/27$, and this solution is equivalent to (3.9).

b.) Secondly, the deflection function of simple supported beam is approximated with a nonorthogonal polynom system. Let us consider a simple supported beam with unique rigidity and unique length. The external load is $q(x) = x - x^4$. The boundary conditions request at the supports that the deflection $u(x)$ and moment function $M(x)$ values are zero:

$$M(0) = 0, \qquad M(1) = 0,$$

$$u(0) = \int_0^x \int_0^x M(\xi) d\xi dx = 0, \quad u(1) = \int_0^x \int_0^x M(\xi) d\xi dx = 0.$$

The nonorthogonal function system is the following:

$$p_0(x) = x \cdot (x - 1), \quad p_1(x) = x^2 \cdot (x - 1), \quad p_2(x) = x^3 \cdot (x - 1). \qquad (3.12)$$

The moment function $M(x)$ is approximated by

$$M(x) \approx \sum_{j=0}^{2} \sum_{i=0}^{2} a_i p_j(x), \quad \mathbf{a} \in R^3 \qquad (3.13)$$

where $a_j = \int_0^1 M(x)p_j(x)dx$.

The second derivatives are

$$\frac{d^2 p_0(x)}{dx} = 2, \quad \frac{d^2 p_1(x)}{dx} = 6x - 2, \quad \frac{d^2 p_2(x)}{dx} = 12x^2 - 6x.$$

The third boundary condition is not satisfied by the polynom system $p_i(x)$, i=0,1,2.

The problem of calculus of variatons based on (3.1) is in the form of

$$\min \int_0^1 \left(\frac{1}{2}(M(x))^2 + u(x)[\frac{d^2 M(x)}{dx^2} - q(x)] + F\int_0^1\int_0^x M(\xi)d\xi dx \right) dx,$$

(3.14)

where F is a constant, mechanically, F is the reaction force.

The approximation of the second derivative of the moment function is

$$\frac{d^2 M(x)}{dx^2} \approx \sum_{j=0}^{2} \sum_{i=0}^{2} a_i \frac{d^2 p_j(x)}{dx}.$$

The second derivatives are given in the basis as follows:

$$\frac{d^2 M(x)}{dx^2} \approx \sum_{j=0}^{2} \sum_{i=0}^{2} a_i c_{ij} p_j(x),$$

where $c_{ij} = \int_0^1 \frac{d^2 p_j(x)}{dx} p_j(x)dx, \quad i,j = 0,1,2.$

The third boundary condition can be expressed by the polynom system in the form of

$$\int_0^1\int_0^x \sum_{j=0}^{2} \sum_{i=0}^{2} a_i p_j(\xi)d\xi dx = \sum_{j=0}^{2} \sum_{i=0}^{2} a_i \int_0^1\int_0^x p_j(\xi)d\xi dx = \sum_{j=0}^{2} \sum_{i=0}^{2} a_i d_j,$$

$$F = \sum_{j=0}^{2} \sum_{i=0}^{2} k_i p_j(x), \quad F\sum_{j=0}^{2} \sum_{i=0}^{2} a_i d_j = \sum_{j=0}^{2} \sum_{i=0}^{2} \sum_{k=0}^{2} \sum_{l=0}^{2} a_i k_k d_l p_j(x).$$

Substituting these approximations for (3.14), the problem of calculus of variations is as follows:

$$\min\{\int_0^1 \left[\frac{1}{2}\left(\sum_{j=0}^2\sum_{i=0}^2 a_i p_j(x)\right)^2 + \sum_{j=0}^2\sum_{i=0}^2 b_i p_j(x)[\sum_{j=0}^2\sum_{i=0}^2 a_i c_{ij} p_j(x) - \sum_{j=0}^2\sum_{i=0}^2 r_i p_j(x)]\right]dx +$$

$$+\sum_{j=0}^2\sum_{i=0}^2\sum_{k=0}^2\sum_{l=0}^2 a_i k_k d_l p_j(x)\}, \tag{3.15}$$

where $u(x) = \sum_{j=0}^2\sum_{i=0}^2 b_i p_j(x)$, the values b_i, i=0,1,2, are variables, and the

given load is expressed in the basis, i.e., $q(x) = x - x^4 = \sum_{j=0}^2 p_j(x)$, thus

$r_i = 1$, $i = 0,1,2$.

Now, the corresponding mathematical programming problem is formulated by introducing the notations $y_{ij} = a_i p_j(x)$, $z_{ij} = b_i p_j(x)$, $i, j = 0,1,2$, is given in the form of

$$\min\left(\frac{1}{2}\left(\sum_{j=0}^2\sum_{i=0}^2 y_{ji}\right)^2 + \sum_{j=0}^2\sum_{i=0}^2 z_{ij}[\sum_{j=0}^2\sum_{i=0}^2 c_{ij} y_{ij} - \sum_{j=0}^2 p_j(x)] + \sum_{j=0}^2\sum_{i=0}^2\sum_{k=0}^2\sum_{l=0}^2 y_{ji} k_k d_l\right)$$

$$\tag{3.16}$$

for every fixed $x \in [0,1]$, where the unknowns are z_{ij}, y_{ij}, and k_j, i,j =0,1,2. By the Lagrange multiplier rule,

$$\sum_{j=0}^2\sum_{i=0}^2 y_{ij} + c_{kl}\sum_{j=0}^2\sum_{i=0}^2 z_{ij} + \sum_{i=0}^2\sum_{j=0}^2 k_i d_j = 0, \quad k,l = 0,1,2,$$

$$\sum_{j=0}^2\sum_{i=0}^2 c_{ij} y_{ij} - \sum_{j=0}^2 p_j(x) = 0,$$

$$\sum_{l=0}^2\sum_{j=0}^2\sum_{i=0}^2 y_{ij} d_l = 0. \tag{3.17}$$

In order to determine the moment function M(x), we use the second equality of (3.17) for every $\mathbf{x} \in [0,1]$, the approximation $y_{ij} = a_i p_j(x)$, i,j=0,1,2, and

the linear independency of the basis functions from which it follows that in our case

$$\begin{bmatrix} c_{00} & c_{10} & c_{20} \\ c_{01} & c_{11} & c_{21} \\ c_{02} & c_{12} & c_{22} \end{bmatrix} \begin{bmatrix} a_0 \\ a_1 \\ a_2 \end{bmatrix} = \begin{bmatrix} 1 \\ 1 \\ 1 \end{bmatrix} \tag{3.18}$$

The system is solved by MATHCAD:

$$p0(x) := x(x-1) \qquad p1(x) := x^2 \cdot (x-1) \qquad p2(x) := x^3 \cdot (x-1)$$

$$dp0(x) := \frac{d}{dx} p0(x) \qquad dp1(x) := \frac{d}{dx} p1(x) \qquad dp2(x) := \frac{d}{dx} p2(x)$$

$$ddp0(x) := \frac{d}{dx} dp0(x) \qquad ddp1(x) := \frac{d}{dx} dp1(x) \qquad ddp2(x) := \frac{d}{dx} dp2(x)$$

$$c_{0,0} := \int_0^1 ddp0(x) \cdot p0(x)\, dx \quad c_{0,1} := \int_0^1 ddp0(x) \cdot p1(x)\, dx \quad c_{0,2} := \int_0^1 ddp0(x) \cdot p2(x)\, d$$

$$c_{1,0} := \int_0^1 ddp1(x) \cdot p0(x)\, dx \quad c_{1,1} := \int_0^1 ddp1(x) \cdot p1(x)\, dx \quad c_{1,2} := \int_0^1 ddp1(x) \cdot p2(x)\, d$$

$$c_{2,0} := \int_0^1 ddp2(x) \cdot p0(x)\, dx \quad c_{2,1} := \int_0^1 ddp2(x) \cdot p1(x)\, dx \quad c_{2,2} := \int_0^1 ddp2(x) \cdot p2(x)\, d$$

$$C := \begin{bmatrix} c_{0,0} & c_{1,0} & c_{2,0} \\ c_{1,0} & c_{1,1} & c_{1,2} \\ c_{2,0} & c_{2,1} & c_{2,2} \end{bmatrix} \qquad b := \begin{pmatrix} 1 \\ 1 \\ 1 \end{pmatrix} \qquad a := C^{-1} \cdot b \qquad a = \begin{pmatrix} -5 \\ 25 \\ -35 \end{pmatrix}$$

$$C = \begin{pmatrix} -0.333 & -0.167 & -0.1 \\ -0.167 & -0.133 & -0.1 \\ -0.1 & -0.1 & -0.086 \end{pmatrix} \qquad C^{-1} = \begin{pmatrix} -15 & 45 & -35 \\ 45 & -195 & 175 \\ -35 & 175 & -175 \end{pmatrix}$$

$$x := 0, 0.1 .. 1$$

$$q(x) := p0(x) + p1(x) + p2(x)$$

$$M0(x) := a_0 \cdot p0(x) + a_0 \cdot p1(x) + a_0 \cdot p2(x) \qquad M1(x) := a_1 \cdot p0(x) + a_1 \cdot p1(x) + a_1 \cdot p2(x$$

$$M2(x) := a_2 \cdot p0(x) + a_2 \cdot p1(x) + a_2 \cdot p2(x) \qquad M(x) := M0(x) + M1(x) + M2(x)$$

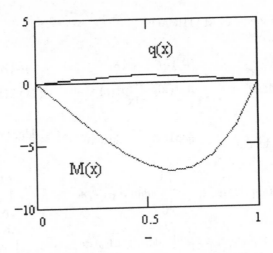

The result is equal with the solution by the finite element method.

References
[1] Hestenes. M.R., Optimization theory, The finite dimensional case, John Wiley & Sons, New York, London, 1975.
[2] Washizu, K., Variational methods in electricity and plasticity, Oxford, Pergamon Press, 1968.
[3] Szabó, J., Roller, B. Anwendung der matrizen rechnung für Stahlwerke. Budapest, Akadémiai Kiadó, 1978.

Parameter Sensitivity of Deterministic and Stochastic Search Methods

K.-J. Böttcher

Federal Armed Forces University Munich, Aero Space Engineering and Technology,

Institute for Mathematics and Computer Sciences, D-85577 Neubiberg / Germany

Keywords. Parameter sensitivity of optimisation methods, stochastic and deterministic search methods, experimental study , measured optimisation time, computer stop watch, importance of time optimal parameter

Abstract. The sensitivity to changes in parameters of 12 common deterministic and stochastic search methods has been investigated by measuring the time they need to approach the known minimum within a preset range of function values. These search methods use or do not use derivatives and depend on 1-5 parameters; their choice largely determines the experimental search time. An example demonstrates that an improper choice of parameters reverses the order of efficiency of search methods: i.e. the fastest method becomes the slowest one. To overcome the difficulty of a proper choice of parameters, the optimisation time was measured as a function of these parameters. Then the optimal parameter vector and optimal search time were determined with regard to the parameter space for any search method and any function separately. The functions of optimisation time versus parameter are presented. Their trend, their structure and position of the optimum are discussed.
The optimal parameter, run times, number of function values, number of decreasing function values, number of iterations required to pass half the distance from the value of the test function at start to it's minimum, etc. at the position of the optimum of optimisation times are discussed being parts of a characterising criterion vector with some 40 components which is typical for each search method and test function.
The practical convergence speed of each search method has been determined by analysing the distribution of function values during iteration of the minimum and their iteration times. The sensitivity of the optimal parameters to changes in the feasibility parameter, the start vector, the distribution of random number vectors in R.S. methods, has been investigated experimentally. In separate programs the function error was investigated for each search method; appropriate termination criteria were selected. By increasing some sample size parameter in many parameter R.S. methods the feasible region could be reached again.

1. Introduction

1.1. Preliminary Notes

There are numerous publications on deterministic and stochastic search methods that

search the optimium of a function, see e.g. [4, 9] and surveys in [10, 14, 15, 17, 18]. They either investigate the advantages or disadvantages of different methods analytically, or experimentally. In order to do so a set of parameters inherent to the search method under investigation is "suitably" chosen. An attempt is made to find one set of parameters "good for a class of functions".
In this article no further survey on search methods will be added to the existing ones. Rather the optimal choice of parameters in each search method - taken from a sample of stochastic and deterministic methods - is examined experimentally. How does this choice affect the "efficiency" of a search method? Does the "efficiency" of deterministic and stochastic methods differ? When does an "efficient method" turn into an "inefficient method"? How many parameters may a search method have that it can practically be handled? Does the same set of parameters hold good for all functions? How fast are these methods in reality? What can the engineer do to find an unknown optimum? These questions result naturally from practical considerations. Nevertheless the outlined parameter problem seems to have been avoided in literature so far due to the large central processor times involved to obtain well established, experimental results and due to the elaborate structure of the testing program that is required.

1.2. Definition of the Problem

The minimisation problem
$$F^*=\min F(x) \quad s.t. \ x \in \Re^n \tag{1}$$
shall be solved with different parameter dependent search methods
$$M_i = M_i(c) \text{ where } c'=(c_0,c_1,c_2,...c_m)\in P_i, \ i=1,2,...12 \tag{2}$$
which are either random or deterministic search methods. Here, P_i denotes the space of admissible parameter vectors for method $M_i(c)$.
Starting from $(x^0,F(x^0))$ they generate a set of iterations
$$(x^0,F^0),(x^1,F^1),(x^2,F^2),...(x^k,F^k),(x^{k+1},F^{k+1}),... \tag{3}$$
that approximate the solution (x^*,F^*) according to some convergence criterion, e.g.
$$|F(x)-F^*|<\varepsilon_f, \ |x-x^*|<\varepsilon_x, \ |F^{k+1}-F^k|<\varepsilon \text{ etc.} \tag{4}$$
The solution (x,F) is obtained after NX iterations, NR generated random No's, NF function values,... NDF calculated gradients,..., NDEMI function values required to pass $F_{1/2}=F^0-0.5\cdot(F^0-F^*))$, NFDOWN decreasing function values, NFEPS calls to the convergence criterion, ..., NFT function values per time, the central processor time RT, required to solve (1), etc. These counting variables may be used as time dependent and time independent components of a criterion vector CR characteristic of method $M_i(c)$:
$$CR = (NX, NR, NF, NDF,...NDEMI, NFDOWN, NFEPS,...NFT,...RT,..) \tag{5}$$
One is tempted to use vector CR as a numerical scale for the performance of a search method. However, practical reasons prevent from doing so, apart from the fact that vector CR has some 40 components to consider: Some components are not present in all search methods. Each component of CR, in general, is a different unknown non-linear function of vector c in $M_i(c)$. To use CR as a measure of performance meant to establish (experimentally), that some sort of vector optimisation problem can be well defined for any method and any function $F(x)$. Hence, the performance of some search method $M_i(c)$ is evaluated by the optimal run time RT* to solve (1):
$$RT^* = \min RT \ s.t. \ c\in P_i \subset \Re^{m+1} \text{ for each } M_i(c) ; \ c^* = \arg \min RT \tag{6}$$
The problem of comparing different search methods will then be decided
a) numerically, by RT*, NX(c*), NF(c*),... and b) by evaluation of the practical

applicability of $M_i(c)$ to solve (1), when their parameter vectors c are changed. Absolute times measured with one computer are required for a). Relative times RT are needed to determine $c*$ and the structure of the RT(c) curve. Comparisons of different $M_i(c)$ on other computers can be done by considering their relative RT(c) relationships. Of course, to apply RT as a measure of performance of some search method in comparison to others entirely depends on the reproducibility of RT by the computer in use.

1.3. Notes on Data Collection

1.3.1. Means and Standard Errors: Choice of the Number NRUN of Repetitions

After NRUN optimisations to solve (1) means μ and their standard errors σ were calculated for each component of criterion vector CR. When the search was deterministic, σ was set equal to 0 in time independent components of CR. Commonly, averaging takes place with data from "successful" runs only, i.e. data from runs that exceeded time or iteration limits are omitted. Hence, μ and σ, -e.g. for RT-, are determined too small. Experimental RT(c) data demonstrated [2] that RT flattened out even if less than 10 from 1000 runs were omitted. Therefore the desired accuracy to obtain the RT(c) function demanded in present investigation all runs to be "successful". Consequently, throughout this study, the stop limits were adequately increased. The number of repetitions NRUN was determined experimentally for each search method for accurate localisation of min RT(c). In particular results from random search methods showed that narrow minima can easily be overlooked for NRUN $\leq$ 100 because standard errors σ of RT are too large. Also, means of random components of vector CR adopt asymptotically with NRUN a constant value: Experimentally, e.g. RT $\approx$ const. for NRUN $\geq$ 500 in this study. Hence, in R.S. methods optimisations were usually repeated 1000 to 30000 times for each value of c. Deterministic methods required substantially less repetitions due to the stability of the central processor time.

1.3.2. Notes on the Choice of ε in Convergence Criterion $|F-F*|<\varepsilon$

The total time was restricted to 150 min's cpu time to run one program. Finally, 12 search methods with their optimal parameters $c*$ set were to be applied successsively to solve (1) for performance comparisons. The demand for accurate data required to have optimisation runs repeated 1000 times for each search method. However, each run should regularly terminate with entry into the feasible region $|F-F*|<\varepsilon$, even for the slowest method. These requirements defined experimentally the lower limit of ε for each test function, e.g. $\varepsilon = 10^{-4}$ for F1 and $\varepsilon = 10^{-6}$ for F6 (see Ch. 1.4). Also, all runs, which measured the RT(c) function in order to determine the minimum position $c*$, had to use the same value ε, which was eventually employed in final comparative runs.

1.3.3. The Use of the Central Processor as a Stop Watch

In this report the time is investigated that some search method needs to localize the minimum of a test function. It had to be verified experimentally that this time is not falsified by undiscovered fluctuations and drifts of time that arise from the workload of the central processor in a job sharing environment. The "time stability" of the computer was investigated in separate programs. The numerical results, see [2], of time stability investigations recorded throughout approximately 70 months demonstrate: The computer had time stable states which lasted sufficiently long for time measurements reported here in order to achieve any degree of accuracy which was practically demanded. Time data RT(c) were reproducible on any day of measurement. As the optimisation time and the scatter from one program were recorded day by day any change in time stability of the central processor was reliably observed. Thus final data were obtained but from a time stable central processor only.

1.4. Relationship Between Run Time and Parameters of Search Methods

In order to determine numerically the relationship between the run time RT and the parameters of a given optimisation method two test functions [6] were used:

F1 Rosenbrock Function, $\varepsilon = 10^{-4}$
$$F(x) = 100(x_1^2 - x_2)^2 + (1 - x_1)^2 \tag{7}$$
$$x^0 = (-1.2,\ 1.0) \quad F(x^0) = 24.2 \quad x^{*'} = (1,1) \quad F^* = 0$$

F6 Engvall Function, $\varepsilon = 10^{-6}$
$$F(x) = x_1^4 + x_2^4 + 2x_1^2 x_2^2 - 4x_1 + 3 \tag{8}$$
$$x^0 = (0.5,\ 2.0) \quad F(x^0) = 19.06 \quad x^{*'} = (1,\ 0) \quad F^* = 0$$

The convergence limit ε was kept constant throughout this study. It's value was determined for each test function separately (see Ch. 1.3.3 and 4.1) to enable fast accurate data collection for all search methods within practical time limits set by the university's computer centre (9000 sec per job). Hence ε differed for F1 and F6. The starting vectors for F1 and F6 remained unchanged, apart from one independent study of their influences (see Ch. 4.2) on optimisation data.

1.5. Search Methods

The iteration process to optimize $F(x)$ is defined by the search method selected from a pool of 12 available deterministic and random search methods:
8 with single, 4 with more than 2 components of their parameter vectors. Table 1.1 summarizes the investigated search methods. M8, M9, M10 are 3, 4, 5 parameters random search methods with modified increments Δx, as it's components are not normalized. Though, the popular names are kept for ease of reference throughout. A brief description will be given; parameter names mentioned here are used to label the axes in graphical data representations.

M8: FSSRS: Fixed Step Size Random Search [13,17]
Starting from a previous best vector XOLD a number of "NXOLD" further reference vectors are consecutively generated. From each one progress is attempted into "NSTEP" random trial directions Δx with different step lengths; components of Δx are N(0,c) distributed. In FSSRS with reversing ("NSTEP < 0") their reversed directions - Δx are tested, too: obviously, a second x value is obtained deterministically, available for comparison, without a time consuming generation of random numbers. The iteration continues into the direction of steepest descent to obtain the next reference point. After generation of this starlike search pattern the convergence criterion is examined.

M9: RANPAT Randomized Pattern Search [7,17]
Additionally to FSSRS in each search star an extended step x - BEF·(XBEST - x) into

Table 1.1. Search Methods

Deterministic Methods $\quad x^{k+1} = x^k + \Delta x^k$

Stochastic Methods $\quad x^{k+1} = \begin{cases} x^k + \Delta x^k & \text{if } F(x^{k+1}) < F(x^k) \\ x^k & \text{if } F(x^{k+1}) \geq F(x^k) \end{cases}$

Symbol	Type	$\Delta x^k = (\Delta x_1^k, \ldots \Delta x_i^k \ldots \Delta x_n^k),$
M 7	Stochastic	$c_o \cdot Z(\omega),\ Z_i(\omega)$: Uniform Distrib. on $[-1.1]$
M 1	Stochastic	$c_o \cdot Z(\omega),\ Z_i(\omega)$: $N(0,1)$ Distrib.
M 2	Stochastic	$c_o \cdot Z(\omega),\ Z_i(\omega)$: $N(0, \lvert \frac{\partial F}{\partial x_i} \rvert)$ Distrib.
M 3	Stochastic	$c_o \cdot Z(\omega),\ Z_i(\omega)$: $N(0, \lvert \frac{\partial F}{\partial x_i} / \frac{\partial^2 F}{\partial x_i^2} \rvert)$ Distrib.
M 4	Deterministic	$- c_o \cdot \nabla F(x^k)$ $\quad$ Gradient Method: Const. Step Width
M 5	Deterministic	$- c_o \cdot (\frac{\partial F}{\partial x_i} / \frac{\partial^2 F}{\partial x_i^2})$ $\quad$ Quasi Newton $H^{-1} \approx ((\frac{\partial^2 F}{\partial x_i^2}))^{-1}$
M 6	Deterministic	$- c_o \cdot (\frac{\partial^2 F}{\partial x^2})^{-1} \cdot \nabla F$ $\quad$ Newton-Method
M 8	Stochastic	3 Parameter Fixed Step Size Random Search, FSSRS
M 9	Stochastic	4 Parameter Random Pattern Search, RANPAT
M 10	Stochastic	5 Parameter Adaptive Step Size Random Search, ASSRS
M 11	Deterministic	$- c_o \cdot \frac{1}{k} \cdot \nabla F(x^k)$ Gradient Method: Variable Step Width
M 12	Deterministic	$- c_o \cdot \rho^* \cdot \nabla F(x^k)$ $\quad \rho^* = \arg \min F[x^k - \rho \nabla F(x^k)]$ (Line Search)

the direction of steepest descent (XBEST - x) is done to check a possible further decrease of F(x). Hence, in RANPAT a 4th parameter, an accelerator BEF > 1 is implemented. Again: for this step no use is made of the random number generator.

<u>M10: ASSRS Adaptive Stepsize Random Search [1,11,12,16]</u>

In principle ASSRS uses a local "radius r" which - in the modified version used here - refers to the $N(0,r)$ distribution of the components of Δx. In case of success radius r is increased by means of an accelerator BEF > 1: r: = r · BEF. In case of failure for KSTEUR ≥ 1 consecutive sample points radius r is decreased by means of a retardation factor VERF < 1: r: = r · VERF. Consequently, - in some simple version [11] - ASSRS has at least 5 parameters to set before the iteration process can start: c_0: the initial "radius", NXOLD, KSTEUR, BEF, VERF.

1.6. Note on Random Numbers

Random numbers for R.S. methods were provided by the UniBw - Unisys A 15 random number generator. It generates uniformly distributed random numbers on [0,1]. For M7 a uniform distribution on [-c, c] is easily derived. R.S. methods apart from M7 used $N(0,1)$ distributed random numbers, which were obtained according to Box-Müller [3]. If

$$X_1 = (-2 \ln Z_1)^{1/2} \cos 2\pi Z_2, \quad X_2 = (-2 \ln Z_1)^{1/2} \sin 2\pi Z_2 \qquad (9)$$

and Z_1, Z_2 are independent and uniform on [0,1], then X_1, X_2 are $N(0,1)$ distributed. For R.S. methods many components of the criterion vector CR (see Ch.1.2), -e.g. the optimisation time- are averages from successful optimisation runs. The random numbers in each of NRUN (see Ch. 1.3.1) repeated optimisation runs were different.

2. Sensitivity of Search Methods to Changes of their Parameter Vectors

2.1. Run Times as a Function of Parameter Vector c

The optimisation problem (1) was solved with each method $M_i(c)$, $i=1,2,...12$ separately for the Rosenbrock function F1 and the Engvall function F6, see Ch. 1.4. Optimisation times $RT(c)$ were measured at gradually closer spaced parameter vectors c, - so to cover the entire permissible parameter space P_i of method $M_i(c)$. In general: for different methods $RT(c)$ depends differently on c. This result refers to the kind of structure, the position of structure within c-space and to convergence or to non-convergence. Also, $RT(c)$ depends on the test function in (1). For any search method and for any test function the dependence of RT on c is not negligible. When experimentally required, any optimisation time $RT \geq RT^*$ could be obtained by properly adjusting parameter c.

2.2. Single Parameter Random Search Methods M7, M1, M2, M3

The shape of the $RT(c)$ function with $c=c_0$, is parabola like, see Fig.s 2.1, 2.2, as well for methods without as with derivatives. The $RT(c_0)$ branch for small values c_0 is steeper than the branch for c_0 beyond the minimum position. $RT(c_0)$ has little to no structure apart from some irregularities of the order of a few standard errors within the minimum region. In particular, for uniformly and normally distributed random variables $Z_i(\omega)$, $i=1,2,...n$ in M7 and M1 (see table 1.1 in Ch. 1.5.) the $RT(c_0)$ functions look very similar, see Fig. 2.1. and Fig. 2.2. Essentially, $RT(c_0)$ for the Rosenbrock function F1 and for the Engvall function don't differ in shape.

2.3. Single Parameter Deterministic Search Methods M4, M5, M6

For different deterministic methods M4, M5 and M6 (see Table 1.1. in Ch. 1.4.) the $RT(c_0)$ functions were different in shape and in structure, see Fig.s. 2.3a, 2.3b, 2.4. They depended considerably on the test functions.
For the gradient method M4 with constant stepsize function values $F(x)$ suddenly exceeded any limit during iteration ("exponent overflow"), once some critical parameter c_R was passed. Also, at isolated c-values beyond c_R optimisation times $RT(c_0)$ became very large unless the runs were stopped, see Fig. 2.3a for F1. $RT(c_0)$ for F6 had several local minima for $c_0 < c_R$, see Fig. 2.3b. When deterministic methods M5 and M6 with second derivatives were applied to solve (1) for test function F6 similar $RT(c_0)$ resulted: $RT(c_0)$ had some relative minima and a sharp rise beyond a parabola type

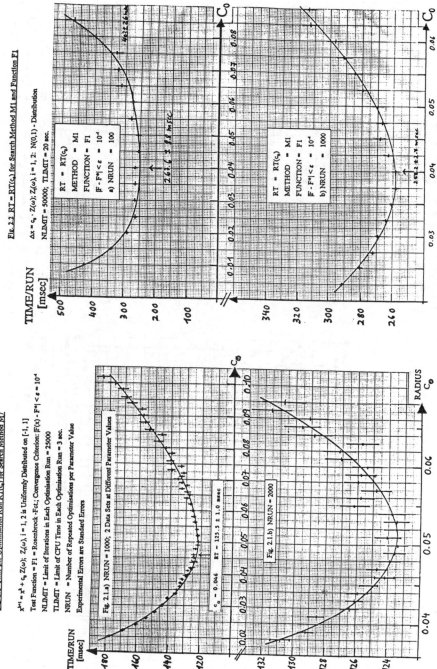

Fig. 2.1 Time per Optimisation Run RT(c_0) for Search Method M7

$x^{k+1} = x^k + c_0 \cdot Z(\omega);$ $Z_i(\omega), i = 1, 2$ is Uniformly Distributed on [-1, 1]

Test Function = F1 = Rosenbrock-Fct.; Convergence Criterion: $|F(x) - F^*| < \varepsilon = 10^{-4}$

NLIMIT = Limit of Iterations in Each Optimisation Run = 25000

TLIMIT = Limit of CPU Time in Each Optimisation Run = 3 sec.

NRUN = Number of Repeated Optimisations per Parameter Value

Experimental Errors are Standard Errors

Fig. 2.1.a) NRUN = 1000; 2 Data Sets at Different Parameter Values

Fig. 2.1.b) NRUN = 2000

$c_0 = 0.046$ RT = 125.5 ± 1.0 msec

TIME/RUN [msec]

RADIUS C_0

Fig. 2.2 RT = RT(c_0) for Search Method M1 and Function F1

$\Delta x = c_0 \cdot Z(\omega);$ $Z_i(\omega), i = 1, 2: N(0,1)$ - Distribution

NLIMIT = 50000; TLIMIT = 20 sec.

RT = RT(c_0)

METHOD = M1

FUNCTION = F1

$|F - F^*| < \varepsilon = 10^{-4}$

a) NRUN = 100

RT = RT(c_0)

METHOD = M1

FUNCTION = F1

$|F - F^*| < \varepsilon = 10^{-4}$

b) NRUN = 1000

TIME/RUN [msec]

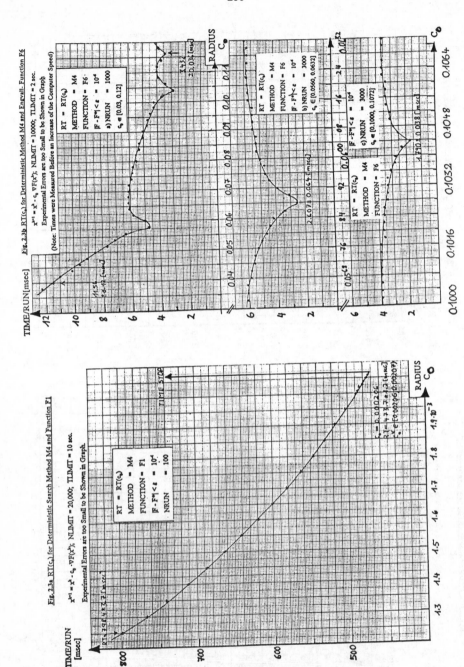

TIME/RUN [msec]

Fig. 2.3b RT(c_e) for Deterministic Method M4 and Enzvsll. Function F6

$x^{i+1} = x^i - c_e \cdot \nabla F(x^i)$; NLIMIT = 10000; TLIMIT = 2 sec.

Experimental Errors are too Small to be Shown in Graph

(Note: Times were Measured Before an Increase of the Computer Speed)

RT = RT(c_e)
METHOD = M4
FUNCTION = F6
|F - F*| < ε = 10^4
a) NRUN = 1000
$c_e \in [0.03, 0.12]$

RT = RT(c_e)
METHOD = M4
FUNCTION = F6
|F - F*| < ε = 10^4
b) NRUN = 3000
$c_e \in [0.0560, 0.0632]$

RT = RT(c_e)
METHOD = M4
FUNCTION = F6
|F - F*| < ε = 10^4
c) NRUN = 3000
$c_e \in [0.1000, 0.1072]$

RADIUS c_e

Fig. 2.3a RT(c_e) for Deterministic Search Method M4 and Function F1

$x^{i+1} = x^i - c_e \cdot \nabla F(x^i)$; NLIMIT = 20,000; TLIMIT = 10 sec.

Experimental Errors are too Small to be Shown in Graph.

TIME/RUN [msec]

RT = RT(c_e)
METHOD = M4
FUNCTION = F1
|F - F*| < ε = 10^4
NRUN = 100

RADIUS c_e

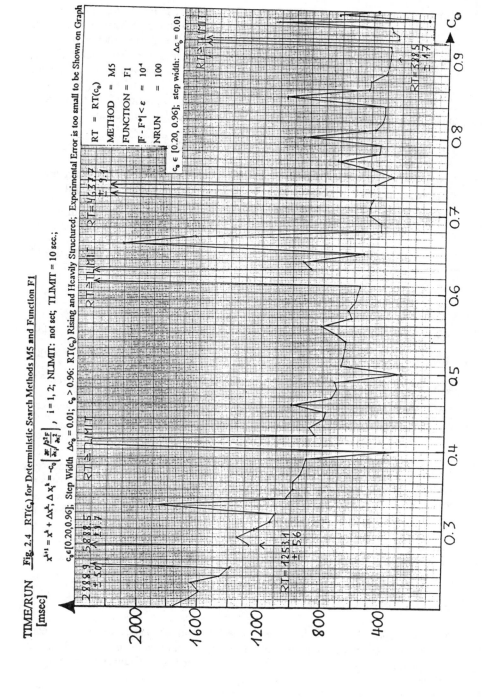

TIME/RUN [msec]

Fig. 2.4 RT(c_0) for Deterministic Search Methods M5 and Function F1

$x^{k+1} = x^k + \Delta x^k$, $\Delta x_i^k = -c_0 \left[\frac{\partial F}{\partial x_i} / \frac{\partial^2 F}{\partial x_i^2} \right]$, $i = 1, 2$; NLIMIT: not set; TLIMIT = 10 sec.;

$c_0 \in [0.20, 0.96]$; Step Width $\Delta c_0 = 0.01$; $c_0 > 0.96$: RT(c_0) Rising and Heavily Structured; Experimental Error is too small to be Shown on Graph

RT = RT(c_0)
METHOD = M5
FUNCTION = F1
$|F - F^*| < \epsilon$ = 10^{-4}
NRUN = 100

$c_0 \in [0.20, 0.96]$; step width: $\Delta c_0 = 0.01$

global minimum. However, when M5 solved (1) for F1 the $RT(c_0)$ function was heavily structured with single abnormal time values within an otherwise decreasing trend curve till some critical parameter c_R, see Fig. 2.4. The severe structure below the critical value c_R became apparent only when the step width Δc_0 in the parameter space was sufficiently reduced. If Δc_0 is too coarse, relative minima in narrow valleys are overlooked. There might be also structure in $RT(c_0)$ for F6 for smaller Δc_0. Because of this difficulty the minimum RT^* and c_0^* were intentionally obtained from the trend of the $RT(c)$ curve, as a replacement of the indeterminable global minimum.

2.4. Single Parameter Deterministic Search Method M11

Deterministic gradient method M11 uses - as in theoretical studies on convergence - variable stepsize c_0/k where k denotes the iteration number, (table 1.1 in Ch. 1.5.).

Table 2.1. Dependence of F(x) for F1 on Number of Iterations $k^{1)}$ and on Time[1]
M11: $x^{k+1}=x^k-(c_0/k)\cdot\nabla F(x^k)$; k = iteration ; $c_0 = 0.006946^{2)}$ $F(x^0) = 24.2$ $F(x^*) = 0$

$k^{1)}$	F(x)	Time [msec]	$k^{1)}$	F(x)	Time [msec]
2	9.915	0.5	462	4.633×10^{-2}	50
5	4.801×10^{-2}	1	927	4.607×10^{-2}	100
43	4.721×10^{-2}	5	4,635	4.548×10^{-2}	500
87	4.695×10^{-2}	10	1,296,684	4.349×10^{-2}	140,000
225	4.659×10^{-2}	25	9,622,299	4.283×10^{-2}	980,000

Notes: 1) The time stop was set to terminate the optimisation, and iteration number k and function value F(x) were recorded.
2) $F(x^0)$ was reduced at most - in given iteration times - with this choice of c_0.

Experimentally, optimisation times $RT(c_0)$ are large and for some ε may practically impede to achieve convergence. E.g. for F6 optimisation times for convergent runs are considerably increased compared to those of M4 with constant step width, see Table 3.3. For the Rosenbrock function F1 iterations got stuck before they reached the feasible region with $\varepsilon = 10^{-4}$, - no matter how much time was made available for the iteration process, see Table 2.1. In order to achieve convergence for final data ε had to be increased to $\varepsilon = 0.05$. If M11 is practically convergent at all convergence takes place within a very narrow range of c_0 values, which easily can be overlooked.

2.5. Many Parameter Deterministic Gradient Method M12

M12 with optimal step width ρ^*, see Table 1.1, requires to activate a second search method in each iteration k to solve a separate optimisation problem (10), with additional parameters p to be set in advance, e.g. some convergence parameter ε_s.
$$x^{k+1} = x^k - c_0\rho^*\nabla F(x^k) \quad G_k(\rho):= F[x^k - \rho\nabla F(x^k)]$$
$$\rho^* = \arg\min G_k(\rho) \qquad G^* = \min G_k(\rho) \text{ s.t. } \rho\in\Re \tag{10}$$
$S_2(p)$ = search method to solve (10) with unknown parameter vector p to optimise.
Over - or under estimation of ρ^* during the main iteration process denoted by index k

will take place. Correction factor c_0 compensates the average over - or under estimate; c_0 has to be optimised for each function $F(x)$.

$RT(c_0)$ has an evolved structure which is discovered only when the grid width Δc_0 for c_0 is sufficiently reduced [2]. Observations show that the $RT(c_0)$ function strongly depends on termination limit ε_s[2]. If ε_s is small, - e.g. $< 10^{-2}$ for F1 with $\varepsilon = 10^{-4}$ or $\varepsilon_s \leq 10^{-1}$ for F6 with $\varepsilon = 10^{-6}$ - $RT(c_0)$ may have a pronounced minimum $RT^*(c_0^*,\varepsilon_s)$. For small increases of c_0 above c_0^* the time RT may rise beyond all limits. If ε_s is too large M12 may diverge: either by too large function values or by not reaching the feasibility region $|F - F^*| < \varepsilon$ no matter how much time is permitted for the search. This may occur at isolated values of the correction factor c_0 or in intervals for c_0.

To determine $RT(c_0)$ for some function and ε in dependence of all variables in the line search routine $S_2(p)$ is very difficult. Much time is needed to optimize $RT(c_0)$ or to improve the performance of M12 in solving (1). This is a great disadvantage of the optimal gradient method M12 compared with gradient method M4 with constant stepwidth. Additionally, M4 was much faster than M12 (see Table 3.3).

2.6. Many Parameter Random Search Methods

2.6.1. R.S. Method M8: FSSRS

FSSRS (see Ch. 1.5) essentially selects in each iteration k the best function value from a sample of NXOLD $\cdot$ NSTEP values, before it proceeds to the next iteration k+1. The run time function RT (c_0) has the parabolalike shape known from M1. It's minimum position for F1 and F6 is almost independent of NXOLD and NSTEP. RT for given NSTEP and c_0 rises with NXOLD $\to$ 0 and NXOLD $\to$ ∞, with some wide ranged minimum in between, see Fig. 2.5. The optimal value of NXOLD depends on the test function. Most important is the dependence of run time RT on the number NSTEP of trial directions from each reference point, see Fig. 2.6. For F1 and F6 time is wasted when several descent directions are tested. Rather one random direction and it's (deterministic) reverse (NSTEP < 0, see Ch. 1.5) shall be examined before progress to the next reference point. The result "NSTEP = - 1 is optimal" is likely to be modified for functions $f(x)$, $x \in \Re^n$ and $n>2$. In essence, FSSRS is faster than R.S. method M1.

2.6.2 R.S. Method M9: RANPAT

RANPAT (see Ch. 1.5) tested here is an improved FSSRS method with some deterministically defined, extended step from each reference point into the direction of steepest descent. Qualitatively, for F1 and F6, the shape of the $RT(c_0)$, RT(NXOLD), RT(NSTEP) dependencies - with the remaining parameters fixed - are similar in RANPAT and FSSRS. However, for both test functions the position c_0^* (BEF) of the minimum of the $RT(c_0)$ function, see Fig. 2.7, and it's range as well as RT^*, depend considerably on accelerator BEF >1 (here NXOLD and NSTEP are kept constant), see Fig. 2.8 and Fig. 2.9 for c_0^*-range and RT^* as functions of BEF. When BEF increases the minimum position of the $RT(c_0)$ function moves to smaller values c_0^*(BEF), see Fig. 2.7 and 2.8, so to compensate the reduction of the $N(0, c_0)$ distributed increments Δx. If BEF becomes too large, - e.g. BEF $\geq$ 150 for F1 and BEF $\geq$ 600 for F6, - function values at iteration points with deterministic increments BEF $\cdot$ Δx become too large and are calculated in vain: run times increase noticeably, and a second minimum

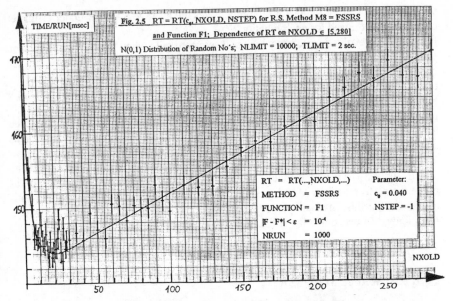

Fig. 2.5 RT = RT(c_0, NXOLD, NSTEP) for R.S. Method M8 = FSSRS
and Function F1; Dependence of RT on NXOLD $\in$ [5,280]
N(0,1) Distribution of Random No's; NLIMIT = 10000; TLIMIT = 2 sec.

Fig. 2.6 - see Fig. 2.5 - Dependence of RT on NSTEP $\in$ [-12,+12]

M8 and F1; N(0,1) Distribution of Random No's; NLIMIT = 10000; TLIMIT = 2 sec.

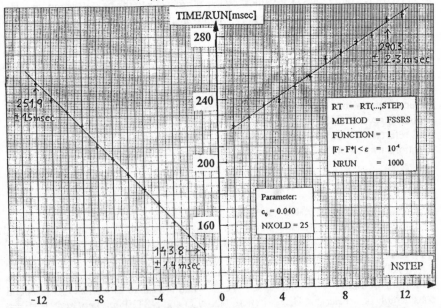

emerges at some position close to that of method FSSRS without any acceleration, see Fig.s 2.8 and 2.9. RANPAT is faster than FSSRS, if BEF is suitably chosen.

2.6.3. R.S. Method M 10: ASSRS

In some program mode the 5 components of parameter vector c in ASSRS (see Ch. 1.5), were changed simultaneously: $c_0 \in \Re^+$, NXOLD $\geq$ 1, BEF > 1, KSTEUR $\geq$ 1,

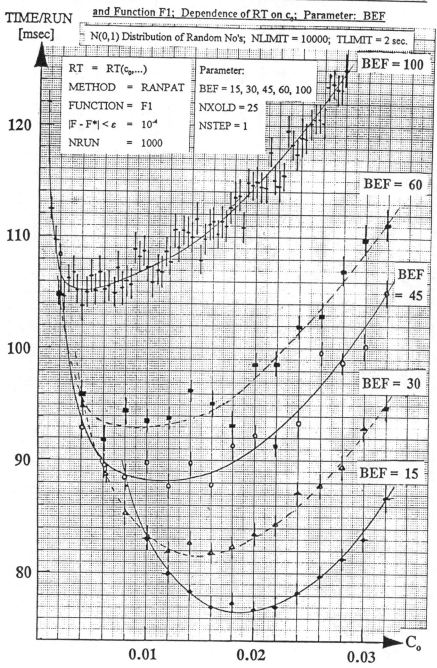

Fig. 2.7 RT = RT(c₀, BEF, NXOLD, NSTEP) for R.S. Method M9 = RANPAT and Function F1; Dependence of RT on c₀; Parameter: BEF

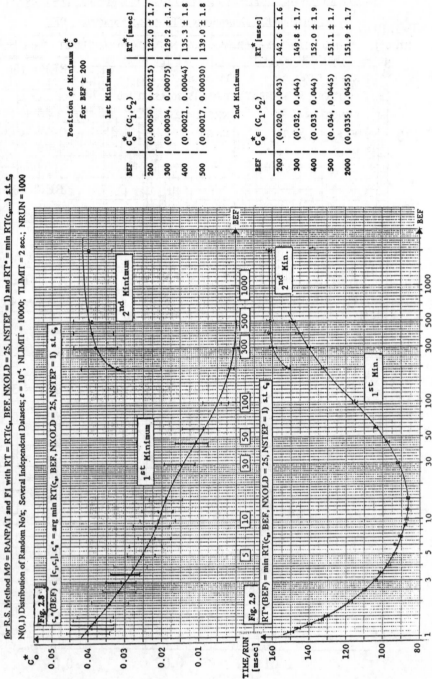

Fig. 2.8, 2.9 Dependence of Optimal Time RT* and Range of Minimum Position $c_0^* \in [c_1, c_2]$ on Parameter BEF ∈ {1,2000} for R.S. Method M9 = RANPAT and F1 with $RT = RT(c_0,$ BEF, NXOLD = 25, NSTEP = 1) and $RT^* = \min RT(c_{...})$ s.t. c_0 N(0,1) Distribution of Random No's; Several Independent Datasets; $\varepsilon = 10^4$; NLIMIT = 10000; TLIMIT = 2 sec.; NRUN = 1000

Fig. 2.8 $c_0^*(\text{BEF}) \in [c_1, c_2]$, $c_0^* = \arg \min RT(c_0,$ BEF, NXOLD = 25, NSTEP = 1) s.t. c_0

Fig. 2.9 $RT^*(\text{BEF}) = \min RT(c_0,$ BEF, NXOLD = 25, NSTEP = 1) s.t. c_0

Position of Minimum C_0^*
for BEF ≥ 200

1st Minimum

BEF	$c_0^* \in (c_1, c_2)$	RT^* [msec]
200	(0.00050, 0.00215)	122.0 ± 1.7
300	(0.00034, 0.00075)	129.2 ± 1.7
400	(0.00021, 0.00044)	135.3 ± 1.8
500	(0.00017, 0.00030)	139.0 ± 1.8

2nd Minimum

BEF	$c_0^* \in (c_1, c_2)$	RT^* [msec]
200	(0.020, 0.043)	142.6 ± 1.6
300	(0.032, 0.044)	149.8 ± 1.7
400	(0.033, 0.044)	152.0 ± 1.9
500	(0.034, 0.0445)	151.1 ± 1.7
2000	(0.0335, 0.0455)	151.9 ± 1.7

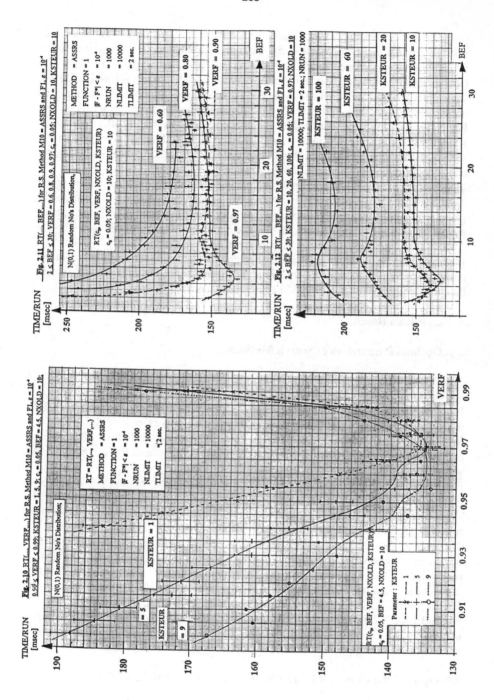

0 < VERF < 1. Measurements of the optimisation time RT(c) demonstrated for each test function F1 and F6 (Ch. 1.4)-, that the initial parameter space could be reduced. RT(c_0) was nearly independent of c_0 in $[10^{-12}, 10^0]$ for F1 and in [0.01, 10.0] for F6. (RT increased by only 25 % of RT* for F6, when $c_0 \in [10^{-4}, 10^3]$). RT(NXOLD) was similar in ASSRS as in FSSRS and in RANPAT. In essence, RT depended critically on the triple BEF, KSTEUR, VERF, see Fig.s 2.10 - 2.12, in a different way for F1 and for F6. Whereas for F6 the local radius should be reduced each time after failure, i.e. KSTEUR $\cong$ 1, function F1 requires only $1 \leq$ KSTEUR ≤ 10 if VERF is optimal for F1, see Fig. 2.10. If VERF had been chosen badly, KSTEUR $\in$ (30, 70) would be a better choice for F1 at the expense of optimisation time.

These few reported observations indicate, that normally an improper parameter vector c will be chosen for ASSRS, which slows down this method in solving (1),- unless the global minimum of RT s.t. $c \in P_{10}$ has been carefully located. E.g. for F1 RT(c) had been measured at some 10,000 values of vector c. In the present form [11], see Ch. 1.5, ASSRS is not well suited to solve (1), but it can easily be improved [2].

In general, a practical limit to the number of components of parameter vector c in any search method appears to be 5. This limit is suggested from the difficulties in finding the global minimum of RT(c) in M 12, in ASSRS and in RANPAT.

3. Numericals Results

3.1. Optimal Parameters of Search Methods

According to (6) RT* is the global minimum of RT(c) on parameter space $P_i \subset \mathfrak{R}^{m_i+1}$ of search method $M_i(c)$, i=1,2,...in (2).

Experimentally, RT* is not well defined and must be replaced [2] by the global minimum of a general trend curve T(c), i.e. RT*:= min T(c) s.t. $c \in P_i$, i=1,2,...12.

R.S. methods: T(c) approximated with least curvature the randomly scattered experimental time data RT(c)$\pm\sigma$(c). In many parameter R.S. methods this approximation was repeated alike for each component of vector c.

Deterministic methods: The experimental scatter of the stochastic time variable RT(c) was but 1 % even at the least run times of this study, RT $\approx$ 0.7 msec. Here the structured experimental RT(c) data are approximated by T(c), which ignores possible narrow relative time minima and resonances at isolated c-values.

Tables 3.1 and 3.2 list the parameter subspaces P_i^* i=1,2,...12 of the observed minimum position RT* for the Rosenbrock-(F1) and Engvall function (F6) respectively. The minimum of optimisation times on P_i^* was determined with independent high accuracy runs with very small step widths Δc. The last column of Tables 3.1 and 3.2 lists c_i^* which corresponds to the minimum; c_i^* is not unique.

Tables 3.1 and 3.2 demonstrate that for one function the same components of the parameter vector, e.g. radius c_0, differ for different methods. Also, optimal parameters for one search method applied to two test functions F1 and F6 are different.

3.2. Optimal Run Times RT* and Number of Function Values NF*

Table 3.3 lists the optimal run times RT* and the optimal number of evaluated function values NF* at c* of Tables 3.1 and 3.2 for F1 and F6.

Table 3.4 shows the influence of derivatives on R.S. methods in solving (1):

Table 1.1. Optimal Parameters for 12 Search Methods and the Rosenbrock Function F1 with $\varepsilon = 10^{-4}$

Method	Symbol	Minimum-Range	Parameter Optimum
Uniform Distrib.	M 7	$c_o \in [0.045,0.060]$	$c_o^* = 0.052$
$N(0,1)$	M 1	$c_o \in [0.025,0.050]$	$c_o^* = 0.040$
$N(0,\|\frac{\partial F}{\partial x_i}\|)$	M 2	$c_o \in [0.040,0.110]$	$c_o^* = 0.075$
$N(0,\|\frac{\partial F}{\partial x_i}/\langle\frac{\partial^2 F}{\partial x_i^2}\rangle\|)$	M 3	$c_o \in [12,28]$	$c_E^* = 21$
VF	M 4	$c_o \in [0.00206,0.00207]$	$c_o^* = 0.002062$
$\frac{\partial F}{\partial x_i}/\frac{\partial^2 F}{\partial x_i^2}$	M 5	$c_o \in [0.68,0.91]$	$c_o^* = 0.90$
$H^{-1}\cdot VF$	M 6	$c_o \in [0.90,1.05]$	$c_o^* = 1.024$
FSSRS	M 8	$c_o \in [0.0325,0.04775]$	$c_o^* = 0.040$
		$NXOLD \in [6,55]$	$NXOLD^* = 25$
		$NSTEP = -2,-1$	$NSTEP^* = -1$
RANFAT	M 9	$c_o \in [0.008,0.032]$	$c_o^* = 0.019$
		$BEF \in [5,30]$	$BEF^* = 15$
		$NXOLD \in [5,50]$	$NXOLD^* = 25$
		$NSTEP = -2,-1$	$NSTEP^* = -1$
ASRS	M 10	$c_o \in [1.8\times10^{-12},1.5]$	$c_o^* = 0.05$
		$BEF \in [2.6,6.8]$	$BEF^* = 4.5$
		$VERF \in [0.9450,0.9775]$	$VERF^* = 0.9650$
		$NXOLD \in [8,60]$	$NXOLD^* = 25$
		$KSTEUR \in [4,18]$	$KSTEUR^* = 5$
$\frac{1}{k}\cdot VF$	M 11	$c_o \in [0.00683,0.00710]$	$c_o^* = 0.00688$ ($\varepsilon = 10^4$:not conv.)
$F[x^k,\rho VF(x^k)]$	M 12	$c_o \in [1.510,1.550]$	$c_o^* = 1.538$ $\varepsilon = 0.05$ was used.
$\to$ min.		$\rho_o = 1.0$ Start:$\varepsilon_{SUB} = 10^3$	

Table 1.2. Optimal Parameters for 12 Search Methods and the Enrell Function F6 with $\varepsilon = 10^{-6}$

Method	Symbol	Minimum-Range	Parameter Optimum
Uniform Distrib.	M 7	$c_o \in [0.009,0.015]$	$c_o^* = 0.012$
$N(0,1)$	M 1	$c_o \in [0.0070,0.0100]$	$c_o^* = 0.0080$
$N(0,\|\frac{\partial F}{\partial x_i}\|)$	M 2	$c_o \in [0.118,0.148]$	$c_o^* = 0.133$
$N(0,\|\frac{\partial F}{\partial x_i}/\langle\frac{\partial^2 F}{\partial x_i^2}\rangle\|)$	M 3	$c_o \in [0.80,1.25]$	$c_o^* = 1.00$
VF	M 4	$c_o \in [0.1042,0.1046]$	$c_o^* = 0.1044$
$\frac{\partial F}{\partial x_i}/\frac{\partial^2 F}{\partial x_i^2}$	M 5	$c_o \in [0.98,1.10]v [1.13,1.18]$	$c_o^* = 1.05$
$H^{-1}\cdot VF$	M 6	$c_o \in [1.02,1.06]v [1.09,1.13]$	$c_o^* = 1.05$
FSSRS	M 8	$c_o \in [0.0070,0.0089]$	$c_o^* = 0.0080$
		$NXOLD \in [10,130]$	$NXOLD^* = 15$
		$NSTEP = -1$	$NSTEP^* = -1$
RANFAT	M 9	$c_o \in [0.0020,0.0030]$	$c_o^* = 0.0025$
		$BEF \in [20,209]$	$BEF^* = 40$
		$NXOLD \in [6,100]$	$NXOLD^* = 15$
		$NSTEP = -2,-1$	$NSTEP^* = -1$
ASRS	M 10	$c_o \in [0.01,14.0]$	$c_o^* = 1.0$
		$BEF \in [3.5,7.5]$	$BEF^* = 4.5$
		$VERF \in [0.54,0.73]$	$VERF^* = 0.62$
		$NXOLD \in [10,24]$	$NXOLD^* = 15$
		$KSTEUR \in [1,4]$	$KSTEUR^* = 1$
$\frac{1}{k}\cdot VF$	M 11	$c_o \in [0.05882,0.05884]$	$c_o^* = 0.05883$
$F[x^k,\rho VF(x^k)]$	M 12	$c_o \in [0.989,0.992]$	$c_o^* = 0.992$
$\to$ min.		$\rho_o = 1.0$ Start:$\varepsilon_{SUB} = 10^5$	

Table 3.3 Optimal Run Times RT* and No. of Function Values NF*

Function		F1: Rosenbrock $\varepsilon = 10^{-4}$		F6: Engvall $\varepsilon = 10^{-6}$	
Method	Symbol	RT* [msec][4]	NF*	RT*[msec][4]	NF*
Uniform Distrib.	M7	122.7 ± 1.1	1268 ± 12	134.1 ±1.8	1354 ± 19
N(0,1)	M1	251.9 ± 2.4	1336 ± 13	222.5 ± 2.5	1,132 ± 14
$N\left(0, \left|\frac{\partial F}{\partial x_i}\right|\right)$	M2	484 ± 14	2384 ± 70	9.187 ± 0.087	44.4 ± 0.4
$N\left(0, \left|\frac{\partial F}{\partial x_i} \middle/ \frac{\partial^2 F}{\partial x_i^2}\right|\right)$	M3	207.0 ± 4.3	1000 ± 14	8.447 ± 0.091	38.3 ± 0.4
∇F	M4	469.15 ± 0.70	4483	0.7283±0.0050	8
$\frac{\partial F}{\partial x_i} \middle/ \frac{\partial^2 F}{\partial x_i^2}$	M5	387.47 ± 0.66	3014	0.7888±0.0034	7
$H^{-1}\nabla F$	M6	2.694 ± 0.02	8	2.013 ± 0.019	6
FSSRS	M8	143.3 ± 1.4	1336 ± 13	130.1 ± 1.5	1149 ± 15
RANPAT	M9	83.55 ± 0.76	809.1 ± 7.3	21.81 ± 0.17	214.6 ± 1.7
ASSRS	M10	134.6 ± 1.9	1151 ± 17	16.61 ± 0.14	141.5 ± 1.2
$\frac{1}{k}\nabla F$	M11[2]	((0.5318 ± 0.0074 6))[1]		28,639 ± 44[3]	264,411
$F[x^k - \rho\nabla F(x^k)] \to$ Min.	M12[2]	1,938.1 ± 1.6	22,029	15.808 ± 0.031	184

Notes: 1) M11 did not converge for F1 with $\varepsilon = 10^{-4}$. Data were obtained with $\varepsilon = 0.05$

2) Data for M1, M2,...M10 were obtained in one program for F1 and for F6;
   Data for M11, M12 for F1, F6 were collected in 4 separate programs

3) Mean values were determined from 1000 optimisation runs apart from method M11
   for F6, which used 300 runs only.

4) Quoted errors are standard errors for R.S. methods and standard deviations for
   deterministic methods, which takes into account the experimental results on time
   stability investigations of the computer, see [2]

Optimal Run Times [msec] for Single Parameter Search Methods.

Table 3.4 : Stochastic Search Methods without and with Derivatives

Symbol	M1	M2	M3
Distribution: see Table 1.1	N(0,1)	$N\left(0, \left|\frac{\partial F}{\partial x_i}\right|\right)$	$N\left(0, \left|\frac{\partial F}{\partial x_i} \middle/ \frac{\partial^2 F}{\partial x_i^2}\right|\right)$
F1: Rosenbrock	252 ± 3	484 ± 14	207 ± 5 msec
F6: Engvall	223 ± 3	9.2 ± 0.1	8.5 ± 0.1 msec

Table 3.5 : Deterministic Search Methods with Derivatives

Symbol	M4	M5	M6
Iteration: see Table 1.1	∇F	$\frac{\partial F}{\partial x_i} \middle/ \frac{\partial^2 F}{\partial x_i^2}$	Newton: $H^{-1} \cdot \nabla F$
F1	469	387	2.7 msec
F6	0.73	0.79	2.0 msec

for F6: the search becomes faster: RT*(M3) < RT*(M2) < RT*(M1);
for F1: the search becomes slower: RT*(M2) > RT*(M1),
     or it becomes faster: RT*(M3) < RT*(M1).
Apparently the introduction of a gradient into stochastic search methods does not necessarily speed up the search for all functions.
Table 3.5. shows the influence of 1st and 2nd derivatives on deterministic methods
for F1: the search becomes faster: RT*(M6) < RT*(M5) < RT*(M4);
for F6: the search becomes slower: RT*(M6) > RT*(M5) > RT*(M4)
I.e. the Newton method M6 was slower than the simple gradient method M4 for F6.
Deterministic gradient methods M11 and M12 are slower than M4, M5, M6;   M11
with step width 1/k is the slowest of all methods in this study.
Noteworthy, too: for F1 stochastic methods M1 and M3 in Table 3.4 are faster than
deterministic methods M4 and M5 in Table 3.5:
for F1: RT*(M3) < RT*(M1) < RT*(M5) < RT*(M4), and RT*(M2) ≈ RT*(M4)
Hence "more information", - i.e. the use of derivatives - did not accelerate the search
in any case. Also, deterministic methods were not always faster than R.S. methods.

Table 3.6 Optimal Run Times [msec] for Many Parameter Random Search Methods

Method	M1	FSSRS	RANPAT	ASSRS
F1: Rosenbrock	252 ± 3	143 ± 2	84 ± 1	135 ± 2 msec
F6: Engvall	223 ± 3	130 ± 2	22 ± 1	17 ± 1 msec

Numerical results in Table 3.6 for many parameter R.S. methods show:
(a) RT* is improved if compared with single parameter method M1 for F1 and F6.
(b) The introduction of an accelerator factor ("BEF") in RANPAT accelerates the search compared to FSSRS (fixed stepsize) for F1 and for F6.
(c) A retardation factor to optimally control the stepsize in ASSRS accelerates the search for F6 but it slows down the search for F1.
Numerical results for M7, with a uniform distribution of random numbers on [-c,c] are:
a) M7 is faster than M1 which uses an N(0,c) distribution of random numbers.
b) M7 is slower than RANPAT. c) M7 is for F1 faster, and for F6 slower than R.S. methods M2,M3,M8,M10.

3.3. Time Optimal Components NFDOWN and NDEMI of Criterion Vector CR

Components NFDOWN and NDEMI of criterion vector CR coarsely describe the progress of iterations relative to NX, the number of generated x-vectors:
NFDOWN   = number of decreasing function values $F(x^{k+1}) < F(x^k)$, k = iteration index
NDEMI   = number of x-vectors required to pass $F_{1/2}$: this number is stored,
     if $F^{k+1} < F^k$ and $F^{k+1} < F_{1/2} = F* + 0.5 \cdot [F(x^0)-F*]$
NFDOWN reveals one major difference between R.S. and deterministic methods:
R.S. methods more often fail to make progress on iteration; for numerical details see [2]
R.S. methods generate decreasing function values with ratios NFDOWN/NX of only
10 % or less for F1 and of about 30 % for F6. Contrarily, deterministic methods
generate mostly decreasing function values with every new iteration vector.
NDEMI, the number of iterated x-vectors to reach $F_{1/2}$ is a rough indicator for a search
method to leave the start position x^0.
R.S. methods: The number of half way iterations NDEMI to arrive at $F_{1/2}$ was different
for test function F1 and F6, see Table 3.7.

For F1 methods M2, M3 (with derivatives) need an average of 8.5 % and 9 % of all iterations in the beginning to decrease $F(x)$ from $F(x^0)$ to $F(x)<F_{1/2}$. Other single and many parameter R.S. methods need at most 6 iterations, which corresponds to less than 0.55 %, to pass the $F_{1/2}$ mark. For F6 between 5 % and 10 % of all iterations reached $F_{1/2}$ after start from x^0.

Table 3.7 NDEMI/NX in % for F1 and F6 with Time Optimal Parameters in M1-M12

Method	M7	M1	M2	M3	M4	M5	M6	M8	M9	M10	M11	M12
F1	0.51	0.43	8.5	9.0	0.09	0.10	14.3	0.39	0.53	0.42	÷	0.05
F6	9.14	10.3	10.0	8.3	14.3	16.7	20.0	10.1	7.8	4.9	~ 0	20.0

Deterministic methods reached $F_{1/2}$ of test function F1 after 1 till 4 iterations and $F_{1/2}$ of F6 after the first iteration, which corresponds to NDEMI/NX $\leq$ 20 %.
Summarily, the convergence rate of all investigated search methods is for both test functions initially very large; it is largest for deterministic methods.

3.4. Practical Convergence Speed

The number of iterations and/or the times required to pass preset function marks on the path of iteration was used as a practical measure of convergence speed. The search methods were applied with their time optimal c^*, see Tables 3.1, 3.2. The distance between $F(x^0)$ at start vector x^0 and F^* was subdivided into 11 intervals JS:
JS = 0: $F^* \leq F < F^* + \varepsilon$; $\varepsilon = 10^{-4}$ for F1 (Rosenbrock); $\varepsilon = 10^{-6}$ for F6 (Engvall)
JS = 1, 2,...10: $F^* + \varepsilon + (JS-1)\Delta F \leq F < F + \varepsilon + JS \cdot \Delta F$, $\Delta F = 0.1 \cdot (F(x^0) - F^*)$
During each run of -in general- 1000 runs iterated function values were sorted into the correct interval JS and the corresponding iteration number and run time were stored. Finally, the averages of the (iteration, run time)-tupels and their standard deviations were determined for each interval JS. Due to the storage and averaging process the increase in time ranged from some 10 % to several 100 %.
As an example only , see [2], Table 3.8 lists the distribution of mean run times with their standard errors versus function interval JS for function F6.
Most of the function decrease takes place during the first iterations, which very fast covered half the distance from $F(x)$ at start x^0 to F^*. This statement has been secured reliably with measurements of the actual distribution of function values to determine the practical convergence speed of a search method. Both deterministic and R.S. methods used up most of the optimisation time and needed most of the iterations shortly before they were successful to enter the feasible region from near by: Convergence speed data showed a jump discontinuity in the number of iterated function values moving from the last 2 function intervals into the ε - environment of F^*. Additionally to convergence speed results of intermediate output of iterated x-vectors and individual function values provide strong support to reduce the local radius c_0 on approach of the minimum during iterations. In this way the probability is increased to enter the feasibility region on approach. A measure for radius decrease may be derived from the history of the iteration process: When the rate of function improvements reduces with an increase of iterations, the radius should decrease.

Table 3.8 Mean Time [msec] Required to Reach Function Interval JS for F6
JS: Definition see Ch. 3.4; NRUN = 1000, apart from M11: NRUN = 1

JS	M7	M1	M2	M3	M4	M5	M6	M8	M9	M10	M11 [2]	M12
10	7.79 ±0.03	8.70 ±0.04	1.31 ±0.35	0.66 ±0.14				7.95 ±0.04	3.80 ±0.07	2.48 ±0.21		
9	14.37 ±0.02	16.79 ±0.04	1.36 ±0.32	1.20 ±0.21				14.51 ±0.03	7.56± 0.026	3.64 ±0.26		
8	21.11 ±0.03	25.06 ±0.04	1.92 ±0.36	1.52 ±0.24				21.15 ±0.03	8.35 ±0.03	4.28 ±0.27		
7	28.70 ±0.03	34.41 ±0.04	1.39 ±0.34	2.20 ±0.29				28.61 ±0.03	9.42 ±0.04	4.86 ±0.25		
6	37.39 ±0.03	45.20 ±0.05	3.10 ±0.47	2.97 ±0.26				37.18 ±0.04	10.55 ±0.04	5.23 ±0.24		
5	47.58 ±0.04	58.04 ±0.05	2.44 ±0.31	3.47 ±0.25	0.13 ±0.00			47.32 ±0.04	11.93 ±0.04	5.58 ±0.22		
4.	60.17 ±0.04	73.72 ±0.05	3.36 ±0.28	4.61 ±0.20				59.77 ±0.04	13.71 ±0.04	7.00 ±0.16		
3	77.06 ±0.04	94.52 ±0.06	4.25 ±0.22	5.44 ±0.16		0.16 ±0.00	0.44 ±0.00	76.40 ±0.04	16.16 ±0.04	8.04 ±0.13		
2	104.32 ±0.05	127.87 ±0.07	5.89 ±0.13	7.17 ±0.09				102.87 ±0.05	20.05 ±0.04	9.07 ±0.09	0.35	
1	174.44 ±0.08	213.25 ±0.10	14.46 ±0.04	14.50 ±0.05	7.06 ±0.01	6.98 ±0.01	7.60 ±0.01	170.10 ±0.08	36.40 ±0.05	21.63 ±0.06	246.5 ±4.2	12.64 ±0.09
[1] s=	32.63	40.44	5.20	4.86	0.64	0.55	0.62	31.28	10.25	8.98	133.6	0.25
0	284.62 ±1.86	359.37 ±2.58	21.32 ±0.13	20.44 ±0.13	8.43 ±0.01	8.23 ±0.01	9.13 ±0.01	267.0 ±1.5	55.88 ±0.26	33.48 ±0.13		22.37 ±0.04

1) Line to demonstrate qualitatively the magnitude of standard deviations of run times in the JS = 1 interval.
2) Iteration times and their scattering were recorded only for iteration 1 (JS = 2) and iterations 2 till 1000 (JS = 1).

3.5. Numerical Performance of Search Methods

For the time optimal choice of parameter vector c in tables 3.1 and 3.2 values RT*, NF* were tabulated before in Table 3.3. Tables 3.9c and 3.9d list for both test functions F1 and F6 the sequence of search methods in the order of increasing optimal run times RT* and of number of calculated function values NF*.
The ordered set of methods with increasing RT*
- differs for both functions F1 and F6;
- differs for one function from the ordered set of methods with rising NF*;
- shows that for F1 several R.S. methods are faster than deterministic methods;
- shows that the fastest and the slowest (=M11) method is deterministic.
For one test function the order of search methods with respect to RT* will change, too, when ε or the start vector x⁰ changes, or when a different component of criterion vector CR (5) were optimized with regard to c.

Table 3.9 Set of Search Methods in the Order of Increasing Run Times RT

or No'of Function Values NF

Table 3.9a Search Methods: Symbols, see Table 1.1 in Ch. 1.5

M7	M1	M2	M3	M4	M5	M6
Uniform Dist.	$N(0,1)$	$N(0,\left|\frac{\partial F}{\partial x_i}\right|)$	$N(0,\frac{\partial F}{\partial x_i}/\frac{\partial^2 F}{\partial x_i^2})$	∇F	$\frac{\partial F}{\partial x_i}/\frac{\partial^2 F}{\partial x_i^2}$	Newton-Method

M8	M9	M10		M11	M12	
FSSRS	RANPAT	ASSRS		$\frac{1}{k}\nabla F$	$F(x^k-\rho\nabla F(x^k))\rightarrow$MIN	

Table 3.9b Function F1: Not Optimal Choice of Parameters in Search Methods

Order	1	2	3	4	5	6	7	8	9	10	11	12
RT	M2	M4	M5	M1	M3	M8	M10	M7	M9	M6	M12	-
NF	M2	M6	M1	M3	M5	M8, M4		M10	M7	M9	M12	-

Table 3.9c Function F1: Time Optimal Parameters in Search Methods, see Table 3.1

RT*	M6	M9	M7	M10	M8	M3	M1	M5	M4	M2	M12	(M11)
NF*	M6	M9	M3	M10	M7	M1	M8	M2	M5	M4	M12	(M11)

Table 3.9d Function F6: Time Optimal Parameters in Search Methods, see Table 3.2

RT*	M4	M5	M6	M3	M2	M12	M10	M9	M8	M7	M1	M11
NF*	M6	M5	M4	M3	M2	M10	M12	M9	M1	M8	M7	M11

Tables 3.9a-3.9d shall serve to demonstrate the importance of the use of optimal parameters in search methods when performance comparisons are made.

For not time optimal parameter vectors in search methods as in table 3.10 of [2] and for test function F1 optimisation times RT and number of calculated function values NF were determined. The order of search methods with increasing run times in Table 3.9b is reversed to the order where parameter vectors are time optimal as in table 3.9c.

To evaluate the performance of a search method also the expense of time to determine the optimal parameter vector c*, it's complexity and applicability for practical problems must be taken into account. The need to determine c* suggests not to use many parameter methods if they have more than 5 parameters to adjust - even if they are fast. One may rather decide to resort to some method with less parameters although it may be slower. The "most efficient" method for any given type of practical optimisation problem must be determined experimentally.

4. Sensitivity of Search Methods to Parameters Other Than c; Search When the Optimum is Not Known

4.1. Dependence of Optimal Variables RT*, NF*, c* on ε

Run times RT were measured dependent on $c = c_0$ for F1 and F6 (Ch. 1.4) using R.S. methods M7 and M1. Only ε was changed and the minimum of $RT(c_0, ε)$ s.t. c_0 was determined as before, see Ch. 2.1; RT^*, c_0^* and NF^* are time optimal. Representations of RT^*, c_0^*, NF^* as a function of $ε \in [10^{-8}, 10^0]$ in a double logarithmic plot were almost straight lines for both methods and for F1 and F6, see Fig.s. 4.1 and 4.2,

$$\log y = \log a + b \log ε \quad y = a \cdot ε^b \quad y: = RT^*, NF^*, c_0^*$$

The functional relationship $RT^*(ε)$, $NF^*(ε)$, $c_0^*(ε)$ is approximately

$$RT^* \sim \frac{1}{\sqrt[3]{ε}} \qquad NF^* \sim \frac{1}{\sqrt[3]{ε}} \qquad c_0^* \sim \sqrt[3]{ε} \quad \text{for M7 and F6}$$

$$RT^* \sim \frac{1}{\sqrt[3.75]{ε}} \qquad NF^* \sim \frac{1}{\sqrt[3.8]{ε}} \qquad c_0^* \sim \sqrt[2.7]{ε} \quad \text{for M7 and F1}$$

The investigation shows that time optimal values c_0^* of parameter vector c for any search method depend strongly on the ε - environment of F^*. For this reason $RT(c,...)$ functions were measured throughout this study with some constant value ε, to determine c^*. If F^* is not known the feasible region $|F - F^*|$ is not defined, and c^* remains undefinable. Similarily, if x^* is not known, the feasible region $|x - x^*| < ε_x$ and consequently some c_x^* are not defined. If both x^* and F^* are unknown optimal parameter vectors with regard to some convergence criterion are difficult to suggest.

4.2 Dependence of Minimum Range, c*, RT*, NF* on Start Vector x^0

The sensitivity of the $RT(c)$ function to changes in the start vector x^0 has been investigated. To do so $\| x^0 \|$ was increased by a factor of 100 to $\| x^{0'} \| = 100 \cdot \| x^0 \|$. Additionly, for test function F1, $ε = 10^{-4}$ and R.S. method M1 with $c = c_0$ the RT (c_0) function was determined at four different start vectors of the same lengths, see Fig.s. 4.3 a-d, that form the corners of a rectangle. In each case the shape of $RT(c_0)$ is similar to $RT(c_0)$ for the original small x^0. However the minimum position was shifted from c_0^* = 0.04 to about 0.11 or 0.30 depending on the components of the start vector. The optimal run times RT^* and NF^* were increased by approximate factors of 6 and 64. Next, the $RT(c_0)$ function was determined for M2 and M4 which started the search from $x^{0'} = 100 \cdot x^0 = (50, 200)$ for F6 and $ε = 10^{-6}$. For M2 and M4 the minimum position was substantially decreased and optimisation times RT^* and the number of calculated function values NF^* were largely increased. The changes were about

M2: $c_0^*(x^{0'}) \approx 1/266 \, c_0^*(x^0)$ $RT^*(x^{0'}) \approx 370 \, RT^*(x^0)$ $NF^*(x^{0'}) \approx 370 \, NF^*(x^0)$
M4: $c_0^*(x^{0'}) \approx 1/8900 \, c_0^*(x^0)$ $RT^*(x^{0'}) \approx 25000 \, RT^*(x^0)$ $NF^*(x^{0'}) \approx 21,600 \, NF^*(x^0)$

Data obtained for F1 with R.S. method M1 (with no derivatives) show that c_0^* must increase when the start vector x^0 is far from x^*. In this way a large distance towards x^* can be covered fast. However, when $F(x)$ is close to F^* the probability to enter the ε-environment reduces severely, and optimisation times must rise. This observation suggests to reduce c_0 for larger iteration numbers. Also, reported data point out at the difficulty to choose parameter c in some search method if x^* is unknown, as from x^0 neither the distance nor the direction to x^* are known.

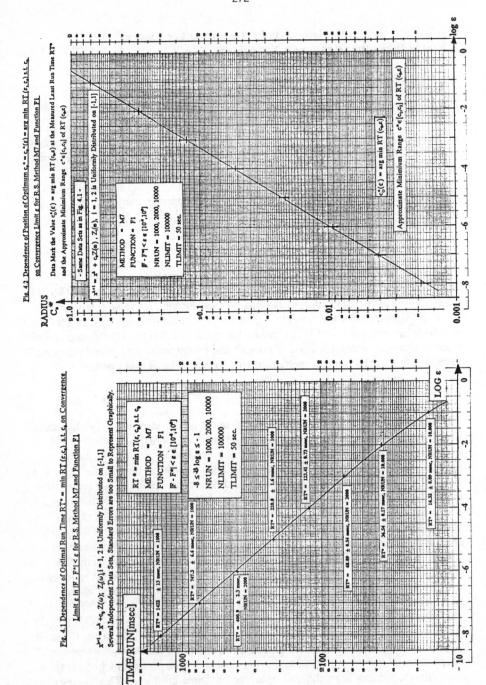

Fig. 4.2 Dependence of Position of Optimum $c_x^* = c_x^*(\varepsilon) = \arg \min RT(\varepsilon, c_x)$ s.t. c_y on Convergence Limit ε for R.S. Method M7 and Function F1

Data Mark the Value $c_x^*(\varepsilon) = \arg \min RT (c_x, \varepsilon)$ at the Measured Least Run Time RT^* and the Approximate Minimum Range $c^* \in [c_x, c_x]$ of $RT (c_x, \varepsilon)$

- Same Data Sets as in Fig. 4.1 -

$x^{n+1} = x^k + c_x Z(\omega), \; Z_i(\omega), \; i = 1, 2$ is Uniformly Distributed on $[-1,1]$

METHOD = M7
FUNCTION = F1
$|F - F^*| < \varepsilon \; [10^{-i}, 10^{0}]$
NRUN = 1000, 2000, 10000
NLIMIT = 100000
TLIMIT = 50 sec.

$c_x^*(\varepsilon) = \arg \min RT (c_x, \varepsilon)$

Approximate Minimum Range $c^* \in [c_x, c_x]$ of $RT (c_x, \varepsilon)$

RADIUS C_x^*

Fig. 4.1 Dependence of Optimal Run Time $RT^* = \min RT(\varepsilon, c_x)$ s.t. c_y on Convergence Limit ε in $|F - F^*| < \varepsilon$ for R.S. Method M7 and Function F1

$x^{n+1} = x^k + c_x Z(\omega), \; Z_i(\omega) \; i = 1, 2$ is Uniformly Distributed on $[-1,1]$

Several Independent Data Sets, Standard Errors are too Small to Represent Graphically.

$RT^* = \min RT(\varepsilon, c_x)$ s.t. c_y
METHOD = M7
FUNCTION = F1
$|F - F^*| < \varepsilon \; [10^{-i}, 10^{0}]$

$-8 \le \log \varepsilon \le -1$
NRUN = 1000, 2000, 10000
NLIMIT = 100000
TLIMIT = 50 sec.

$RT^* = 1423 \pm 13$ msec, NRUN = 1000

$RT^* = 747.3 \pm 6.6$ msec, NRUN = 1000

$RT^* = 400.9 \pm 2.3$ msec, NRUN = 2000

$RT^* = 228.0 \pm 1.6$ msec, NRUN = 1000

$RT^* = 123.42 \pm 0.73$ msec, NRUN = 2000

$RT^* = 62.89 \pm 0.54$ msec, NRUN = 2000

$RT^* = 34.54 \pm 0.17$ msec, NRUN = 10,000

$RT^* = 16.33 \pm 0.09$ msec, NRUN = 10,000

TIME/RUN [msec]

LOG ε

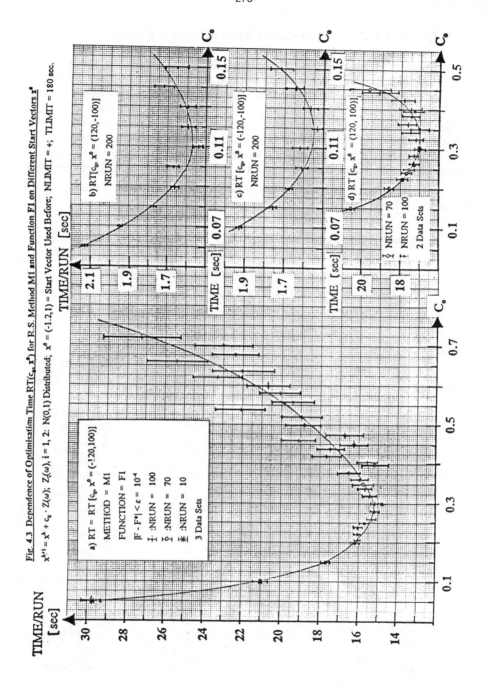

Fig. 4.3 Dependence of Optimisation Time RT(c₀, x⁰) for R.S. Method M1 and Function F1 on Different Start Vectors x⁰

$x^{k+1} = x^k + c_0 \cdot Z(\omega); \ Z_i(\omega), i = 1, 2: N(0,1)$ Distributed; $x^0 = (-1,2,1) =$ Start Vector Used Before; NLIMIT = +; TLIMIT = 180 sec.

4.3. Run Times RT(c) for Normalized Random Number Vectors

How does the distribution of random numbers in R.S. methods affect the search for the optimum of a function? Investigations reported in Ch.2 have shown that the $RT(c_0)$ functions for method M7 (with uniform distribution) and for M1 (with normal distribution of random numbers) are both parabola like in shape, see Fig. 2.1 and 2.2. Their minima were at $c_0{}^* = 0.052$ (M7) and $c_0{}^* = 0.040$ (M1) for test function F1 and at $c_0{}^* = 0.012$ (M7) and $c_0{}^* = 0.0080$ (M1) for F6. Also, M7 was faster than M1 (Ch.3.2). Additionly, the influence of normalized random number vectors in R.S. search methods on optimisation times $RT(c_0)$ has been investigated. These random number

Fig 4.4 $RT = RT(c_0)$ for R.S. Methods M1, M2 with Normalized Random Number

Vectors $Z(\omega)/\|Z(\omega)\|$ and $RT(c_0)$ for M1 with Not Normalized $Z(\omega)$

Function = F1; $|F - F^*| < \varepsilon = 10^{-4}$; NRUN = 100;

No's Inserted Are Number of Runs Stopped with RT > TLIMIT Among 100 Runs.

These Runs are Not Counted to Determine the Mean Optimisation Time at Some Value c_0.

— $\frac{\text{I}}{\text{I}}$ — M1: $x^{k+1} = x^k + c_0\, Z(\omega)/\|Z(\omega)\|$; $Z_i(\omega)$: N(0,1) Distributed; TLIMIT = 20 sec.

— $\frac{\text{I}}{\text{X}}$ — M2: $x^{k+1} = x^k + c_0\, Z(\omega)/\|Z(\omega)\|$; $Z_i(\omega)$: N(0, $\left|\frac{\pi}{2k_i}\right|$) Distrib.; TLIMIT = 50 sec.

- - o - - M1: $x^{k+1} = x^k + c_0 Z(\omega)$; $\qquad Z_i(\omega)$: N(0,1) Distributed; TLIMIT = 4 sec.

TIME/RUN[msec]

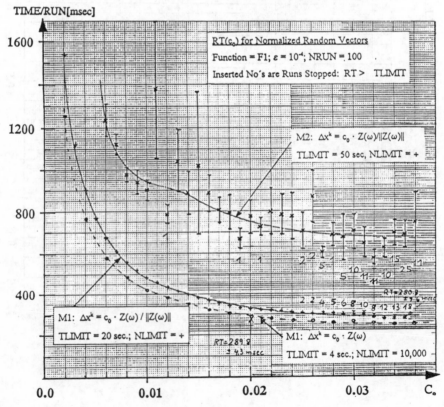

vectors are uniformly distributed on a sphere with radius c_0 [18]. This type of distribution has been suggested by several authors - e.g. [1,7,12,13,17]. As an example the $RT(c_0)$ functions for M1, M2 and F1 are represented in Fig. 4.4. $RT(c_0)$ for M3 has not been included in Fig. 4.4 as $RT(c_0)$ would lie in between both $RT(c_0)$ functions and obstruct the readability of that figure. Comparison of the $RT(c_0)$ functions for M1, and for M2 with $RT(c_0)$ for M1, which uses not normalized random numbers, shows:
For $c_0 \leq 0.26$ $RT(c_0)$ is similar in shape with slightly increased run times; for $c_0 \geq 0.26$ $RT(c_0)$ remains flat. The position c_0^* of min $RT(c_0)$ s.t. c_0 for M1, M2, M3 for normalized random number vectors is approximately $c_0^* \sim 0.024$, if runsets were considered that have only "successful" runs and no run stops due to non converging runs. Position $c_0^* \sim 0.024$ differs from c_0^* in M1, M2, M3 of table 3.1 with not normalized random number vectors. Run times RT* and number of function values NF* are approximately larger by factors 1.2 for M1, 1.6 for M2, and by 2.0 for M3. If random vectors have constant lengths larger than approximately $c_0 \approx 0.24$, an increasing number of runs failed to enter the feasible region, as was qualitatively expected, see e.g. [11].

4.4. Change of Convergence Criterion from $|F^k - F^*| < \varepsilon$ to $|F^{k+1} - F^k| < \varepsilon$

4.4.1 Distance F - F* for New Convergence Criterion.

If F* is known $|F - F^*| < \varepsilon$ is a suitable convergence criterion for any method to stop the search for x*. Investigations reported before had the aim to make statements on the optimal parameter vector c* in 12 search methods: If x* were known an analogous research were to be carried out with $|x-x^*| < \varepsilon_x$. If neither F* nor x* are known the ε-environment of F* qr of x* is not defined. The problem rises where to stop the search. Some convergence criterion must be applied that is related to the history of the search for the optimum, e.g. $\min\{F(x^k), k=1,2,...n\}$ s.t. $k = 1,2,...n$ or $|F^{k+1} - F^k| < \varepsilon$.
If $F^{k+1} < F^k$ and $|F^k - F^*| < \varepsilon$ then follows $|F^{k+1} - F^*| < \varepsilon$, but the reverse is not true! As an example the distance from the known optimum F* had been investigated using previous time optimal c*, when the search was terminated with:

$$|F^{k+1} - F^k| < \varepsilon. \tag{11}$$

The distances are listed in Table 4.1: With (11) some deterministic methods still achieved convergence. However, most methods stopped far ahead of F* with the search.

4.4.2 Dependence of F - F* at Stop as a Function of Sample Parameter NXOLD in Many Parameter R.S. Methods.

The main disadvantage of single parameter search methods to approximate F* when (11) is used, is the examination of convergence criterion (11) every time after two successive iterations. Real search problems with unknown optimum (x^*,F^*) are best solved with many parameter methods that group the iterated function values in sets of a sufficiently large number of NSET sample points. Then the minimum of two consecutive sets may be compared:

$$\left.\begin{array}{l} F^k = \min\ \{F(x_{j+1}), F(x_{j+2}),...F(x_{j+NSET})\} \\ \\ F^{k+1} = \{F(x_{r+1}, F(x_{r+2}),...F(x_{r+NSET})\} \end{array}\right\} \rightarrow "|F^{k+1} - F^k| < \varepsilon\ ?\ " \tag{12}$$

Table 4.1 Distance $F^{k+1} - F^*$ when the Search is Terminated with $|F^{k+1}-F^k| < \varepsilon$
Search Methods M1 - M12 Use Optimal Parameter c^* of Tables 3.1, 3.2

Function		F1: Rosenbrock	F6: Engvall
Method	ε	10^{-4}	10^{-6}
M7	uniform distrib.	$[1831.2 \pm 219.0] \times 10^{-4}$	$[16,203 \pm 8091] \times 10^{-6}$
M1	$N(0,1)$	$[1435.1 \pm 216.7] \times 10^{-4}$	$[13,994 \pm 10,912] \times 10^{-6}$
M2	$N\left(0, \left\|\frac{\partial F}{\partial x_i}\right\|\right)$	$[303.9 \pm 71.1] \times 10^{-4}$	$[30.8 \pm 18.5] \times 10^{-6}$
M3	$N\left(0, \left\|\frac{\partial F}{\partial x_i}\bigg/\frac{\partial^2 F}{\partial x_i^2}\right\|\right)$	$[462.2 \pm 96.5] \times 10^{-4}$	$[2.34 \pm 0.33] \times 10^{-6}$
M4	∇F	426.3×10^{-4}	0.29×10^{-6}
M5	$\frac{\partial F}{\partial x_i}\bigg/\frac{\partial^2 F}{\partial x_i^2}$	300.1×10^{-4}	0.000058×10^{-6}
M6	$H^{-1}\nabla F(x)$	0.38×10^{-4}	0.0011×10^{-6}
M8	FSSRS	$[7.73 \pm 1.43] \times 10^{-4}$	$[1.496 \pm 0.094] \times 10^{-6}$
M9	RANPAT	$[6.54 \pm 0.91] \times 10^{-4}$	$[0.488 \pm 0.024] \times 10^{-6}$
M10	ASSRS	$[84.12 \pm 43.89] \times 10^{-4}$	$[0.051 \pm 0.011] \times 10^{-6}$
M11	$\frac{1}{k}\nabla F$	491.5×10^{-4}	1216×10^{-6}
M12	$F[x^k - \rho\nabla F(x^k)] \to \text{Min.}$	1282.4×10^{-4}	0.020×10^{-6}

Table 4.2 Dependence of $|F^{k+1}-F^*|$ and of RT on NXOLD with Termination Criterion $|F^{k+1}-F^k| < \varepsilon$:
Test Function = F1; $F^* = 0$, $\varepsilon = 10^{-4}$

Notes: Between F^{k+1} and F^k $2*$ NXOLD sample points where generated;
         Search methods are: M8 = FSSRS, M9 = RANPAT, M10 = ASSRS

	Method	NXOLD = 5	10	15	20	25
$\frac{\|F^{k+1}-F^*\|}{\varepsilon}$	M8	563.5 ± 132.9	77.4 ± 39.0	18.5 ± 3.1	45.7 ± 38.7	7.7 ± 1.4
	M9	236.9 ± 71.4	31.4 ± 3.1	18.1 ± 2.4	8.1 ± 0.8	6.5 ± 0.9
	M10	125.0 ± 33.5	84.9 ± 18.7	72.7 ± 19.5	83.7 ± 39.3	84.1 ± 43.9
Run Time [msec]	M8	179.2 ± 6.1	182.1 ± 5.8	199.2 ± 6.6	207.9 ± 6.9	209.9 ± 6.9
	M9	60.21 ± 0.91	68.1 ± 1.1	74.2 ± 1.0	81.1 ± 1.7	85.6 ± 1.8
	M10	89.2 ± 1.1	95.5 ± 1.3	97.0 ± 1.3	102.1 ± 1.4	106.6 ± 1.6

		NXOLD = 30	40	60	80	100
$\frac{\|F^{k+1}-F^*\|}{\varepsilon}$	M8	9.1 ± 1.7	3.2 ± 0.5	1.60 ± 0.27	0.77 ± 0.06	0.70 ± 0.05
	M9	4.2 ± 0.6	1.6 ± 0.2	0.64 ± 0.07	0.31 ± 0.03	0.19 ± 0.02
	M10	30.3 ± 10.5	9.2 ± 1.1	4.30 ± 0.73	2.20 ± 0.52	0.96 ± 0.16
Run Time [msec]	M8	197.9 ± 5.1	215.3 ± 6.4	226.9 ± 5.9	258.4 ± 7.7	258.9 ± 7.2
	M9	90.6 ± 2.0	94.6 ± 1.4	115.4 ± 3.6	120.4 ± 2.6	150.0 ± 5.3
	M10	112.4 ± 1.8	115.8 ± 1.7	131.4 ± 1.9	145.5 ± 1.9	157.7 ± 2.0

To demonstrate the effect of the sample size, R.S. methods M8 (FSSRS), M9 (RANPAT), M10 (ASSRS) have been run at their time optimal parameters of Tables 3.1 and 3.2 apart from a varying sample size NSET, called NXOLD (see Ch. 1.5), to solve (1) with (12). Table 4.2 lists the distance F - F* for test function F1 as a function of NXOLD, when methods M8, M9, M10 stop the search because condition $|F^{k+1} - F^k| < \varepsilon$ is met. Experimentally follows, the ε-environment of F* is reached: for M8 with NXOLD $\geq$ 80, for M9 with NXOLD $\geq$ 60, for M10 with NXOLD $\geq$ 100.

Method	M8	M9	M10
table 3.3:	RT*(25)=143.3±1.4[msec]	RT*(25)= 83.55±0.76[msec]	RT*(25)=134.6±1.9[msec]
table 4.2:	RT (80) = 258.4±7.7[msec]	RT (60)= 115.4±3.6[msec]	RT(100)=157.7±2.0[msec]
time factor	1.80	1.38	1.17

These data show, many parameter methods can be successful to reach $|F - F^*| < \varepsilon$ at the expense of optimisation time: Though, the retardation was here only less than 2.
They must have a sampling facility for function values in order to artificially delay the test of the termination criterion. R.S. methods will considerably accelerate the search if at any random iteration step a function value is calculated too in the deterministic opposite direction in order to make the best choice for progress. These methods should keep some record on previous iterations (directions of descent) and adapt their stepsize, i.e. the "radius" locally: e.g. by activation of an accelerator on progress and of a reduction factor after successive failures to move on.

5. Conclusions

Investigations have proven that the computer can be used as a stop watch for optimisation times of the order measured in this study. It's time fluctuations and it's time drifts do not impede an accurate time data collection: The time stability of the computer was monitored reliably at any instant.
The experimental results have shown that optimisation times RT needed to solve (1) depend severely on the choice of parameter vector c in the applied search method $M_i(c)$, i = 1,2,...12. Convergence or non-convergence (i.e. time excess or divergence with too large function values) are determined by parameter vector c. For different methods the RT(c) functions are different. RT(c) may have a few relative minima, or may heavily be structured with low time coordinates and resonances. In general, the RT(c) functions for stochastic methods were little structured.
If time optimal parameters c* were chosen for each search method one definite order of search methods with rising run times RT* can be found. This order is different for each function. The order - related to one particular function - was altered, if one component in parameter vector c, e.g. "radius" c_0, was set constant in all search methods. Indeed, by changing parameter vector c that order may be reversed manipulating some slow method to become "more efficient" than a fast one in solving (1). The order of methods with regard to rising optimal run times differs from the order that follows for the number of calculated function values.
Once the search method to use in some given optimisation problem has been decided upon, it's parameter vector should be changed tentatively to select some practical "good" value c for further fast optimisations. The need to do so is based on the observed strong c-dependence of optimisation times in any search method. The search

must be carried out with different sample sizes to control the stability of the final (x, F) position at termination by convergence criterion $|F^{k+1} - F^k| < \varepsilon$. Eventually the search for an unknown optimum must be repeated with decreasing feasibility parameters ε in order to locate the unknown optimum F^* more accurately.

References

[1]   Beltrami, F.J. and Indusi, J.P. "An Adaptive Random Search Algorithm for Constraint Minimisation" IEEE Trans C-21, (1972) 1004

[2]   Böttcher, K.-J. "Efficiency Comparison and Parameter Sensitivity of Deterministic and Stochastic Search Methods"; Universität der Bundeswehr München, Forschungsschwerpunkt Simulation und Optimierung Deterministischer und Stochastischer Dynamischer Systeme, June 1996

[3]   Box, G.E.P. and Müller, M.E. "A Note on the Generation of Random Normal Deviates" Ann. Math. Stats 29 (1958) 610

[4]   Brooks, S.H. "A Discussion of Random Methods for Seeking Maxima" Comp.J. Vol. 6 (1958) No 2

[5]   de Graag, D.P. "Parameter Optimisation Techniques for Hybrid Computers" 6th AICA/IFIP Conf. on Hybrid Computation

[6]   Himmelblau, D.M. "A Uniform Evaluation of Unconstrained Optimisation Techniques" in Numerical Methods for Nonlinear Optimisation Acad. Press, London 1972, p. 69-97

[7]   Lawrence, J.P. and Steiglitz, "Randomized Pattern Search" IEEE Trans. C21 (1972) 382

[8]   Matyas, J. "Das zufällige Optimierungsverfahren und seine Konvergenz" 5th Intern Analogue Comput. Meeting 1968, p. 540-544

[9]   Matyas, J. "Random Optimisation" Automation and Remote Control 26 (1965) No 2

[10]   Müller, P.H., Nollau, V., and Polovinkin, A.J., "Stochastische Suchverfahren", Harry Deutsch Verlag 1986

[11]   Neumann, P. Programs: FSSRS, RANPAT, ASSRS with Minor Corrections Universität München, Juli 1979

[12]   Oppel, V.G., Neumann P., "Random Search and Evolution", Bericht Universität München (LMU) 1980

[13]   Rastrigin, L.A. " The Convergence of the Random Search Method in the Extremal Control of a Many Parameter System", Automation and Remote Control 24 (1963) 133

[14]   Schittkowski, K. "Nonlinear Programming Codes", Lecture Notes in Economics and Mathematical Systems No 183 (1980)

[15]   Schwefel, H.P. "Numerische Optimierung von Computer Modellen mittels der Evolutions Strategie", Birkhäuser Verlag, Basel 1977

[16]   Schumer, M.A. and Steiglitz "Adaptive Stepsize Random Search" IEEE Transactions on Automatic Control AC 13 (1968) No 3

[17]   White Jr., R.C "A Survey of Random Methods for Parameter Optimisation", TH-Report 70-E-16, p. 1-20, Dept. of Electrical Engineering, Technical University Eindhoven, Netherlands

[18]   Zielinski, R. and Neumann, P. "Stochastische Verfahren zur Suche nach dem Minimum einer Funktion", 1983 Akad. Verlag Berlin

Regression Estimators Related to Multinormal Distributions: Computer Experiences in Root Finding

István Deák[1]

[1]Operations Research Group, Dept. of Differential Equations
Technical University of Budapest
H-1111 Budapest, XI. Muegyetem rkp. 3.
email: deak@inf.bme.hu

Several simple regression estimators can be constructed to approximate the distribution function of the m-dimensional normal distribution along a line. These functions can be used to find the border points of the feasible region of probability constrained stochastic programming models. Computer experiences show a fast and robust behaviour of the root finding techniques.

Keywords: multinormal distribution, stochastic programming, regression estimators, quantile computation

1   Introduction

Consider the m-dimensional normal distribution with expected value $\mathbf{0}$ and correlation matrix $\mathbf{R}$. Its distribution function and density function are given as

$$\Phi(\mathbf{h}) = \int_{-\infty}^{h_1} \cdots \int_{-\infty}^{h_m} \phi(\mathbf{z})d\mathbf{z}, \tag{1}$$

$$\phi(\mathbf{z}) = (2\pi)^{(-m/2)}|\mathbf{R}|^{-\frac{1}{2}}\exp\{-\frac{1}{2}\mathbf{z}\mathbf{R}^{-1}\mathbf{z}\}.$$

Computation of the function values $\Phi(\mathbf{h})$ is required in numerical optimization procedures of stochastic programming problems, when the random variables of the model have a joint normal distribution. This is the case in solution procedures of the STABIL stochastic programming model [11] and the two-stage model [9]. Other problems, where computation of (1) is required can be found in diverse areas of statistics and engineering ([1], [6], [4]).

There are three Monte Carlo methods for efficiently computing values of $\Phi(\mathbf{h})$ in higher ($m > 5$) dimensions. One of them is the method of orthonormalized estimators [2], the other one is based on Boole-Bonferoni inequalities [13], and the latest technique is a hybrid method [8]. These Monte Carlo methods give realizations $\eta_1, \ldots, \eta_n$ of a random variable η, where $E(\eta) = \Phi(\mathbf{h})$, so in numerical computations the unbiased estimator

$$y = \frac{1}{n} \sum_{i=1}^{n} \eta_i$$

is used to approximate the value $\Phi(\mathbf{h})$. If $D^2(\eta) = \sigma^2$, then $D^2(y) = \sigma^2/n$, so the standard deviation of the method – which is usually called the error of the Monte Carlo computation – is $\sigma/\sqrt{n}$. Invoking the central limit theorem η (and y) is assumed to have a truncated normal distribution $\eta \in \tilde{N}(\Phi(\mathbf{h}), \sigma)$, where $0 < \Phi(\mathbf{h}) - 3\sigma < \eta < \Phi(\mathbf{h}) + 3\sigma < 1$.

A problem, frequently arising in optimization procedures is to find the intersection of a line with the boundary of the set of feasible solutions. Assume, that for a given point $\mathbf{z}$ and a direction $\mathbf{d}$ the halfline $\mathbf{z} + x(\mathbf{d} - \mathbf{z})$, $x > 0$, intersects the surface given by $\Phi(\mathbf{h}) = p$, then using the notation $f(x) = \Phi(\mathbf{z} + x(\mathbf{d} - \mathbf{z}))$, the intersection point can be found by solving the equation

$$f(x) = \Phi(\mathbf{z} + x(\mathbf{d} - \mathbf{z})) = p \tag{2}$$

for the unknown x, the halfline is assumed to contain the root. Solving (2) is called root finding, which basically amounts to finding the p-quantile of the function f. For the solution of (2) Szántai [14] presented a heuristic algorithm, that moves up and down along a line, increasing the accuracy of function evaluation and decreasing the steplength. This seems to be unsatisfactory, since the new approximate root is determined depending only on the last two or three function values, completely disregarding previous function values – our technique takes into consideration all previously computed function values, that hopefully increases the efficiency.

A solution of the problem (2) can be obtained in the following way: (i) using regression techniques a function, approximationg $f(x)$ is constructed, (ii) then the root of this approximating function is computed, (iii) if this approximate root is sufficiently accurate, we can stop, otherwise by using more points the approximating regression function is recomputed, and the whole procedure is repeated.

In the next section the basic linear regression techniques are described, – some more estimators, full details of the estimators and other possible uses are given in [4], [5]. In Section 3. a preliminary algorithm of root finding is outlined, together with some details and considerations about the parameters. In Section 4. the final algorithm is described, while in Section 5. computer experiences and numerical results are presented. Finally some remarks and conjectures are drawn in the last section.

2 Least squares regression estimators

Consider the one-dimensional function $f(x) = \Phi(z + x(d - z))$, for fixed vectors z and d, and assume for now, that d is an increasing direction, that is $f(x_1) < f(x_2)$, if $x_1 < x_2$. Since $\Phi(\cdot)$ is logconcave (see [10],[11]), then $f(x)$ is also logconcave. It is assumed, that we have a Monte Carlo method for computing approximate values of the function f, that is for arbitrary points $x_i \in R^1, i = 1, \ldots, n$ realizations $p_i, i = 1, \ldots, n$ of a random variable ξ_i can be computed, where $\xi_i \in \tilde{N}(f(x_i), \sigma)$, where $\tilde{N}$ is a truncated normal distribution, restricted to $(0,1)$. To find approximations to $f(\cdot)$ regression techniques will be used, that is for a given set $S = \{x_i, p_i\}_{i=1}^n$ the parameters of an estimator $t(x)$ will be computed. Several types of the function $t(x)$ are suggested in this section.

2.1 Initial estimator – approximation of $f(x)$ by quadratic regression

Let us look for the estimator $t_1(x)$ in the form of a quadratic function $g_1(x) = a_1 x^2 + b_1 x + c_1$. To determine the best fit of type $g_1(x)$ to the function $f(x)$ for a given set $S = \{x_i, p_i\}_{i=1}^n$ the problem

$$\min_{a_1, b_1, c_1} \sum_{i=1}^n [p_i - (a_1 x_i^2 + b_1 x_i + c_1)]^2. \tag{3}$$

is to be solved for the unknown parameters a_1, b_1, c_1. The first order optimality criteria of this minimization problem can be obtained by differentiating (3) with respect to the parameters, and setting the derivatives equal to zero; a system of three linear equations emerges, from where the parameters can be expressed easily (for the corresponding explicit formulas see [7], [4]).

Since $f(x)$ is monotone increasing, only the increasing part of the function $g_1(x)$ can be used as an estimator to $f(x)$, so by defining the solution of the equation $g_1'(x_t) = 0$ to be the truncation point (where $x_t = -b_1/(2a_1)$) the estimator is given as

$$t_1(x) = \begin{cases} g_1(x), & \text{if } -\infty < x \le x_t \\ g_1(x_t) = c_1 - b_1^2/(4a_1), & \text{if } x_t < x < \infty. \end{cases} \tag{4}$$

2.2 Logarithmic estimator – approximation of $\log f(x)$ by quadratic regression

Instead of approximating $f(x)$ directly, now a quadratic function $g_2(x) = a_2 x^2 + b_2 x + c_2$ will be used to approximate the function $\log f(x)$, that is the logarithmic estimator $t_2(x)$ of the form $\exp(g_2(x))$ will be used to approximate $f(x)$. This is still a linear estimator – the parameters of $t_2(x)$ can be expressed from a system of linear equations. To find the best – least square – approximation of the form $g_2(x)$ to $\log f(x)$ the following problem is to be solved

$$\min_{a_2,b_2,c_2} \sum_{i=1}^{n} [q_i - (a_2x_i^2 + b_2x_i + c_2)]^2 w_i, \tag{5}$$

where $q_i = \log p_i$ and w_i are some weights, given below. The parameters a_2, b_2, c_2 can be expressed from (5) the same way, as it was done for the initial estimator. The weights w_i are introduced to counteract the distorting effect of the logarithmic transformation: $\log(x+\delta) - \log x$ depends on x also, not only on δ. For a given (x_i, p_i) the quantity $w_i^* = (p_i + t^*(x_i))^2/4, i = 1,\ldots, n$ is defined, where $t^*(x)$ is any previously computed estimator. After normalization the final weights are obtained as $w_i = w_i^*/\sum_j w_j^*$. Truncation point is $x_t = -b_2/(2a_2)$, and so the logarithmic estimator becomes

$$t_2(x) = \begin{cases} g_2(x), & \text{if } -\infty < x \leq x_t \\ g_2(x_t) = c_2 - b_2^2/(4a_2), & x_t < x < \infty. \end{cases} \tag{6}$$

2.3 Reverse – logarithmic estimator – approximation of $\log(1 - f(x))$ by quadratic regression

Now the function $\log(1 - f(x))$ is approximated with a quadratic function $g_3(x) = a_3x^2 + b_3x + c_3$ that is a function $t_3(x)$ is constructed in the form $1 - \exp(g_3(x))$ to approximate the original function $f(x)$. This is a linear estimator, too, its parameters can be obtained by solving the problem

$$\min_{a_3,b_3,c_3} \sum_{i=1}^{n} [r_i - (a_3x_i^2 + b_3x_i + c_3)]^2 w_i, \tag{7}$$

where $r_i = 1 - \log p_i$ and the weights w_i are computed by normalizing the values $w_i^* = (1 - p_i + \exp(g_3^*(x)))^2/4$, where $g_3^*(x)$ is a previously computed version of g_3. Obviously, instead of $\exp(g_3^*(x))$ the function $1 - t^*(x)$ can be used with any previous estimator t^* of $f(x)$.

Here we have $1 - f(x)$ as a monotone decreasing function, so the monotone decreasing portion of $g_3(x)$ is to be used in the approximation only. Correspondingly, taking $x_t = -b_3/(2a_3)$ for the truncation point the estimator t_3 is defined by the following

$$t_3(x) = \begin{cases} 1 - \exp(g_3(x_t)) = 1 - \exp(c_3 - b_3^2/(4a_3)), & \text{if } -\infty < x \leq x_t \\ 1 - \exp(g_3(x)), & x_t < x < \infty. \end{cases} \tag{8}$$

3   Considerations on root finding strategies

Now we return to our problem of finding a solution x_r which approximates the real root $\overline{x}_r$ of the problem

$$f(\overline{x}_r) = p, \tag{9}$$

which corresponds to the problem of finding an approximate solution to the equation $\Phi(\mathbf{z} + x(\mathbf{d} - \mathbf{z})) = p$, for given vectors $\mathbf{z}, \mathbf{d}$ and reliability level p. An approximate root x_r can be obtained by solving the equation $t(x) = p$, where t is any of the previously described estimators.

Our strategy for finding $\overline{x}_r$ is realized by an iterative procedure, where the estimator $t(x)$ is succesively made more and more accurate in the neighbourhood of the approximate root. This is achieved by constructing a sequence of shrinking intervals $[\alpha_i, \beta_i]$ around the value p, and the corresponding intervals $[a_i, b_i]$ around the approximate roots x_r of $t^i(x_r) = p$ (here $t^i(a_i) = \alpha_i, t^i(b_i) = \beta_i, t^i$ is the actual estimator in the i-th step of the algorithm). The length of the intervals is decreased in each step of the procedure. To keep notations simple, the algorithm is described for $t_1(x) = g_1(x) = a_1 x^2 + b_1 x + c_1$, (after the necessary small modifications the estimators of t_2 and t_3 can be equally well used), and to further simplify notations the indices of t_1, g_1, a_1, b_1, c_1 are also dropped.

3.1   Preliminary version of the algorithm

0. [Initialization.] Assume that we have a set of points $S_0 = \{x_{0j}, p_{0j}\}_{j=1}^{K}$, an initial function value interval $[\alpha_0, \beta_0]$ and iteration counter $i = 0$. Using S_0 the function $g(x)$ can be constructed. Set $\sigma_0 = 3\sigma$ as the half length of the initial function value interval, where σ is the standard deviation of the values p_i.

1. [Computation of the function value interval $[\alpha_i, \beta_i]$.] Increase the iteration counter $i = i + 1$ and compute $\sigma_i = \varrho\sigma_{i-1}$, where $\varrho < 1$ is a reduction factor. Let $\alpha = \max(p - \sigma_i, 0), \beta = \min(p + \sigma_i, 1)$.

2. [Computation of the interval $[a_i, b_i]$.] Compute $a_i = g^{-1}(\alpha), b_i = g^{-1}(\beta)$.

3. [Recomputing the estimator $g(x)$.] Determine K new points $x_{ij}, j = 1, \ldots, K$ in the interval $[a_i, b_i]$ (for example uniformly) and using a Monte Carlo technique compute the approximate values $p_{ij} \sim f(x_{ij}), j = 1, \ldots, K$. Let $S = S_0 \cup \cdots \cup S_i$, where $S_i = \{x_{ij}, p_{ij}\}_{j=1}^{K}$, and recompute $g(x)$ using the set of points S.

4. [Termination criteria.] If $x_r = g^{-1}(p)$ is close enough to $\overline{x}_r$ according to some convergence criteria, then stop, otherwise go back to Step 1.

In the following subsections details of the basic version are considered and using results of the computer experiences some suggestions are made concerning the details of the final algorithm and the values of the constants.

3.2 The initial interval $[a_0, b_0]$

The interval $[a_0, b_0]$ should contain the main bulk of $f(x)$, that is around $\bar{x}_r$, in order to make the first approximation more or less be near to $f(x)$. Given $f(\cdot)$ and a suitable Monte Carlo method, by halving and doubling the steplength an initial interval $[a_0, b_0]$ can be selected, which contains $\bar{x}_r$ with large probability (for example assume, that $f(a_0) \le p - 3\sigma, f(b_0) \ge p + 3\sigma$ holds, near equality is preferred). The initial points $x_{0j}, j = 1, \ldots, K$ can be selected at equal distance (previously computed pairs $\{x_i, p_i\}$ can be included), and they might be dispersed on a wider interval as well, that is $f(a_0) \sim \delta, f(b_0) \sim 1 - \delta, \delta = 0.01 - 0.1$ is acceptable as well.

We have to avoid cases when several points x_{0j} produce very small (near to 0.01) or very great (near to 0.99) function values, because this would result in a very poor initial approximation $g(x)$, causing very slow convergence of the root finding algorithm. Generally we have some previous knowledge of the function, since some feasible solutions of the optimization problem – for which f takes on values greater than p – are available and this information helps in constructing the initial interval. In short, the initial interval should be wide enough to contain the meaningful part of f, and small enough to facilitate convergence.

An easy and practical way to ensure, that the initial interval is well located (more or less symmetric around the root, see Subsection 3.4) is to count the number of function values greater than p, that should be the same (or almost the same) as the number of function values smaller than p.

3.3 The reduction factor ϱ

This constant governs the speed of decrease of the intervals (first the decrease of the function value interval $[\alpha_i, \beta_i]$, than by using the transformation g^{-1} that of $[a_i, b_i]$). There are two, somewhat contradictory aims to realize in determining ϱ and the intervals. First we would like to decrease $[\alpha_i, \beta_i]$ as fast as possible, so it would be very small around p, so that $[a_i, b_i]$ would be small around $\bar{x}_r$, and thus new points x_{ij} would be near to the real root, because this is the point where we want to make our approximation $g(x)$ to be very close to $f(x)$. Secondly, a very fast decrease ($\varrho \sim 0.1$) would very soon result in a pointlike interval $[a_i, b_i]$ which with large probability would not contain $\bar{x}_r$, so consecutive intervals $[a_i, b_i]$) would jump up and down along our line, trying to locate the root, thus we are loosing stability and speed.

The best choice is to have a sequence of intervals $[a_i, b_i] \supset [a_{i+1}, b_{i+1}] \supset \ldots \supset \bar{x}_r$. According to this consideration the choice $\varrho \sim 0.9$ is preferred. Furthermore, the value of ϱ is related to the number K of newly generated points. Since K new point ideally would reduce the error of the estimator around $\bar{x}_r$ by about a factor $1/\sqrt{K}$, so $\varrho\sqrt{K}$ should be a constant, greater than 1 (to preserve stability). Computer experiences suggest to take $\varrho\sqrt{K} \sim 2$.

3.4 Symmetrization of $[a_i, b_i]$

By generating points not symmetrically, the appproximation would be loopsided - accurate in regions not interesting for us, so the best strategy seems to determine $[a_i, b_i]$ symmetrically around $\bar{x}_r$. Since the real root is not known at the time of the computation, we determine $[a_i, b_i]$ symmetrically around x_r, the actual approximation, also taking into account the derivative of the function g. Compare the differences in the function values by evaluating the fraction $o = (p - g(x_r - \delta))/(g(x_r + \delta) - p)$, where δ is the probable halflength of the next interval $x_r - a_{i+1} \sim b_{i+1} - x_r$ (we can set $\delta = 0.4(x_r - a_i)$), and determine the new function value interval as $\alpha_{i+1} = o\sigma_i$, $\beta_{i+1} = \sigma_i$.

So the determination of the intervals is done by first making a symmetrical interval around the root, then determining the function value interval and finally computing the interval, where new points are generated.

3.5 First and second phase

To safeguard against the unpleasant consequences of having a very small interval $[a_i, b_i]$ it might be very useful to implement a second phase in the algorithm: if the function value interval becomes very small, we do not decrease the length of the function value interval any more (or we change it much slower). Assume, that we want to determine a root with accuracy ϵ, that is $|f(x_r) - p| < \epsilon$. Then in case of $\beta_i - \alpha_i < \epsilon$, we change the value of the reduction factor by setting $\varrho = 1$ (or $\varrho = 0.99$). From this point on α_i and β_i remain the same, but $a_i = g^{-1}(\alpha_i), b_i = g^{-1}(\beta_i)$ are still updated (since the function g is changed in each iteration).

3.6 Stability of the estimator

Unfortunately the concavity of the approximating function can not be always guaranteed, due to possible large errors in the values of p_i. So g can become a convex function, a full computer implementation of the root-finding algorithm has to be prepared to handle this case too. Generally it happens in the first or second step of the iterative procedure, when the number of points x_{ij} is rather small.

This "flipping over" is also an indication, that the function f, compared to the error σ is very flat, so increased accuracy - greater K, or greater ϱ - is required. This involves checking the negativity of the constant a, and if $a > 0$ then the greater root of the quadratic function g has to be accepted (the truncation in this case would give the right hand side of a convex function). This phenomena may occur also in the case of a "bad" initial interval.

3.7 Truncation in case of high reliability

In cases, when the reliability level p is very high $p \sim 0.95 - 0.999$ (or very small, $p \sim 0.05 - 0.001$) the truncation of the quadratic function should be changed. The original truncation – replacing the decreasing part by a constant – can severely restrict the upward movement of the approximate root. A new truncation rule could be the following. Select a point x_{tt} a little smaller, than the original truncation point, that is let $x_{tt} = x_t - \epsilon$, where $\epsilon \sim (b_i - a_i)/10 \cdots (b_i - a_i)/100$ and use a linear increasing function in the truncated part, which is the tangent line of the function $g(x)$ at the point x_{tt}, until it reaches the values 1 (this kind of truncation is called linear truncation, as opposed to the previous constant truncation). That is let the line be given by $a_t x + b_t$. then we have the equations $a_t x_{tt} + b_t = g(x_{tt})$, $2a_t x_{tt} + b_t = a_t$, from which the constants a_t and b_t can be expressed. The final form of $t(x)$ (for the initial estimator) becomes

$$t_1(x) = \begin{cases} g_1(x), & \text{if } -\infty < x \leq x_{tt} \\ \min(1, a_t x + b_t), & \text{if } x_{tt} < x < (1 - b_t)/a_t. \end{cases} \tag{10}$$

3.8 Updating the estimator

The computation of $f(x)$ is done by a Monte Carlo method, so its error (standard deviation) σ determines the necessary amount of work in evaluating $f(x)$. The root finding algorithm's computationally most demanding part is the determination of the estimator $g(x)$, when using old and new points in S the function $g(x)$ is recomputed. We show, that – neglecting the first few steps of the algorithm – a simple updating procedure can be used, saving thus a considerably amount of time. This modification is described for the logarithmic estimator $t_2(x)$; for the reverse-logarithmic estimator and the initial estimator the same steps are to be made.

In evaluating the parameters of $t_2(x) = \exp(a_2 x^2 + b_2 x + c_2)$ we have expressions like $\overline{q0} = \frac{1}{n} \sum_{i=1}^{n} q_i w_i, \ldots, \overline{x3} = \frac{1}{n} \sum_{i=1}^{n} x_i^3 w_i$ (for details see [7], [4]). Assume, that in the last iteration of the root finding procedure we had I points in S, and K points $\{x_{ij}, p_{ij}\}, j = 1, \ldots, K$ are additionally determined. Then the new expressions $\overline{q0}^*, \ldots \overline{x3}^*$ apppearing in the formulas for a_2, b_2, c_2 can be obtained as

$$\overline{q0}^* = \frac{I}{I + K}\overline{q0} + \frac{1}{I + K} \sum_{j=1}^{K} q_{ij}, \ldots, \overline{x4}^* = \frac{I}{I + K}\overline{x4} + \frac{1}{I + K} \sum_{j=1}^{K} x_{ij}^4.$$

This updating scheme is possible, because the weights are basically the same, if the interval, where the values x_{ij} are coming from is small enough (in the following root finding algorithm this will be true from the second or third iteration on already).

4 Root finding algorithm – p-quantile determination

The algorithm is described for the estimator $t_1(x)$; to apply the same algorithm to the logarithmic and reverse-logarithmic estimators just some minor changes are to be made (p_i is replaced by q_i or r_i, the formula for the ratio o is to be changed, conditions on the roots being smaller than the truncation point should be reversed for reverse logarithmic estimators, etc.)

0. [Initialization.] Assume, that we have an initial interval $[a_0, b_0]$ and K points x_{0j} in it (given maybe at equal distances) and the "noisy" fuction values $p_{0j} \sim f(x_{0j})$, furthermore the estimator $g(x)$ using the points $S_0 = \{x_{0j}, p_{0j}\}_{j=1}^{K}$ is already computed. Set the initial values $\sigma_0 = \sigma/\varrho$, i=0, steplength $stl = (b_0 - a_0)/K$.

1. [Computation of the length of the function value interval.] Increase the iteration counter $i = i + 1$, compute $\sigma_i = \varrho\sigma_{i-1}$ and the truncation point $x_t = -b/(2a)$.

2. [Computation of the preliminary interval $[a_i, b_i]$.] Determine the approximate roots x_r from the equation $g(x_r) = p$. If $a < 0$, then accept the root, which is smaller than x_t (or set $x_r = x_t$ if there is no solution to the equation). If $a > 0$, then accept the root, which is greater then x_t (or set $x_r = x_t$ if there is no solution to the equation). Compute $\delta = 2 * stl$ and let $x^- = x_r - \delta, x^+ = x_r + \delta$. The ratio $o = (p - g(x^-)/(g(x^+) - p)$ indicates the relation of the left and righthandside derivatives approximately. Let $\alpha_i = \max(p - o\sigma_i, 0.0001), \beta_i = \min(p + \sigma_i, 0.9999)$.

3. [Computation of the interval $[a_i, b_i]$.] Compute the values of $a_i = g^{-1}(\alpha_i), b_i = g^{-1}(\beta_i)$. If $a < 0$, select those roots, that are smaller than x_t, if $a > 0$, then select the greater roots (if no solution exist, then take $a_i = x_t, b_i = a_i + 0.1\delta$).

4. [Updating $g(x)$.] Select K new points in $[a_i, b_i]$ by letting $x_{ij} = (b_i - a_i)(j - 1)/(K - 1) + a_i, j = 1, \ldots, K$ and compute by a Monte Carlo method the noisy function values $p_{ij} \sim f(x_{ij})$, $Ep_{ij} = f(x_{ij})$. Let the set of new points be $S_i = \{x_{ij}, p_{ij}\}_{j=1}^{K}$, and $S = \cup_{l=0}^{i} S_l$. Compute the new approximation $g(x)$ using all points in S, and determine the new approximate root $x_r^* = g^{-1}(p)$ (if $a < 0$, then choose the smaller root, otherwise the greater one and set $x_r = x_t$, if no solution exists).

5. [Test of convergence.] If $N_0 < I = (i + 1)K$ then stop, where N_0 is a prescribed number of function evaluations. Otherwise set $x_r = x_r^*, stl = 2(b_i - a_i)/K$ and go back to Step 1.

The second phase is not built in into this version, so the modification described in subsection 3.5 are to be incorporated in case of need. Also, the new truncation rule suggested for large p was omitted. For the sake of clarity the updating scheme proposed in Subsection 3.8 was also left out, but is should be incorporated whenever speed is important. Instead of the content of Step 5. other stopping rules can be used as well; the simplest one is described above to stop the algorithm after a prescibed number of iterations. The necessary number N_0 of iterations can be determined from the equation $\epsilon = \sigma/\sqrt{(N_0 + 1)K}$, where σ is the standard deviation in the Monte Carlo computation of one function evaluation, ϵ is the required precision of the result measured in function value difference $|f(x_r) - f(\overline{x}_r)|$ (this formula is based on computer experiences, see Section 6.). Another possibility for the stopping rule is to take $|g(x_r) - g(x_r^*)| < \epsilon_1$ (or $|x_r - x_r^*| < \epsilon_2$, with some prescribed precisions ϵ_1, ϵ_2) and stop, if the condition is satisfied.

5 Computational results

In the computer tests a crude Monte Carlo technique was used for computing the values $\Phi(\cdot)$ (that is $f(\cdot)$) of the distribution function of the m-dimensional normal distribution: 100 normally distributed vectors were generated and then tested if they lie in the domain of integration (this is the simplest importance sampling, see [2],[3]). The results had a standard deviation $\sqrt{p^*(1 - p^*)}/10$, if the function value to be evaluated was p^*, so for $p^* = 0.5$ this gave an error of 0.05. The number of points in each iteration was set to $K = 10$, the reduction factor to $\varrho = 0.6$ (see 3.3), the number of iterations was kept fixed at 9, so altogether 100 function value evaluations of $f(\cdot)$ were performed in each run, that were used to compute the final regression estimator.

The errors were measured in function value differences, since values of parameter x can be changed by scaling. The entry "error" in the tables was calculated as $|f(x_r) - f(\overline{x}_r)|$. The "true root" was computed by the same algorithm, with increased accuracy and more sample points, so it is an approximation only. A great number of computer runs were performed during the testing, here only a small portion of the results are presented, these result are not the best, neither the worst, they show just average behaviour. The description of the four main examples are given in the Appendix.

The sequence of approximate roots produced by the root finding procedure depends on the specific estimator used in the root finding algorithm (that could be any of the initial, logarithmic and reverse-logarithmic estimators), this will be called the main estimator and its name is given in boldface. Using the points x_i and function values p_i determined by this main estimator, the other estimators can be fitted to $S = \{x_i, p_i\}_{i=1}^N$, and approximate roots computed for the other two estimators as well. To give an introductory picture, a set of results is shown in the first table.

There seems to be no discernable difference between the performance of the main estimator, and the results obtained from the other estimators. Also, there seems to be no difference in using different estimators for the main estimator. To illustrate this statement the following set of results, concerning Example 4 is offered in the second table. Three different runs were performed, the main estimator's name is given in boldface; here we have m=50 dimensions, uncorrelated components, reliability level $p = 0.95, \sigma = 0.05$, the true root is $\overline{x}_r = 6.368$, the initial interval was $[a_0, b_0] = [3.0, 10.2]$, for which we have $f(3.0) = 0.79, f(10.2) = 0.97$.

No. exmp	Dimens. m	Rel. p	True root $\overline{x}_r$	Estim.	Root x_r	Error
1.	2	0.8	1.1114	**Init**	1.0996	0.0025
				Log	1.1001	0.0035
				R-log	1.1310	0.0048
1.	2	0.9	1.8358	**Init**	1.8193	0.0013
				Log	1.8679	0.0023
				R-log	1.9201	0.0057
2.	2	0.8	2.5131	**Init**	2.5012	0.0013
				Log	2.4887	0.0027
				R-log	2.5741	0.0063
3.	10	0.8	2.510	**Init**	2.479	0.0043
				Log	2.468	0.0055
				R-log	2.478	0.0043
3.	10	0.9	3.910	Init	3.936	0.0010
				Log	3.852	0.0024
				R-log	3.910	0.0005
3.	10	0.95	5.895	Init	5.7241	0.0024
				Log	5.8273	0.0010
				R-log	5.9986	0.0012
4.	50	0.9	4.276	Init	4.201	0.0038
				Log	4.164	0.0050
				R-log	4.2158	0.0029

No. run	Final interv. $[a_9, b_9]$	$f(a_9)$	$f(b_9)$	Estim.	Root x_r	Error
1.	[6.56, 6.69]	$p + 0.0024$	$p + 0.0039$	**Init**	6.560	0.0024
				Log	6.551	0.0022
				R-log	6.845	0.0057
2.	[6.62,6.71]	p+0.0031	p+0.0041	Init	6.687	0.0030
				Log	6.676	0.0037
				R-log	6.963	0.0066
3.	[6.79,6,92]	p+0.0050	p+0.0063	Init	6.736	0.0044
				Log	6.727	0.0043
				R-log	6.849	0.0056

Next to show, how results differ from each other in case of different initial intervals we give the results for Example 4. on the next page (with $p = 0.9, \sigma =$

0.05, the true root being $\overline{x}_r = 4.2716$) of four different runs, the main estimator was always the initial estimator $t_1(x)$.

No. run	Init. interv. $[a_0, b_0]$	$f(a_0), f(b_0)$	Final interv. $[a_9, b_9]$	Estim.	Root x_r	Error
1.	[2.5,7.0]	0.62, 0.95	[4.23,4.25]	**Init**	4.275	0.0001
				Log	4.272	0.0001
				R-log	4.351	0.0032
2.	[1.0,5.5]	0.002, 0.93	[4.03,4.05]	**Init**	4.060	0.0103
				Log	4.057	0.0102
				R-log	4.139	0.0060
3.	[1.0,8.2]	0.002, 0.96	[4.28,4.30]	**Init**	4.295	0.0009
				Log	4.308	0.0015
				R-log	4.437	0.0067
4.	[1.0,8.2]	0.002,0.96	[4.32,4.35]	**Init**	4.335	0.0026
				Log	4.336	0.0027
				R-log	4.443	0.0069

Reliability p	True root $\overline{x}_r$	Estimator	Root x_r	Error $|f(x_r) - f(\overline{x}_r)|$
0.5	0.8915	**Init**	0.8806	0.0036
		Log	0.8786	0.0043
		R-log	0.8854	0.0021
0.5	0.8915	Init	0.8780	0.0045
		Log	0.8734	0.0060
		R-log	0.8843	0.0024
0.5	0.8915	Init	0.8888	0.0009
		Log	0.8860	0.0019
		R-log	0.8959	0.0014
0.8	2.513	**Init**	2.495	0.0020
		Log	2.492	0.0023
		R-log	2.545	0.0034
0.8	2.513	Init	2.5265	0.0014
		Log	2.5214	0.0008
		R-log	2.5633	0.0052
0.8	2.513	Init	2.5063	0.0008
		Log	2.5004	0.0013
		R-log	2.5468	0.0035
0.95	5.8907	**Init**	5.655	0.0035
		Log	5.635	0.0034
		R-log	6.307	0.0052

In this third table of results (at the top) it would be worthwhile to have a closer loook at the first and the second run, where all the estimators gave final roots outside the final interval, showing that the algorithm is self-correcting. Note the outstanding performance of the first run, which is due to the choice of the good initial interval $[a_0, b_0]$ (it is almost symmetrical around the root $\overline{x}_r$, but this does not show in the function values ($f(a_0) = 0.62$ and $f(b_0) = 0.95$ is not symmetric around $f(\overline{x}_r) = 0.9$). The second run indicates, what can a badly placed inital interval do to the final result.

Numerical results strongly support the need for the initial interval to contain the main bulk of the density function around the real root.

The last set of results shown (previous page bottom) is given for Example 3. to illustrate, that the performance of the method does not depend on the probability p to be computed ($m = 10, \sigma = 0.05$, initial interval varying, for the first run $[a_0, b_0] = [0.0, 1.8], f(a_0) = 0.05, f(b_0) = 0.75$).

6   Remarks

Presently the behaviour of the algorithm is not fully explained yet, the remarks below are conjectures, based on extensive numerical computations. The most important conjecture is that with more or less properly set parameters the error of approximation around $\overline{x}_r$ (that is the difference of the function values $|f(x_r) - f(\overline{x}_r)|$) is decreasing with $\sigma/\sqrt{N}$, where N is the number of all points in the set S and σ is the error of one function evaluation (so in the numerical examples we registered errors less than $0.005 = 0.05/10$ almost everywhere. Probably this phenomena is due to that property of the algorithm, that it very fast focuses on a small interval, where the function we are estimating is rather flat, compared to the error σ.

In other words if we want to receive a root with an error ϵ, then instead of trying to compute with error ϵ the function $f(x)$ for some fixed x values, trying to figure out the location of the root $\overline{x}_r$, we should use the root finding algorithm, where function values are computed with error 10ϵ, and after 10 iterations, in each of them 10 points; then in light of the computer experiences this root finding algorithm produces an approximate root with deviation ϵ in the neighbourhood of the root, and it is accomplished at the cost of computing only one function value with an error ϵ.

Furthermore, this root finding procedure gives an approximation $t(x)$, that has about the same error on and around the final interval, not only in a point, so it becomes very easy to safeguard against leaving the region of feasible solutions (e.g. this property can be utilized in barrier methods). We can compute for example a "safe" root x_s, for which $f(x_s) = p + \delta$, where δ is a prescribed safety margin. If the error ϵ of the root finding procedure is less than δ, then x_s is in the feasible region with high probability.

The root finding algorithm is self-correcting in the following sense. Assume, that in the i-th iteration an interval $[a_i, b_i]$ was determined, which does not contain the true root $\overline{x}_r$. If in the next iteration(s) more points are added to

S in $[a_i, b_i]$, then after a while the approximation becomes sufficiently close to the function $f(x)$, and will generate another interval, nearer to the root, or containing the root.

The algorithm can be applied for non-increasing directions as well, just the selection of the proper root has to be apppropriately modified, dependent on the problem we have at hand.

7 Appendix

The details of the numerical examples used in the computer experimentations are given here. Two of the four listed examples describe 2-dimensional normal distributions, one is a 10-dimensional example and the last one is a 50-dimensional distribution. The examples are described as follows: for a given distribution function $\Phi(\cdot)$ of the m-dimensional normal distribution the function $f(x)$ is used in the numerical computations, where

$$f(x) = \Phi(\mathbf{z} + x(\mathbf{d} - \mathbf{z})).$$

Example 1. Dimension $m = 2$, $\mathbf{z} = (0, 1.5)$, $\mathbf{d} = (1, 0)$, the correlation matrix $\mathbf{R}$ is given by $r_{12} = r_{21} = -0.9, r_{11} = r_{22} = 1.0$.

Example 2. Dimension $m = 2$, $\mathbf{z} = (0.8, 1.5)$, $\mathbf{d} = (2.4, 1.6)$, the correlation matrix $\mathbf{R}$ is given by $r_{12} = r_{21} = 0.95, r_{11} = r_{22} = 1.0$.

Example 3. Dimension $m = 10$, $\mathbf{z} = (1.1, 0.3, 0, 1.1, -0.3, 0.8, 1.2, 3, 1, 1.6)$, $\mathbf{d} = (1.2, 2, 0.5, 1.9, 3, 2.1, 2, 3.1, 1.9, 1.8)$, the correlation matrix R is given by $r_{12} = r_{21} = 0.9, r_{34} = r_{43} = -0.95, r_{56} = r_{65} = 0.5, r_{78} = r_{87} = -0.9$, $r_{9,10} = r_{10,9} = 0.8, r_{ii} = 1.0$ for $i = 1, \ldots, 10$, all other elements of $\mathbf{R}$ are zero.

Example 4. Dimension $m = 50$, $\mathbf{z} = (1.1, 0.3, 0, 1.1, -0.3, 0.8, 1.2, 3, 1, 1.6,$ $-3, 1, 1.2, 2.1, -1.1, -0.8, -0.2, 0.3, 1.3, 1.8, 1.2, 0.4, 0.6, 2.1, -0.6, -0.3, -2.1,$ $2.2, 3.1, 0.4, 0.3, 0.7, 0.8, 1.3, -1.3, -1.1, -0.4 - 0.1, 0.5, 0.8, 0.2, 2, 2.1, -0.6, 0.5,$ $0.9, 1.1, 1.3, 6, 1)$, $\mathbf{d} = (1.2, 2, 0.5, 1.9, 3, 2.1, 2, 3.1, 1.9, 1.8, -1, 1.3, 3, 2.9, 0, 0.5, 2,$ $1.2, 2.1, 2.8, 4.6, 2.4, 2.1, 4, 1.21.4, -0.4, 4, 6, 3, 2, 2.2, 2.8, 2.2, 2.1, 1, 1.4, 0.9, 2.2,$ $2.3, 1.9, 2.5, 4.1, 1.2, 2.5, 2.4, 1.9, 1.8, 9, 3)$, the correlation matrix is the identity matrix, that is $r_{ii} = 1.0, i = 1, \ldots, 50$, and $r_{ij} = 0$, if $i \neq j$.

Remark. This work was supported by grants from the National Scientific Research Fund, Hungary, T13980 and T16413, and the "Human Capital and Mobility of the European Union ERB CJHR XTC 930087.

References

[1] Bjerager, P.: On computation methods for structural reliability analysis, Structural Safety 9 (1990) 79-96.

[2] Deák, I.: Three digit accurate multiple normal probabilities, Numerische Math. 35 (1980) 369-380.

[3] Deák, I.: Random Number Generators and Simulation, in: Mathematical methods of Operations Research (series ed. A.Prékopa), Akadémiai Kiadó, Budapest, 1990, pp.342.

[4] Deák, I.: Simple regression estimators for the multinormal distributions, lecture at the XIII. Int. Conf. on Mathematical Programming, Mátraháza, Hungary, 1996, manuscript.

[5] Deák, I.: Linear regression estimators for multinormal distributions in optimization of stochastic programming models, European Journal of Operational Research, (submitted).

[6] Ditlevsen, O. J., Bjerager, P.: Plastic reliability analysis by directional simulation, J. Engineering Mechanics, V.115. (1989) 1347-1362.

[7] Ezekiel, M., Fox, K.A.: Methods of correlation and regression analysis, Wiley, 1959, pp. 548.

[8] Gassmann, H.: Conditional probability and conditional expectation of a random vector, in: Numerical techniques for stochastic optimization, ed. Y. Ermoliev, R. Wets, Springer series in comp. math., 1988, 237-254.

[9] Kall, P., Ruszczyński, A., Frauendorfer, K.: Approximation techniques in stochastic programming, in: Numerical techniques for stochastic optimization, ed. Y. Ermoliev, R. Wets, Springer series in comp. math., 1988, 33-64.

[10] A.Prékopa, A.: Stochastic Programming, in: Mathematics and its Applications 324, Kluwer, 1995.

[11] Prékopa, A., Ganczer, S. Deák, I., Patyi, K.: The STABIL stochastic programming model and its experimental application to the electrical energy sector of the Hungarian economy, in:Proc. of the International Symp. on Stochastic Programming, ed. M.Dempster (1980) Academic Press 369-385.

[12] Sen, A., Srivastava, M.: Regression analysis – theory, methods, and applications, Springer, 1990, pp. 347.

[13] Szántai, T.: Evaluation of a special multivariate gamma distribution function, Math. Programming Study 27, (1986) 1-16.

[14] Szántai, T.: A computer code for solution of probabilistic constrained stochastic programming problems, (in: Numerical techniques for stochastic optimization, ed. Yu. Ermoliev, R.J.B.Wets) Laxenburg, 1988. 229-235.

Some Aspects of Algorithmic Differentiation of Ordinary Differential Equations

Peter Eberhard[1] and Christian Bischof[2]

[1] Institute B of Mechanics, University of Stuttgart, 70550 Stuttgart, Germany, Email: pe@mechb.uni-stuttgart.de
[2] Argonne National Laboratory, Mathematics and Computer Science Division, Argonne, IL 60439-4844, U.S.A., Email: bischof@mcs.anl.gov

Summary: Many problems in mechanics may be described by ordinary differential equations (ODE) and can be solved numerically by a variety of reliable numerical algorithms. For optimization or sensitivity analysis often the derivatives of final values of an initial value problem with respect to certain system parameters have to be computed. This paper discusses some subtle issues in the application of Algorithmic (or Automatic) Differentiation (AD) techniques to the differentiation of numerical integration algorithms. Since AD tools are not aware of the overall algorithm underlying a particular program, and apply the chain rule of differential calculus at the elementary operation level, we investigate how the derivatives computed by AD tools relate to the mathematically desired derivatives in the presence of numerical artifacts such as stepsize control in the integrator. As it turns out, the computation of the final time step is of critical importance. This work illustrates that AD tools compute the derivatives of the program employed to arrive at a solution, not just the derivatives of the solution that one would have arrived at with strictly mathematical means, and that, while the two may be different, high-level algorithmic insight allows for the reconciliation of these discrepancies.

Keywords: Algorithmic Differentiation, Sensitivity Analysis, Multibody Systems, Automatic Differentiation, Differential Equations, Numerical Integration

1 Ordinary Differential Equations and Multibody Systems

Ordinary differential equations appear in many fields of science and technology. Although the modeling and description of systems may be very different, the frequently resulting differential equations can be divided into a few major categories, e.g. ordinary differential equations or partial differential equations. Only for some relatively simple types of differential equations can one provide a closed-form analytical solution. Even if this is possible, the mathematical derivation can be difficult and cumbersome. Therefore, differential equations are usually solved by numerical integration algorithms. Many research groups have developed reliable and efficient algorithms and a large body of literature is devoted to this subject, see e.g. [14], [5], [12].

In this paper we restrict ourselves to the solution of nonlinear ordinary differential equations. The resulting initial value problem may then be formulated as follows:

For a given value of system parameters $p \in I\!R^h$, find the trajectories $x(t, p), x \in I\!R^n$ for $t^0 \leq t \leq t^1$ where x is the state vector, t the time, t^0 the initial time and t^1 the final time, respectively. The states are determined by the solution of the initial value problem

$$\dot{x} = f(x, p, t), \quad x(t = t^0, p) = x^0, \tag{1}$$

where f is the vector of (usually nonlinear) state derivatives and x^0 is the initial state.

Our interest in the differentiation of integration algorithms arose during investigations related to the optimization of multibody systems. The multibody system approach can be applied successfully to the description and analysis of mechanical systems where the individual parts undergo large translational and rotational displacements, whereas the deformation of the parts themselves is neglected [13]. The basic elements of a multibody system model are rigid bodies, coupling elements including springs, dampers or active force elements, and joints such as bearings or ideally position controlled elements, see Fig.1.

The generalized coordinates describing e.g. translations or rotations of bodies are summarized in a vector $y \in R^f$. Analogously, the translational and angular velocities of the individual bodies are described by generalized velocities $z \in R^g$, where we have $f = g$ for holonomic multibody systems. The position and velocity for the whole time history are then determined by the differential equations of motion and the implicit initial conditions

$$\begin{aligned}
\dot{y} &= v(t, y, z, p), & \Phi(t^0, y^0, p) &= 0, \\
M(t, y, p)\dot{z} + k(t, y, z, p) &= q(t, y, z, p), & \dot{\Phi}(t^0, y^0, z^0, p) &= 0
\end{aligned} \tag{2}$$

which can be found from Newton's law, Euler's law, and d'Alembert's principle, see [13]. Here M is the mass matrix of the system, q the vector of applied forces, k the vector of Coriolis forces, and v the kinematics vector.

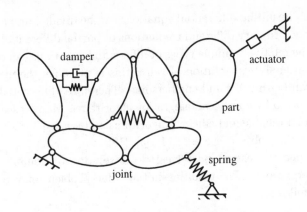

Fig. 1: Basic elements of a multibody system

The set of ordinary differential equations can be solved numerically by an appropriate numerical integration scheme, for example, by Runge-Kutta methods [12] or the Shampine-Gordon algorithm [14]. The final time t^1 is determined by an implicit final condition

$$H^1(t^1, y^1, z^1, p) = 0. \tag{3}$$

In the software system NEWOPT/AIMS [7], the fully symbolical generation of the differential equations even for very large multibody systems is supported by the NEWEUL approach [11].

The multibody systems modelling approach shown above is typical of many other approaches in mechanics. However, in the sequel, we deal with the notationally simpler and more general notation (1) instead of (2). No special use is made from the structure of (2) throughout this paper.

2 Algorithmic Differentiation

To illustrate the basic ideas of Algorithmic (or Automatic) Differentiation, we discuss the simple formula $f = x_1 - e^{2x_2}$, where f is the dependent variable and x_1, x_2 are the independent variables with respect to differentiation. The goal is to compute the total derivative df/dx^T with $x = [x_1, x_2]^T$. Differentiating by hand or using a formula manipulation program like Maple [6], one can easily find the solution $df/dx^T = [1, -2e^{2x_2}]^T$.

On the other hand, to evaluate the function, i.e. to compute *numerical values* for the results, a compiler would transform the whole function into a sequence of simple elementary operations as shown in the middle of Fig.2. This view of a computer program is usually called a 'computational graph', and one can interpret automatic differentiation as an augmentation procedure of such a computational graph. This viewpoint, and generalizations thereof, are discussed in more detail in [4].

The total derivative may be evaluated using the chain rule, i.e.

$$\frac{dx_j}{dx_i} = \sum_{P \in M_P(x_i, x_j)} \left(\prod_{e \in P} (\text{partial derivative along arc } e) \right) \qquad (4)$$

with

M_P ... set of all paths between x_j and x_i,

P ... a single path from M_P,

e ... a single arc along P.

Note that the required partial derivatives along the arcs can easily be computed, because at each node only basic operations are performed. The chain rule in (4) can be evaluated using different computation sequences, in the literature often called 'modes', see [10]. The different evaluation sequences of course have to yield the same results, but this freedom can be utilized to lower computational or storage complexity. The exploration of these tasks is a field of very active research, see [1].

The two traditional modes of automatic differentiation, the forward mode and the reverse mode, are both displayed in Fig.2 for our simple example.

- Using the forward mode, a gradient *vector* ∇x_i is assigned to each intermediate node x_i, representing the total derivatives of this node with respect to all *independent* variables. Each arc is assigned the partial derivative of the 'output node' with respect to the 'input node'. The gradient associated with a particular node is computed using the chain rule, at the same time as the value of the node is computed, resulting in a computational cost that increases linear with the number of independent variables.

- Using the reverse mode, only a *scalar* gradient $\bar{x}_i$ is assigned to each intermediate node. This scalar contains the total derivative of the *dependent* variable with respect to the intermediate node x_i. To calculate the required final gradient the computation sequence must be reversed, i.e. first the values of *all* intermediate nodes have to be computed and then the gradients are computed from the dependent to the independent variables and *not* from the independent to the dependent variable as in the forward mode. This requires storing or recomputing the values of all intermediate nodes, but the computation time is not depending on the number of independent variables any longer.

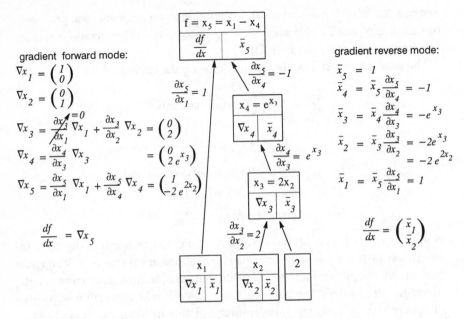

Fig. 2: Forward and reverse mode for the function: $f = x_1 - e^{2x_2}$

Automatic differentiation tools perform this derivative augmentation in a completely mechanical fashion, applying and generalizing the principles outlined above to programs of arbitrary length and complexity. It should be noted, however, that the implementation approaches[1] taken may differ considerably from the graph-oriented view outlined above. In our experiments, we employed the ADIFOR [3] tool.

Note that AD tools do not have an understanding of the global behavior of the code they differentiate. Derivatives are computed for all variables whose value depends on the independent variables and impact the dependent variables (such variables are called *active*). So, in particular, AD tools do not know the mathematical considerations that a programmer went through in constructing a particular piece of code.

3 Differentiation of Numerical Integration Algorithms

In optimization of mechanical systems often total derivatives of some final values of an initial value problem with respect to some system parameters have

---

[1]see http://www.mcs.anl.gov/Projects/autodiff/AD_Tools for an overview of current AD tools

to be computed. Thus, it is interesting to differentiate a numerical integration algorithm with AD techniques and investigate whether and under which circumstances this 'differentiated integration algorithm' really computes the correct total derivatives.

Other, highly specialized algorithms, e.g. the adjoint variable method [2], [8] to compute sensitivities of integral type performance criteria in multibody dynamics exist for similar purposes, but the development of such codes requires usually a lot of insight and always a lot of time. Therefore, the first goal applying AD tools to numerical integration algorithms is to get the correct results and save 'man time', but not necessarily to save a lot of CPU-time.

From the ODE (1) one usually cannot derive solutions $x(t)$ analytically. Therefore, numerical integration algorithms have to discretize and solve the problem in an appropriate way. If we restrict ourselves to single-step algorithms, this can be written as

$$x_{i+1}(p) = x_i(p) + h_i(p)\overline{\dot{x}_i}(p),\tag{5}$$

where the index i denotes a value at time t_i, h_i is the actual stepsize and $\overline{\dot{x}_i}$ is a slope estimation. If $h_i = h = \text{const}$, we have an algorithm without stepsize control. Usually the slope estimation is an average of evaluations for different times and approximations. For constant stepsize and $\overline{\dot{x}_i} = \dot{x}_i = f_i$ we have the simplest integration scheme, the explicit Euler algorithm. Note that the trajectory x_i as well as the slopes $\overline{\dot{x}_i}$ and the stepsize h_i are all dependent on the system parameters p. Fig.3 shows a simplified description of the time-stepping loop of an typical explicit integration algorithm with stepsize control, where g is some function that adjusts the time step. Methods I and II are two integration methods of different order. For simplicity, we ignored the fact that the time step will be adjusted upwards if there is a good fit.

If, for a a given p, we are interested in $\partial x/\partial p^T|_{t=t^1}$, we can employ an AD tool to differentiate this code with respect to p. If we differentiate with respect to p, and use $\nabla \bullet$ to denote $d \bullet /dp^T$, the chain rule of differential calculus now implies that

$$\nabla h_{i+1} = \frac{\partial g}{\partial h_i}\nabla h_i + \frac{\partial g}{\partial \delta}\nabla \delta.\tag{6}$$

Clearly, $\nabla \delta \neq 0$ in general, as δ depends on x, which in turn depends on p. Thus we have the interesting situation that, when $\partial g/\partial \delta \neq 0$, the computational equivalent of time will have a nonzero derivative with respect to the parameter p. Viewed from an analytical perspective, this is nonsense — the values of time and the parameter are not related. From a computational perspective however, it does make sense — depending on the value of the

Given: parameter p, current time t_i, current solution $x_{ci} \approx x_i(t_i, p)$,
        suggested time step h_i
1) Compute $x_1 \approx x(t_i + h_i, p)$ using method I
2) Compute $x_2 \approx x(t_i + h_i, p)$ using method II
3) Compute $\delta = \|x_1 - x_2\|$ for some norm $\| \cdot \|$
4) If $\delta <$ some given threshold
      accept the higher-order of x_1 and x_2 as x_{i+1}
      and update $t_{i+1} \leftarrow t_i + h_i$
   else
      $h_i = g(h_i, \delta)$
      goto 1)
   endif

Fig. 3: Simplified description of a numerical integration algorithm

parameter, we may choose a different time discretization. Thus, what we really *compute* as the final value $x^1(p)$ is

$$x^1(p) = x(t(p), p)|_{t(p)=t^1} \tag{7}$$

(note the dependence of t on p). Thus, we obtain

$$\nabla x_{t=t^1} = \frac{\partial x}{\partial t}\Big|_{t=t^1} \nabla t_{t=t^1} + \frac{\partial x}{\partial p^T}, \tag{8}$$

and with Eq.(1)

$$\nabla x_{t=t^1} = f(x^1, p, t^1)\nabla t_{t=t^1} + \frac{\partial x}{\partial p^T}\Big|_{t=t^1}. \tag{9}$$

Note that ∇x and ∇t will have been computed by the AD-generated derivative code. We observe the following:

(i). Depending on how the time discretization was chosen, we will obtain different values for $\nabla t_{t=t^1}$ and thus for $\nabla x_{t=t^1}$. Most certainly, we will *not* obtain $\partial x/\partial p^T|_{t=t^1}$ which is the result desired by most users.

(ii). If Δt would have been zero at every step, we would have $\nabla t_{t=t^1} = 0$ and thus $\nabla x_{t=t^1} = \partial x/\partial p^T|_{t=t^1}$, as desired by the user. By default, this happens in methods using a fixed step size.

(iii). Independent of how the time discretization was chosen, we can recover the desired solution as

$$\partial x/\partial p^T|_{t=t^1} = \nabla x_{t=t^1} - f(x^1, p, t^1)\nabla t_{t=t^1}. \tag{10}$$

These issues are discussed in more detail in a forthcoming paper [9].

Note that approaches (ii) and (iii) are really geared toward the sophistic-ated AD user. When an integrator code is written, it is probably feasible to indicate the places where the next time step is assigned and to indicate that an AD tool should treat this statement as constant with respect to differenti-ation, resulting in the assignment of a zero gradient. At any rate, unless the developer of the integrator provides this information, the considerable sophist-ication of these codes makes it difficult for others to extract this information from the code.

While one might take the attitude that this was not really an issue given the 'fix' (iii), this is not really the case. Even when $\partial x / \partial p^T$ is well behaved, ∇t and ∇x can become very large and can overflow. Furthermore, the user of an AD tool may well be unaware of these issues, or may not be able to localize the problem since the integrator may be buried under other layers of software. However, as shown in the next section, if the final time is handled appropriately, we are likely to obtain $\nabla t_{t=t^1} = 0$ and everything works out; we suspect that this situation has happened in quite a few AD applications.

We note that while (ii) and (iii) will result in the right derivatives $\partial x / \partial p^T$, there is no guarantee that the derivatives will be obtained at the same accuracy as the solution x, since the guard of the if-statement governing acceptance or rejection of a step will *not* be augmented by AD, and thus still will be only governed by the behavior of x. Thus, the derivatives obtained by Eq.(10) will be consistent, but they may not be as accurate as those obtained by solving the sensitivity equations, which are obtained by direct differentiation of Eq.(1):

$$\frac{d}{dp^T}(\dot{x}) = \frac{d}{dp^T}\left(\frac{dx}{dt}\right) = \frac{\partial f}{\partial x^T}\frac{dx}{dp^T} + \frac{\partial f}{\partial p^T}, \tag{11}$$

$$\rightarrow \quad \frac{d}{dt}\left(\frac{dx}{dp^T}\right) = \frac{d}{dt}(\nabla x) = \frac{\partial f}{\partial x^T}\nabla x + \frac{\partial f}{\partial p^T}. \tag{12}$$

It is easy to add the norm of $\nabla \delta$ to the guard for stepsize control, but an AD tool cannot be expected to do so without user guidance.

The sensitivity equation (12) can also be used to obtain the required gradi-ents using the undifferentiated integration algorithm, but usually it requires a lot of work to derive the sensitivity equations by hand or with formula manip-ulation programs. Often the integration algorithm is only one part of a bigger program. Then it is not trivial to extract the ODE without major changes in the original program flow to derive the sensitivity equations and to reintegrate them back into the program. In [2] an approach similar to Eq.(12) is used to combine hand-derived sensitivity equations with automatically generated partial derivatives. AD tools on the other side allow to compute the gradients

without much additional efforts as soon as the initial value problem itself can
be formulated and solved correctly.

4   Handling of Final Time

So far we intensionally ignored the treatment of the final integration step, but
we will see that this is an essential task. Several integration algorithms have
been investigated and by manual changes like (10) we are able to compute
correct gradients.

It is somewhat surprising that with suitable handling of the final time these
corrections are not required and nevertheless one arrives at the right results.
In Fig.4a the final time handling is shown for an algorithm which limits the
last step size such that the last step satisfies $t_i = t^1$. In comparison with
the previous described algorithm we have to add only the computation of the
proposed next time step h_{i+1}. The most important information is contained in
the limitation of the last step size $h_i = t^1 - t_i$ which leads in the differentiated
code to $\nabla h_i = -\nabla t_i$ if the final time t^1 does not depend on the parameters.
From the time step update $t_{i+1} = t_i + h_i$ we get $\nabla t_{i+1} = \nabla t_i + \nabla h_i$ and
because for the last step (and only for the last step) we have $\nabla h_i = -\nabla t_i$, it
follows that $\nabla t_{i+1} = -\nabla t_i + \nabla t_i = 0$. Therefore, if the algorithm guarantees
that the last step is performed at the final time the correction from (10) is not
required and it is not even necessary to modify the automatically generated
code for the gradients. Of course, the correct computation of ∇t_i for the
whole time interval is still essential.

Another frequently used approach in numerical integration codes to handle
the final time is shown in Fig.4b, where the integration algorithm continues
the time integration until the the final time has been passed. Then the values at the desired final time are computed by interpolation. This has the
advantage that no costly additional evaluation of the system is required, only
a computationally cheaper interpolation. The problem, however, is that the
gradients ∇x and ∇t are interpolated as well and usually we have $\nabla t^1 \neq 0$.
A correction using (10) still leads to the correct results, but not without user
manipulation of the generated code.

Note that the time discretization and step size control are not necessarily
the only numerical artifacts. Features such as variable order polynomial in-
terpolations and projections also depend on the input quantities and influence
the dependent variables and thus correspond to active variables. Their associated gradients will then also contribute to the finally computed gradients.
In this case, the correction (10) may then not be sufficient. Thus, if an unknown new integration algorithm has to be differentiated, one should carefully

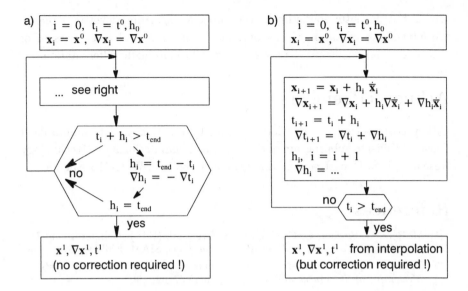

Fig. 4: Handling of the final time

study which numerical artifacts impact the final solution and correct for their influence in the AD-generated derivatives.

5 Conclusions

In this paper it has been shown that Algorithmic Differentiation techniques can be successfully applied to numerical integration algorithms by taking into account the influence of numerical artifacts such as numerical stepsize control in an a-posteriori correction of the AD-computed results. Often, AD-computed results will even be correct without any manual correction. Thus, AD allows to compute gradients for solutions of many initial value problems without manually writing a large amount of code.

However, it must be emphasized that at least a basic understanding of the problem formulation and the numerical algorithms is required to avoid subtle pitfalls and sources of errors arising from the discrepancy of the algorithm arrived at the solution and the mathematical formulation of the solution. High-level, but in-depth, knowledge of the underlying algorithm is necessary to account for internal (numerical and physical) dependencies to obtain correct results. The impact of the computation of the final time step on the overall gradients showed the criticality of this issue.

At the moment, due to the relative novelty of general-purpose AD tools for Fortran and C, the discovery of these internal dependencies may not be trivial, but this situation is likely to improve as algorithm developers realize the advantages of automatically deriving a sensitivity-enhanced version of their code, and document the issues relevant for automatic differentation.

Acknowledgements

C.B. was supported by the Mathematical, Information, and Computational Sciences Division subprogram of the Office of Computational and Technology Research, U.S. Department of Energy, under contract W-31-109-Eng-38.

References

[1] M. Berz, C. Bischof, G. Corliss, and A. Griewank (Eds.). *Computational Differentiation: Techniques, Applications, and Tools*, SIAM, Philadelphia, 1996.

[2] D. Bestle and P. Eberhard. Analyzing and Optimizing Multibody Systems. *Mechanics of Structures and Machines*, 20(1):67–92, 1992.

[3] C. Bischof, A. Carle, P. Khademi, and A. Mauer. ADIFOR 2.0: Automatic differentiation of Fortran 77 programs, *IEEE Computational Science & Engineering*, 3(3):18–32, 1996.

[4] C. Bischof and M. Haghighat. *On Hierarchial Differentiation*. In [1], pp. 83–94.

[5] J.C. Butcher. *The Numerical Analysis of Ordinary Differential Equations (Runge-Kutta and General Linear Methods)*. John Wiley and Sons, New York, 1987.

[6] B.W. Char, K.O. Geddes, G.H. Gonnet, M. Monagan, and B.M. Watt. *First Leaves, A Tutorial Introduction to Maple*. Watcom Publications, Waterloo, 1989.

[7] P. Eberhard. *Zur Mehrkriterienoptimierung von Mehrkörpersytemen*. Reihe 11, No. 227. VDI-Verlag, Düsseldorf, 1996.

[8] P. Eberhard. *Adjoint Variable Method for Sensitivity Analysis of Multibody Systems Interpreted as a Continuous, Hybrid Form of Automatic Differentiation*. In [1], pp. 319–328.

[9] P. Eberhard and C. Bischof. *Automatic Differentiation of Numerical Integration Algorithms*. (to appear).

[10] A. Griewank. On automatic differentiation. In *Mathematical Programming: Recent Developments and Applications*, pp. 83–108, Amsterdam, 1989. Kluwer Academic Publishers.

[11] E. Kreuzer and G. Leister. *Programmsystem NEWEUL 92, AN-32*. University, Institute B of Mechanics, Stuttgart, 1991.

[12] W.H. Press, S.A. Teukolsky, W.T. Vetterling, and B.P. Flannery. *Numerical Recipes in Fortran: The Art of Scientific Computing*. Cambridge University Press, 2nd edition, 1992.

[13] W.O. Schiehlen (Ed.). *Advanced Multibody System Dynamics*. Kluwer Academic Publishers, Dortrecht, 1993.

[14] L.F. Shampine. *Numerical Solution of Ordinary Differential Equations*. Chapman & Hall, New York, 1994.

Refinement Issues in Stochastic Multistage Linear Programming[1]

K. Frauendorfer and Ch. Marohn

University of St. Gallen, Institute of Operations Research, Holzstrasse 15, CH-9010 St. Gallen

Summary. Linear stochastic multistage programs are considered with uncertain data evolving as a multidimensional discrete-time stochastic process. The associated conditional probability measures are supposed to depend linearly on the past. This ensures convexity of the problem and allows application of barycentric scenario trees. These approximate the discrete-time stochastic process, and provide inner and outer approximation of the value functions.

The main issue is to refine the discretization of the stochastic process efficiently, using the nested optimization and integration of the dynamic, implicitly given value functions. We analyze and illustrate how errors evolve across nodes of the scenario trees.

Keywords. approximation, discretization, refinement scheme

1 Introduction

In various applications decision makers are required to take uncertain future into account. In particular, today's decisions have to be taken without knowing prices or resources.

Let the uncertain evolvement of data over a finite planning horizon $[0, T]$ be described as a multidimensional discrete-time stochastic process $(\omega_t, t = 0, \ldots, T)$ on a common Borel space $(\Omega, \mathcal{B}^M)$ with $\Omega \subset \mathbb{R}^M$ compact. In many problem statements both prices and demand or supply of resources are stochastic and induce specific structural properties of the value functions involved. This motivates us to decompose the multidimensional discrete-time stochastic process $(\omega_t, t = 0, 1, \ldots, T)$ into two stochastic parts, one referring to prices $(\eta_t, t = 0, 1, \ldots, T)$, the second $(\xi_t, t = 0, 1, \ldots, T)$ to demand or supply of resources, i.e. $\omega_t = (\eta_t, \xi_t)$.

Let P represent the (regular) joint probability distribution of $\omega := (\omega_0, \ldots, \omega_T)$. The associated regular conditional distributions with respect to ω_t are

---

[1] Research of this report was supported by *Schweizerischer Nationalfonds* Grant Nr. 21-39'575.93

denoted $P_t(\cdot|\omega_{t-1})$ for $t = 1, \ldots, T$. All these have compact support. The time points $t = 0, 1, \ldots, T$ at which decisions $u_t \in \mathbb{R}^n$ are taken, are supposed to be predefined. Setting $\eta := (\eta_0, \ldots, \eta_T)$ and $\xi := (\xi_0, \ldots, \xi_T)$, a corresponding mathematical program is written formally as

$$
\begin{aligned}
\min \quad & \int_\Omega \left[\sum_{t=0}^T \rho_t(u_t, \eta_t) \right] dP(\eta, \xi) \\
\text{s.t.} \quad & f_t(u_{t-1}, u_t) \leq h(\xi_t), \quad t = 0, 1, \ldots, T.
\end{aligned}
\tag{1}
$$

The convention is that negative subscripts of variables indicate decisions of the past, negative subscripts of the stochastic data indicate data of the past, both of which represent the input data at the present stage $t = 0$; in particular, stochastic data with subscript 0 are currently observed data and, hence, deterministic.

In (1), the cost $\rho_t(\cdot)$ at t are determined by observation η_t and decision u_t. The feasibility region for $u_t \in \mathbb{R}^n$ is supposed to depend on u_{t-1} and on the observed outcome ξ_t; $f_t(u_{t-1}, \cdot)$ is a vector-valued function in u_t and represents the demand for resources, the components of $h(\xi_t)$ are the supply components of the resources at t. Decisions u_t have to be selected at t after (η_t, ξ_t) is observed, but prior to the observations $(\eta_{t+1}, \ldots, \eta_T)$ and $(\xi_{t+1}, \ldots, \xi_T)$. According to this rule, *nonanticipative* or *measurable* decisions have to be determined, which minimizes the expected value of the overall cost and which satisfies the constraints.

The feasibility set, viewed as a constraint multifunction in ξ, is supposed to be *strictly nonanticipative* and convex compact-valued with a nonempty interior for every ξ. This ensures that for any nonanticipative and feasible decision u_t there exist interior feasible and nonanticipative decisions $u_{t+1}, \ldots, u_T$ for any sequence of outcomes $(\xi_{t+1}, \ldots, \xi_T)$ (see Rockafellar and Wets 1976/1978 [45],[46] and Frauendorfer 1996 [23]).

Standard dynamic programming arguments yield optimal value functions $\phi_t(\cdot)$ corresponding to periods $t = 0, \ldots, T$ (see e.g. Bertsekas 1995 [2], [3]). This allows to write the stochastic multistage programm as a sequence of nested two-stage programs. Start at period T with $\phi_{T+1}(\cdot) := 0$ and define backwards for $t = T, \ldots, 0$:

$$
\phi_t(u_{t-1}, \eta_t, \xi_t) := \min \rho_t(u_t, \eta_t) + \int_{\Omega_{t+1}} \phi_{t+1}(u_t, \omega_{t+1}) dP_{t+1}(\omega_{t+1}|\eta_t, \xi_t)
$$

$$
\text{s.t.} \quad f_t(u_{t-1}, u_t) \leq h(\xi_t).
\tag{2}
$$

In case $\rho_t(u_t, \eta_t)$ are continuous convex-concave saddle functions, $f_t(\cdot)$ is convex vector-valued, $h_t(\cdot)$ is linear-affine in ξ_t, and in case the conditional probability distributions $P_{t+1}(\cdot|\eta_t, \xi_t)$ depend linearly on (η_t, ξ_t) and are unaffected by the decisions taken, then it has been proven that for $t = 1, \ldots, T$ the value functions $\phi_t(u_{t-1}, \eta_t, \xi_t)$ are continuous saddle functions, convex in (u_{t-1}, ξ_t) and concave in η_t with respect to their domains. Under these conditions, referred to as *the convex case* below, primal and dual solvability of the convex programs is ensured in (2). For details it is referred to Rockafellar and Wets 1976 [45], Frauendorfer 1994/1996 [22] and [23].

For the ease of exposition it will be helpful to use for (2) the notation

$$\phi_t(u_{t-1}, \eta_t, \xi_t) := \min\ [\rho_t + \mathrm{E}_t \phi_{t+1}](u_t, \eta_t, \xi_t). \tag{3}$$

One major challenge with (1) is the nested minimizations and multidimensional integrations of implicitly given value functions. Unlike many control type formulations, no analytical expressions can be expected within the stochastic multistage setting due to the non-smoothness of the value functions. For overcoming these difficulties numerically in the convex case we discretize the stochastic processes with respect to their outcomes taking into account the subdifferentiability of the value functions.

Let the conditional probability measures $P_t(\cdot|\omega_{t-1})$ be successively discretized for $t = 0, \cdots, T$, yielding discrete conditional probability measures $Q_t(\cdot|\omega_{t-1})$ with corresponding support $A_t(\omega_{t-1})$. This way, a *scenario tree* A and the associated path probabilities $Q(\omega)$ are given according to

$$
\begin{aligned}
A &:= \{\omega \in \Omega | \omega_t \in A_t(\omega_{t-1}),\ \forall t \geq 1\}, \\
Q(\omega) &:= \prod_{t=1}^{T} Q_t(\omega_t|\omega_{t-1}).
\end{aligned}
\tag{4}
$$

Let the projections of A onto $[0, t]$ be denoted A^t. Clearly, any stochastic process, which is discrete in both time and state, is representable as scenario tree. It is easily seen when the conditional probability measure $P_{t+1}(\omega_{t+1}|\eta_t, \xi_t)$ is discrete with finite support, that the stochastic two-stage programs in (2) has block structure. Then (1) may be written as a mathematical program with a dynamic block structure and high sparsity, whose size depends on the number of scenarios within the tree. This indicates why the solvability of stochastic multistage programs strongly benefits from sophisticated algorithms. Recent works include Rockafellar and Wets 1991 [47], Wets 1989 [51], Gassmann 1990 [27], Birge 1985/1994/1995 [5],[7],[6], Mulvey and Ruszczyński 1995 [38], Mulvey, Vanderbei and Zenios 1995 [40], Zenios 1991 [52], Robinson 1991 [44], Ruszczyński 1993 [49],[48], Dantzig 1990/1993 [13], [14], Edirisinghe and Ziemba 1994/96 [18], [17], [19], Kall, Ruszczyński and Frauendorfer 1988 [32], Infanger 1992/1994 [30],[31], Ermoliev and Wets 1988 [20], Kall and Wallace 1994 [34], Kall and Stoyan 1982 [33], Wets 1989 [51], Hiller and Eckstein 1994 [29].

However, one has to be aware of the fact that discretizing the conditional probability measures is a simplification of the real dynamics, which may have severe impact on the goodness of the surrogate problem. It is not difficult to define conditions based on which the discretization has to be refined, to ensure weak convergence of the discrete measures to the real probability measure and, hence, epi-convergence of the minimizers. But, due to the fact that the number of scenarios grows exponentially, the quality of discretizations has to be monitored carefully. For example, applying Monte-Carlo simulation to the conditional probability measures with a sample size of 10^5 in each of 6 stages results in a scenario tree with 10^{30} possible scenarios. Suffering from the *curse of dimensionality* the corresponding deterministic equivalent program is numerically unsolvable. In the convex case, *barycentric approximation* allows design distinguished scenario trees that provide information on how accurate the real dynamics are approximated.

We discuss issues on how to refine the scenario trees taking into account the nested optimization and integration of the value functions. Hence, this work provides improvements within algorithmic procedures that solve stochastic multistage linear programs.

The paper is organized as follows: Section 2 introduces a multistage finance problem and releases its structural properties, which may be exploited within the refinement process. Section 3 reviews briefly *barycentric scenario trees* and the approximation of the value functions which help overcome the difficulties with nested minimizations and multidimensional integrations. Section 4 represents the major part of this work and investigates how the error evolves along the scenario tree backwards in time, and measures to what extent error arises from integration. Section 5 concludes.

2 Application in banking

An example from banking is given to illustrate the application of multistage stochastic linear programming. Besides linearity, further structural properties are inherent in many financial problem statements and, hence, facilitate their solvability. Herein, we consider funding of non-fixed rate mortgages which face interest rate and prepayment risk. This constitutes an important problem for corporate financial managers of Swiss banks due to the risks these managers are exposed to.

Non-fixed rate mortgages offer the clients to finance their mortgages at current market rates which strongly correlate with the current interest rate level. It should be noted that the national bank not only monitors the mortgage rate but also defines some cap in case the market interest rates are too high. This is intended to protect the clients and, on one hand, reduces the credit risk for the bank, on the other hand, additional risk is faced with respect to the funding mechanism. In Switzerland, it happened over a rather long period within the last decade that the mortgage rate the clients had to pay was even less than the one-year-borrowing rate at which the bank funded a considerable part of their business. In such an interest rate environment the clients clearly prefer the variable rate mortgages. This increases the volume and the funding costs and therefore the risk the bank is exposed to. If the interest rate level is low, it has been observed that the mortgage rate is kept above a certain floor, causing fixed-rate mortgages to fall even below the non-fixed ones. As a natural consequence, clients change their liabilities to the fixed-rate mortgages, causing the non-fixed rate volume to decrease in an environment where banks would have the possibility to fund their non-fixed rate business at low costs. This type of risk is known as prepayment risk. Summarizing these dynamics from the view point of banking industry, one observes that the profitability of non-fixed rate mortgages suffers from a high interest level as well as from a low interest level.

The challenge of funding non-fixed rate mortgages is seen in optimizing the monthly funding activities taking into account the stochasticity of interest rate dynamics and mortgage volume. Depending on the asset and liability structure of a bank, the non-fixed rate mortgages are funded to a certain extent with bonds of different maturities. The range of these maturities may vary between one month and 10 years. Taking into account the liquidity with respect to which the various maturities are traded, it is observed that short term bonds, say bonds with a maturity of up to 1 year, may be borrowed in a considerably larger volume than bonds with a maturity beyond 2 years. In addition, the short term rates are much more volatile than long term rates.

The funding problem, referred to below, has been introduced in Frauen-dorfer 1996 [24] and is discussed with respect to interest rate models in Frauendorfer and Schürle 1997 [26]. Herein, this problem is investigated with respect to its solvability.

Let $\mathcal{D} := \{1, 2, \ldots, D\}$ be the set of prescribed times at which bonds mature. Taking into account the liquidity within the various maturities, only a subset $\mathcal{D}^S \subset \mathcal{D}$ is regarded for funding. Furthermore, it is assumed that the bonds are held until maturity, so that changes in the price of a bond during the holding period may be relaxed. The volume of bonds borrowed at time $t = 0, \ldots, T$ with maturity $d \in \mathcal{D}$ is denoted $v_t^{d,+}$; v_t^d represents the total volume of bonds with maturity d at time t. Clearly, v_t^d is determined by

$$v_t^d = v_{t-1}^{d+1} + v_t^{d,+} \quad t = 0, 1, \ldots, T; \ \forall d \in \mathcal{D}^S,$$
$$v_t^d = v_{t-1}^{d+1} \quad t = 0, 1, \ldots, T; \ \forall d \notin \mathcal{D}^S.$$

The total funding volume at t is given with

$$x_t = \sum_{d \in \mathcal{D}} v_t^d \quad t = 0, \ldots, T.$$

Its evolvement over time is determined with the stochastic change $\xi_t \in \mathbb{R}$,

$$x_t = x_{t-1} + \xi_t \quad t = 1, 2, \ldots, T.$$

The term structure dynamics may be represented by a finite-dimensional discrete-time stochastic process $\eta_t \in \mathbb{R}^K; t = 0, \ldots, T$. Let the accrued interest payments of an unit borrowed with maturity d be denoted $\rho_t(\eta_t, d)$. Setting $v_t := (v_t^d; d \in \mathcal{D}), v_t^+ := (v_t^{d,+}; d \in \mathcal{D}^S)$ allows for writing the accrued interest paymenets as inner product within $\mathbb{R}^{|\mathcal{D}^S|}$:

$$< \rho_t(\eta_t), v_t^+ > = \sum_{d \in \mathcal{D}^S} \rho_t(\eta_t, d) \cdot v_t^{d,+}$$

The stochastic multistage linear program which minimizes the expected present value of interest payments reads as

$$\min \int_\Omega \sum_{t=0}^{T} <\rho_t(\eta_t), v_t^+> dP(\eta, \xi)$$

$$
\begin{aligned}
v_t^d - v_{t-1}^{d+1} - v_t^{d,+} &= 0 \quad \forall t, \forall d \in \mathcal{D}^S \\
v_t^d - v_{t-1}^{d+1} &= 0 \quad \forall t, \forall d \notin \mathcal{D}^S \\
x_t - \sum_{d \in \mathcal{D}} v_t^d &= 0 \quad \forall t \\
x_t - x_{t-1} &= \xi_t \quad \forall t \\
v_t^{d,+} \geq 0; v_t^d, x_t \in \mathbb{R} \text{ nonanticipative} & \quad \forall t, \forall d \in \mathcal{D}.
\end{aligned}
\tag{5}
$$

Observe that the left-hand side of the constraints (5) is deterministic. This fact requires that the stochastic interest payments, that clearly depend on the funding decision, has to be incorporated in the dynamis of ξ. To preserve convexity one has to relax the stochastic dependency of the volume change on the decisions. Then probability measures must be unaffected by the decisions taken. The stochastic interest payments appear in the objective, only. If, in addition, the conditional probabilities $P_t(\cdot | \eta_{t-1}, \xi_{t-1})$ depend linearly on the observation in period $t-1$, the convexity of the stochastic multistage program (5) is preserved and the value function of stage t associated with (5) are saddle functions concave-convex in (η_t, ξ_t) (see Frauendorfer 1994/1996 [22], [23]).

The underlying saddle structure of the value functions motivates the application of the barycentric approximation (see Figure (1) and section 3), which optimizes the discretization of the stochastic interest rate and volume processes.

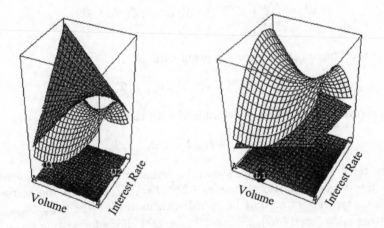

Figure 1: Bilinear approximation of the value function

An illustration of a scenario tree is given in Figure 2.

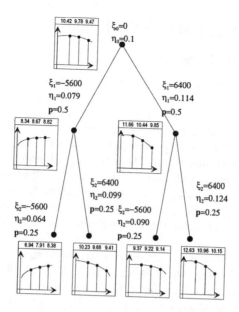

Figure 2: Scenario tree for 2-stage funding problem with interest rate curve being driven by three maturities.

It is further noted that $+1$ and -1 are the only non-zero coefficients of the matrix on the left-hand side of (5). The submatrices associated with the various stages $t = 1, \ldots, T$ remain unchanged and are of high sparsity. In addition to the low dimensions of the term structure representation, the dynamics of the stochastic right-hand side is characterized by a one-dimensional discrete-time process. All these properties increase the solvability of the underlying problem considerably, as these are exploited by sophisticated algorithms.

Due to the achieved progress in the methodological developments within mathematical programming stochastic programming has received increasing attention in finance. At this place it is referred to the successful and valuable contributions of Dantzig 1990/1993 [13], [14], Ziemba 1975/1986/1992/1994/1997 [57], [36], [55], [9], [56], [42], Dempster 1997 [11], Klaassen 1997 [35], Dupačová 1991/1992/1997 [15], [16] [1], Mulvey 1992/1994/1996/1997 [41], [37], [39], [56], [42], Zenios 1991/1992/1993/1995/1996 [52], [53],[54], [55], [12] [28], [50], [43], Hiller and Eckstein 1994 [29], and Cariño et al. 1994/1995/1997 [9], [10], [8].

3 Barycentric scenario trees

In the dynamic formulation (2),

$$\int \phi_t(u_{t-1}, \eta_t, \xi_t) dP_t(\eta_t, \xi_t | \omega_{t-1}). \tag{6}$$

has to be evaluated within the nested optimization, which encounters a serious challenge. As said, in the convex case the value functions are saddle functions in (1). This motivates the application of the *barycentric approximation technique*, which not only helps carry out the nested optimization, but also provides information on how accurate the real dynamics are mapped.

In this section, the *barycentric approximation technique* is shortly reviewed, illustrating how the multidimensional integration is bounded from above and below. For details it is referred to Frauendorfer 1992/1996 [21], [23].

Since supports are compact, the stochastic outcomes $\omega_t := (\eta_t, \xi_t)$ may be covered by a $\times$-simplex[2] $\Omega_t := \Theta_t \times \Xi_t$ for $t = 1, \dots, T$. Taking into account the dependency of (η_t, ξ_t) on (η_{t-1}, ξ_{t-1}), one may select the $\times$-simplicial coverage conditioned on $\omega_{t-1} = (\eta_{t-1}, \xi_{t-1})$, $\Omega_t(\omega_{t-1}) := \Theta_t(\omega_{t-1}) \times \Xi_t(\omega_{t-1})$.

To a given scenario tree A of the form (4) and its projections A^t for $t = 1, \dots, T$, a $\times$-simplicial partition of the support of ω_t may be selected for any $\omega_{t-1} \in A_{t-1}(\omega_{t-2})$:

$$C_t(\omega_{t-1}) := \{\Omega_{t,i_t}(\omega_{t-1}) | \cup_{i_t=1}^{I_t} \Omega_{t,i_t}(\omega_{t-1}) = \Omega_t(\omega_{t-1}) \supset \text{supp } P_t(\cdot|\omega_{t-1}),$$

$$\Omega_{t,i'_t}(\omega_{t-1}) \cap \Omega_{t,i''_t}(\omega_{t-1}) = \emptyset, \forall\, i'_t \neq i''_t, \; i_t = 1, \dots, I_t\}. \tag{7}$$

We call the resulting set process $C := (C_t(\omega_{t-1}); t = 1, \dots, T)$ a $\times$-*simplicial coverage process* consistent with A iff $A_t(\omega_{t-1}) \subset \Omega_t(\omega_{t-1})$ for all $\omega_{t-1} \in A_{t-1}(\omega_{t-2})$, $t = 1, \dots, T$.

For the ease of exposition, we first consider a partition $C_t(\omega_{t-1})$ consists of one $\times$-simplex, i.e.: $C_t(\omega_{t-1}) := \Omega_t(\omega_{t-1}) := \Theta_t(\omega_{t-1}) \times \Xi_t(\omega_{t-1})$. $\Theta_t(\omega_{t-1}), \Xi_t(\omega_{t-1})$ are regular simplices in $\mathbb{R}^K$, $\mathbb{R}^L$ with associated vertices $a_{t,\nu}$ and $b_{t,\mu}$, $\nu = 0, \dots, K$, $\mu = 0, \dots, L$. Below, $\lambda_\mu(\eta_t)$ and $\tau_\nu(\xi_t)$ denote the barycentric weights of η_t and ξ_t with respect to the simplices $\Theta_t(\omega_{t-1})$ and $\Xi_t(\omega_{t-1})$.

The conditional probability measure $P_t(\cdot|\omega_{t-1})$ induces mass distribution $\mathcal{M}_{t,\nu}$ with corresponding generalized barycenter

$$\xi_{t,\nu} := \sum_\mu b_{t,\mu} \frac{\int \lambda_\mu(\eta_t) \cdot \tau_\nu(\xi_t) dP_t(\eta_t, \xi_t | \omega_{t-1})}{\int \tau_\nu(\xi_t) dP_t(\eta_t, \xi_t | \omega_{t-1})} \tag{8}$$

on the (L-dimensional) simplices $\{a_{t,\nu}\} \times \Xi_t(\omega_{t-1})$.[3] For $\nu = 0, \dots, K$ let

---

[2] A $\times$-simplex is a set whose closure is representable as a Cartesian product of simplices.
[3] In the notation it is omitted for simplicity that the vertices and generalized barycenters depend on ω_{t-1}.

the mass

$$\mathcal{M}_{t,\nu}(\{a_\nu\} \times \Xi_t(\omega_{t-1})) := \int \tau_\nu(\xi_t) dP_t(\eta_t, \xi_t | \omega_{t-1}) \tag{9}$$

be assigned to $(a_{t,\nu}, \xi_{t,\nu})$. As for $\nu = 0, \dots, K$ the mass distributions $\mathcal{M}_{t,\nu}$ add up to a conditional probability distribution, one gets a discrete conditional probability measure $Q_t^l(\cdot | \omega_{t-1})$ on $\Theta_t(\omega_{t-1}) \times \Xi_t(\omega_{t-1})$.

Due to symmetry, the conditional probability measure $P_t(\cdot | \omega_{t-1})$ induces mass distributions $\mathcal{M}_{t,\mu}$ with associated generalized barycenters

$$\eta_{t,\mu} := \sum_\nu a_{t,\nu} \frac{\int \lambda_\mu(\eta_t) \cdot \tau_\nu(\xi_t) dP_t(\eta_t, \xi_t | \omega_{t-1})}{\int \lambda_\mu(\eta_t) dP_t(\eta_t, \xi_t | \omega_{t-1})} \tag{10}$$

on the (K-dimensional) simplices $\Theta_t(\omega_{t-1}) \times \{b_{t,\mu}\}$. Again, for $\mu = 0, \dots, L$ let the mass

$$\mathcal{M}_{t,\mu}(\Theta_t(\omega_{t-1}) \times \{b_{t,\mu}\}) := \int \lambda_\mu(\eta_t) dP_t(\eta_t, \xi_t | \omega_{t-1}) \tag{11}$$

be assigned to the points $(\eta_{t,\mu}, b_{t,\mu})$. Analogously, for $\mu = 0, \dots, L$ the mass distributions $\mathcal{M}_{t,\mu}$ add up to a conditional probability distribution, yielding a discrete conditional probability measure $Q_t^u(\cdot | \omega_{t-1})$ on $\Theta_t(\omega_{t-1}) \times \Xi_t(\omega_{t-1})$.

As it becomes obvious from (8),(9),(10) and (11), the advantagous feature from a computational viewpoint is that generalized barycenters and their probabilities are completely determined by the first moments of η_t and ξ_t, and by the cross moments $E(\eta_t \cdot \xi_t)$.

It has been proven in Frauendorfer 1992 [21] that in the convex case the expression in (6) is approximated from below and above by

$$(E_t^l \phi_t)(u_{t-1}, \omega_{t-1}) \quad := \quad \int \phi_t(u_{t-1}, \omega_t) dQ_t^l(\omega_t | \omega_{t-1}) \le$$
$$\le (E_t \phi_t)(u_{t-1}, \omega_{t-1}) \quad := \quad \int \phi_t(u_{t-1}, \omega_t) dP_t(\omega_t | \omega_{t-1}) \le \tag{12}$$
$$\le (E_t^u \phi_t)(u_{t-1}, \omega_{t-1}) \quad := \quad \int \phi_t(u_{t-1}, \omega_t) dQ_t^u(\omega_t | \omega_{t-1})$$

This technique has been termed *barycentric approximation* and can easily be applied to a partition of the form (7). For details it is referred to Frauendorfer 1992/1994/1996 [21], [22] and [23].

Let be defined

$$A_t^l(\omega_{t-1}) := \text{supp } Q_t^l(\cdot | \omega_{t-1}),$$
$$A_t^u(\omega_{t-1}) := \text{supp } Q_t^u(\cdot | \omega_{t-1}). \tag{13}$$

Clearly, $A_0^l = A_0^u = \omega_0$. Starting with a partition $C_1^l(\omega_0) = C_1^u(\omega_0)$, the associated barycentric approximation yields $Q_1^l(\omega_0)$ and $Q_1^u(\omega_0)$ and the associated supports $A_1^l(\omega_0)$ and $A_1^u(\omega_0)$. Hence, the partitions $C_t^l(\omega_{t-1})$ and $C_t^u(\omega_{t-1})$ are selected with respect to the support of $Q_t^l(\cdot | \omega_{t-1})$ and

$Q_t^u(\cdot|\omega_{t-1})$ in an inductive manner. This way, the *barycentric scenario trees* and the associated path probabilities are given in analogy to (4):

$$A^l := \{\beta^l|\beta_t^l \in A_t^l(\beta_{t-1}^l),\ \forall t \geq 1\}, \tag{14}$$

$$Q^l(\beta^l) := \prod_{t=1}^{T} Q_t^l(\beta_t^l|\beta_{t-1}^l), \tag{15}$$

and

$$A^u := \{\beta^u|\beta_t^u \in A_t^u(\beta_{t-1}^u),\ \forall t \geq 1\}, \tag{16}$$

$$Q^u(\beta^u) := \prod_{t=1}^{T} Q_t^u(\beta_t^u|\beta_{t-1}^u). \tag{17}$$

$A^{t,l}, A^{t,u}$ denotes the projection of A^l, A^u onto $[0,t]$, and $\beta^l = (\beta_0^l, \ldots, \beta_T^l)$, $\beta^u = (\beta_0^u, \ldots, \beta_T^u)$ the associated barycentric scenarios. Note that due to the construction the barycentric scenario trees, A^l, A^u and the associated ×-simplicial coverage processes C^l, C^u are consistent.

Substituting in (1) the probability measure P by its discretizations Q^l and Q^u yields the associated stochastic multistage programs

$$\begin{aligned} \min\ &\int \left[\sum_{t=0}^{T} \rho_t(u_t, \eta_t)\right] dQ^l(\eta, \xi) \\ s.t.\ &f_t(u_{t-1}, u_t) \leq h(\xi_t), \quad t = 0, 1, \ldots, T, \end{aligned} \tag{18}$$

and

$$\begin{aligned} \min\ &\int \left[\sum_{t=0}^{T} \rho_t(u_t, \eta_t)\right] dQ^u(\eta, \xi) \\ s.t.\ &f_t(u_{t-1}, u_t) \leq h(\xi_t), \quad t = 0, 1, \ldots, T. \end{aligned} \tag{19}$$

The associated value functions are given through the dynamic formulation

$$\begin{aligned} \psi_t(u_{t-1}, \eta_t, \xi_t) &:= \min \rho_t(u_t, \eta_t) + \int \psi_{t+1}(u_t, \omega_{t+1}) dQ_t^l(\omega_{t+1}|\eta_t, \xi_t) \\ s.t.\ &f_t(u_{t-1}, u_t) \leq h(\xi_t), \end{aligned} \tag{20}$$

and

$$\begin{aligned} \Psi_t(u_{t-1}, \eta_t, \xi_t) &:= \min \rho_t(u_t, \eta_t) + \int \Psi_{t+1}(u_t, \omega_{t+1}) dQ_t^u(\omega_{t+1}|\eta_t, \xi_t) \\ s.t.\ &f_t(u_{t-1}, u_t) \leq h(\xi_t), \end{aligned} \tag{21}$$

with $\psi_{T+1}(\cdot) = \Psi_{T+1}(\cdot) := 0$.

It has been proven in Frauendorfer 1994 [22] that (20) and (21) provide lower and upper approximates for the value function $\phi_t(u_{t-1}, \eta_t, \xi_t)$ with $\psi_t(u_{t-1}, \eta_t, \xi_t)$ and $\Psi_t(u_{t-1}, \eta_t, \xi_t)$, i.e:

$$\psi_t(u_{t-1}, \eta_t, \xi_t) \leq \phi_t(u_{t-1}, \eta_t, \xi_t) \leq \Psi_t(u_{t-1}, \eta_t, \xi_t). \tag{22}$$

Solving (18) and (19) yields policies $u^l := (u_0^l, u_1^l, \ldots, u_T^l)$ and $u^u :=$ $(u_0^u, u_1^u, \ldots, u_T^u)$ where u_t^l, u_t^u denote the decisions made after $\beta^{t,l} \in A^{t,l}$, $\beta^{t,u} \in A^{t,u}$, $(t = 1, \ldots, T)$ is observed. Hence, upper and lower approximations of the value functions may correspond to different policies. Both $\psi_t(\cdot)$ and $\Psi_t(\cdot)$ epiconverge to $\phi_t(\cdot)$ in case the weak convergence of the conditional discrete probability measures $Q_t^l(\cdot|\beta_{t-1}^l)$, $Q_t^u(\cdot|\beta_{t-1}^u)$ to $P_t(\cdot|\beta_{t-1}^l)$, $P_t(\cdot|\beta_{t-1}^u)$, respectively, is ensured for $t = 1, \ldots, T$. This requires that the sub-×-simplices of the coverage processes become arbitrarily small with respect to their diameters. In the following section we discuss in what way this convergence can be monitored, taking into account that lower and upper approximation refer to different policies and different coverage processes.

4 Error analysis

As mentioned in the introduction, the difficulty in solving stochastic linear multistage programs is seen in the nested optimization and multidimensional integration of implicitly given value functions. Discretizing the conditional probability measures in the barycentric sense helps to overcome these difficulties and, in the convex case, provides approximate policies for the underlying problem including bounds on the optimal expected value. However, lower and upper approximation of the value functions refer to different scenario trees and associated consistent ×-simplicial coverage processes, which only coincide with respect to the data at $t = 0$. For assessing how the inaccuracy of the approximation evolves over time, based on which the refinement process can be monitored efficiently, additional information has to be determined.

For the ease of exposition, the following notations are used:

$$(\mathrm{E}_t^l \Psi_t)(u_{t-1}, \omega_{t-1}) := \int \Psi_t(u_{t-1}, \omega_t) dQ_t^l(\omega_t|\omega_{t-1})$$

$$(\mathrm{E}_t^u \Psi_t)(u_{t-1}, \omega_{t-1}) := \int \Psi_t(u_{t-1}, \omega_t) dQ_t^u(\omega_t|\omega_{t-1})$$

$$(\mathrm{E}_t^l \psi_t)(u_{t-1}, \omega_{t-1}) := \int \psi_t(u_{t-1}, \omega_t) dQ_t^l(\omega_t|\omega_{t-1}) \qquad (23)$$

$$(\mathrm{E}_t^u \psi_t)(u_{t-1}, \omega_{t-1}) := \int \psi_t(u_{t-1}, \omega_t) dQ_t^u(\omega_t|\omega_{t-1})$$

Let the scenario trees A^l, A^u and the associated consistent coverage processes C^l, C^u be and the associated stochastic multistage programs (18) and (19) be solved. Given a scenario $\beta^{t,l} = (\beta_1^l, \ldots, \beta_t^l) \in A^{t,l}$ up to t, decision u_t^l and its value $\psi_t(u_{t-1}^l, \beta_t^l)$ is available. The stochastic evolvement during the next period is approximated by the conditional barycentric discretization $\beta_{t+1}^l \in A_{t+1}^l(\beta_t^l)$ with the associated minimal value $\psi_{t+1}(u_t^l, \beta_{t+1}^l)$ (see Figure 3); analogoulsy, for $\beta^{t,u} \in A^{t,u}$ the decision u_t^u and the value $\Psi_t(u_{t-1}^u, \beta_u^t)$ is known. The stochastic evolvement during the next period is approximated by the conditional barycentric discretization $\beta_{t+1}^u \in A_{t+1}^u(\beta^u)$ with the associated minimal values $\Psi_{t+1}(u_t^u, \beta_{t+1}^u)$ (see Figure 4). Due to (22), the following

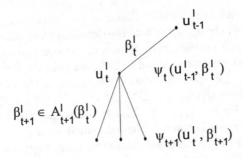

Figure 3: Scenario tree corresponding to the lower approximation

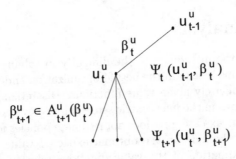

Figure 4: Scenario tree corresponding to the upper approximation

inequalities hold :

$$\psi_t(u^l_{t-1}, \beta^l_t) \leq \phi_t(u^l_{t-1}, \beta^l_t) \tag{24}$$

$$\phi_t(u^u_{t-1}, \beta^u_t) \leq \Psi_t(u^u_{t-1}, \beta^u_t). \tag{25}$$

The missing upper bound for $\phi_t(u^l_{t-1}, \beta^l_t)$ is available with the evaluation of $\Psi_t(u^l_{t-1}, \beta^l_t)$, the missing lower bound for $\phi_t(u^u_{t-1}, \beta^u_t)$ can be obtained with $\Psi_t(u^u_{t-1}, \beta^u_t)$. In both cases, a $(T-t)$-stage multistage program must be solved with respect to the barycentric scenario trees conditioned on β^l_t, β^u_t, respectively, providing the inner approximation, the outer approximation, respectively. Due to the convexity of the value function ϕ_t in the decision u_t, we focus on the minimizer of the outer approximation: This becomes evident from Figure 5: Evaluating the inner approximation at the minimizer u^l_t of the outer approximation provides a further bound on $\phi_t(u^l_{t-1}, \beta^l_t)$, i.e.:

$$\phi_t(u^l_{t-1}, \beta^l_t) \leq [\rho_t + \mathrm{E}^u_t \Psi_{t+1}](u^l_t, \beta^l_t) \tag{26}$$

Note that u^l_t solves (20) and is feasible for (21) given u^l_{t-1} and β^l_t (see Figure 6).

To the contrary, the value of the outer approximation at the minimizer u^u_t of the inner approximation does not necessarily bound $\phi_t(\cdot)$ from below

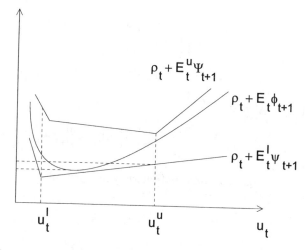

Figure 5: Lower and upper approximation for $\rho_t + E_t \phi_{t+1}$

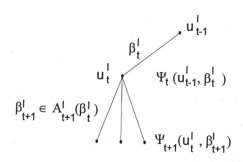

Figure 6: Evaluation of the upper approximation $\Psi_t(u_{t-1}^l, \beta_t^l)$ corresponding to each node of the scenario tree A^l.

(see Figure 5), i.e.:

$$[\rho_t + E_t^l \psi_{t+1}](u_t^u, \beta_t^u) \not\leq \phi_t(u_{t-1}^u, \beta_t^u) = \min[\rho_t + E_t \phi_{t+1}](u_t, \beta_t^u) \qquad (27)$$

Note that u_t^u solves (21) and is feasible for (20) given u_{t-1}^u and β_t^u.

Above observations let us focus on the barycentric scenario tree which corresponds to the lower approximation (see Figure 7), for which immediately error bounds can be obtained at any node of that tree by

$$\epsilon_t(\beta_t^l) \quad := \quad \Psi_t(u_{t-1}^l, \beta_t^l) - \psi_t(u_{t-1}^l, \beta_t^l). \qquad (28)$$

The evaluation of the error at each period provides a measure for the goodness of the lower approximation (see Figure 8). If the error at some node in period t is zero, the approximation of the value function $\phi_t(\cdot)$ and their minimizers is exact and does not need to be improved beyond this node.

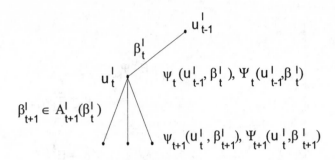

Figure 7: Information known in period t

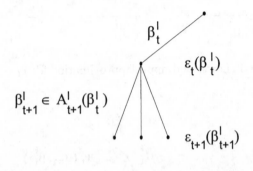

Figure 8: Error bounds at the nodes of β_t^l

Clearly, in case that $\epsilon_t(\beta_t^l)$ is positive at some node, than the error bounds prior to that node are positive, too.

By definition $\epsilon_{T+1} \equiv 0$. Moving backwards in time, the error bounds increase. The integration error occurs first and, hence, causes increasing error within the nested minimization and integration procedure. We shall investigate here how the error bound $\epsilon_t(\cdot)$ evolves. For this purpose, we summarize the inequalities (24) and (26) according to

$$[\rho_t + E_t^l \psi_{t+1}](u_t^l, \beta_t^l) = \psi_t(u_{t-1}^l, \beta_t^l) \le \phi_t(u_{t-1}^l, \beta_t^l) \le$$

$$\le \Psi_t(u_{t-1}^l, \beta_t^l) = [\rho_t + E_t^u \Psi_{t+1}](\hat{u}_t^u, \beta_t^l) \le \qquad (29)$$

$$\le [\rho_t + E_t^u \Psi_{t+1}](u_t^l, \beta_t^l)$$

where $\hat{u}_t^u$ solves (21) given u_{t-1}^l and β_t^l. Obviously,

$$[\rho_t + E_t^l \psi_{t+1}](u_t^l, \beta_t^l) \le [\rho_t + E_t^l \Psi_{t+1}](u_t^l, \beta_t^l), \qquad (30)$$

however

$$[\rho_t + E_t^l \Psi_{t+1}](u_t^l, \beta_t^l) \not\le [\rho_t + E_t^u \Psi_{t+1}](u_t^l, \beta_t^l), \qquad (31)$$

as the approximate value function $\Psi_{t+1}(\cdot)$ need not necessarily be a saddle function. For the evaluation of $[E_t^l \Psi_{t+1}](u_t^l, \beta_t^l)$ and $[E_t^u \Psi_{t+1}](u_t^l, \beta_t^l)$ it is referred to Figures 9 and 10. Therefore,

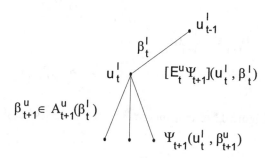

Figure 9: Evaluation of $[E_t^u \Psi_{t+1}](u_t^l, \beta_t^l)$

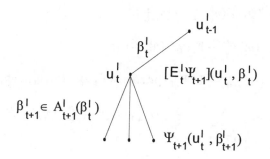

Figure 10: Evaluation of $[E_t^l \Phi_{t+1}](u_t^l, \beta_t^l)$

$$\delta_t(\beta_t^l) := [E_t^u \Psi_{t+1}](u_t^l, \beta_t^l) - [E_t^l \Psi_{t+1}](u_t^l, \beta_t^l), \tag{32}$$

may also become negative. Accepting $\delta_t(\beta_t^l)$ as error estimate associated with the integration of $\Psi_{t+1}(\cdot)$ (see Figure 11), we conclude that in case $\delta_t(\beta_t^l) \gg 0$, the current discretization has caused an inaccurate integration of $\Psi_{t+1}(\cdot)$ and, hence, of $\phi_{t+1}(\cdot)$. In case that $\delta_t(\beta_t^l)$ is even negative, the current approximation of the saddle function $\phi_{t+1}(\cdot)$ by $\Psi_{t+1}(\cdot)$ cannot be accepted as sufficiently accurate, as $\Psi_{t+1}(\cdot)$ does not even satisfy the saddle property.

We derive in what way $\delta_t(\beta_t^l)$ contributes to $\epsilon_t(\beta_t^l)$. This comes immediate

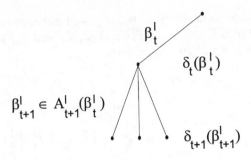

Figure 11: Error increments at the nodes of A^l

from

$$\epsilon_t(\beta_t^l) = \Psi_t(u_{t-1}^l, \beta_t^l) - \psi_t(u_{t-1}^l, \beta_t^l) \leq$$

$$\leq [\rho_t + E_t^u \Psi_{t+1}](u_t^l, \beta_t^l) - [\rho_t + E_t^l \psi_{t+1}](u_t^l, \beta_t^l) =$$

$$= E_t^u \Psi_{t+1}(u_t^l, \beta_t^l) - E_t^l \psi_{t+1}(u_t^l, \beta_t^l) =$$

$$= E_t^u \Psi_{t+1}(u_t^l, \beta_t^l) - E_t^l \Psi_{t+1}(u_t^l, \beta_t^l) + E_t^l \Psi_{t+1}(u_t^l, \beta_t^l) - E_t^l \psi_{t+1}(u_t^l, \beta_t^l) =$$

$$= \delta_t(\beta_t^l) + \sum_{\beta_{t+1}^l \in A_{t+1}^l(\beta_t^l)} \epsilon_{t+1}(\beta_{t+1}^l) \cdot Q_{t+1}^l(\beta_{t+1}^l | \beta_t^l) =: \Delta_t(\beta_t^l).$$

$$(33)$$

According to the above relation, $\delta_t(\beta_t^l)$ is an upper bound for the error increment from period $t+1$ to period t conditioned on β_t^l. Note that

$$\epsilon_t(\beta_t^l) = \delta_t(\beta_t^l) + \sum_{\beta_{t+1}^l \in A_{t+1}^l(\beta_t^l)} \epsilon_{t+1}(\beta_{t+1}^l) \cdot Q_{t+1}^l(\beta_{t+1}^l | \beta_t^l) = \Delta_t(\beta_t^l) \quad (34)$$

holds in case u_t^l is a minimizer of $[\rho_t + E_t^u \Psi_{t+1}](\cdot, \beta_t^l)$. This implies that the error increment is due to integration of $\Psi_{t+1}(u_t^l, \beta_t^l)$. If, additionaly, $\Psi_{t+1}(u_t^l, \beta_t^l)$ is bilinear, then $\delta_t(\beta_t^l) = 0$ and

$$\epsilon_t(\beta_t^l) = \sum_{\beta_{t+1}^l \in A_{t+1}^l(\beta_t^l)} \epsilon_{t+1}(\beta_{t+1}^l) \cdot Q_{t+1}^l(\beta_{t+1}^l | \beta_t^l) = \Delta_t(\beta_t^l). \quad (35)$$

The above relations are illustrated on a 3-stage funding problem for the ease of understanding.

The optimal value of the lower approximation is 5906.92 and of the upper approximation is 6694.96, the accuracy is 11.77% (see table 1). Three refinements have been performed which have decreased the relative error to 5.81%. It is observed that the upper bound remains unchanged with respect to these

# ref.	upper bound	lower bound	accuracy
0	6694.96	5906.92	11.77 %
1	6694.96	6268.61	6.37 %
2	6694.96	6269.41	6.36 %
3	6694.96	6305.73	5.81 %

Table 1: Lower and upper bounds for the first 3 refinements

refinements. This is due to fact that in the underlying funding problem the inner approximation $\Psi_t(\cdot)$ of the value function $\phi_t(\cdot)$ has been bilinear over $t = 2, 1, 0$, for which case the nested integration and minimization of $\Psi_t(\cdot)$ is exact, and, hence, the upper bound and the corresponding minimizers remain unchanged.

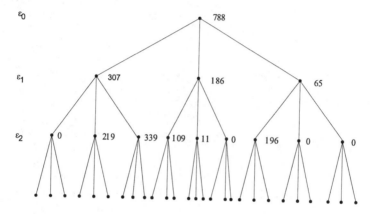

Figure 12: Evolvement of the error $\epsilon_t(\beta_t^l)$ backwards in time

The evolvement of $\epsilon_t(\beta_t^l), \delta_t(\beta_t^l)$ and $\Delta_t(\beta_t^l)$ is illustrated in figures 12, 13, and 14. Nodes in 12 at which $\epsilon_t = 0$ indicate that inner and outer approximation of the value functions coincide and are bilinear. Nodes in 13 at which $\delta_t = 0$ indicate that the inner approximation is bilinear. Nodes at which $\epsilon_t = \Delta_t$ indicate that the error increment from stage $t + 1$ to t is due to inaccurate integration of the inner approximation, with no impact on the minimizer of the inner and outer approximation. This way, critical nodes may be assessed beyond which the corresponding coverage process should be refined. Given a node of A^l conditioned on which the partitions are refined, one is faced with the two-stage situation. Theoretically, the various refinement schemes that have been developed by Kall and Stoyan 1982 [33], Birge and Wets 1986 [4], Frauendorfer and Kall 1988 [25], Kall, Ruszczyński and Frauendorfer 1988 [32], Edirisinghe and Ziemba 1994/1996 [18], [17], [19], are applicable. However, note that contrary to the two-stage stochastic programs, the recourse function of the nested two-stage formulation within

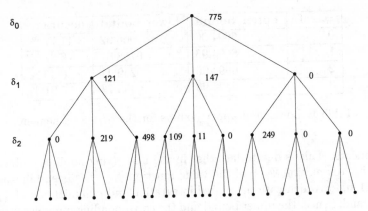

Figure 13: Error increments $\delta_{t-1}(\beta_{t-1}^l)$ at the nodes (backwards in time)

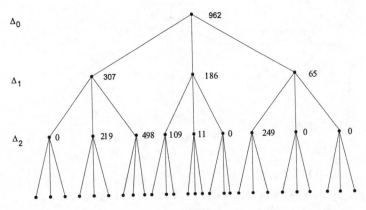

Figure 14: Error bounds $\Delta_{t-1}(\beta_{t-1}^l)$ of 3-stage scenario tree of lower approximation

the multistage problem is only approximately available by the implicitly given inner and outer approximation. This fact certainly has to be taken into account and requires further investigations.

5 Conclusions

This work contributes to the solvability of stochastic multistage linear programs, which suffers from the nested optimization and multidimensional integration of implicitly given value functions. In the convex case, which holds if the conditional probability distributions depend linearly on the past and remain unaffected by the desicions taken, structural properties help to overcome numerical difficulties to a certain extent. In particular, applying the barycentric approximation technique yields distinguished scenario trees. The solution of the associated deterministic equivalent programs provides lower and upper bounds of the minimal expected costs given the entire planning horizon $[0, T]$. The approximate policies refer to different scenario trees and, hence, may differ so that the inaccuracies beyond $t \geq 1$ cannot be assessed immediately. The inner and outer approximation of the value function are implicitly given and may only be compared as long they refer to the same history. Due to the convexity of the value function with respect to the decisions, it has been realized that the outer approximation ensures the minorization of the minimum values subject to the stages $t = 0, 1, \ldots, T$. This has been the basis to focus on the corresponding scenario tree A^l, and, then to evaluate the upper bound with respect to each node of that tree. This way the error can be assessed with respect to any history within A^l. It has been observed that the error caused by integration of the inner approximation is mainly responsible for the error increments backwards in time. For determining the critical nodes beyond which the approximation should be refined, relations have been derived which characterize the total error at stages $t, t+1$ and the integration error that arises from $t+1$ to t. Given a history (i.e., a node) of A^l conditioned on which the coverage process has to be refined, one is faced with the two-stage situation. However, for applying the various refinement schemes that have been developed for stochastic two-stage programs, it has to be taken into account that the recourse functions of the nested two-stage formulation within the multistage setting are only approximately available by the implicitely given value functions of the surrogate problems. This certainly opens further research activities which hopefully help increase the solvability and, hence, the applicability of stochastic multistage programs.

References

[1] BERTOCCHI, M., DUPAČOVÁ, J., AND MORIGGIA, V. Postoptimality for scenario based financial planning models with an application to bond portfolio management. In *Worldwide Asset and Liablility Management*, W. T. Ziemba and J. M. Mulvey, Eds. Cambridge University Press, 1997.

[2] BERTSEKAS, D. P. *Dynamic Programming and Optimal Control*, vol. 1. Athena Scientific, 1995.

[3] BERTSEKAS, D. P. *Dynamic Programming and Optimal Control*, vol. 2. Athena Scientific, 1995.

[4] BIRGE, J., AND WETS, R. J.-B. Designing approximation schemes for stochastic optimization problems, in particular for stochastic programs with recourse. *Mathematical Programming Study 27* (1986), 54–102.

[5] BIRGE, J. R. Decomposition and partitioning methods for multistage stochastic linear programs. *Operations Research 33*, 5 (1985), 989–1007.

[6] BIRGE, J. R. Current trends in stochastic programming computation and applications. Working paper, Department of Industrial and Operations Engineering, The University of Michigan, Ann Arbor, August 1995.

[7] BIRGE, J. R., AND ROSA, C. H. Parallel decomposition of large-scale stochastic nonlinear programs. Working paper, Department of Industrial and Operations Engineering, The University of Michigan, Ann Arbor, 1994.

[8] CARIÑO, D. R., AND KENT, T. Multistage planning for asset allocation. In *Worldwide Asset and Liablility Management*, W. T. Ziemba and J. M. Mulvey, Eds. Cambridge University Press, 1997.

[9] CARIÑO, D. R., KENT, T., MYERS, D. H., STACY, C., SYLVANUS, M., TURNER, A. L., WATANABE, K., AND ZIEMBA, W. T. The Russell-Yasuada kasai finanical planning model: An asset/liability model for a japanese insurance company using multistage stochastic programming. In *Worldwide Asset and Liablility Management*, W. T. Ziemba and J. M. Mulvey, Eds. Cambridge University Press, 1997.

[10] CARIÑO, D. R., MYERS, D. H., AND ZIEMBA, W. T. Concepts, technical issues, and uses of the Russel-Yasuda kasai financial planning model. Working paper, Frank Russell Company, May 1995.

[11] CONSIGLI, G., AND DEMPSTER, M. A. H. Dynamic stochastic programming for asset-liability management. In *Worldwide Asset and Liablility Management*, W. T. Ziemba and J. M. Mulvey, Eds. Cambridge University Press, 1997.

[12] DAHL, H., MEERAUS, A., AND ZENIOS, S. A. Some financial optimization models: I. risk management. In *Financial Optimization*, S. A. Zenios, Ed. Cambridge University Press, 1993, pp. 3–36.

[13] DANTZIG, G. B., AND GLYNN, P. W. Parallel processors for planning under uncertainty. *Annals of Operations Research 22* (1990), 1–21.

[14] DANTZIG, G. B., AND INFANGER, G. Multi-stage stochastic linear programs for portfolio optimization. *Annals of Operations Research 45* (1993), 59–76.

[15] DUPAČOVÁ, J. Stochastic programming models in banking. Working paper, International Institute for Applied Systems Analysis (IIASA), 1991.

[16] DUPAČOVÁ, J. Portfolio optimization under uncertainty. *submitted for Annals of Operations Research* (1992).

[17] EDIRISINGHE, N. C. P., AND ZIEMBA, W. T. Bounding the expectation of a saddle function with application to stochastic programming. *Mathematics of Operations Research 19*, 2 (1994), 314–340.

[18] EDIRISINGHE, N. C. P., AND ZIEMBA, W. T. Bounds for two-stage stochastic programs with fixed recourse. *Mathematics of Operations Research 19*, 2 (1994), 292–313.

[19] EDIRISINGHE, N. C. P., AND ZIEMBA, W. T. Implementing bounds-based approximations in convex-concave two-stage stochastic programming. *Mathematical Programming 75*, 2 (1996), 295–326.

[20] ERMOLIEV, Y., AND WETS, R. J.-B., Eds. *Numerical Techniques for Stochastic Optimization.* Springer-Verlag, Berlin, 1988.

[21] FRAUENDORFER, K. *Stochastic Two-Stage Programming.* Lecture Notes in Economics and Mathematical Systems 392. Springer-Verlag, Berlin, 1992.

[22] FRAUENDORFER, K. Multistage stochastic programming: Error analysis for the convex case. *Zeitschrift für Operations Research 39* (1994), 93–122.

[23] FRAUENDORFER, K. Barycentric scenario trees in convex multistage stochastic programming. *Mathematical Programming 75*, 2 (1996), 277–294.

[24] FRAUENDORFER, K. Stochastic multistage programming in financial decision making. *Zeitschrift für Angewandte Mathematik und Mechanik 76* (1996), 21–24.

[25] FRAUENDORFER, K., AND KALL, P. A solution method for SLP recourse problems with arbitrary multivariate distribution - the independent case. *Problems of Control and Information Theory 17* (1988), 177–205.

[26] FRAUENDORFER, K., AND SCHÜRLE, M. Barycentric approximation of stochastic interest rate processes. In *Worldwide Asset and Liablility Management*, W. T. Ziemba and J. M. Mulvey, Eds. Cambridge University Press, 1997.

[27] GASSMANN, H. MSLIP: A computer code for the multistage stochastic linear programming problem. *Mathematical Programming* (1990), 407–423.

[28] GOLUB, B., HOLMER, M., MCKENDALL, R., POHLMAN, L., AND ZENIOS, S. A. A stochastic programming model for money management. *European Journal of Operational Research* (1995), 282–296.

[29] HILLER, R. S., AND ECKSTEIN, J. Stochastic dedication: Designing fixed income portfolios using massively parallel benders decomposition. *Management Science 39*, 11 (1994), 1422–1438.

[30] INFANGER, G. Monte carlo (importance) sampling within a benders decomposition algorithm for stochastic linear programs. *Annals of Operations Research 39* (1992), 69–95.

[31] INFANGER, G. *Planning under Uncertainty, Solving Large-Scale Stochastic Linear Programs*. The Scientific Press Series, Danvers, Massachusetts, 1994.

[32] KALL, P., RUSZCZYŃSKI, A., AND FRAUENDORFER, K. Approximation techniques in stochastic programming. In *Numerical Techniques for Stochastic Optimization*, Y. Ermoliev and R. J.-B. Wets, Eds. Springer-Verlag, 1988, pp. 33–64.

[33] KALL, P., AND STOYAN, D. Solving stochastic programming problems with recourse including error bounds. *Mathematische Operationsforschung und Statistik Series Optimization 13* (1982), 431–447.

[34] KALL, P., AND WALLACE, S. W. *Stochastic Programming*. Wiley and Sons Ltd., Chichester, 1994.

[35] KLAASSEN, P. Financial asset-pricing theory and stochastic programming models for asset/liability management: a synthesis. In *Worldwide Asset and Liablility Management*, W. T. Ziemba and J. M. Mulvey, Eds. Cambridge University Press, 1997.

[36] KUSY, M. I., AND ZIEMBA, W. T. A bank asset and liability management model. *Operations Research 34* (1986), 356–376.

[37] MULVEY, J. M. Multi-stage financial planning. In *Operations Research Models in Quantitative Finance*, R. L. D'Ecclesia and S. A. Zenios, Eds. Physica-Verlag, 1994, pp. 18–35.

[38] MULVEY, J. M., AND RUSZCZYŃSKI, A. A new scenario decomposition method for large scale stochastic optimization. *Operations Research 43* (1995), 477–453.

[39] MULVEY, J. M., AND THORLACIUS, E. The tower perrin global capital market scenario generation system: Cap-link. In *Worldwide Asset and Liablility Management*, W. T. Ziemba and J. M. Mulvey, Eds. Cambridge University Press, 1997.

[40] MULVEY, J. M., VANDERBEI, R., AND ZENIOS, S. Robust optimization of large scale systems. *Operations Research 43*, 2 (1995), 264–281.

[41] MULVEY, J. M., AND VLADIMIROU, H. Stochastic network programming for financial planning problems. *Management Science 38* (1992), 1642–1664.

[42] MULVEY, J. M., AND ZIEMBA, W. T. Asset and liability modeling: discussion of issues. In *Worldwide Asset and Liablility Management*, W. T. Ziemba and J. M. Mulvey, Eds. Cambridge University Press, 1997.

[43] NIELSEN, S. S., AND ZENIOS, S. A. A stochastic programming model for funding single premium deferred annuities. *Mathematical Programming 75*, 2 (1996), 177–200.

[44] ROBINSON, S. M. Extended scenario analysis. *Annals of Operations Research 31* (1991), 385–398.

[45] ROCKAFELLAR, R. T., AND WETS, R. J.-B. Nonanticipativity and L^1-martingales in stochastic optimization problems. *Mathematical Programming 6* (1976), 170–187.

[46] ROCKAFELLAR, R. T., AND WETS, R. J.-B. The optimal recourse problem in discrete time: L^1-multipliers for inequality constraints. *SIAM Journal on Control and Optimization 16*, 1 (1978), 16–36.

[47] ROCKAFELLAR, R. T., AND WETS, R. J.-B. Scenarios and policy aggregation in optimization under uncertainty. *Mathematics of Operations Research 16*, 1 (1991), 119–147.

[48] RUSZCZYŃSKI, A. Interior point methods in stochastic programming. Working paper, International Institute for Applied Systems Analysis, Laxenburg, Februar 1993.

[49] RUSZCZYŃSKI, A. Parallel decomposition of multistage stochastic programming problems. *Mathematical Programming 58*, 2 (1993), 201–228.

[50] VASSIADOU-ZENIOU, C., AND ZENIOS, S. A. Robust optimization models for managing callable bond portfolios. *Interfaces 24* (1994), 29–49.

[51] WETS, R. J.-B. The aggregation principle in scenario analysis and stochastic optimization. In *Algorithms and Model Formulations in Mathematical Programming*, S. W. Wallace, Ed. Springer, Berlin, Heidelberg, 1989, pp. 91–113.

[52] ZENIOS, S. A. Massively parallels computations for financial planning under uncertainty. In *Very Large Scale Computing in the 21-st Century*, J. Mesirov, Ed. SIAM, Philadelphia, 1991.

[53] ZENIOS, S. A. Asset/liability management under uncertainty: The case of mortgage-backed securities. Technical report, Department of Decision Sciences, The Wharton School, University of Pennsylvania, Philadelphia, 1992.

[54] ZENIOS, S. A., Ed. *Financial Optimization*. Cambridge University Press, 1992.

[55] ZENIOS, S. A., AND ZIEMBA, W. T. Financial modeling. *Management Science 38* (1992).

[56] ZIEMBA, W. T., AND MULVEY, J. M., Eds. *Worldwide Asset and Liability Management*. Cambridge University Press, 1997.

[57] ZIEMBA, W. T., AND VICKSON, R. G., Eds. *Stochastic Optimization Models in Finance*. Academic Press, New York, 1975.

On Solving Stochastic Linear Programming Problems

P. Kall and J. Mayer
IOR, University of Zurich, Moussonstr. 15, CH–8044 Zürich

Abstract. Solving a stochastic linear programming (SLP) problem involves selecting an SLP solver, transmitting the model data to the solver and retrieving and interpreting the results. After shortly introducing the SLP model classes in the first part of the paper we give a general discussion of these various facets of solving SLP problems. The second part consists of an overview on the model–solver connection as implemented in SLP–IOR, our model management system for SLP. Finally we summarize the main features and capabilities of the solvers in the collection of solvers presently connected to SLP–IOR.

Keywords. 90C15 (1991 MSC)

1 Stochastic linear programs

In this section we briefly summarize the stochastic linear programming (SLP) model classes which will be considered in this paper. For a detailed introduction see Kall [11], Kall and Wallace [19] and Prékopa [28].

SLP with fixed recourse

$$(1.1) \qquad \begin{cases} \min\{c^\mathrm{T}x + E_\omega Q(x,\omega)\} \\ \text{s.t.} \quad Ax \quad \propto b \\ \qquad x \quad \in [l,u], \end{cases}$$

where

$$(1.2) \qquad \begin{cases} Q(x,\omega) = \min q^\mathrm{T}(\omega)y \\ \quad \text{s.t.} \quad Wy \quad \propto h(\omega) - T(\omega)x \\ \qquad y \quad \geq 0. \end{cases}$$

The symbol $\propto$ stands for any one of $=, \leq, \geq$, row-wise. The following classes of recourse problems will be considered:

— W (arbitrary) fixed recourse.

— W complete fixed recourse, i.e. $\{z \mid z = Wy, y \geq 0\} = \mathbb{R}^{m_2}$ and $\{u \mid W^T u \leq q(\omega)\} \neq \emptyset$ w.p. 1. Note that these assumptions imply that the recourse problem (1.2) has a feasible solution $\forall x \, \forall \omega$ and an optimal solution for $\forall x$ w.p. 1.

— W simple recourse, i.e. $W = (I, -I)$ and $T(\omega) \equiv T$, $q(\omega) \equiv q$. In this case the expected value term in the objective of (1.1) becomes separable w.r. to the components of $\chi = Tx$.

In the model above $\omega \in \Omega$, $(\Omega, \mathcal{F}, P)$ is a probability space; $q(\omega), h(\omega)$ are random vectors and $T(\omega)$ is a random matrix. These stochastic parts are assumed to be given by the following affine relations:

$$
(1.3) \quad
\begin{cases}
q(\omega) &= q^0 + \sum_{j=1}^r q^j \xi_j(\omega) \\
h(\omega) &= h^0 + \sum_{j=1}^r h^j \xi_j(\omega) \\
T(\omega) &= T^0 + \sum_{j=1}^r T^j \xi_j(\omega)
\end{cases}
$$

with $\xi_1(\omega), \ldots, \xi_r(\omega)$ being random variables with a known joint probability distribution. The stochastic independence assumption for this type of models will mean the stochastic independence of $\xi_1(\omega), \ldots, \xi_r(\omega)$.

Under mild assumptions the SLP problem with fixed recourse (1.1) is a convex programming problem, see e.g. Kall and Wallace [19].

SLP with a joint chance constraint

$$
(1.4) \quad
\begin{cases}
\min c^T x \\
P(\{\omega \mid Tx \geq h(\omega)\}) \geq \alpha \\
Ax \propto b \\
x \in [l, u],
\end{cases}
$$

with $0 \leq \alpha \leq 1$ being some (high) probability level. The probability distribution of the random vector $h(\omega)$ is assumed to be known.

Problem (1.4) is also called an SLP model with a probabilistic constraint. We gave the model formulation with a single probabilistic constraint because this case will be considered later on. Let us notice however that the general formulation may involve several constraints of this type.

For a broad class of multivariate probability ditributions with an existing density function (including the nondegenerate multinormal distribution) the SLP problem with a joint chance constraint (1.4) is a convex programming problem, see e.g. Kall and Wallace [19] and Prékopa [28].

Notice that problem (1.4) has been formulated with a deterministic technology matrix T. The reason is that under a random technology matrix $T(\omega)$ the problem becomes in general nonconvex even for a multinormal distribution.

SLP with separate chance constraints

$$(1.5) \quad \begin{cases} \min c^T x \\ P(\{\omega \mid t_i^T(\omega)x \geq h_i(\omega)\}) \;\geq\; \alpha_i, \; \forall i \\ \qquad\qquad Ax \;\propto\; b \\ \qquad\qquad x \;\in\; [l, u], \end{cases}$$

with $0 \leq \alpha_i \leq 1, \forall i$, given (high) probability levels and $t_i^T(\omega)$ denoting the i-th row of $T(\omega)$. The joint probability distribution of $(t_i^T(\omega), h_i(\omega))$ is assumed to be given $\forall i$ and we also assume that these random vectors are stochastically independent.

Notice that from the purely theoretical point of view (1.5) could be considered as a special case of (1.4) with several joint chance constraints and random technology matrices $T(\omega)$. The separately chance constrained problem (1.5) is however practically the single numerically tractable subclass of jointly chance constrained problems with a random technology matrix. This is the reason for introducing chance constrained models as they stand above.

For certain distributions (including multinormal) and certain probability levels the SLP problem with separate chance constraints (1.5) turns out to be a convex programming problem, see e.g. Kall and Wallace [19] and Marti [23].

All three model classes involve in general multidimensional integrals. In the recourse case an additional difficulty is rooted in the fact that the integrand is only implicitly given as the optimal value function of a parametric LP. Because of this feature SLP models are numerically hard problems.

If we replace the random variables in the above models by their expected values (assuming their existence), deterministic LP models result. We will call them *underlying LP's* in the sequel.

2   Algebraic equivalents

In some special cases it is possible to reformulate SLP models as mathematical programming problems involving only functions explicitly given by algebraic formulas. We call these equivalent MP problems *algebraic equivalents* in order to distinguish them from the so called *deterministic equivalents* (for the latter see e.g. Kall and Wallace [19]). Below we discuss those algebraic equivalents which will be addressed later on in the paper.

Recourse problems, discrete distribution

For a discrete distribution the expected value in the objective of (1.1) becomes a sum and the optimal value function can be eliminated on the cost of introducing an additional vector variable for each one of the realizations.

Let us assume that (q^k, h^k, T^k), $k = 1, \ldots, N$, are the joint realizations of $(q(\omega), h(\omega), T(\omega))$ with corresponding probabilities p_k, $k = 1, \ldots, N$. The algebraic equivalent will be the following LP problem:

$$(2.6) \quad \begin{cases} \min \left[c^T x + p_1 (q^1)^T y^1 + \ldots + p_N (q^N)^T y^N \right] & \\ A x & = b \\ T^1 x + W y^1 & = h^1 \\ \quad \vdots \qquad \qquad \ddots & \quad \vdots \\ T^N x + \qquad\qquad\qquad W y^N & = h^N \\ x & \geq 0 \\ & y^i \geq 0 \; \forall i. \end{cases}$$

This LP problem has a so–called dual block angular structure. Considering r=9 independent random variables the number of diagonal blocks of the problem is:

- $N = 3^9 = 19683$ for 3 realizations for each random variable;

- $N = 5^9 = 1953125$ for 5 realizations for each random variable.

Thus the algebraic equivalent LP's are typically large scale problems having a special structure. The fact that such an equivalent LP exists does not imply that all recourse problems with a discrete distribution can be solved by just solving the equivalent LP. As the above example illustrates the size of the problem rapidly grows with an increasing number of realizations of the components of the random vector and with an increasing number of random variables. The size of the problem may easily grow to an extent where it is impossible to generate the equivalent LP not to speak of solving it.

Separate chance constraints

In the special case when in (1.5) only the right hand side is stochastic, the problem can obviously be reformulated as an LP based on quantiles.

In the general case for certain multivariate distributions and probability levels convex programming algebraic equivalents exist. For the case of multinormal distributions and $\alpha_i \geq 0.5$ see e.g. Kall and Wallace [19].

3 Solving SLP problems

The solution phase plays a crucial role in the modeling life-cycle of SLP models, see Kall and Mayer [15]. In this section we shortly summarize the various facets of solving SLP problems. In the sequel by a solver we mean a computer implementation of a solution algorithm.

Access to solvers

For solving an SLP model first of all access to a solver is needed. For moderately sized recourse problems with a discrete distribution the problem may be solved by formulating the algebraic equivalent LP (2.6) and by solving it e.g. by a readily available commercial LP solver. The same is true for those separately chance constrained problems where an algebraic equivalent exists. When only the right hand size is stochastic then again a general purpose LP solver can be utilized, otherwise an NLP solver is needed.

For realistically sized recourse problems and for jointly chance constrained problems specialized SLP solvers are needed. The difficulty is that according to our knowledge there do not exist commercial SLP solvers and the existing SLP solvers are located at various academic institutions. One of the purposes of this paper is to provide information how SLP solvers can be accessed.

Selecting an approriate solver

Let us first emphasize that unlike in the LP case, in the SLP case there does not exist a general SLP solver capable to solve all of the various SLP model types.

The main problem features on which the selection of an appropriate solver for a given SLP model instance depends, are the following: The type of the model, the fact which parts of the problem are stochastic, the stochastic dependency structure and probability distribution of the random variables and the dimensionality limitations imposed by the solver and by the computing environment. This implies that selecting an SLP solver usually presupposes some technical knowlege on solver capabilities.

Solver input dataformat

Let us assume that a solver has been selected. The next issue is to transform model data into the solver's input dataformat. Notice that this conversion may involve also a model conversion when the solver aims at an algebraic equivalent.

A standard input dataformat, S–MPS, exists for recourse problems including also multistage models, see Birge et al. [1]. This is an extension of the well–known linear programming dataformat MPS. An SLP model instance can be specified in three text files. The first one serves for specifying the underlying LP, the second one for pinpointing the random entries and fixing their probability distribution whereas the third file defines the stages. The first one of these files is basically an MPS file, for writing it an algebraic modeling language like GAMS (see Brooke et al. [2]) can be utilized.

For solvers not endowed with the capability of reading S–MPS or for solvers aiming at models not included into the S–MPS dataformat data must be formatted according to the specific input requirements of the solver. The data specification for a solver can become in both cases quite a problem for large scale models.

Considering large scale problems an additional difficulty is to find data errors: Such problems must usually be debugged like a computer program. Debugging is extremely difficult with data being in the solver's input format (including S–MPS). Besides debugging repeated runs also occur when solving variants of a model instance. This implies that repeated conversion between a "readable" dataformat like a spreadsheet or an algebraic modeling language and the solver's input format should be supported.

Solver parameters

As already mentioned SLP problems are numerically hard. SLP algorithms either treat them as large scale LP's or directly face the problem of dealing with multidimensional integrals. This implies that solving an SLP problem may involve several runs with various settings of the solver parameters (e.g. various tolerances); "tuning" the parameters plays an important role. We experienced this problem even with SLP solvers which solve the algebraic equivalent LP.

Selecting an appropriate setting of the solver parameters is especially important for the stochastic algorithms; the performance of the corresponding solvers largely depends on the parameter settings. It is very important to provide some guidance to parameter selection for the users of these algorithms.

Specification of solver parameters is either implemented as command line parameters or by employing "SPECS" or "OPTIONS" files. The format of these files largely depends on the solver, this being true even for commercial LP solvers.

Output of results

Solvers usually write the solution into an output file which is in most cases "readable" meaning that the solution is tabulated for the sake of easy comprehension. For large scale models huge tables arise which must be further processed for judging the solution or for analyzing it. This means that the user might wish to load the solution into his own working environment for further processing. The difficulty is that the different solvers may write quite differently formatted output tables, i.e. again data format conversion is involved which should be automated.

Assessing the quality of a solution

As mentioned above solving an SLP problem may involve several runs, and it is very important to judge the quality of the current solution.

Let us consider recourse problems first. For large scale problems it may be very costly or even be impossible to compute a single exact objective function value. In the case of the stochastic algorithms it is epecially important to judge the quality of the solution. Successive discrete approximation methods play an important role in this respect: They provide lower and upper bounds on the optimal objective value.

For jointly chance constrained problems the difficulty is to compute the probability involved in the chance constraint. In this case Boole-Bonferroni type inequalities can be used for the purpose of assessing a solution.

For the bounds mentioned above see Kall and Wallace [19] and Prékopa [28].

4 SLP–IOR: The solver interface

This section is devoted to an overview on the solver interface of SLP–IOR, a model management system for SLP developed by the authors, see Kall and Mayer [13], [14], [16]. Presently we work on a further development of the system by including also multistage models. SLP–IOR is freely available for academic purposes; the current version is for IBM PC/AT 486 (or higher) computers running under MSDOS.

As we discussed in the previous section one of the main difficulties in dealing with different solvers is the handling of the various input/output solver dataformats. The *main idea* in the design of the solver interface of SLP–IOR is the following: We utilize the well documented solver interface of the algebraic modeling system GAMS, see Brooke et al. [2], for connecting the SLP solvers to SLP–IOR.

A solver run consists of the following steps:

- SLP–IOR writes the model instance in the GAMS modeling language. As GAMS does not include facilities for representing random variables we use our own conventions for specifying the random variables data.

- GAMS reads the model instance and subsequently outputs LP data according to the GAMS interface format and random variable data according to our format convention. The LP data correspond either to an algebraic equivalent or otherwise to the underlying LP problem.

- In the next step these data are converted by SLP–IOR to the input dataformat of the solver.

- The solver is started up.

- After solver termination results are converted by SLP–IOR to the GAMS format.

- GAMS reads the results and writes a listing file which documents the model instance as well as the run characteristics and the results. For the sake of easy retrieval the solution is also written into a separate text file.

- SLP–IOR retrieves the results.

The single deviation from this scheme is for solvers with input data in S–MPS format: For efficiency reasons in this case the S–MPS files are directly generated by SLP–IOR.

An obvious advantage of the outlined approach is that this way a uniform interface arises: All solvers have to be interfaced according to the GAMS interface format. Regarding only the uniformity issue admittedly a much simpler user interface could be built by defining and using an own solver interface format.

Besides uniformity our approach has however further advantages: This way we have immediate access to the powerful general purpose MP solvers of GAMS, which can e.g. be used for solving algebraic equivalents for comparative purposes. Another advantage is that a documentation of the model and of the computational results is automatically available in the modeling language GAMS. A further reason which does not concern solvers is the following: SLP models frequently arise as stochastic versions of an underlying LP. According to present day modeling standards the user should be provided with the important facility to formulate this LP in an algebraic modeling language and import it afterwards into SLP–IOR for subsequently building the stochastic variants. We employ GAMS also for this purpose.

The main features of the solver interface of SLP–IOR can be summarized as follows:

- A wide variety of solvers is connected to SLP–IOR.

- Selecting an appropriate solver is supported by providing a list of appropriate solvers for the current model instance.

- Model data are automatically transformed to the solver's input format.

- Setting the solver's parameters and repeated runs occur in an interactive menu driven fashion.

- Solver output results are automatically retrieved e.g. for further analysis.

Technical note: All solvers connected to SLP–IOR are instances of a general solver class in the object oriented sense. SLP problems are themselves instances of model classes. Selecting an appropriate solver for a model instance is implemented as follows: The solver instances in turn inspect the current model instance and send a message to the model manager component whether they consider themselves appropriate for solving that model instance. This feature facilitates connecting solvers to SLP–IOR.

5 SLP–IOR: Connecting solvers

Solvers can be connected to SLP–IOR according to the following categories:

An *external tool* is an external software (besides a solver it can e.g. be a text editor) which is just started up by SLP–IOR and after its termination control simply goes back to SLP–IOR. Connecting such a tool can be performed in a menu driven fashion.

An *external solver* is a solver which receives the data of the current model instance but after performing its tasks no computational results are returned to SLP–IOR (*solver* is to be understood in this context quite broadly, it may e.g. be a model analysis system). This facility can also be used to loosely connect an SLP solver to SLP–IOR provided that the solver data input format belongs to one of the available formats in SLP–IOR (e.g. S–MPS). The connection can again be established in a menu driven way.

A *GAMS solver* is one of the solvers of the user's GAMS system, connecting it to SLP–IOR is again guided by menus. This is a close connection, the GAMS solvers participate in all system operations in the same way as the internal solvers.

An *internal solver* is a solver connected to SLP–IOR in the closest possible way. Connecting a solver this way may involve minor changes in the source of the solver and sometimes also in the source of SLP–IOR.

The rest of this section is devoted to discussing the connection of internal solvers and is intended for the technically interested reader.

We only connect solvers to SLP–IOR as internal solvers when the source of the code is available.

One of the reasons for making minor changes in the solver source is measuring elapsed time. In order to perform reasonable comparative computational studies the elapsed time returned by the solvers should have the same meaning for all solvers. This is unfortunately usually not the case: Some solvers measure e.g. preprocessing time or I/O time separately, others just return total elapsed time. When comparing solver performance the total elapsed time is of interest whereas comparing algorithms requires comparison of elapsed time of the solution procedure part. Let us notice that the latter is not unambiguous, sometimes important solver specific transformations in the preprocessing phase are not measured as part of the solution time of an algorithm.

Below we summarize the main points concerning changing the source code of the solver or of SLP–IOR.

— *Solver input format:* If this is not S–MPS, then we first look into the solver code to find out whether the data input part can safely be changed to read data in one of the already available formats in SLP–IOR. If this is the case the change is carried out, otherwise as a last resort the code of SLP–IOR is changed: The list of available dataformats is augmented by the new dataformat.

— *Solver output:* A small piece of code is inserted into the solver source for outputting termination status, elapsed time, solution etc. If this is not possible because of the complexity of the code this information is being cut out from the solution listing (i.e. the code of SLP–IOR must be changed).

— *Solver parameters:* In SLP–IOR solver parameters can be specified by the user before a solver run in an interactive fashion. This is implemented as follows: In the case when solver parameters are implemented as command line parameters these are simply offered for the user for selection. For solvers employing option files a default option file is offered for editing before the run.

Let us emphasize that for solvers not developed by ourselves for changes in the solver source we ask for the author's permission.

6   SLP–IOR: Solvers connected

In this section we first give a list of solvers which are either connected to SLP–IOR or which are planned to be connected in the near future. For each solver the underlying solution algorithm is listed first followed by the name of the solver, its developers and references. For details see the specified references, Kall and Wallace [19] and the following survey papers: Kall [12], Kall, Ruszczyński and Frauendorfer [17], Mayer [25], Prékopa [27] and Wets [34].

Fixed recourse, algebraic equivalent

- L-shaped method, Van Slyke and Wets [33], further developed by Gassmann [7]. **MSLiP**, Gassmann 1992. The solver capabilities include also multistage problems. Connecting a new version is in progress.

- Basis reduction method, Strazicky [31].

- Regularized decomposition, Ruszczyński [29].
  QDECOM, Ruszczyński 1985; a new version **DECOMP**, Ruszczyński and Świętanowski [30].

- General purpose simplex solvers. **XMP**, Marsten [21], and the GAMS solvers **CONOPT, MINOS5, OSL, ZOOM**.

- General purpose interior point solvers:
  BPMPD, Mészáros [26]; **HOPDM**, Gondzio [8]; **(R)OB1**, Marsten et al. [22] and the GAMS solver **OSL**.

Complete recourse, original problem

- Successive discrete approximation, Kall and Stoyan [18], Frauendorfer and Kall [5], Frauendorfer [4]. **DAPPROX**, Kall and Mayer 1994.

- Stochastic Quasigradient method, Ermoliev [3], Gaivoronski [6].

- Stochastic decomposition, Higle and Sen [9], [10]. **SDECOM**, Kall and Mayer 1993[1].

Simple recourse, original problem

- Successive discrete approximation, Kall and Stoyan [18]. **SRAPPROX**, Kall and Mayer 1992.

- Convex hull method of van der Vlerk [20], for integer recourse. **SIRD2SCR**, van der Vlerk and Mayer 1993.

Joint chance constraints, multinormal distribution

- Supporting hyperplane type method, Szántai [32]. **PCSP**, new implementation, Szántai, 1996, (also Dirichlet and Gamma distributions); **PCSPIOR**, Mayer, 1995.

- Reduced gradient type method, Mayer [24]. **PROCON**, new implementation, Mayer 1995.

- Central cutting plane type method, Mayer, 1995. **PROBALL**, Mayer, 1995.

In the tables on the next page the main characteristics of those solvers are tabulated which are presently connected to SLP–IOR.

Table 6.1 gives a summary of the solver characteristics. The first column shows appropriate model types, the second column indicates model parts which may be stochastic (for the denotation here see Section 1). The third column shows allowed probability distributions. Although in this column **cd** generally stands for continuous distributions please notice that in the present version this means uniform, normal or exponential distributions. The last column shows the availability of the code. GAMS in this column indicates a commercial GAMS solver, available with GAMS. SLP–IOR stands for an internal solver of SLP–IOR which is distributed along with SLP–IOR in an executable form. A literature reference in this column indicates that the solver is licensed, i.e. for using it a license from the author is needed.

In Table 6.2 an asterisk indicates that the solver is appropriate for the corresponding model type. This table does not contain possible independence requirements for the random variables, for this information see Table 6.1.

---

[1]Implemented with generous support by Higle and Sen.

Table 6.1 Solver characteristics

	Models	St. Parts	Distr.	Avail.
BPMPD	FR, SC	T, h, q	dd	SLP–IOR
CONOPT	FR, SC	T, h, q	dd	GAMS
DAPPROX	CR	T, h	i, dd, cd	SLP–IOR
HOPDM	FR, SC	T, h, q	dd	[8]
MINOS5	FR, SC	T, h, q	dd	GAMS
MSLiP	FR	T, h, q	dd	[7]
OB1	FR, SC	T, h, q	dd	[22]
OSL	FR, SC	T, h, q	dd	GAMS
PROBALL	JC	h	nd	SLP–IOR
PROCON	JC	h	nd	SLP–IOR
PCSPIOR	JC	h	nd	SLP–IOR
QDECOM	FR	T, h, q	dd	SLP–IOR
SDECOM	CR	T, h	i, dd, cd	SLP–IOR
SIRD2SCR	SIR	h	dd	SLP–IOR
SRAPPROX	SR	h	dd, cd	SLP–IOR
XMP	FR, SC	T, h, q	dd	[21]
ZOOM	FR, SC	T, h, q	dd	GAMS

Table 6.2 Solvers versus models

	FR dd	CR dd	CR cd	SR dd	SR cd	SIR dd	JC nd	SC cd
BPMPD	*	*	–	*	–	–	–	*
CONOPT	*	*	–	*	–	–	–	*
DAPPROX	–	*	*	*	*	–	–	–
HOPDM	*	*	–	*	–	–	–	*
MINOS5	*	*	–	*	–	–	–	*
MSLiP	*	*	–	*	–	–	–	*
OB1	*	*	–	*	–	–	–	*
OSL	*	*	–	*	–	–	–	*
PROBALL	–	–	–	–	–	–	*	–
PROCON	–	–	–	–	–	–	*	–
PCSPIOR	–	–	–	–	–	–	*	–
QDECOM	*	*	–	*	–	–	–	–
SDECOM	–	*	*	*	*	–	–	–
SIRD2SCR	–	–	–	–	–	*	–	–
SRAPPROX	–	–	–	*	*	–	–	–
XMP	*	*	–	*	–	–	–	*
ZOOM	*	*	–	*	–	–	–	*

FR, CR, SR and **SIR**: fixed, complete, simple continuous and simple integer recourse. **JC** and **SC**: joint and separate chance constraints. **dd, cd, nd**: discrete, continuous, normal distributions; **i**: independence.

References

[1] J. R. Birge, M. A. H. Dempster, H. Gassmann, E. Gunn, A. J. King, and S. W. Wallace. A standard input format for multiperiod stochastic linear programs. Working Paper WP-87-118, IIASA, 1987.

[2] A. Brooke, D. Kendrick, and A. Meeraus. *GAMS. A User's Guide, Release 2.25*. Boyd and Fraser/The Scientific Press, Danvers, MA, 1992.

[3] Y. Ermoliev. Stochastic quasigradient methods and their application to systems optimization. *Stochastics*, 9:1–36, 1983.

[4] K. Frauendorfer. Solving SLP recourse problems with arbitrary multivariate distributions - the dependent case. *Mathematics of Operations Research*, 13:377–394, 1988.

[5] K. Frauendorfer and P. Kall. A solution method for SLP recourse problems with arbitrary multivariate distributions — the independent case. *Problems of Control and Information Theory*, 17:177–205, 1988.

[6] A. Gaivoronski. Stochastic quasigradient methods and their implementation. In Y. Ermoliev and R.J-B. Wets, editors, *Numerical Techniques for Stochastic Optimization*, pages 313–351. Springer Verlag, 1988.

[7] H.I. Gassmann. MSLiP: A computer code for the multistage stochastic linear programming problem. *Mathematical Programming*, 47:407–423, 1990.

[8] J. Gondzio. HOPDM (version 2.12) – A fast LP solver based on a primal-dual interior point method. *European Journal of Operational Research*, 85:221–225, 1995.

[9] J. L. Higle and S. Sen. Stochastic decomposition: An algorithm for two-stage linear programs with recourse. *Mathematics of Operations Research*, 16:650–669, 1991.

[10] J. L. Higle and S. Sen. *Stochastic decomposition. A statistical method for large scale stochastic linear programming*. Kluwer Academic Publishers, 1996.

[11] P. Kall. *Stochastic linear programming*. Springer Verlag, 1976.

[12] P. Kall. Computational methods for solving two-stage stochastic linear programming problems. *Zeitschrift für angewandte Mathematik und Physik*, 30:261–271, 1979.

[13] P. Kall and J. Mayer. SLP-IOR: A model management system for stochastic linear programming — system design —. In A.J.M. Beulens and H.-J. Sebastian, editors, *Optimization-Based Computer-Aided Modelling and Design*, pages 139–157. Springer Verlag, 1992.

[14] P. Kall and J. Mayer. A model management system for stochastic linear programming. In P. Kall, editor, *System Modelling and Optimization*, pages 580–587. Springer Verlag, 1992.

[15] P. Kall and J. Mayer. Computer support for modeling in stochastic linear programming. In K. Marti and P. Kall, editors, *Stochastic Programming: Numerical Techniques and Engineering Applications*, pages 54–70. Springer Verlag, 1995.

[16] P. Kall and J. Mayer. SLP–IOR: An interactive model management system for stochastic linear programs. *Math. Programming*, 75:221–240, 1996.

[17] P. Kall, A. Ruszczyński, and K. Frauendorfer. Approximation techniques in stochastic programming. In Y. Ermoliev and R.J-B. Wets, editors, *Numerical Techniques for Stochastic Optimization*, pages 33–64. Springer Verlag, 1988.

[18] P. Kall and D. Stoyan. Solving stochastic programming problems with recourse including error bounds. *Mathematische Operationsforschung und Statistik, Ser. Optimization*, 13:431–447, 1982.

[19] P. Kall and S. W. Wallace. *Stochastic programming*. John Wiley & Sons, 1994.

[20] W. K. Klein Haneveld, L. Stougie, and M. H. van der Vlerk. On the convex hull of the simple integer recourse objective function. Research memorandum 516, IER, University of Groningen, 1993.

[21] R. E. Marsten. The design of the XMP linear programming library. *ACM Transactions on Mathematical Software*, 7:481–497, 1981.

[22] R. E. Marsten, M. J. Saltzman, D. F. Shanno, G. S. Pierce, and J. F. Ballintijn. Implementation of a dual affine interior point algorithm for linear programming. *ORSA Journal on Computing*, 4:287–297, 1989.

[23] K. Marti. Konvexitätsaussagen zum linearen stochastischen Optimierungsproblem. *Zeitschrift für Wahrscheinlichkeitstheorie und verw. Geb.*, 18:159–166, 1971.

[24] J. Mayer. Probabilistic constrained programming: A reduced gradient algorithm implemented on PC. Working Paper WP-88-39, IIASA, 1988.

[25] J. Mayer. Computational techniques for probabilistic constrained optimization problems. In K. Marti, editor, *Stochastic Optimization: Numerical Methods and Technical Applications*, pages 141–164. Springer Verlag, 1992.

[26] Cs. Mészáros. The augmented system variant of IPMs in two–stage stochastic linear programming computation. Working Paper WP-95-11, MTA SzTAKI, Budapest, 1995.

[27] A. Prékopa. Numerical solution of probabilistic constrained programming problems. In Y. Ermoliev and R.J-B. Wets, editors, *Numerical Techniques for Stochastic Optimization*, pages 123–139. Springer Verlag, 1988.

[28] A. Prékopa. *Stochastic programming*. Kluwer Academic Publishers, 1995.

[29] A. Ruszczyński. A regularized decomposition method for minimizing a sum of polyhedral functions. *Mathematical Programming*, 35:309–333, 1986.

[30] A. Ruszczyński and A. Świętanowski. On the regularized decomposition method for two stage stochastic linear problems. Working Paper WP-96-014, IIASA, 1996.

[31] B. Strazicky. On an algorithm for solution of the two-stage stochastic programming problem. *Methods of Operations Research*, XIX:142–156, 1974.

[32] T. Szántai. A computer code for solution of probabilistic-constrained stochastic programming problems. In Y. Ermoliev and R.J-B. Wets, editors, *Numerical Techniques for Stochastic Optimization*, pages 229–235. Springer Verlag, 1988.

[33] R. Van Slyke and R. J-B. Wets. L-shaped linear program with applications to optimal control and stochastic linear programs. *SIAM J. Appl. Math.*, 17:638–663, 1969.

[34] R. J-B. Wets. Stochastic programming: Solution techniques and approximation schemes. In A. Bachem, M. Grötschel, and B. Korte, editors, *Mathematical programming: The state of the art*, pages 566–603. Springer Verlag, 1983.

On an On/Off Type Source with Long Range Correlations

Ryszard Antkiewicz and Arkadiusz Manikowski

Institute of Mathematics and Operations Research,
Military University of Technology, Warsaw, Poland

Abstract. In this paper we study the ON/OFF model of telecommunication traffic source. The time duration of the state ON of this model has heavy-tailed probability distribution with infinity variance. We prove, that Index of Dispersion for Counts of traffic generated by such source is unbounded for t increasing to infinity. It means, that this traffic possesses long-range dependency.

Keywords. Heavy-tailed distribution, long-range dependency, Index of Dispersion for Counts

1 Introduction

Meaurements on a LAN-network at Bellcore [1] have shown, that LAN-traffic has long-range dependency. There have been proposed a few models covering this fenomena. Norros [4] has applied Fractional Brownian Motion to model LAN arrival process. Veitch [5] proposed model, which could be named "Fractional renewal process ". J. Le Boudec [3] has used five stage semi-Markov process for modelling self-similar data traffic. Very interesting and simple model was developed by M. Villen [6]. This model is based on Poisson arrival of bursts, which duration times are random variables with infinite variance.

All mentioned above models differ from that used up till now. We will try to show, in this paper, that traditional ON/OFF model could capture long-range dependency of LAN traffic. We only assume, that state ON has heavy-tailed distribution with infinity variance. We use here Pareto distribution. We use

Pareto distribution, because it is heavy-tailed and it was suggested in [8], that such distribution well describe state ON duration time.

2 Description of the model

It is considered ON/OFF source, which generates traffic with intensity $\lambda(t)$:

$$\lambda(t) = d \cdot X(t) \tag{1}$$

where d is a peak rate of source, and $X(t)$ is a stochastic process defined as follows:

$$X(t) = \begin{cases} 1 \text{ if source is active } (\text{in state ON}) \text{ at moment t;} \\ 0 \text{ if source is not active } (\text{in state OFF}) \text{ at moment t.} \end{cases}$$

The example of process $X(t)$ trajectory is presented in Figure 1:

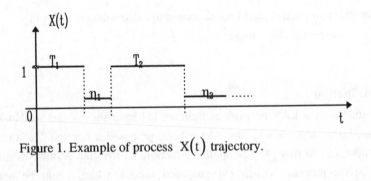

Figure 1. Example of process $X(t)$ trajectory.

We assume, that (T_i) and (η_i) are sequences of i. i. d. random variables with distribution functions:

$$F(t) = P\{T_i \langle t\}$$

$$G(t) = P\{\eta_i \langle t\} \ , i = 1,2,....$$

where:

$$F(t) = \begin{cases} 0 & t \le a \\ 1 - \left(\dfrac{a}{t}\right)^\alpha & ,t \rangle a, \alpha \rangle 0, \end{cases} \qquad (2)$$

It means, that duration time of state ON has Pareto distribution. We suppose, that $\alpha \in (1,2)$, thus the mean value of random variables $(T_i)_i$ is :

$$\theta_1 = a + \frac{a}{\alpha - 1} = \frac{a\,\alpha}{\alpha - 1} \qquad (3)$$

and their variance is infinite. About $(\eta_i)_i$ we assume, that they have finite expected value :

$$\theta_2 = \int_0^\infty G(t)\,dt \langle + \infty$$

Let $Y(t)$ be amount of traffic arrived in an interval $(0,t)$. Taking into account (1) we have:

$$Y(t) = \int_0^t \lambda(y)\,dy = d \cdot \int_0^t X(y)\,dy \qquad (4)$$

We assume that process $X(t)$ is stationary and $X(0) = 1$. It implies, that [7]:

$$F_1(t) = \frac{1}{\theta_1} \int_0^t (1 - F(y))\,dy$$

We would like to know, if process $Y(t)$ possesses long-range dependence. Index of Dispersion for Counts (IDC) will be derived for $Y(t)$, in order to check it. IDC is defined as follows [1]:

$$IDC(t) = \frac{D^2 Y(t)}{E Y(t)} \qquad (5)$$

It is known, that:

$$EY(t) = E\left(\int_0^t dX(y)dy\right) = d\int_0^t EX(y)dy = \int_0^t P\{X(y) = 1\}dy$$

and

$$EY(t) = d \cdot p_{ON} \cdot t$$

where: $p_{ON} = P\{X(t) = 1\} = \dfrac{\theta_1}{\theta_1 + \theta_2}$ for stationary renewal process.

Now, will be derived variance of $Y(t)$. It is known, that:

$$D^2Y(t) = d^2 \, D^2\left(\int_0^t X(y)dy\right) = d^2 \cdot 2\int_0^t (t-\tau)K_x(\tau)d\tau \qquad (6)$$

where:

$$K_x(\tau) = E\left(\left(X(t+\tau) - E(X(t+\tau))\right)\left(X(t) - EX(t)\right)\right) \qquad (7)$$

By assumptions, process $X(t)$ is stationary. It follows that:

$$E(X(t+\tau)) = E(X(t)) = p_{ON}$$

thus:

$$K_X(t) = E X(t+\tau)X(t) - p_{ON}^2 \qquad (8)$$

It remains to derive $EX(t+\tau)X(t)$. From definition of $X(t)$ we obtain:

$$EX(t+\tau)X(t) = P\{X(t+\tau) = 1, X(t) = 1\} = \qquad (9)$$

$$= P\{X(t+\tau) = 1 | X(t) = 1\} \cdot P\{X(t) = 1\}$$

and from [7]:

$$P\left\{X(t+\tau)=1\middle|X(t)=1\right\}=\frac{1}{\theta_1}\cdot\int_\tau^\infty R(y)dy+\int_0^\tau R(\tau-v)d\,H_2\,(v)$$

where : $\qquad R(\tau)=1-F(\tau)$

and $H_2(\tau)$ is expected number of process $X(t)$ transitions from state 0 to 1 in an interval $(0,t)$, given that $X(0)=1$.

We now find approximations for two parts of equation (9) right side. We know from assumptions, that $R(t)=a^\alpha\cdot t^{-\alpha}$ for $t\rangle a$. Hence, for $\tau\rangle a$:

$$R_1\left(\tau\right)=\frac{1}{\theta_1}\cdot\int_\tau^\infty R(y)dy=\frac{1}{\theta_1}\cdot\int_\tau^\infty a^\alpha\cdot y^{-\alpha}\,dy=\frac{a^\alpha}{\theta_1}\cdot\frac{1}{\alpha-1}\cdot\tau^{-\alpha+1}\quad(10)$$

From Smith' theorem [7] it is known, that:

$$\lim_{\tau\to\infty}\int_0^\tau R(\tau-v)d\,H_2\,(v)=\frac{1}{\theta}\int_0^\infty R(y)dy$$

where $\theta=\theta_1+\theta_2$. It is clear, that:

$$\frac{1}{\theta}\int_0^\infty R(y)dy=\lim_{\tau\to\infty}\frac{1}{\theta}\int_0^\tau R(y)dy$$

Therefore, we can use the following approximations for $\tau\to\infty$:

$$\int_0^\tau R\left(\tau-v\right)d\,H_2\,(v)\sim\frac{1}{\theta}\int_0^\tau R(y)dy=$$

$$=\frac{a}{\theta}+\frac{a^\alpha}{\theta}\cdot\frac{1}{\alpha-1}\cdot\left(a^{-\alpha+1}-\tau^{-\alpha+1}\right)\qquad(11)$$

From (9) - (11) we obtain for $\tau\to\infty$:

$$P\{X(t+\tau)=1 \mid X(t)=1\} \sim$$

$$\sim \frac{a^{\alpha}}{\theta_1} \cdot \frac{1}{\alpha-1} \cdot \tau^{-\alpha+1} + \frac{a}{\theta} + \frac{a^{\alpha}}{\theta} \cdot \frac{1}{\alpha-1}\left(a^{-\alpha+1} - \tau^{-\alpha+1}\right) =$$

$$= \frac{a^{\alpha}}{\alpha-1} \cdot \tau^{-\alpha+1}\left(\frac{1}{\theta_1} - \frac{1}{\theta}\right) + \frac{a \cdot \alpha}{\theta(\alpha-1)}$$

Finally:

$$P\{X(t+\tau)=1 \mid X(t)=1\} = A \cdot \tau^{-\alpha+1} + p_{ON} \qquad (12)$$

where :

$$A = \frac{a^{\alpha}}{\alpha-1} \cdot \left(\frac{1}{\theta_1} - \frac{1}{\theta}\right)$$

Thereby, from (8) and (9) we have, that

$$K_X(\tau) \sim \left(A \cdot \tau^{-\alpha+1} + p_{ON}\right) \cdot p_{ON} - p_{ON}^2 = \qquad (13)$$

$$= A \, p_{ON} \cdot \tau^{-\alpha+1} = A' \cdot \tau^{-\alpha+1}, \qquad as \ \tau \to \infty$$

where : $A' = A \cdot p_{ON}$. According to (6) and (13) we conclude, that:

$$D^2 Y(t) = d^2 \cdot 2 \cdot \int_0^t (t-\tau) K_X(\tau) d\tau \sim$$

$$\sim d^2 \cdot 2 \cdot \int_0^t (t-\tau) \cdot A' \cdot \tau^{-\alpha+1} \cdot d\tau \qquad as \quad t \to \infty$$

Because

$$\int_0^t (t-\tau) A' \cdot \tau^{-\alpha+1} d\tau = A' \cdot t^{-\alpha+3}\left(\frac{1}{2-\alpha} - \frac{1}{3-\alpha}\right)$$

we finally obtain:

$$D^2 Y(t) \sim A'' \cdot t^{-\alpha + 3} \quad , \quad \text{as} \quad t \to \infty \tag{14}$$

where

$$A'' = 2d^2 \cdot A' \cdot \left(\frac{1}{2-\alpha} - \frac{1}{3-\alpha} \right)$$

Now, we can evaluate IDC (t) for $t \to \infty$:

$$IDC(t) = \frac{D^2 Y(t)}{E\,Y(t)} \sim \frac{A'' \cdot t^{-\alpha+3}}{d \cdot P_{ON} \cdot t} = \frac{A''}{d \cdot P_{ON}} \cdot t^{-\alpha+2} \tag{15}$$

We see, that IDC(t) $\to + \infty$, as $t \to \infty$ because $\alpha \in (1,2)$. Thus process Y(t) possesses long-range dependence. It is possible to evaluate the Hurst parameter H.

For the second-order self-similar process Z is [4]:

$$D^2 Z(\beta \cdot t) = \beta^{2H} \cdot D^2 Z(t) \tag{16}$$

In our model according to (14) we have

$$D^2 Y(\beta \cdot t) \sim A'' \cdot t^{-\alpha+3} \cdot \beta^{-\alpha+3} \sim \beta^{-\alpha+3} \cdot D^2 Y(t) \tag{17}$$

and relation between H and α is following:

$$2H = -\alpha + 3$$

thus $H = \dfrac{-\alpha + 3}{2}$ and $\alpha = 3 - 2H$. \hfill (18)

3 Model verification

In order to verify the proposed model some simulation experiments have been done. It has been generated traffic according to proposed ON/OFF source model. We assume that parameters of considered source have the following values:

- peak rate d=10Mbits/s;
- time duration of state ON is random variable with Pareto distribution, Hurst parameter H=0.8 and this implies, that parameter α=1.4;
- the minimal time duration of state ON a=55µs, it implies from minimal packet size in Ethernet (72 bytes) and peak rate;
- time duration of state OFF is a random variable exponentially distributed with expected value θ_2=40ms.

Results of simulation are presented in the following Figures.

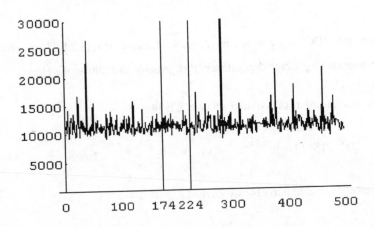

Fig.2a. Time unit=100s (Pareto)

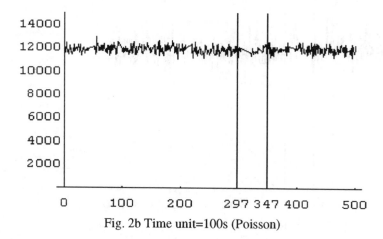

Fig. 2b Time unit=100s (Poisson)

Fig. 3a. Time unit=10s (Pareto)

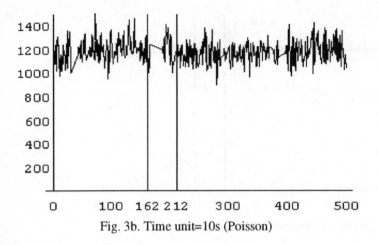

Fig. 3b. Time unit=10s (Poisson)

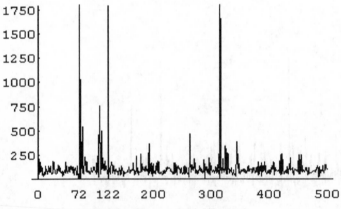

Fig. 4a. Time unit=1s (Pareto)

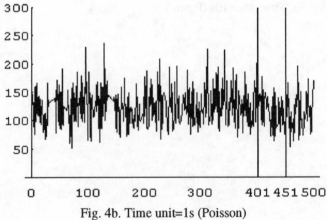

Fig. 4b. Time unit=1s (Poisson)

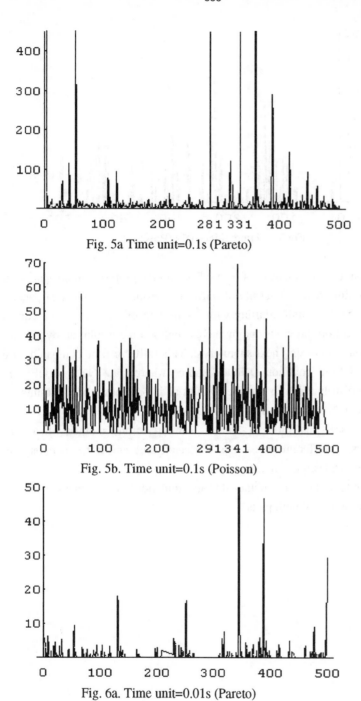

Fig. 5a Time unit=0.1s (Pareto)

Fig. 5b. Time unit=0.1s (Poisson)

Fig. 6a. Time unit=0.01s (Pareto)

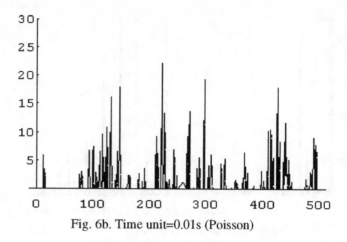

Fig. 6b. Time unit=0.01s (Poisson)

Figures 2a.-6a. present intensity of traffic from source proposed in this paper for five different time scales. Intensite of traffic is measured in number of packets (ATM cells) per time unit. Starting with a time unit of 100s, each subsequent plot is obtained from previous one by increasing the time resolution by a factor 10 and randomly choosing next subinterval. This traffic can be compared with one generated from common ON/OFF source where state ON and OFF are exponentialy distributed with expected values as in previous model. Results for the second model are presented in Figures 2b.-6b.

It is simply to see, that plots 2a.-6a. are „similar" to one another, when plots 2b.-6b. differs for different time scales. For small time unit traffic is bursty but for higher time scales it is too „smooth".

Figures 7a. and 7b. show IDC obtained from simulation and analyticaly.

It is seen consistency of both plots

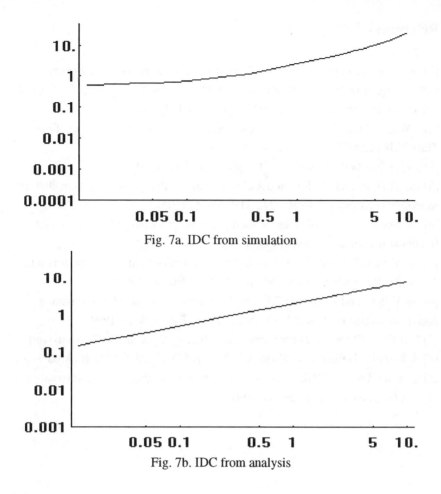

Fig. 7a. IDC from simulation

Fig. 7b. IDC from analysis

4 Conclusion

We proposed some modification of known ON/OFF model of LAN traffic source. We assume, that time duration of state ON has Pareto distribution with infinite variance. Such assumption implies, that traffic generated by source possesses long-range dependency. The advantages of presented model are the following:

- simplicity compared with the others models of LAN traffic proposed in last a few years;
- fact, that ON/OFF models are commonly used to study ATM networks.

References

[1] H.j. Fowler and W.E. Leland, "Local Area Network Traffic Characteristics with Implication for Broadband Network Congestion Management" , IEEE J. Select. Areas Commun., vol. 9, NO. 7, pp. 1139-1149,Sept. 1991;

[2] W.E. Leland   ," On the Self -Similar Nature of Ethernet Traffic",IEEE/ACM
Trans. on Networking, vol. 2, NO. 1,pp. 1-15, Febr. 1994;

[3] S. Robert and J. Le Boudec,"Can self-similar traffic be modeled by Markovian processes ", COST 242 TD(095)26, 1995;

[4] I. Norros , " Studies on a model for connectionless traffic, based on fractional Brownian motion ", COST 242D(92)041, 1992;

[5] D. Veitch, " Novel Models of Broadband Traffic", in Proc. 7th Australian Teletraffic Research Seminar , Murray River, Australia, 1992;

[6] M.Villen and J. Gamo,"A simple, tentative model for explaining the statistical characteristics of LAN traffic ", COST 242(94)28, 1994;

[7] D.R.Cox ,"Renewal Theory",Sovetskoe Radio, Moskva, 1967 (in Russian).

[8] K.S. Meier-Hellstren, P.Wirth, Y-L.Yan and D. Hoeflin, "Traffic Models for ISDN Data Users": Office Atomation Application, Proc. 13 th International Teletraffic Congress, Copenhagen 1991.

Optimization Methods in Structural Reliability

K. Breitung[1], F. Casciati[1] and L. Faravelli[1]

[1] Dipartimento di Meccanica Strutturale, Università di Pavia
Via Ferrata 1, I-27100 Pavia, Italy

Abstract. In this paper a method for constrained minimization of functions is outlined. This method is similar to the method developed by Hasofer/Lind and Rackwitz/Fiessler; but firstly it can be generalized to problems with several constraints and secondly under slight regularity conditions its convergence can be demonstrated.

Further it is shown that the sequential quadratic programming schemes, which produce an approximate Hessian of the Lagrangian, can be used easily for calculating SORM approximations, since the determinant of this Hessian divided by the squared length of the gradient of the limit state function is the inverse of the square of the SORM correction factor.

Keywords. Structural reliability, constrained minimization, SORM approximations, Lagrange multipliers, asymptotic approximations.

1   Introduction

In the usual structural reliability formulation, the state of a structure is modelled by a random vector $\boldsymbol{X}$ denoting the basic random variables which describe the loads, the material properties and the geometry. Let $g(\boldsymbol{x})$ be the limit state function, then the failure domain is given by

$$F = \{\boldsymbol{x}; g(\boldsymbol{x}) \leq 0\}. \tag{1}$$

We have now to calculate the probability of failure given by

$$\mathrm{P}(F) = \int\limits_{g(\boldsymbol{x}) \leq 0} f(\boldsymbol{x}) \, d\boldsymbol{x}. \tag{2}$$

with $f(\boldsymbol{x})$ the joint p.d.f. of the random vector $\boldsymbol{X}$.

All analytic methods as well as methods which use analytic approximations as a starting point, need a numerical algorithm for finding points on the limit surface $G = \{\boldsymbol{x}; g(\boldsymbol{x}) = 0\}$ where a function is minimal with respect to this surface. In the case of a standard normal distribution, it is necessary to find the points with minimal distance to the origin, i.e. the x^*'s, where

$$|x^*| = \min_{x \in G} |x|. \tag{3}$$

In the general case we consider the log-likelihood of the joint p.d.f. defined by $l(x) = \ln(f(x))$, i.e. we seek points, where

$$l(x^*) = \max_{x \in G} l(x). \tag{4}$$

In both cases this is equivalent to maximize the joint p.d.f. $f(x)$ of the random vector X on the limit state surface. Concepts how to calculate then the asymptotic approximations for the failure probabilities are outlined in [2] and [3].

Liu and Der Kiureghian [6] compared several algorithms without coming to a conclusive result which one is the best. This is not surprising; depending on the structure of the reliability problem, for different cases different methods may be preferable. All algorithms use only information about the function and its gradient.

In sequential quadratic programming methods this information is used to approximate the Hessian of the Lagrangian by an updating scheme. Therefore here it is necessary to store an $n \times n$ matrix and to solve a linear equation system with $n + 1$ equations at each step.

Often used structural reliability is the method proposed by Hasofer/Lind and Rackwitz/Fiessler [10], in the following abbreviated HL-RF method.

The HL-RF method computes the next point x_{k+1} by linearizing the function $g(x)$ at x_k and computing the point on the hyperplane $g_L(x) = 0$ with minimal distance to the origin. This point is then the next point x_{k+1}, i.e.

$$x_{k+1} = |\nabla g(x_k)|^{-2} \left(x_k^T \nabla g(x_k) \right) - g(x_k) \right) \nabla g(x_k) \tag{5}$$

If we define the unit vector $\alpha^{(k)}$ by

$$\alpha^{(k)} = |\nabla g(x_k)|^{-1} \nabla g(x_k) \tag{6}$$

we can write this as

$$x_{k+1} = \left(x_k^T \alpha^{(k)} - \frac{g(x_k)}{|\nabla g(x_k)|} \right) \alpha^{(k)}. \tag{7}$$

To improve the convergence properties, a modification of the HL-RF method was proposed by Veneziano et al. ([12]; [4], p. 156). At each iteration step the unit vector $\alpha^{(k)}$ was replaced by the vector

$$\alpha_{im}^{(k)} = \frac{\left(\psi \alpha^{(k-1)} + (1 - \psi) \alpha^{(k)} \right)}{|\psi \alpha^{(k-1)} + (1 - \psi) \alpha^{(k)}|}, \tag{8}$$

where the value of ψ was chosen different for even and odd step numbers to avoid periodic jumps. The values were $\psi = 0.4$ for odd and $\psi = 0.39$ for even step numbers.

Other modifications were proposed in the paper by Hohenbichler et al. [5] and in Liu and Der Kiureghian [6]. But for both of them, the original method and the modifications, some convergence problems remained.

2 The linearization method of Pshenichnyj

In this section we will describe a minimization method developed by Pshenichnyj [9]. This book is a translation of a Russian/Ukrainian book which appeared in 1983. A paper by the same author [8] outlining the method was already mentioned in [1]. There it was also noted that it the HL-RF method appears to be a simplified form of this algorithm. Pshenichnyj developed this method especially for the minimization of functions under constraints. It can be applied for equality and inequality constraints.

In the following we will consider only the case of one equality constraint $g(\boldsymbol{x}) = 0$, i.e.

$$\min f(\boldsymbol{x}) \quad \text{under} \quad g(\boldsymbol{x}) = 0. \tag{9}$$

The first step is to linearize the functions f and g around the starting point $\boldsymbol{x}_0$ of the search. They are replaced by

$$f_L(\boldsymbol{x}) = f(\boldsymbol{x}_0) + \nabla f(\boldsymbol{x}_0)^T (\boldsymbol{x} - \boldsymbol{x}_0) \tag{10}$$

$$g_L(\boldsymbol{x}) = g(\boldsymbol{x}_0) + \nabla g(\boldsymbol{x}_0)^T (\boldsymbol{x} - \boldsymbol{x}_0), \tag{11}$$

We have the problem to find a vector $\boldsymbol{d}$ such that

$$f_L(\boldsymbol{x}_0 + \boldsymbol{d}) = \min_{g_L(\boldsymbol{x}_0 + \boldsymbol{d}) = 0} f_L(\boldsymbol{x}_0 + \boldsymbol{d}). \tag{12}$$

Using now the method of Lagrange multipliers, we obtain the following equation system

$$\nabla f(\boldsymbol{x}_0) + \lambda \nabla g(\boldsymbol{x}_0) = \boldsymbol{o} \tag{13}$$

$$\nabla g(\boldsymbol{x}_0)^T \boldsymbol{d} = -g(\boldsymbol{x}_0) \tag{14}$$

But this problem will have a solution only if the gradients are parallel.

The modification proposed by Pshenichnyj is to add a term in the function to be minimized to make the equation system solvable and to avoid too large steps in the algorithm. The additional term is $|\boldsymbol{d}|^2/2$, i.e. the squared norm of the step divided by two. So the problem is now to find the minimum of

$$f(\boldsymbol{x}_0) + \nabla f(\boldsymbol{x}_0)^T \boldsymbol{d} + |\boldsymbol{d}|^2/2 \tag{15}$$

under the constraint $g_L(x_0 + d) = 0$. Using again Lagrange multipliers, we find

$$\nabla f(x_0) + d + \lambda \nabla g(x_0) = \mathbf{o} \tag{16}$$
$$\nabla g(x_0)^T d = -g(x_0) \tag{17}$$

This gives

$$d = -(\nabla f(x_0) + \lambda \nabla g(x_0)) \tag{18}$$
$$\nabla g(x_0)^T d = -g(x_0) \tag{19}$$

Replacing the vector d in the last equation by the right hand side in (18) gives for λ

$$-g(x_0) = -\nabla g(x_0)^T (\nabla f(x_0) + \lambda \nabla g(x_0)) \tag{20}$$
$$\lambda = |\nabla g(x_0)|^{-2} \left(g(x_0) - \nabla g(x_0)^T \nabla f(x_0) \right) \tag{21}$$

So the solution vector is

$$d = -\left[\nabla f(x_0) + |\nabla g(x_0)|^{-2} \left(g(x_0) - \nabla g(x_0)^T \nabla f(x_0) \right) \nabla g(x_0) \right]. \tag{22}$$

Now there comes an additional modification. If we calculate the new point $x_0 + d$, we do not know if this point is "better" than the starting point. But if at a point x^* the function $f(x)$ has a constrained minimum under $g(x) = 0$, the function $f(x) + N|g(x)|$, the augmented Lagrangian, has an unconstrained minimum if $N > 0$ is large enough (for details see [9], chap. 1.2.8). So if we check the value of the augmented Lagrangian

$$H_N(x) = f(x) + N|g(x)| \tag{23}$$

is less than at the starting point, we see if the new point is "better". If the value has not decreased, instead of taking as next point $x_0 + d$, the step length is decreased by multiplying repeatedly it by 0.5 until we find a point $x_0 + \alpha d$ with $H_N(x_0 + \alpha d) < H_N(x_0)$. In the figure the iteration step length is $0.5d$ to give a decrease in the augmented Lagrangian function.

The difference between this method and the HL-RF algorithm is that the latter minimizes directly the function $|x|^2$ under the constraint $g_L(x) = 0$ without checking if the computed new point is in some sense "better" than the last one. This can lead to a non-convergent behavior. The method of Pshenichnyj instead ensures by the addition of the term $|d|^2$ in the function to be minimized that the step length does not become too large and further the check of the value of the augmented Lagrangian guarantees convergence towards a stationary point of the Lagrangian under slight regularity conditions (see [9], p. 45-9).

It should be noted that the direction used in this method is the same as Liu/Der Kiureghian [6] obtain by introducing a merit function to improve

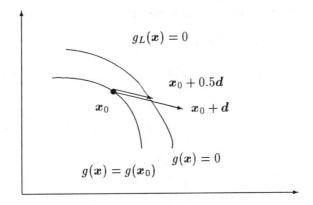

Figure 1: An iteration step of the linearization algorithm

the robustness of the HL-RF method. The difference is that as outlined in the next paragraph the line search which is done here is to get a decrease in the augmented Lagrangian and not in the merit function.

3   The algorithm

The proposed method for minimizing a function $f(x)$ under the constraint $g(x) = 0$ can be summarized as follows:

1. Set $k = 1$. Choose a starting value N for the augmented Lagrange function $H_N(x)$ and a value ϵ. Choose a starting point x_k.

2. For the point x_k calculate the coefficient λ_k by

$$\lambda_k = |\nabla g(x_k)|^{-2} \left(g(x_k) - \nabla g(x_k)^T \nabla f(x_k) \right) \tag{24}$$

and the new search direction d_k by

$$d_k = - \left[\nabla f(x_k) + \lambda_k \nabla g(x_k) \right]. \tag{25}$$

3. Set $\alpha_k = 1$ and then divide α_k by two until the inequality

$$\begin{aligned} f(x_k + \alpha_k d_k) \quad &+ \quad N|g(x_k + \alpha_k d_k)| \\ \leq f(x_k) \quad &+ \quad N|g(x_k)| - \epsilon \alpha_k |d_k|^2 \end{aligned} \tag{26}$$

is satisfied.

4. Take as the next point in the iteration

$$x_{k+1} = x_k + \alpha_k d_k. \tag{27}$$

5. If we have for the factor N of the augmented Lagrangian that

$$N \leq \lambda_k, \tag{28}$$

set

$$N = 2\lambda_k. \tag{29}$$

6. Set $k = k + 1$ and return to step 2.

If this method is used for finding the beta point, the function to be minimized is $f(x) = |x|^2$ and its gradient is $\nabla f(x) = 2x$.

How to choose N and ϵ? Since N should be larger than the Lagrange multiplier $\beta |\nabla g(x^*)|^{-1}$ at the beta point, it should be taken large enough at the beginning, for example $N = 2 \times |x_0|/|\nabla g(x_0)|$. If it is too small, it will be increased automatically in step 5. As value for ϵ should be taken as a number between 0.1 and 0.5. In the scheme above no stopping criterion is included. Usually the algorithm should be stopped if the steplengths become too small.

Modifications of this algorithm for the case of several limit state functions are given in [9], chap. 2. There are given also some modifications similar to the quadratic programming schemes outlined in the next paragraph to improve the convergence velocity in the final steps of the algorithm, but they do not make a reconstruction of the modified Hessian.

4 Quadratic programming and the SORM-factor

In this paragraph we will consider now a different topic, i.e. how to calculate the SORM factor in an easy way. In sequential quadratic programming methods a point (x^*, λ^*) is sought where

$$\begin{aligned}
\nabla f(x^*) + \lambda^* \nabla g(x^*) &= \mathbf{o} \\
g(x^*) &= 0,
\end{aligned} \tag{30}$$

i.e. a stationary point of the Lagrangian. This is done by trying to find approximately the Jacobian of these functions, i.e. the matrix

$$L(x, \lambda) = \begin{pmatrix} H_f(x) + \lambda H_g(x) & \nabla g(x) \\ \nabla g(x)^T & 0 \end{pmatrix}. \tag{31}$$

and then using Newton's method to calculate the step d_k from a point x_k to the next point x_{k+1} by

$$
\begin{pmatrix} x_{k+1} \\ \lambda_{k+1} \end{pmatrix} = \begin{pmatrix} x_k \\ \lambda_k \end{pmatrix} + \begin{pmatrix} d_k \\ \kappa_k \end{pmatrix}
$$
$$
= \begin{pmatrix} x_k \\ \lambda_k \end{pmatrix} + L(x_k, \lambda_k)^{-1} \begin{pmatrix} -\nabla f(x_k) \\ -g(x_k) \end{pmatrix}. \tag{32}
$$

Here $\kappa_k = \lambda_{k+1} - \lambda_k$ is the change in the Lagrange multiplier. To ensure convergence the steplength is usually modified by making some line search along the direction defined by d_k, i.e some point $x_k + \alpha_k d_k$ with $0 \le \alpha_k \le 1$ is taken as x_{k+1}.

The characteristic of these methods is that the matrix is not computed directly, but reconstructed approximately from the differences of the gradients at the various steps of the algorithm. So the numerical calculation of second derivatives is avoided. A widely used algorithm of this form was developed by Schittkowski [11].

If we compare the computation of the new step in these methods with the linearization method, we see that the linearization method for one equality constraint is obtained from the sequential quadratic programming scheme by taking instead of the matrix $H_f(x) + \lambda H_g(x)$ the unity matrix.

The SORM approximation for the failure probability takes into account the additional information of the second derivatives of the limit state function at the design point, i.e. the failure probability is approximated by

$$
P(F) \sim \Phi(-\beta) \prod_{i=1}^{n-1} (1 - \beta \kappa_i)^{-1/2}. \tag{33}
$$

with the κ_i's the main curvatures of G at the beta point x^*.

One objection against using SORM approximations is that it requires the numerical calculation of second derivatives and then an eigenvalue analysis for the modified Hessian at the beta point.

But if a sequential quadratic programming scheme is used to find the beta point, we can use this approximate Hessian of the Lagrangian for calculating the SORM factor. Writing the SORM factor as a function of the main curvatures is intuitively appealing, since it gives a clear geometric meaning, but it is not well suited for a numerical analysis. This factor $\prod_{i=1}^{n-1}(1 - \beta \kappa_i)^{-1/2}$ can be written as (see [2], [3])

$$
\prod_{i=1}^{n-1} (1 - \beta \kappa_i)^{-1/2}
$$
$$
= \det(\underbrace{(I_n - P)H(I_n - P) + P)}_{=H^*(x^*)})^{-1/2} \tag{34}
$$

with

$$H = \left(\delta_{ij} + \frac{\beta}{|\nabla g(\boldsymbol{x}^*)|} g_{ij}(\boldsymbol{x}^*) \right)_{i,j=1,\dots,n} \tag{35}$$

and $\boldsymbol{P}$ the projection matrix onto the gradient vector $\nabla g(\boldsymbol{x}^*)$, i.e.

$$\boldsymbol{P} = |\nabla g(\boldsymbol{x}^*)|^{-2} \nabla g(\boldsymbol{x}^*) \nabla g(\boldsymbol{x}^*)^T. \tag{36}$$

By a suitable rotation we can achieve always that the beta point is $\boldsymbol{x}^* = (0,\dots,0,\beta)^T$. In this case $\boldsymbol{H}^*(\boldsymbol{x}^*)$ has the form

$$\begin{pmatrix} 1 + \lambda^* g_{11}(\boldsymbol{x}^*) & \cdots & \lambda^* g_{1,n-1}(\boldsymbol{x}^*) \\ \lambda^* g_{21}(\boldsymbol{x}^*) & \cdots & \lambda^* g_{2,n-1}(\boldsymbol{x}^*) \\ \vdots & \ddots & \vdots \\ \lambda^* g_{n-1,1}(\boldsymbol{x}^*) & \cdots & 1 + \lambda^* g_{n-1,n-1}(\boldsymbol{x}^*) \end{pmatrix}.$$

with $\lambda^* = |\nabla g(\boldsymbol{x}^*)|^{-1}\beta$. Now the gradient is $\nabla g(\boldsymbol{x}^*) = |\nabla g(\boldsymbol{x}^*)|(0,\dots,0,1)^T$ and the modified Hessian $\boldsymbol{L}(\boldsymbol{x}^*,\lambda)$ at $(\boldsymbol{x}^*,\lambda^*)$ is

$$\begin{pmatrix} 1 + \lambda^* g_{11}(\boldsymbol{x}^*) & \cdots & \lambda g_{1n}(\boldsymbol{x}^*) & 0 \\ \lambda^* g_{21}(\boldsymbol{x}^*) & \cdots & \lambda^* g_{2n}(\boldsymbol{x}^*) & 0 \\ \vdots & \ddots & \vdots & \vdots \\ \lambda^* g_{n1}(\boldsymbol{x}^*) & \cdots & 1 + \lambda^* g_{nn}(\boldsymbol{x}^*) & g_n(\boldsymbol{x}^*) \\ 0 & \cdots & g_n(\boldsymbol{x}^*) & 0 \end{pmatrix}.$$

Expanding the determinant of this matrix with respect to the last row and column we see that

$$\begin{aligned} \det(\boldsymbol{L}(\boldsymbol{x}^*,\lambda^*)) &= -g_n^2(\boldsymbol{x}^*)\det(\boldsymbol{H}^*(\boldsymbol{x}^*)) \\ &= -|\nabla g(\boldsymbol{x}^*)|^2 \det(\boldsymbol{H}^*(\boldsymbol{x}^*)). \end{aligned} \tag{37}$$

So we obtain

$$\prod_{i=1}^{n-1}(1-\beta\kappa_i)^{1/2} = \frac{|\det(\boldsymbol{L}(\boldsymbol{x}^*,\lambda^*))|^{1/2}}{|\nabla g(\boldsymbol{x}^*)|}. \tag{38}$$

In this form the SORM factor can be computed directly from the matrix which is used in the sequential quadratic programming method. Due to the rotational symmetry of the standard normal density, this result is independent of the chosen coordinate system and therefore we get

$$\prod_{i=1}^{n-1}(1-\beta\kappa_i)^{-1/2} = \frac{|\nabla g(\boldsymbol{x}^*)|}{\sqrt{|\det(\boldsymbol{L}(\boldsymbol{x}^*,\lambda^*))|}}. \tag{39}$$

5 Numerical examples

Example 1
We consider the second example in the paper of Liu and Der Kiureghian [6]. Here a limit state function is given in the form

$$g(x) = x_1 + 2(x_2 + x_3) + x_4 - 5(x_5 + x_6) + 0.001 \sum_{i=1}^{6} \sin(100x_i). \qquad (40)$$

The random variables $X_1, \ldots, X_6$ are independent and all have a lognormal distribution. The variables X_1 to X_4 have mean 120 and standard deviation 12. The variable X_5 has mean 50 and standard deviation 15. X_6 has mean 40 and standard deviation 12. The terms in the sum produce noise in the function.

Liu/Der Kiureghian ([6]) found as beta point the point with the coordinates $(-.228, -.400, -.400, -.228, 1.75, 1.12)$ and record as β the value 2.3482 which is not the norm of this vector. The reason might be some rounding error. We found as beta point a point $(-.167, -.330, -.330, -.167, 1.75, 1.19)$ with $\beta = 2.18$. Since even for the limit state function without noise we had convergence to this point, we assume that it is the correct beta point.

The algorithm was stopped if the step length was less than $1.E - 4$. The linearization method needed six evaluations of the gradient and additionally 123 evaluations of the function g. This is better than all recorded values for the methods compared in [6].

This result shows that this algorithm is convergent even in the presence of noise in the limit state function and that the computation effort in this example is less than for other algorithms. Due to the presence of noise in the limit state function it is here not useful to calculate a SORM factor.

Example 2
A further study covers the example 5.6, p. 97-101 in Madsen et al. [7]. Here a plane frame structure is considered. Plastic hinge mechanisms are considered for causing the failure of the structure.

The plastic moment capacities $X_1, \ldots, X_5$ are lognormally distributed random variables with means 134.9 kNm and standard deviations 13.49 kNm. The load X_6 is also a random variable with a lognormal distribution having mean 50 kNm and standard deviation 15 kNm. The load X_7, also lognormally distributed, has mean 40 kNm and standard deviation 12 kNm. They are all independent of each other. The limit state function is

$$g(x_1, \ldots, x_7) = x_1 + 2x_3 + 2x_4 + x_5 - h \cdot x_6 - h \cdot x_7. \qquad (41)$$

We take $h = 5m$. The limit state function can be written directly as a function of standard normal random variables $U_1, \ldots, U_7$.

By running the sequential quadratic programming algorithm of Schittkowski [11] as described in Liu/Der Kiureghian [6], but with a slight modification to ensure numerical stability (i.e. only the projection of the Hessian on the tangential space was reconstructed), we find the beta point

$$\boldsymbol{x}^* = (-.22, 0, -.43, -.43, -.22, 2.39, 1.45). \tag{42}$$

The FORM approximation is 1.97×10^{-3}.

We found as SORM factor, by evaluating the modified Hessian at the beta point the value $1/\sqrt{0.546} = 1.35$. The value obtained from taking the reconstructed Hessian from the sequential quadratic programming algorithm was $1/\sqrt{0.531} = 1.37$. This gives as SORM approximation 2.66×10^{-3} with the exact Hessian and 2.7×10^{-3} with the approximate Hessian. The exact failure probability for this example is 2.69×10^{-3} (see [7], p. 116).

6 Summary and conclusions

In this paper a method for constrained optimization for determining the design point in structural reliability problems was outlined. The method is similar to the Hasofer/Lind-Rackwitz/Fiessler algorithm, but it can be shown that under slight regularity conditions it will converge to a stationary point of the Lagrangian. This convergence is achieved, since in each step of the search the step length is varied such that the Lagrangian function of the problem decreases.

In principle the most promising approach for calculating the design point and the approximation for the failure probability appears to be a hybrid method. First to locate approximately the position of the point by using the linearization method whose convergence velocity is linear and then to start a sequential quadratic programming algorithm with quadratic convergence to get the exact position of the point and, thence, also the SORM factor.

Acknowledgement

The co-operation of the first author was made possible within the exchange program supported by the European Union grant ERBCHXCT940565 (Human Capital Mobility Network on Stochastic Mechanics).

References

[1] K. Breitung. Asymptotic approximations for multinormal integrals. *Journal of the Engineering Mechanics Division ASCE*, 110(3):357–366, 1984.

[2] K. Breitung. *Asymptotic Approximations for Probability Integrals.* Springer, Berlin, 1994. Lecture Notes in Mathematics, Nr. 1592.

[3] K. Breitung and L. Faravelli. Response surface methods and asymptotic approximations. In F. Casciati and J.B. Roberts, editors, *Mathematical Models for Structural Reliability*, chapter 5, pages 237–298. CRC, Boca Raton, FL, 1996.

[4] F. Casciati and L. Faravelli. *Fragility Analysis of Complex Structures.* Research Studies Press Ltd., Taunton, UK, 1991.

[5] M. Hohenbichler, S. Gollwitzer, W. Kruse, and R. Rackwitz. New light on first-and second-order reliability methods. *Structural Safety*, 4:267–284, 1987.

[6] P.-L. Liu and A. Der Kiureghian. Optimization algorithms for structural reliability. *Structural Safety*, 9(3):161–177, 1991.

[7] H. Madsen, S. Krenk, and N.C. Lind. *Methods of Structural Safety.* Prentice-Hall Inc., Englewood Cliffs, N.J., 1986.

[8] B.N. Pshenichnyj. Algorithms for general mathematical programming problems. *Cybernetics*, 6(5):120–125, 1970. Translation of the journal "Kibernetika".

[9] B.N. Pshenichnyj. *The Linearization Method for Constrained Optimization.* Springer, Berlin, 1994.

[10] R. Rackwitz and B. Fiessler. Structural reliability under combined random load sequences. *Computers and Structures*, 9:489–494, 1978.

[11] K. Schittkowski. On the convergence of a sequential quadratic programming method with an augmented Lagrangian line search function. *Math. Operationsforschung und Statistik, Ser. Operationsforschung*, 14:197–216, 1983.

[12] D. Veneziano, F. Casciati, and L. Faravelli. Method of seismic fragility for complicated systems. In *Proc. 2nd Specialist Meeting on Probabilistic Methods in Seismic Risk Assessment for Nuclear Power Plants*, pages 67–88, Livermore, 1983.

Mathematical Aspects of the Boundary Initial Value Problems for Thermoelasticity Theory of Non-simple Materials with Control for Temperature

Jerzy Gawinecki and Lucjan Kowalski

Institute of Mathematics and Operations Research, Faculty of Cybernetics, Military University of Technology, 01-489 Warsaw, Poland

Abstract. We consider the initial–boundary value problem for the hyperbolic partial differential equations of thermoelasticity theory for non-simple materials. The new approach is based on the fact that we consider the initial–boundary value problem for these equations with control for temperature. We formulate the control for termperature in the terms of maximal monotone set. Existence, uniqueness and regularity of the solution to this initial–boundary value problems are proved in Sobolev space. In our proof, we use the semigroup theory and the method of Hilbert space.

Keywords: control for temperature, optimal control, non-simple thermoelastic materials, boundary–initial value problem, Hilbert space, Sobolev space, semigroup theory, stochastic equations

1. Introduction

The heat conduction problem with control for temperature was formulated and solved by Duvaut and Lions (cf. [7]). This problem was also studied in [3, 9, 21]. The linear thermoelasticity equations (for hyperbolic system), with control for termperature was investigated in [11]. We extend our considerations in order to solve the initial–boundary value problem with control for temperature for linear thermoelasticity theory of non-simple materials. We consider non-simple materials whose local state is characterized by the temperature, its gradient, the time rate of change of temperature, the deformation gradient and its gradient (cf. [6, 14]). Existence and uniqueness of the solution for linear thermoelasticity for non-simple materials (without control for temperature) were considered in ([14]). Using the method of Sobolev spaces and the method of semigroup theory we prove that the solution of the boundary–initial value problem with the control for temperature for the equation describing non-simple materials exists and is unique.

In order to formulate the control for temperature (in generalized form) we used the theory of maximal monotone operators.

Our paper is organized as follows.

In section 2 the statement of the problem is given. Section 3 is devoted to the proof of the theorem of existence and uniquenes of the solution to the initial–boundary value problem with control for temperature for the equation describing non-simple thermoelastic materials. Finally, in the last section some concluding remarks are given.

2. Statement of the problem

Let Ω be a domain in $\mathbb{R}^3$ with the smooth boundary $\Gamma = \partial\Omega$. The time variable t takes values from $[0, T] \subset \mathbb{R}$.

We consider the thermoelastic, anisotropic medium, characterized by temperature, its gradient, the time rate of change of tempearature, the deformation gradient and its gradient i.e. so called non-simple materials (cf. [6]) occupying domain $\bar{\Omega}$. For $x \in \Omega$ and $t \in [0, T]$ we denote by $u = u(x, t)$ the displacement vector field of the medium and by $\Theta = \Theta(x, t)$ the temperature of the medium. Below, we consider boundary–initial value problem for thermoelasticity of non-simple materials with control for temperature. Now we describe the meaning of the control for temperature.

We need (in many applications in technology) the temperature of the medium to take the value from the interval $[\Theta_1(x), \Theta_2(x)]$ for any $t \in (0, T)$.

How to reach this?

In order to obtain this aim we must control the voluminal heat source satisfying the role of the temperature controller with intensity $\hat{g}$.

Let the intensity of this heat source belong to the interval $[g_1, g_2]$ (we assume that $0 \in [g_1, g_2]$, g_i $(i = 1, 2)$ is also equal to the intensity).

We control the intensity of the additional source as follows:

$1°$ If $\Theta(x, t) \in [\Theta_1(x), \Theta_2(x)]$, then $\hat{g} = 0$

$2°$ If $\Theta(x) \notin [\Theta_1(x), \Theta_2(x)]$, then we lead the heat which is directly proportional to the difference between the temperature $\Theta(x; t)$ and the interval $[\Theta_1(x), \Theta_2(x)]$. So, we have.

$$\Theta(x, t) > \Theta_2(x) \Rightarrow \begin{cases} -\hat{g} = k_2(\Theta - \Theta_2) & \text{if } k_2(\Theta - \Theta_2) \leqslant g_2 \\ -\hat{g} = g_2 & \text{if } k_2(\Theta - \Theta_2) > g_2 \end{cases}$$

$$\Theta(x, t) < \Theta_1(x) \Rightarrow \begin{cases} -\hat{g} = k_1(\Theta - \Theta_1) & \text{if } k_1(\Theta - \Theta_1) \geqslant g_1 \\ -\hat{g} = g_1 & \text{if } k_1(\Theta - \Theta_1) < g_1 \end{cases}$$

We can write these cases as follows

$$-\hat{g} = \beta(\Theta)$$

where

$$\beta(\Theta) = \begin{cases} g_1 & \text{if } \Theta \leqslant \Theta_1 + \dfrac{1}{k_1} g_1 \\[2mm] k_1(\Theta - \Theta_1) & \text{if } \Theta_1 + \dfrac{1}{k_1} g_1 \leqslant \Theta \leqslant \Theta_1 \\[2mm] 0 & \text{if } \Theta_1 \leqslant \Theta \leqslant \Theta_2 \\[2mm] k_2(\Theta - \Theta_2) & \text{if } \Theta_2 \leqslant \Theta \leqslant \Theta_2 + \dfrac{1}{k_2} g_2 \\[2mm] g_2 & \text{if } \Theta \geqslant \Theta_2 + \dfrac{1}{k} g_2 \end{cases}$$

We can describe the control for temperature more generally in term of maximal montone operator i.e.

$$-\hat{g} \in \beta(\Theta)$$

where: $\beta : \Omega \times \mathbb{R} \to \mathbb{R}$ is multivaleed operator, such that $\forall x \in \Omega$, $\Theta \to \beta(x, \Theta)$ is maximal monotone operator. Below, we give some examples of the operator $\beta(\Theta)$:

Example 2.1. Jumping control:

Let $k_1 = k_2 = +\infty$, $-\infty < g_1$, $g_2 < +\infty$. In this case the intensity of the heat source is bounded operator $\beta(\cdot)$ and has the following form:

$$\beta(\Theta) = \begin{cases} g_1 & \text{if } \Theta < \Theta_1 \\ [g_1, 0] & \text{for } \Theta = \Theta_1 \\ 0 & \text{for } \Theta_1 \leqslant \Theta \leqslant \Theta_2 \\ [0, g_2] & \text{for } \Theta = \Theta_2 \\ g_2 & \text{for } \Theta > \Theta_2 \end{cases}$$

for given Θ_1, Θ_2 $(\Theta_1 < \Theta_2)$; g_1, g_2 $(\Theta \in (g_1, g_2))$

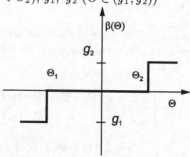

Fig. 2. 1.

Example 2.2. Upper restriction for the temperature. In this case we have

$$\beta(\Theta) = \begin{cases} 0 & \text{if } \Theta < \Theta_2 \\ [0, +\infty) & \text{if } \Theta = \Theta_2 \\ \varnothing & \text{if } \Theta > \Theta_2 \end{cases}$$

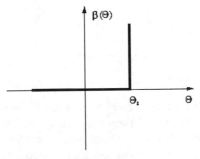

Fig. 2. 2.

Example 2.3. Lower restriction for the temperature

$$\beta(\Theta) = \begin{cases} \varnothing & \text{if } \Theta < \Theta_1 \\ (-\infty, 0] & \text{if } \Theta = \Theta_1 \\ 0 & \text{if } \Theta > \Theta_1 \end{cases}$$

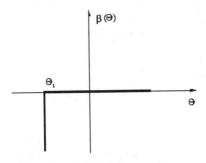

Fig. 2. 3.

Example 2.4. Lower and upper restriction for the temperature

$$\beta(\Theta) = \begin{cases} \varnothing & \text{if } \Theta < \Theta_1 \\ (-\infty, 0] & \text{if } \Theta = \Theta_1 \\ 0 & \text{if } \Theta < \Theta \leqslant \Theta_2 \\ [0, +\infty) & \text{if } \Theta = \Theta_2 \\ \varnothing & \text{if } \Theta > \Theta_2 \end{cases}$$

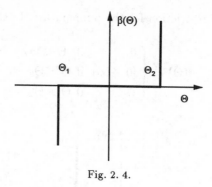

Fig. 2. 4.

We assume that $\mathbb{D}(\beta) = \mathbb{R}$

We will be seeking the displacement vector field u and the temperature Θ which satisfy the equations of thermoelasticity theory for non-simple materials [6, 11]. A linear theory of thermoelasticity for non-simple materials based upon an entropy production inequality proposed by Green and Laws [12] was devised by Ieşan [14].

In that theory the local state of non-simple materials is characterised by the temperature, its gradient, the time rate of change of temperature, the deformation gradient and its gradient. Below, we consider the linear theory of thermoelasticity for non-simple materials as it is established in [14]. The basic equations in that theory are as follows:

— the equation of motion

$$\tau_{ji,i} - \mu_{sji,sj} + f_i = \rho \ddot{u}_i \tag{2.1}$$

— the equation of energy

$$\rho T_0 \dot{\eta} = q_{i,i} + S \tag{2.2}$$

— the constitutive equations

$$
\begin{aligned}
\tau_{ij} &= A_{ijrs} e_{rs} + B_{ijpqr} \varkappa_{pqr} + a_{ij} + (\Theta + \alpha \dot{\Theta}), \\
\mu_{ijk} &= B_{rsijk} e_{rs} + C_{ijkmnr} \varkappa_{mnr} + c_{ijk}(\Theta + \alpha \dot{\Theta}), \\
\rho \eta &= a + d\Theta + h\dot{\Theta} - b_i \Theta_{,i} - a_{ij} e_{ij} - c_{ijk} \varkappa_{ijk}, \\
q_i &= T_0(b_i \dot{\Theta} + k_{ij} \Theta_{,j}),
\end{aligned}
\tag{2.3}
$$

— the kinematic relations

$$2e_{ij} = u_{i,j} + u_{j,i}, \qquad \varkappa_{ijk} = u_{k,ij}. \tag{2.4}$$

In these relations we used the following notations: ρ — the constant mass density; T_0 — the constant absolute temperature of the body in its reference state; u_i — the components of the displacement vector; Θ — the temperature variation measured from the constant temperature T_0; e_{ij} and $\varkappa_{ijk}$ — the kinematic characteristics of the body; τ_{ij} and μ_{ijk} — the components of the hyperstress tensors; η — the specific entropy; q_i — the components of the heat flux vector; f_i — the components of the body force per unit volume; A_{ijrs}, B_{ijpqr}, C_{ijkmnr}, c_{ijk}, a_{ij}, k_{ij}, b_i, a, d, h and α are characteristic constants of the material and they obey the symmetry relations:

$$A_{ijrs} = A_{rsij} = A_{jirs}, \quad B_{ijpqr} = B_{jipqr} = B_{ijqpr},$$
$$C_{ijkpqr} = C_{pqrijk} = C_{jikpqr}, \quad c_{ijk} = c_{jik}, \quad a_{ij} = a_{ji} \quad k_{ij} = k_{ji}. \tag{2.5}$$

The entropy inequality implies

$$(d\alpha - h)\dot{\Theta}^2 + 2b_i\dot{\Theta}\Theta_{,i} + k_{ij}\Theta_{,i}\Theta_{,j} \geqslant 0.$$

Substituting now (2.3) and (2.4) into relations (2.1) and (2.2) using symmetry relation (2.5) and putting instead of equality (=) the relation $\in$ in the equation (2.2) and adding the term $\beta(x, \Theta)$ on the right hand side of this relation, we obtain, the following system of coupled equations:

$$\partial_t^2 u_i = A_{jirs}u_{r,sj} + B_{jipqr}u_{r,pqj} + a_{ji}(\Theta_{,j} + \alpha\partial_t\Theta_{,j}) +$$
$$- B_{mnsji}u_{m,nsj} - C_{sjimnr}u_{r,mnsj} - C_{sji}(\Theta_{,sj} + \alpha\partial_t\Theta_{,sj}) + f_i$$
$$i = 1, 2, 3 \tag{2.7}$$

$$\partial_t^2\Theta \in -d\partial_t\Theta + 2b_i\partial_t\Theta_{,i} + a_{ij}\partial_t u_{i,j} + c_{ijk}\partial_t u_{k,ij} + k_{ij}\Theta_{,ij} + S + \beta(x, \Theta) \tag{2.8}$$

(without loss of generality we assume that $\rho = 1$ and $h = 1$, $T_0 = 1$).

We adjoin the folowing boundary conditions to the system of field equations (2.7), (2.8):

$$
\begin{aligned}
u_i(x, t) &= 0 & \text{for} \quad (x, t) &\in \Gamma \times (0, T) \\
\Theta(x, t) &= 0 & \text{for} \quad (x, t) &\in \Gamma \times (0, T) \\
\frac{du_i(x, t)}{dn} &= 0 & \text{for} \quad (x, t) &\in \Gamma \times (0, T), \quad i = 1, 2, 3
\end{aligned}
\tag{2.9}
$$

where $\frac{du_i(x,t)}{dn} = u_{i,j}n_j$ and n_j are the direction cosines of the outward normal to the boundary $\Gamma = \partial\Omega$ and with the initial conditions:

$$\Theta(x,0) = \Theta^0(x), \qquad \partial_t \Theta(x,0) = \Theta^1(x) \qquad \text{for} \quad x \in \Omega$$
$$u(x,0) = u^0(x), \qquad \partial_t u(x,0) = u^1(x) \qquad \text{for} \quad x \in \Omega. \tag{2.10}$$

REMARK. Relation (2.8) describes the problem with the control for temperature for the non-simple thermoelastic materials.

The following assumption on the material properties will be used throughout this paper:

$$\alpha > 0, \quad h = 1, \quad \rho = 1, \quad T_0 = 1$$
$$(d\alpha)\zeta^2 + 2\zeta b_i \xi_i + k_{ij}\xi_i\xi_j > k_0(\zeta^2 + \xi_i\xi_i), \quad k_0 > 0, \tag{2.11}$$

for all arbitrary ζ and ξ_i;

$$A_{ijrs}\xi_{ij}\xi_{rs} + 2B_{ijmnr}\xi_{ij}\chi_{mnr} + C_{ijkpqr}\chi_{ijh}\chi_{pqr} \geqslant$$
$$\geqslant a_0(\xi_{ij}\xi_{ij} + \chi_{ijk}\chi_{ijk}), \qquad a_0 > 0, \tag{2.12}$$

for all arbitirary,

$$\xi_{ij} = \xi_{ji}, \qquad \chi_{ijk} = \chi_{jik}. \tag{2.13}$$

The above assumptions are in agreement with the restrictions imposed in [14]. The assumption (2.11) represents a considerable strengthening of the consequence (2.6) of the entropy production inequality. The condition (2.12) was used in [13] in order to obtain the existence and uniqueness of solution for the boundary value problems in some non-simple theorem of elastostatics.

In the next section we prove the existence and uniqueness of the solution to the problem (2.7) (2.8) – (2.9) (2.10).

3. Existence and uniquenes of the weak solutions

Let $\mathbb{X}$ be Hilbert space:

$$\mathbb{X} = \{\mathbb{W} = (u, v, \Theta, \psi); \quad u \in [\mathbb{H}_0^1(\Omega) \cap \mathbb{H}^2(\Omega)]^3,$$
$$v \in [\mathbb{H}_0(\Omega)]^3, \quad \Theta \in \mathbb{H}^1(\Omega), \quad \psi \in \mathbb{H}_0(\Omega)\} \tag{3.1}$$

Now, we define the following operators;

$$A_i(\mathbb{W}) = v_i$$

$$B_i(\mathbb{W}) = A_{jirs}u_{r,sj} + B_{jipqr}u_{r,pqj} + a_{ij}(\Theta_{,j} + \alpha\psi_{,j}) +$$
$$- B_{mnsji}u_{m,nsj} - C_{sjimnr}u_{r,mnsj} - C_{sji}(\Theta_{,sj} + \alpha\psi_{,sj})$$

$$C(\mathbb{W})) = \psi$$

$$\mathbb{D}(\mathbb{W}) = -d\psi + 2b_i\psi_{,i} + a_{ij}v_{i,j} + C_{ijk}v_{k,ij} + k_{ij}\Theta_{,ij} \tag{3.2}$$

and the operator:

$$A(\mathbb{W}) = (A(\mathbb{W}), B(\mathbb{W}), C(\mathbb{W}), D(\mathbb{W})) \tag{3.3}$$

with the domain:

$$D(\mathcal{A}) = \{\mathbb{W} \in \mathbb{X} : A(\mathbb{W}) \in \mathbb{X}, v = \frac{\partial v}{\partial n} = 0 = \psi = 0 \text{ on } \Gamma = \partial\Omega\}$$

Let $\bar{\beta} \subset (\mathbb{H}^1(\Omega) \times \mathbb{H}^1(\Omega)$ be the operator defined as follows:

$$\bar{\beta}(\Theta) = \{v \in \mathbb{H}^1(\Omega) : v \in \beta(\Theta(x)) \ e.a. \ x \in \Omega\} \tag{3.4}$$

$\bar{\beta}$ is also maximal monotone operator.

Now, we define the operator $\mathbb{B}$ as follows:

$$\mathbb{B} = \mathbb{X} \to \mathbb{X}$$
$$\mathbb{B} = [B_1, B_2, B_3, B_4] \tag{3.5}$$

where:

$$\begin{aligned}
B_1\mathbb{W} &= 0 \\
B_2\mathbb{W} &= 0 \\
B_3\mathbb{W} &= \bar{\beta}(\Theta) \\
B_4\mathbb{W} &= 0 \\
D(\mathbb{B}) &= [L^2(\mathbb{W})]^7 \times D(\bar{\beta})
\end{aligned} \tag{3.6}$$

From (3.3), (3.6) and [3] it follows that $\mathbb{B}$ is maximal monotone operator.

Introducing the vector of the form $W = (u, \partial_t u, \Theta, \partial_t \Theta)$ we can convert initial–boundary value problem (2.1)–(2.6) with the control for temperature into evolution equation in Hilbert space $\mathbb{X}$:

$$\partial_t W + \mathcal{A}W + \mathbb{B}W \ni' \mathbb{F} \quad \text{for } t \in (0, T), \ W(0) = W^0 \tag{3.7}$$

where: $\mathbb{F} = (0, f, 0, S)$ and

$$W^0 = (u^0, u^1, \Theta^0, \Theta^1) \tag{3.8}$$

The symbol $\in'$ has the following meaning:

$$(x_1, \cdots, x_7, x_8) \in' A_1 \times \cdots \times A_7 \times A_8 \Leftrightarrow A_i \in \{x_i\} \text{ for } i = 1, 2, \cdots, 6, 8, x_7 \in A_7$$

Now, we use the result of the simigroup theorem in order to obtain the existence and uniqueness of the solution to the problem (3.7)–(3.8).

It is easy to see that the operator $\mathcal{A}$ has the following properties:

$$
\begin{aligned}
&1) \quad \mathbb{D}(\mathcal{A}) \text{ is dense in } \mathbb{X}, \\
&2) \quad <\mathcal{A}W, W >\geqslant 0 \ \forall W \in \mathbb{D}(\mathcal{A}) \\
&3) \quad R(\lambda I - \mathcal{A}) = \mathbb{X}, \ \forall \lambda > 0.
\end{aligned}
\tag{3.9}
$$

Basing on the properties (3.9) (1–3) we deduce that the operator $\mathcal{A}$ is maximal monotone operator in $X \times X$ (cf. [21]).

Also, operator $\mathbb{B}$ is maximal monotone.

If $\mathcal{D}(\mathcal{A}) \cap \operatorname{Int}\mathcal{D}(\mathbb{B}) \neq 0$, then from Rockafaller-Moreau theorem it follows that, the operator $K = \mathcal{A} + \mathbb{B}$ generates a $\mathcal{C}_0$ semigroup of contraction in the Hilbert space $\mathbb{X}$. So, (cf. [3], p. 131, 136) we can prove that there exists a unique solution (weak) to the problem (3.7)–(3.8). It means that the following theorem is true:

Theorem 2.1. *If* $W^0 \in \mathcal{D}(K)$, $\mathbb{F} \in \mathbb{W}^{1,1}((0,T), \mathbb{X})$, *then there exists a unique solution of the problem*

$$
\begin{aligned}
\partial_t W + \bar{K}W &\ni' \mathbb{F} \\
W(0) &= W^0
\end{aligned}
\tag{3.10}
$$

with properties

$$
W \in \mathbb{W}^{1,\infty}((0,T); \mathbb{X}) \quad {}^* \tag{3.11}
$$

i.e.

$$
(u, \partial_t u, \Theta, \partial_t \Theta) \in \mathbb{W}^{1,\infty}((0,T), \mathbb{X})
$$

---

* 1° $\mathbb{W}^{k,2}(I, V)$; $k \in \mathbb{N}$ denotes the space of measurable functions $f : I \to V$ with $\mathrm{d}^n f/dt^n \in L^2(I, V)$ for $0 \leqslant n \leqslant k$ (derivatives in the weak sence); the norm in $\mathbb{W}^{k,2}(I, V)$ is given by: $\|f\|_k^2 = \sum_{k=0}^{n} \int_0^t \|\mathrm{d}^n f(\cdot, t)| dt^n\|_V^2 \, dt$

2° $\mathbb{W}^{1,\infty}(I, V) = \{f : I \to V, \ \mathrm{d}^n f | dt^n \in L^\infty(I, V) \text{ where } I = (0,T), 0 \leqslant n \leqslant 1\}$

4. Concluding remarks

Remark 4.1. Applying the method of the convex analysis one can investigate in the similar way the problem with control for the temperature on the boundary $\Gamma = \partial\Omega$

Remark 4.2. The problem (3.11) has the solution of the form:

$$W(t) = T(t)W^0 + \int_0^t T(t-s)\mathbb{F}(s)\mathrm{d}s$$

for

$$\forall t \in (0,T) \qquad\qquad (4.1)$$

where: $T(t)$ is C_0 — semigroup of contractions on $\mathbb{X}$ generated by the operator K.

Remark 4.3. In order to obtain more regular solution to the problem of optimal control of the temperature we can substitute the maximal monotone operator β by smooth function β_0.

If we take into account the maximal monotone operator $\beta(\Theta)$ (cf. section 2) we can accept Friedrichs mollifier of the continuous function $\beta_c(\Theta)$ of the form:

$$\beta_c(\Theta) = \begin{cases} g_1 & \text{if } \Theta < \Theta_1 - \varepsilon \\ \dfrac{g_1}{2\varepsilon}(\Theta_1 + \varepsilon - \Theta) & \text{if } \Theta_1 - \varepsilon \leqslant \Theta < \Theta_1 - \varepsilon \\ 0 & \text{if } \Theta_1 + \varepsilon \leqslant \Theta < \Theta_2 - \varepsilon \\ \dfrac{g_2}{2a}(\Theta + \varepsilon - \Theta_2) & \text{if } \Theta_2 - \varepsilon \leqslant \Theta \leqslant \Theta_2 + \varepsilon \\ g_2 & \text{if } \Theta > \Theta_2 + \varepsilon \end{cases} \qquad (4.2)$$

for fixed, but arbitrary small $\varepsilon > 0$.

Then the initial condition has very important influence on the reqularity of the solutions. Under some assumption we get the solution with required regularity:

For example let us denote:

$$\mathcal{D}(K^k) = [v \in \mathcal{D}(K^{k-1}), \quad Kv \in \mathcal{D}(K^{k-1})] \qquad (4.3)$$

for $k \geqslant 2$, $k \in \mathcal{Z}$, $\mathcal{D}(K^k)$ is the Hilbert space with the inner product

$$(u, v)_{\mathcal{D}(K^k)} = \sum_{j=0}^{k} (K^j u, K^j v) \tag{4.4.}$$

So, we have (cf. [5]):

Let $W^0 \in \mathcal{D}(\mathbb{W}^k)$, $k \geqslant 2$, than the initial value problem (3.11) has the unique solution W given by formula (4.1) (for $\mathbb{F}$ — being $\mathbb{C}^k$ — function) with properties:

$$W \in \mathbb{C}^{k-j}\left([0,T]; \mathcal{D}(K^j)\right) \quad \text{for} \quad j = 0, 1, \cdots, k \tag{4.5}$$

Remark 4.4. We can also extent our results to the stochastic equation describing thermoelastic non-simple materials for example, we can consider in the right hand side of the equation (3.9) the term $\frac{\partial W}{\partial t}$ which is the weak derivative of Wiener process. The intial value W^0 is a random variable. Using the method of Hillbert space we can obtain the behaviour of the probability distribution μ_t of $W(t, \cdot)$.

References

[1] R. A. Adams, *Sobolev Spaces*, Academic Press, New York 1975.

[2] G. Ahmadi, K. Firoozbakhsh, *First strain-gradient theory of thermoelasticity*, Int. J. Solids Structures, 11 (1975), 339–345.

[3] V. Barbu, *Nonlinear semigroups and differential equations in Banach spaces*, Noordhoff, Int. Publ. Leyden 1976.

[4] R. C. Batra, *Thermodynamics of non-simple elastic materials*, J. of Elasticity, 6 (1976), 451–456.

[5] H. Brezis, *Operateurs maximaux monotones et semigroupes de contractions dans les espaces de Hilbert*, Math. Studies. 5, North Holland, Amsterdam 1973.

[6] S. Chirita, G. H. Rusu, *On existence and uniqueness in thermoelasticity of non-simple materials*, Rev. Rum. Sci. Techn. Mec. Appl., 30, 1 (1985), pp. 21–30.

[7] G. Duvaut, J. L. Lions, *Les inéqauations en mécanique et en physique*, Dunod, Paris 1972.

[8] A. Falqués, *Estabilidad asiniótica en la termoelasticidad generalizada*, Thesis. Faculdad de Fisica, Universidad de Barcelona 1982.

[9] A. FALQUÉS, *Thermoelasticity and heat conduction with memory effects*, J. Thermal Stresses, 5 (1982), 145–160.

[10] G. FICHERA, *Existence theorems in elasticity*, in *Handbuch der Physik*, Vol. VIa/2, Springer-Verlag, Berlin 1972.

[11] J. GAWINECKI, *Mathematical aspects of initial-boundary value Problems for Thermoelasticity Theory with Control for Temperature*, Lecture on 3r-d GAMM|IFIP–Workshop end Tutorial Stochastic Optimization Numerical Methods and Technical Applications, June 17–20, 1996, Münich, Germany.

[12] A. E. GREEN, N. LAWS, *On the entropy production inequality*, Arch. Rational Mech. Anal., 45 (1972), 47–59.

[13] I. HLAVÁČEK, M. HLAVÁČEK, *On the existence and uniqueness of solution and some variational principles in linear theories of elesticity with couple-stresses*, Aplikace Matematiky, 14 (1949), 411–427.

[14] D. IEŞAN, *Thermoelasticity of non-simple materials*, Journal of Thermal Stresses, 6 (1983), pp. 167–188.

[15] R. D. MINDLIN, *Microstructure in linear elasticity*, Arch. Rational Mech. Anal., 16 (1964), 51–78.

[16] R. D. MINDLIN, N. N. ESHEL, *On first strain-gradient theories in linear elasticity*, Int. J. Solids Structures, 4 (1968), 109–124.

[17] S. MIZOHATA, *The theory of partial differential equations*, Cambridge Univ. Press, London 1973.

[18] J. NAUMANN, *Einführung in die Theorie parabolischer Variationsungleichungen*, Teubner–Texte zur Mathematik, No 64, Leipzig 1984.

[19] C. B. NAVARRO, R. QUINTANILLA, *On existence and uniqueness in incremenyal thermo-elasticity*, (to appear).

[20] P. PANAGIOTOPOULUS, *Inequality Problems in Mechanics and Applications*, Birkhäuser, Boston–Stuttgart 1985.

[21] S. PAZY, *Semigroups of Linear Operators and Applications to Partial Differential Equations*, Springer–Verlag, New York, Berlin, Heidelberg, Tokyo 1983.

[22] R. A. TOUPIN, *Elastic materials with couple-stresses*, Arch. Rational Mech. Anal., 11 (1962), 385–414.

[23] K. YOSIDA, *Functional Analysis*, Springer, Berlin, Heidelberg, New York 1978.

Stochastic Trajectory Planning for Manutec r3 with Random Payload

Shihong Qu *

Institute of Mathematics and Computer Science
Federal Armed Forces University Munich
85577 Neubiberg/Munich, Germany

Key Words: trajectory planning, stochastic uncertainty

1. Introduction

Optimal trajectory planning for robots is a basic tool for improving manufacturing processes.

A robot is a multi-body mechanical system. The bodies are driven by torques and forces generated by the motors of the robot. The relationship between the input (torques and forces generated by motors) and output (position of the robot in configuration space) can be defined by the dynamic equation [1]:

$$\sum_{j=1}^{n} J_{ij}(q,p)\ddot{q}_j + \sum_{j,k=1}^{n} D_{ijk}(q,p)\dot{q}_j\dot{q}_k + G_i(q,p) = \tau_i, \ 1 \le i \le n, (1)$$

where the following notations are used:

q : vector of configuration coordinates of the robot,

p : vector of the model parameters,

τ_i : torques and forces generated by motors,

J_{ij} : elements of the inertia matrix of the robot,

D_{ijk} : coefficients of the centrifugal and Coriolis forces,

G_i : gravity forces.

Introducing a geometric path parameter s, we can represent the trajectory $q = q(t)$ as a function of s, $q = q(s)$, while t and s are connected by means of the function $s = s(t)$. This yields then

$$\dot{q}_i(s) = q_i'\dot{s}, \tag{2}$$

---
* Supported by DFG-Schwerpunktprogramm
"Echtzeit-Optimierung großer Systeme".

$$\ddot{q}_i(s) = q_i'\ddot{s} + q_i''\dot{s}, \tag{3}$$

$$\ddot{s}(s) = \frac{1}{2}\frac{d}{ds}(\dot{s}^2), \tag{4}$$

where $(\ )' = \dfrac{d}{ds}$ and $(\dot{\ }) = \dfrac{d}{dt}$ mean the derivative with respect to path parameter s and time t, respectively. With these definitions we can rewrite the dynamic equation as follows:

$$a_i(s,q,q',p)\beta' + b_i(q,q',q'',p)\beta + c_i(q,p) = \tau_i,\ 1 \le i \le n, \tag{5}$$

where

$$\beta = \dot{s}^2 \tag{5.1}$$

$$a_i = \frac{1}{2}\sum_{j=1}^{n} J_{ij}(q,p)q_j',\ 1 \le i \le n, \tag{5.2}$$

$$b_i = \sum_{j=1}^{n} J_{ij}(q,p)q_j'' + \sum_{j,k=1}^{n} D_{ijk}(q,p)q_j'q_k',\ 1 \le i \le n, \tag{5.3}$$

$$c_i = G_i(q,p),\ 1 \le i \le n. \tag{5.4}$$

The problem of trajectory planning for robots is then to find the functions $\beta = \beta(s)$ and $q = q(s)$ such that a given objective function is optimized, the available limits for torques and forces τ and allowable limits for position q and velocity $\dot{q}$ are not exceeded, and the robot must move from a prespecified initial position in its configuration space to a certain given terminal point.

Mathematically the problem can be described as follows [2, 3, 4]:

$$\min_{\beta,q} \int_0^{s_e} f_0(s,q,q',q'',\beta,\beta')ds \tag{6}$$

subject to

$$\beta(0) = \beta(s_e) = 0,\ \beta(s) \ge 0,\ 0 \le s \le s_e, \tag{6.1}$$

$$q(0) = q_0,\ q(s_e) = q_e, \tag{6.2}$$

$$q_{min} \le q \le q_{max},\ \dot{q}_{min} \le q'\sqrt{\beta} \le \dot{q}_{max},\ 0 \le s \le s_e, \tag{6.3}$$

$$\tau_{min,i} \le a_i(q,p)\beta' + b_i(q,p)\beta + c_i(q,p) \le \tau_{max,i}, \tag{6.4}$$

$$1 \le i \le n,\ 0 \le s \le s_e,$$

where $q_{min}, q_{max}, \dot{q}_{min}, \dot{q}_{max}, \tau_{min,i}$ and $\tau_{max,i}$ describe the bounds for configuration coordinates, their derivatives with respec to time t and for torques and forces generated by the motors, respectively.

As an objective function we consider a linear combination of the criterions of the time optimal and the energy minimal problem:

$$f_0 = \frac{\kappa_t}{\sqrt{\beta}} + \kappa_e \sum_{i=1}^{n} \frac{\tau_i^2}{\sqrt{\beta}}. \tag{7}$$

If $\kappa_t = 1$ and $\kappa_e = 0$, we get the time optimal problem, else a mixed problem.

2. Substitute Problems

Very often the model parameter p is not exactly known. To get a more reliable solution of the planning problem, one should utilize the available statistical information about the uncertainty. Hence, for the trajectory problem with random parameter disturbances, we propose the following substitute problems [2,4]:

2.1. Chance Constrained Substitute Problem

$$\min_{\beta, q} \int_0^{s_e} E_p \, f_0(s, q, q', q'', \beta, \beta', p) ds \tag{8}$$

subject to:

$$\beta(0) = \beta(s_e) = 0, \ \beta(s) \geq 0, \ 0 \leq s \leq s_e. \tag{8.1}$$

$$q(0) = q_0, \ q(s_e) = q_e, \tag{8.2}$$

$$q_{min}(s) \leq q(s) \leq q_{max}(s), \tag{8.3}$$

$$\dot{q}_{min}(s) \leq q'(s) \sqrt{\beta(s)} \leq \dot{q}_{max}(s), \ 0 \leq s \leq s_e, \tag{8.4}$$

$$P\left(\tau_{min,i} \leq a_i \beta' + b_i \beta + c_i \leq \tau_{max,i}\right) \geq \alpha_i, \tag{8.5}$$

$$1 \leq i \leq n, \ 0 \leq s \leq s_e.$$

Here, we consider the expected value of the objective function, and we demand that the stochastic conditions (6.4) are fulfilled separately with given minimum probabilities α_i.

2.2. Cost Constrained Substitute Problem

$$\min_{\beta, q} \int_0^{s_e} E_p \, f_0(s, q, q', q'', \beta, \beta', p) ds \tag{9}$$

subject to:

$$\beta(0) = \beta(s_e) = 0, \ \beta(s) \geq 0, \ 0 \leq s \leq s_e. \tag{9.1}$$

$$q(0) = q_0, \ q(s_e) = q_e, \tag{9.2}$$

$$q_{min}(s) \leq q(s) \leq q_{max}(s), \tag{9.3}$$

$$\dot{q}_{min}(s) \leq q'(s) \sqrt{\beta(s)} \leq \dot{q}_{max}(s), \ 0 \leq s \leq s_e, \tag{9.4}$$

$$E_p \Gamma_i \left(a_i \beta' + b_i \beta + c_i, \tau_{min,i}, \tau_{max,i} \right) \leq \delta_i, \tag{9.5}$$

$$1 \leq i \leq n, \ 0 \leq s \leq s_e$$

where Γ_i are given cost functions for the evaluation of the violation of the restrictions (6.4). In this substitute problem we consider the expected value of the objective function and demand that given upper bounds δ_i for the expected values of the cost functions should not be exceeded.

3. Numerical Methods for the Substitute Problem

3.1. Calculation of the Probabilities

To solve the chance constrained problem (8),(8.1-5) we have to compute the probabilities (8.5) at first. Since many model parameters appear linearly in dynamic equations of robots and lie in some given intervals, we consider the following model:

$$y = \vec{\xi} \cdot \vec{a} + \xi_0, \tag{10}$$

where a_i, i=1,2,...n, are uniformly distributed in $[c_i, d_i]$ and independent on each other. For the random variable y we have therefore

$$P\{y \leq \eta\} = \frac{1}{n! \prod_{i=1}^{n}(\Delta_i |\xi_i|)} \left(\sum_{j=1}^{N} (-1)^{\tilde{I}_n(\vec{a}^j - \vec{a}^1)} (\eta - y_j)_+^n \right) \tag{11}$$

with $\Delta_i = d_i - c_i$ and $\tilde{I}_n(\vec{a}) = \dfrac{a_1}{\Delta_1} + \dfrac{a_2}{\Delta_2} + ... + \dfrac{a_n}{\Delta_n}$,

where $\vec{a}^{(j)}$, $j = 1, 2, ...2^n =: N$, are the vertex points of $[c_1, d_1] \times [c_2, d_2] \times ... \times [c_n, d_n]$ and $y_j = \vec{\xi} \cdot \vec{a}^{(j)} + \xi_0$, $j = 1, 2, ...N$.

For

$$y = a_1 + 2a_2 + 3a_3, \tag{12}$$

where a_i, $i = 1, 2, 3$, are independent and uniformly distributed in $[0, 1]$, we get

$$P\{y \leq \eta\} = \begin{cases} \dfrac{\eta^3}{36} & \text{for } 0 \leq \eta < 1 \\[2mm] \dfrac{1}{36} - \dfrac{\eta}{12} + \dfrac{\eta^2}{12}, & \text{for } 1 \leq \eta \leq 2 \\[2mm] \dfrac{1}{4} - \dfrac{5\eta}{12} + \dfrac{\eta^2}{4} - \dfrac{\eta^3}{36}, & \text{for } 2 \leq \eta < 4 \\[2mm] -\dfrac{55}{36} + \dfrac{11\eta}{12} - \dfrac{\eta^2}{12}, & \text{for } 4 \leq \eta < 5 \\[2mm] -5 + 3\eta - \dfrac{\eta^2}{2} + \dfrac{\eta^3}{36}, & \text{for } 5 \leq \eta \leq 6 \end{cases} \tag{13}$$

3.2. Calculation of the Expected Cost Functions

The computation of the expected values of objective function (8), (9) and cost functions (9.5) can be very difficult, because the objective function and the cost functions are very complicated functions with respect to model parameter p in general. If the derivatives of the functions with respect to p up to a certain order exist, the expected values of these functions can be calculated approximatively by means of the Taylor expansion of the functions at the expectation $\bar{p}$ of p [4]. In this case we need only some central moments but not the whole distribution of the stochastic variable p.

For a function $u = u(p, s)$ having derivatives with respect to the random vector p up to the (K+1)-th order, by means of Taylor expansion of u at the expectation $\bar{p}$ of p we get

$$\begin{aligned} u &= u(\bar{p}, s) + \sum_{k=1}^{K} \frac{1}{k!} \frac{\partial^k u}{\partial p^k}(\bar{p}, s) \cdot (p - \bar{p})^k \\ &+ \frac{1}{(K+1)!} \frac{\partial^{(K+1)} u}{\partial p^{(K+1)}}(\tilde{p}, s) \cdot (p - \bar{p})^{(K+1)} \\ &\approx u(\bar{p}, s) + \sum_{k=1}^{K} \frac{1}{k!} \frac{\partial^k u}{\partial p^k}(\bar{p}, s) \cdot (p - \bar{p})^k \end{aligned} \tag{14}$$

where $\tilde{p}$ is a point between p and $\bar{p}$. Let μ_k be the system of k-th central moments of p. Given the moments of the stochastic variable p up to K-th order, $E_p u(p, s)$ can be calculated approximatively by

$$E_p u(p, s) \approx u(\bar{p}, s) + \sum_{k=1}^{K} \frac{1}{k!} \frac{\partial^k u}{\partial p^k}(\bar{p}, s) \cdot \mu_k . \tag{15}$$

3.3. Discritization

The solution of the substitute problems are the functions $q = q(s)$ and $\beta = \beta(s)$. Describing the functions $q_i = q_i(s)$ and $\beta = \beta(s)$ as linear combinations of known functions, $B_j(s)$, $1 \leq j \leq J$, and $B_j^0(s)$, $1 \leq j \leq J_0$,

$$q_i(s) = \sum_{j=1}^{J} \gamma_j^{(i)} B_j(s), \quad 1 \leq i \leq n \tag{16}$$

$$\beta(s) = \sum_{j=1}^{J_0} \gamma_j^{(0)} B_j^0(s), \tag{17}$$

and putting these into the equations (2)–(2.3) or (3)–(3.3), we get then a standard parameter optimization problem. This problem can then be solved by means of numerical optimization methods. In our situation we have made use of SQP-type algorithms.

The solution of the problem depends, of course, on the choice of the basis $B_j(s)$, $1 \leq j \leq J$, and $B_j^0(s)$, $1 \leq j \leq J_0$. Because of the well known excellent properties, we choose cubic and quadratic B-Splines $B_j(s)$ and $B_j^0(s)$

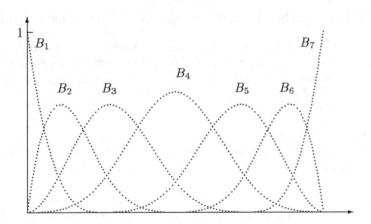

Fig. 1. Cubic B-Spline functions

The functions $B_j(s)$ and $B_j^0(s)$ have the following properties

1. $B_1(0) = 1$, $B_j(0) = 0$, $j = 2, ...J$
   $B_1^0(0) = 1$, $B_j^0(0) = 0$, $j = 2, ...J_0$

2. $B_J(s_e) = 1$, $B_j(s_e) = 0$, $j = 1, ...J - 1$
   $B_{J_0}^0(s_e) = 1$, $B_j^0(s_e) = 0$, $j = 1, ...J_0 - 1$

3. $B_j \geq 0$, $j = 1, \ldots J$ and $B_j^0 \geq 0$, $j = 1, \ldots J_0$

4. for every $s \in (0, s_e)$ there are at most 4 nonvanishing B_j and 3 nonvanishing B_j^0.

According to these properties, condition (8.2) or (9.2) will be fulfilled automatically by taking q_0, q_e, resp., as the first, last coefficient in the representation (16) of $q(s)$. Choosing the first and last coefficients of β as zero and the other nonnegative guarantees condition (8.1) and (9.1). The local Property 4 is very useful for the effective computation of the Jacobian matrix.

4. Applications on Manutec r3

To demonstrate our idea, we apply the method described above to solve the optimal trajectory planning problem for robot Manutec r3. The robot Manutec has in fact six rotary joints; the first three are mainly responsible for the position of the end effector and the last three for the orientation of the hand. To simplify our problem, the last three joints are fixed, so we have only three degrees of freedom. The dynamic equations and the model parameters can be found in [5]. We suppose that the payload m_l is a random variable being uniformly distributed on the interval $[0, 15]$.

4.1. Solution without Position and Velocity Restrictions

The time optimal problem is considered at first and the restrictions for the configuration coordinates and their derivatives are neglected. The initial position and the end position are chosen as $q_0 = (0, -1.5, 0)$ and $q_e = (1, -1.95, 1)$, and the robot should move the payload from the initial position to the end position as fast as possible.

This problem is solved for the reliabilities $\alpha_i = 0.99$, $\alpha_i = 0.75$, and we solve this problem also deterministically by replacing m_l by its expected value. The total run time are 0.5429 seconds for $\alpha_i = 0.99$, 0.5191 sec. for $\alpha_i = 0.75$ and 0.4925 sec. for the deterministic case, respectively. Obviously, the run time increases, if the stochastic uncertainty is considered. Moreover, we get a longer run time, if the constraints are required to hold with greater reliability α.

In the following figures we describe the configuration coordinates q_2 and q_3 as functions of q_1 and compare the results for three cases. It is shown that one get smaller values of q_2 and q_3 in case of stochastic uncertainty. And for higher reliability the functions q_2 and q_3 are decreased. Since q_2 is the angle between the upper arm and the vertical axis and q_3 is the angle between upper and lower arms, in this situation a smaller value of q_2 means a closer position of the elbow to the base of the robot and a larger value of q_3 leads to a higher position of the payload. This means that in order to guarantee a higher reliability, the robot tries to pull the elbow to itself and to prevent a too high position of the payload.

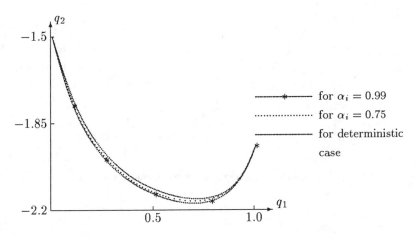

Fig. 2. Trajectories in configuration space without position restrictions

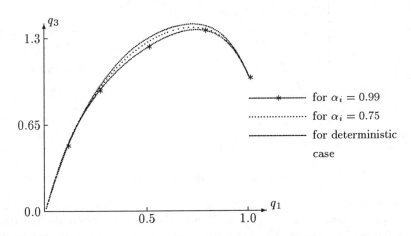

Fig. 3. Trajectories in configuration space without position restrictions

4.2. Solution with Position and Velocity Restrictions

Now let us consider the situation with position and velocity constraints. The bounds are also the same as given in [5]. The total perform time are 0.5715 seconds for $\alpha_i = 0.99$, 0.5501 sec. for $\alpha_i = 0.75$ and 0.5298 sec. for the deterministic case.

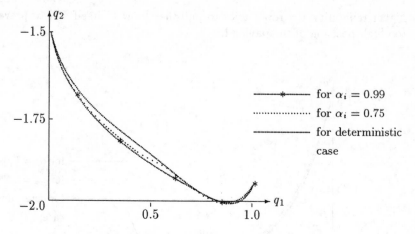

Fig. 4. Trajectories in configuration space under position restrictions

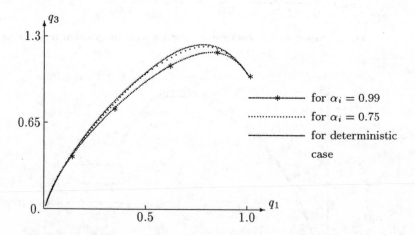

Fig. 5. Trajectories in configuration space under position restrictions

In the same way we show the configuration coordinates q_2 and q_3 as functions of q_1 in Figure 4 and Figure 5. Due to the restrictions on the position and velocity, the feasible domain for the optimal solution becomes smaller, hence, the optimal trajectories for different cases are closer to each other. But one can still see, for a higher reliability the robot tries to pull the elbow to itself and to prevent the payload from being held too high.

4.3. Planning Problem with a Mixed Objective Function

In next examples we study the influence of the weight factors in the objective function on the results of the planning problem and the computing time.

For the same initial and end positions as given above we solve the planning problem with $\kappa_t = 1$ and different values of κ_e. The results are shown in the following tables:

	$\kappa_e = 0$	$\kappa_e = 0.001$	$\kappa_e = 0.0001$
t_c	$920sec$	$145sec$	$374sec$
t_f	0.571528	0.578460	0.571605
e_f	38.72099	31.33955	37.81237
z_f	0.571528	0.609800	0.575386

Table 4.1. Stochastic case with $\alpha_i = 0.99$

	$\kappa_e = 0$	$\kappa_e = 0.001$	$\kappa_e = 0.0001$
t_c	$1358sec$	$120sec$	$397sec$
t_f	0.529847	0.541252	0.531638
e_f	48.61951	32.19380	45.90686
z_f	0.529847	0.573445	0.536229

Table 4.2. Deterministic case

where t_c means the computing time, and t_f, e_f and z_f are the total perform time, the energy consumption and the value of the objective function, respectively.

These results show that by an appropriate choice of κ_e the computing time may be reduced tremendously. To study this phenomenon, we perform the computations for different tasks (different initial and end positions) and get the following results, where Table 4.3 and Table 4.4 present the results for $q_0 = (0, -1.5, 0)$, $q_e = (1, -1.5, 0)$ and Table 4.5 and Table 4.6 for $q_0 = (0.228, 0.31, 1.75)$ and $q_e = (-0.147, 0.73, -1.82)$:

	$\kappa_e = 0$	$\kappa_e = 0.001$	$\kappa_e = 0.01$
t_c	$430sec$	$98sec$	$52sec$
t_f	0.607482	0.612947	0.644437
e_f	43.95982	33.64664	24.20151
z_f	0.607482	0.646594	0.886452

Table 4.3. Stochastic case with $\alpha_i = 0.99$

	$\kappa_e = 0$	$\kappa_e = 0.001$	$\kappa_e = 0.01$
t_c	$432sec$	$69sec$	$66sec$
t_f	0.559192	0.571750	0.611204
e_f	52.31024	31.35354	25.06793
z_f	0.559192	0.603104	0.861883

Table 4.4. Deterministic case

	$\kappa_e = 0$	$\kappa_e = 0.001$	$\kappa_e = 0.01$
t_c	$1075sec$	$497sec$	$143sec$
t_f	1.024160	1.033095	1.101216
e_f	57.78957	45.49765	23.56912
z_f	1.024160	1.078592	1.336907

Table 4.5. Stochastic case with $\alpha_i = 0.99$

	$\kappa_e = 0$	$\kappa_e = 0.001$	$\kappa_e = 0.01$
t_c	$925sec$	$230sec$	$163sec$
t_f	0.873024	0.884234	0.991281
e_f	72.14502	51.29047	23.95391
z_f	0.873024	0.935525	1.230821

Table 4.6. Deterministic case

From these results we can see that by means of an appropriate choice of κ_e, the computing time is cut down, while the performing time increases only slightly. The influence of the weight factors on the optimal trajectories is shown in the following:

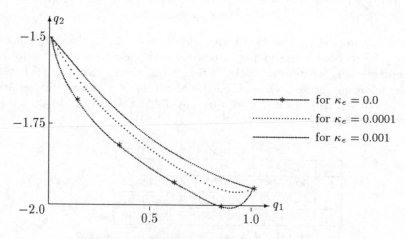

Fig. 6. Solutions in the q_2 and q_1 space for different weight factors

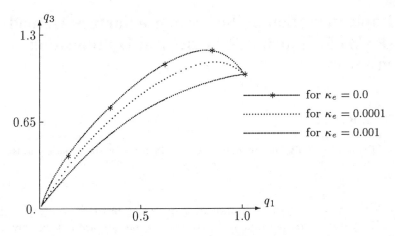

Fig. 7. Solutions in q_3 and q_1 space for different weight factors

These figures show that the robot try to prevent to move too violently, if the energy is also considered.

References

1 Fu, K.S. and Gonzalez, R.C.: Robotics, Control, Sensing, Vision and Intelligence, McGraw-Hill, New York, 1987
2 Marti, K. and Qu, S.: Optimal Trajectory Planning for Robot Considering Stochastic Parameters and Disturbances -Computation of an Efficient Open-Loop Strategy, J. Intelligent and Robotic Systems, Vol. 15, pp.19-23, 1996
3 Pfeiffer, F. and Johanni, R.: A Concept for Manipulator Trajectory Planning, IEEE J. Robot. Automat. RA-3(3), pp 115-123, 1987
4 Qu, S.: Optimale Bahnplanung unter Beruecksichtigung stochastischer Parameterschwankungen, VDI-Verlag, Duesseldorf, 1995
5 Tuerk, M.: Zur Modellierung der Dynamik von Robotern mit rotatorischen Gelenken, VDI-Verlag, Duesseldorf, 1990

Implementation of the Response Surface Method (RSM) for Stochastic Structural Optimization Problems

Dr. J. Reinhart [1]

[1]Fakultät für Luft– und Raumfahrttechnik, Institut für Mathematik und Rechneranwendungen, Universität der Bundeswehr München, D–85577 Neubiberg

Abstract. In structural optimization one often has the situation that some of the parameters like material data or load factors are not exactly known, but they have a well–known statistical behaviour. There exists a lot of possibilities to reformulate the minimization problem such that the stochastic variation of the parameters can be taken into account already in the planning phase. One way to solve the resulting so–called substituting problems is the *response surface method* (RSM) with an adaptive matrix step size control. To get a fast and stable implementation of this algorithm, the setting of some control parameters is very critical. The topic of this paper is to present RSM as a technique capable to solve stochastic optimization problems, but to show also the difficulties in the selection of the control parameters in a proper implementation of the procedure.

Keywords. stochastic optimization, response surface method, stochastic approximation, matrix stepsize control

1 Basic problem

An often used formulation for a stochastic optimization problem is

$$\min_{\underline{x} \in D} \ \mathcal{E} \, f(\underline{x}, \underline{a}(\omega)) \quad (=: F(\underline{x})), \tag{1}$$

where the design variables $\underline{x} \in \mathbb{R}^r$ lie in a relatively simple region D and $\underline{a} \in \mathbb{R}^s$ is a vector of independently distributed random variables. There exists an abstract probability space Ω and a corresponding probability measure P such that the expectation $\mathcal{E}$ in (1) is well defined and finite. Examples for regions D, defined by simple constraints, are:

- box constraints, i.e., $D = \{\underline{x} \in \mathbb{R}^r \mid \underline{x}_l \leq \underline{x} \leq \underline{x}_u\}$ or
- linear "deterministic" restrictions, which means that no expensive evaluation of the expectation is needed within the constraints, i.e.,

$$D = \{\underline{x} \in \mathbb{R}^r \mid \underline{x}_l \leq \underline{x} \leq \underline{x}_u \text{ and } A\underline{x} \geq \underline{b}\}.$$

2 Solution method

The presented algorithm in principal is a stochastic approximation method, which uses estimations of first and second order derivatives of the object-ive function F. Deterministic constraints are taken into account by using a projection operator, such that the main rule of the algorithm is

$$\underline{x}_{n+1} = \pi_D(\underline{x}_n - \rho_n \underline{S}_n), \quad n = 1, 2, \ldots, \tag{2}$$

where $\underline{x}_1 \in D$ is chosen arbitrarily. In (2), $\rho_n > 0$ denotes the step size and $\underline{S}_n \in \mathbb{R}^r$ the search direction.

To get feasible iteration points, in every iteration step, a simple optimization problem has to be solved, namely

$$\pi_D(\underline{y}) := \arg\min_{\underline{x} \in D} \|\underline{y} - \underline{x}\|^2. \tag{3}$$

The projection is a relatively inexpensive operation because of the assumed simple structure of D. In the following we consider appropriate selections of the step size ρ_n and the search direction $\underline{S}_n$.

2.1 Gradient approximation by using RSM

The basic idea of stochastic approximation methods is that the whole iteration process is based on *estimates* of functions and derivatives. Thus, a gradient approximation method on the basis of only a few function evaluations of f is needed. One method of that type is the well–known response surface method (RSM).

Following the Taylor expansion of F in $\underline{\hat{x}}$,

$$F(\underline{x}) = \sum_{k=0}^{\nu} \frac{1}{k!} D_{\underline{x}-\underline{\hat{x}}}^{(k)} F(\underline{\hat{x}}) + R^{(\nu)}(\underline{x}, \underline{\hat{x}}), \tag{4}$$

where

$$D_{\underline{h}}^{(k)} F(\underline{\hat{x}}) := \left. \frac{\partial^k F(\underline{\hat{x}} + t \cdot \underline{h})}{\partial t^k} \right|_{t=0}, \quad k = 0, 1, \ldots, \tag{5}$$

and

$$R^{(\nu)}(\underline{x}, \underline{\hat{x}}) := \frac{1}{(\nu + 1)!} \cdot \left. \frac{\partial^{\nu+1} F(\underline{\hat{x}} + t (\underline{x} - \underline{\hat{x}}))}{\partial t^{\nu+1}} \right|_{t=\delta} \quad \text{with } \delta \in (0, 1), \tag{6}$$

a symmetrical polynomial model for F is used to approximate the behaviour of F near $\underline{\hat{x}}$:

$$\widehat{F}_\nu(\underline{x}) := \sum_{k=0}^{\nu} \frac{1}{k!} \sum_{i_1, \ldots, i_k=1}^{r} b_{i_1 \cdots i_k} \cdot (x_{i_1} - \hat{x}_{i_1}) \cdots (x_{i_k} - \hat{x}_{i_k}) \tag{7}$$

Then, a suitable gradient estimation is just the gradient of the model function $\widehat{F}_\nu$ in $\widehat{\underline{x}}$:

$$\widehat{\nabla} F := \nabla \widehat{F}_\nu(\widehat{\underline{x}}) \approx \nabla F(\widehat{\underline{x}}). \tag{8}$$

To estimate values for the unknown model parameters $b_{i_1 \cdots i_k}$, unbiased estimations of F are taken at p design points $\underline{x}^{(i)}$, where $p \geq \binom{r+\nu}{\nu} =: r_\nu$ which is the number of involved parameters up to the νth degree. An important example is given by

$$u^{(i)} := \frac{1}{K} \sum_{l=1}^{K} f(\underline{x}^{(i)}, \underline{a}(\omega^{(i,l)})) \approx F(\underline{x}^{(i)}), \quad i = 1, \ldots, p, \tag{9}$$

where $\underline{a}(\omega^{(i,l)})$ are independent realizations of the stochastic parameter vector. According to those estimations, the model function $\widehat{F}_\nu$ is fitted by the least square estimation method:

$$\min_{b_{i_1 \cdots i_k}} \sum_{i=1}^{p} (u^{(i)} - \widehat{F}_\nu(\underline{x}^{(i)}))^2. \tag{10}$$

It is possible to write the resulting gradient approximation as the product of a certain matrix L_ν and the vector of function estimations $\underline{u} = (u^{(1)}, .., u^{(p)})^T$, i. e.

$$\widehat{\nabla} F = L_\nu \underline{u}. \tag{11}$$

The involved matrix L_ν contains some combinations of the difference vectors $\underline{d}^{(i)} := \underline{x}^{(i)} - \widehat{\underline{x}}, i = 1, \ldots, p$. For an explicit formula, cf. [5].

2.2 Error analysis

Under some standard assumptions on the function estimations $\underline{u}$ defined in (9), the mean approximation error for the gradient can be written as

$$\underline{e}^{det} := \mathcal{E}(\widehat{\nabla} F - \nabla F(\widehat{\underline{x}})) = L_\nu \underline{R}_\nu \tag{12}$$

where $\underline{R}_\nu := (R^{(\nu)}(\underline{x}^{(1)}, \widehat{\underline{x}}), \ldots, R^{(\nu)}(\underline{x}^{(p)}, \widehat{\underline{x}}))^T$ is the vector of remainders in the Taylor expansion (6).

A measure for the stochastic approximation error is the trace of the covariance matrix of the difference between gradient and gradient approximation:

$$e^{stoch} := \mathrm{tr}\,[\mathcal{C}ov\,(\widehat{\nabla} F - \nabla F(\widehat{\underline{x}}))] \tag{13}$$

If

$$\sigma_i^2 := \mathcal{E}\,(f(\underline{x}^{(i)}, \underline{a}(\omega)) - F(\underline{x}^{(i)}))^2, \quad i = 1, \ldots, p, \tag{14}$$

we have that

$$e^{stoch} = \mathrm{tr}\,[\,L_\nu \cdot diag(\frac{\sigma_i^2}{K}) \cdot L_\nu^T\,]. \tag{15}$$

Hence for the deterministic error we find

$$\|\underline{e}^{det}\|^2 = \|L_\nu \cdot \underline{R}_\nu\|^2 \leq \|L_\nu L_\nu^T\| \cdot \|\underline{R}_\nu\|^2 \tag{16}$$

$$\leq \|L_\nu L_\nu^T\| \cdot \frac{1}{((\nu+1)!)^2} \sup_{\underline{x}\in\widehat{D}} \|\nabla^{\nu+1}F(\underline{x})\|^2 \sum_{i=1}^{p} \|\underline{d}^{(i)}\|^{2(\nu+1)}$$

with $\widehat{D} = \{\widehat{\underline{x}} + t\underline{x}^{(i)}, i \in \{1,\ldots,p\}, t \in [0,1]\} \cap D$, and an upper bound for the stochastic error is given by

$$|e^{stoch}| \leq \|L_\nu L_\nu^T\| \cdot \frac{r}{K} \max_{i=1,\ldots,p} \sigma_i^2; \tag{17}$$

note that both error estimates contain the factor $\|L_\nu L_\nu^T\|$.

Now the main goal of an efficient algorithm is to decrease these two bounds, (16), (17), simultaneously, step by step. This task is related to the research field called *experimental design* [1].

2.3 Experimental design

The accuracy of the gradient approximation mainly depends on the position of the design points $\underline{x}^{(i)}, i = 1,\ldots,p$ with respect to $\widehat{\underline{x}}$. First we assume that the distance of the design points to $\widehat{\underline{x}}$ is fixed. Then there are, dependent on the degree ν of the model function, lower bounds r_ν for the number p of design points. According to this number there exist a few possibilities to choose "good" design directions $\underline{d}^{(i)}$.

If we choose a **linear** approximation, i. e. $\nu = 1$, then a *regular simplex* (Fig. 1) is the best solution in a certain sense, but *(fractional) factorial designs* (Fig. 2) are good selections, too. Especially, if we change during the iteration process to a **quadratic** approach, then we can use some results from the linear approximation, for example in a *central composite design* (Fig. 4).

Usually p is greater than r_ν, but the additional design points could be used, for example, to decide whether a linear or quadratic approach is suitable. Another possible design is the so called *Koshal design* (Fig. 3). Here the number of design points equals exactly the lower bound, but the design neither is orthogonal nor rotational, which are two important features of a suitable design.

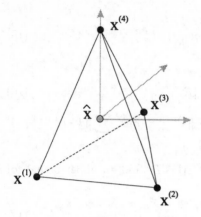

Figure 1: Regular simplex

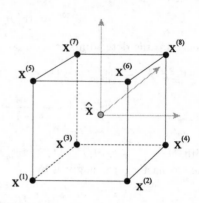

Figure 2: Factorial design

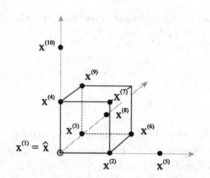

Figure 3: Koshal design

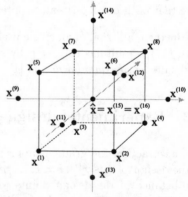

Figure 4: Central composite design

2.4 Influence of the distance $\|\underline{x}^{(i)} - \widehat{\underline{x}}\|$

Suppose that the design vector $\underline{d}^{(i)} := \underline{x}^{(i)} - \widehat{\underline{x}}$ is replaced by $\widetilde{\underline{d}}^{(i)} := \mu \cdot \underline{d}^{(i)}$, $\mu > 0$, such that the new design points are

$$\widetilde{\underline{x}}^{(i)} = \widehat{\underline{x}} + \widetilde{\underline{d}}^{(i)} = \widehat{\underline{x}} + \mu\,(\underline{x}^{(i)} - \widehat{\underline{x}}) = \mu\underline{x}^{(i)} + (1 - \mu)\widehat{\underline{x}}. \tag{18}$$

Then it is easy to prove that

$$\widetilde{L}_\nu = \frac{1}{\mu}\,L_\nu \tag{19}$$

and therefore

$$\|\widetilde{L}_\nu \widetilde{L}_\nu^T\| = \frac{1}{\mu^2}\,\|L_\nu L_\nu^T\|, \tag{20}$$

where $\widetilde{L}_\nu$ results from L_ν by $\underline{d}^{(i)} \to \widetilde{\underline{d}}^{(i)}, i = 1, \ldots, p$.

According to the new design points there is also a change in the bound for the remainder term in the Taylor expansion, cf. (6) and (16),

$$\|\widetilde{\underline{R}}_\nu\|^2 \le \mu^{2(\nu+1)} \cdot \frac{1}{((\nu+1)!)^2} \sup_{\underline{x} \in \mu \widehat{D}} \|\nabla^{\nu+1} F(\underline{x})\|^2 \sum_{i=1}^{p} \|\underline{d}^{(i)}\|^{2(\nu+1)}. \quad (21)$$

Consequently, (20), (21), resp., yield a new bound for the deterministic error:

$$\|\widetilde{\underline{e}}^{det}\|^2 \le \mu^{2\nu} \cdot \|L_\nu L_\nu^T\| \cdot \sup_{\underline{x} \in \mu \widehat{D}} \|\nabla^{\nu+1} F(\underline{x})\|^2 \frac{1}{((\nu+1)!)^2} \sum_{i=1}^{p} \|\underline{d}^{(i)}\|^{2(\nu+1)}. \quad (22)$$

There is no principal rule, how the factor σ_i^2 (see (14)) is influenced by reducing or enlarging the distance of $\underline{x}^{(i)}$ relative to $\widehat{\underline{x}}$. Thus, the new bound for the stochastic approximation error is given by

$$|\widetilde{\underline{e}}^{stoch}| \le \frac{1}{\mu^2} \|L_\nu L_\nu^T\| \cdot \frac{r}{K} \max_{i=1,\ldots,p} \widetilde{\sigma}_i^2, \quad (23)$$

where $\widetilde{\sigma}_i^2$ is an upper bound of the variance σ_i^2 under consideration. Regarding the inequalities (22) and (23), there is a discrepancy between the behaviour of the deterministic and stochastic approximation error, resp., with respect to a reduction of the factor $\mu < 1$; while the deterministic error is lowered, the stochastic error is enlarged. A factor $\mu > 1$ would cause the contrary.

2.5 Hessian approximation

An essential feature of the present algorithm is, that it makes use of second order information, too. Hence, it is necessary to approximate also the Hessian of the objective function F, again based on only a few function estimations.

Starting with a positive definite matrix H_0, the matrix $H_n \approx \nabla^2 F(\underline{x}_n)$ is updated by

$$
\begin{aligned}
H_n^{(0)} &:= H_{n-1} \\
H_n^{(i)} &:= H_n^{(i-1)} + \frac{1}{n} (\underline{h}_n^{(i)} - H_n^{(i-1)} \underline{e}_{l(n,i)}) \underline{e}_{l(n,i)}^T, \quad i = 1, \ldots, l_n, \\
H_n &:= H_n^{(l_n)}
\end{aligned}
\quad (24)
$$

using the *stochastic central differential quotient*

$$\underline{h}_n^{(i)} := \frac{1}{2 c_n} (\underline{Y}_n^{(i)} - \underline{Y}_n^{(i+l_n)}), \quad i = 1, \ldots, l_n, \quad (25)$$

where
$$\underline{Y}_n^{(i)} := \widehat{\nabla} F(\widehat{\underline{x}}_n^{(i)}), \quad i = 1, \ldots, 2l_n, \tag{26}$$
are approximations of the gradient of F in special points, namely
$$\begin{aligned}
\widehat{\underline{x}}_n^{(i)} &:= \underline{x}_n + c_n \, \underline{e}_{l(n,i)}, \\
\widehat{\underline{x}}_n^{(i+l_n)} &:= \underline{x}_n - c_n \, \underline{e}_{l(n,i)}, \quad i = 1, \ldots, l_n,
\end{aligned} \tag{27}$$
where

- $l(n, i), i = 1, \ldots, l_n, \ n = 1, 2, \ldots$ runs through $\{1, \ldots, r\}$ cyclically,
- $\underline{e}_k \in \mathbb{R}^r, k = 1, \ldots, r$, denotes the k-th unit vector,
- and c_n is defined by $c_n := \dfrac{c_0}{n^\mu}$ with certain $c_0 > 0$ and μ.

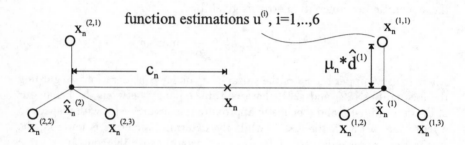

Figure 5: Design points for Hessian approximation

According to (8), we get the approximations $\underline{Y}_n^{(i)}$ of the gradient using the design points, cf. (9),
$$\underline{x}^{(i,j)} := \widehat{\underline{x}}^{(i)} + \mu_n \cdot \widehat{\underline{d}}^{(j)}, \quad i = 1, \ldots, 2l_n, j = 1, \ldots, p, \tag{28}$$
with fixed design vectors $\widehat{\underline{d}}^{(j)}$ and a factor $\mu_n := \dfrac{\mu_0}{n^{\frac{2\mu}{\nu}}}$.

In Fig. 5 the situation is shown for $r = 2$, $l_n = 1$ and $p = 3$. Thus, using a regular simplex design in every iteration step, the function f has to be evaluated $6 \cdot K_n$-times.

To prove convergence, the approximation of the expectation in (9) has to become more accurate, if the iteration number grows. Thus, the number K_n of function evaluations at each design point $\underline{x}^{(i,j)}$ is chosen in the following way
$$K_n := \lceil K_0 \cdot n^{\frac{4\mu}{\nu}} \rceil, \tag{29}$$
with a suitable initial value K_0.

2.6 Implementation of the algorithm

During the development of a suitable stochastic approximation method which makes use of second order information, it became clear, that the quality of a Hessian approximation isn't very high in the starting phase of the algorithm. Thus, a partition into several phases is necessary, in order to get a robust method.

2.6.1 Phase 1

There is only little information about the objective function during the first few iterations. Hence, a simple gradient step with a plain step size rule is sufficient: It holds

$$
\begin{aligned}
\underline{S}_n &:= \widehat{\nabla} F_n \quad \text{(gradient approximation, acc. to Section 2.1)} \\
\rho_n &:= \frac{\rho_0}{n} \quad \text{with a suitable initial step size } \rho_0.
\end{aligned}
\tag{30}
$$

We change to the second phase, if either a fixed number n_1 of iterations is achieved or the relative distance of two successive iteration points is smaller than a given value ϵ_1.

2.6.2 Phase 2

After some iterations within the first phase, it is more probable, that the iteration points lie already in a certain neighbourhood of the optimum. Hence, one can try adaptive step size rules. At that stage of the algorithm, the Hessian is approximated for the first time. Therefore the gradient approximations in $\widehat{\underline{x}}_n^{(i)}, i = 1, \ldots, 2l_n$, cf. (26), are sufficient for a gradient approximation in $\underline{x}_n$. We choose

$$
\begin{aligned}
\underline{S}_n &:= \frac{1}{2l_n} \sum_{i=1}^{2l_n} \underline{Y}_n^{(i)} \\
\rho_n &:= \widehat{\rho}_{n-1} Q_n (\widehat{\rho}_{n-1}, \delta_n, \Gamma_n) \frac{\widehat{\zeta_{n-1}}}{\zeta_n} \eta_n.
\end{aligned}
\tag{31}
$$

In the step size rule we have $\widehat{\rho}_{n-1} := \min\{\overline{r}_n, \rho_{n-1}\}$, e.g. with $\overline{r}_n = \overline{q}\frac{\delta_n}{\Gamma_n}$ and an arbitrary $\overline{q} \in (0, 1)$. Furthermore, a convergence factor Q_n is selected according to

$$
\begin{aligned}
Q(\cdot, \delta, \Gamma) : [0, \overline{q}\frac{\delta}{\Gamma}] &\to (0, 1] \\
r &\to \frac{1}{1 + r\delta - r^2\Gamma};
\end{aligned}
\tag{32}
$$

here, second order information is used in form of estimations of the eigenvalues of the approximation H_n of the Hessian, namely $\delta_n \approx 2\lambda_{\min}(H_n)$,

$$\delta_n := \begin{cases} \dfrac{2}{\text{tr}[H_n^{-1}]}, & \text{if } \det(H_n) > 0, \text{ tr}[H_n] > 0, \text{and } \delta_n \in [\underline{\delta}, \overline{\delta}] \\ \delta_0, & \text{otherwise,} \end{cases} \tag{33}$$

and $\Gamma_n \approx \lambda_{\max}^2(H_n)$, e.g.

$$\Gamma_n := \begin{cases} (\text{tr}[H_n])^2, & \text{if } \Gamma_n \leq \overline{\Gamma} \\ \Gamma_0, & \text{otherwise.} \end{cases} \tag{34}$$

A further part of the step size rule is an estimate of the stochastic error

$$\widehat{\zeta}_n := \begin{cases} \zeta_l, & \text{if } \zeta_n < \zeta_l \\ \zeta_n, & \text{if } \zeta_n \in [\zeta_l, \zeta_u] \\ \zeta_u, & \text{if } \zeta_n > \zeta_u, \end{cases} \tag{35}$$

with

$$\zeta_n := \frac{n-1}{n}\zeta_{n-1} + \frac{1}{n}\|\frac{1}{2l_n}\sum_{i=1}^{l_n}(\underline{Y}_n^{(i)} - \underline{Y}_n^{(i+l_n)})\|^2, \tag{36}$$

and finally we have a control parameter $\eta_n := \exp(\kappa_n/n^2)$, where κ_n is given e. g. by

$$\kappa_n := \langle \underline{x}_n - \underline{x}_{n-1}, \underline{x}_{n-1} - \underline{x}_{n-2}\rangle. \tag{37}$$

Hence, a zigzag–course leads to small values of η_n and straight courses yield larger step sizes. We change to the third phase, if both

$$k_{\text{change}} := \text{number of changes from Phase 3 back to Phase 2} \tag{38}$$

doesn't exceed a fixed number $k_{\text{change}}^{\max}$ and the relative difference between two successive Hessian approximations is less than a given value ϵ_2, because then we can suppose that the approximation is sufficiently stable to use H_n directly in the search direction.

2.6.3 Phase 3

As already mentioned, now the search direction is directly based on H_n, such that we get a kind of quasi–newton search directions. In this case the convergence proof requires that

$$\lim_{n\to\infty} n\rho_n = 1, \quad \text{a.s.}$$

Hence, we get

$$\begin{aligned} \underline{S}_n &:= \widehat{H}_n^{-1}\frac{1}{2l_n}\sum_{i=1}^{2l_n}\underline{Y}_n^{(i)} \\ \rho_n &:= \frac{\rho_{n-1}}{1+\rho_{n-1}}. \end{aligned} \tag{39}$$

For the sake of robustness we use the approximation H_n of the Hessian only if it fulfills a certain regularity condition, i. e.

$$\hat{H}_n := \begin{cases} H_n, & \text{if } H_n \in \mathcal{H}_0 \\ H_0, & \text{otherwise} \end{cases} \tag{40}$$

where e.g.

$$\mathcal{H}_0 := \{H \in \mathbb{R}^{r \times r} \mid \det(H) \neq 0 \text{ and } r^2 \max_{1 \leq i,k \leq r} ((H^{-1})_{ik})^2 < \frac{1}{\alpha_0^2}\} \tag{41}$$

with a prescribed number α_0.

If the optimum of problem (1) lies on the boundary of the feasible region D, then there exists a convergence proof only if the search direction is a gradient approximation. Thus, we change back to the second phase, if $\underline{x}_n \in \partial D$.

3 Conditions for convergence and parameter selection

There are a lot of conditions, which have to be fulfilled for guaranteeing convergence of the algorithm. Some of them are standard in optimization theory, some of them are problem dependent, as the proper selection of lower and upper bounds for the estimation of the eigenvalues of the Hessian, cf. (33). A complete list of convergence conditions is given in [5]. Here, only the most important are quoted:

- $\lim\limits_{n \to \infty} \sqrt{n}\, c_n^{1+\overline{\mu}} = 0$ with a problem dependent constant $\overline{\mu} > 0$,

- $\lim\limits_{n \to \infty} \sqrt{n}\, \mu_n^{\nu} = 0$ for the convergence of the deterministic error and

- $\dfrac{1}{\mu_n^2 K_n} \leq C < \infty.$ to bound the stochastic error.

These conditions prescribe a special selection of c_n, μ_n, K_n, resp. A suitable choice is given by

- $c_n := \dfrac{c_0}{n^\mu}$ with $\max\{\dfrac{1}{4}, \dfrac{1}{2(1+\overline{\mu})}\} < \mu < \dfrac{1}{2}$

- $\mu_n := \dfrac{\mu_0}{n^{\frac{2\mu}{\nu}}}$

- $K_n := \lceil K_0 \cdot n^{\frac{4\mu}{\nu}} \rceil,$

which contain the tuning parameters c_0, μ_0 and K_0. An important result from the research in this field is, that the iteration course is greatly influenced by the choice of those parameters. In the next sections some hints for the proper selection of the tuning parameters are given.

3.1 Initial step size ρ_0

From the two contourplots in Fig. 6, Fig. 7, resp., the consequences of an inappropriate selection of the initial step size ρ_0 in Phase 1 can be seen. To

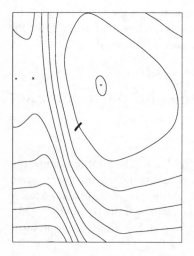

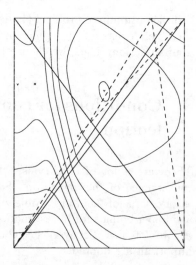

Figure 6: Initial step size ρ_0 too small

Figure 7: Initial step size ρ_0 too large

get a proper choice, a first condition has to be fulfilled, which is not very restrictive. Namely, the bounds for the box constraints $\underline{x}_l$ and $\underline{x}_u$ have to be selected such that the distance between the optimum $\underline{x}^*$ and the initial point $\underline{x}_1$ is of the same magnitude as the distance between $\underline{x}_l$ and $\underline{x}_u$. This supposition is justified by the fact that the initial point of the stochastic optimization problem often results from a previous deterministic optimization run. Hence, we define

$$\rho_0 := \widehat{\rho}_0 \cdot \min_{1 \leq i \leq r} \frac{(\underline{x}_u - \underline{x}_l)_i}{|(\underline{S}_1)_i|} \tag{42}$$

with $\widehat{\rho}_0 \approx 0.2$ and $\underline{S}_1$ defined in (30). Then, it holds

$$|(\underline{x}_2 - \underline{x}_1)_i| = |(\rho_0 \cdot \underline{S}_1)_i| \leq \widehat{\rho}_0 \cdot (\underline{x}_u - \underline{x}_l)_i, \tag{43}$$

hence, that we may expect that $\underline{x}_2$ doesn't leave the box.

A similar idea could be found in [2], but there one has no distinction between the different components, which, however, allows the handling of rectangles with great width–to–height relations.

3.2 The parameter μ_0

A satisfactory iteration run requires good gradient approximations above all. Hence, a very important parameter is the length of the design vector $\underline{d}^{(i)}$, $i = 1, \ldots, p$, cf. Sect. 2.4, used for the RSM–estimation. An obvious aspect is that μ_0 has to be chosen relatively to the distance between $\underline{x}_l$ and $\underline{x}_u$:

$$\mu_0 := \widehat{\mu}_0 \, \|\underline{x}_u - \underline{x}_l\| \quad \text{with } \widehat{\mu}_0 \approx 0.05. \tag{44}$$

In contrast to deterministic algorithms, the relative distance $\widehat{\mu}_0$ within stochastic approximation methods has to lie between about 0.01 and 0.5 and not around values like 10^{-4}.

Another important result is obtained from Fig. 8, which shows the relative error of the first gradient approximation as a function of $\widehat{\mu}_0$ for different

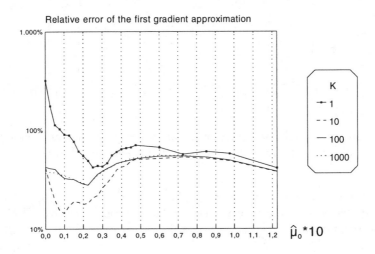

Figure 8: Influence of $\widehat{\mu}_0$ on the approximation error for different values of K

numbers K of realizations of the random variables. The difference between the given curves is large on the left hand side and small on the right hand side.

This reflects the portion of the deterministic, cf. (12), and the stochastic error, cf. (13), in the total approximation error. Greater values of $\widehat{\mu}_0$ reduce the stochastic error, as mentioned in (23), such that on the right hand side mainly the deterministic error is figured, which is independent of K. Lower values of $\widehat{\mu}_0$ yield a lower deterministic error, cf. (22), and enlarge the stochastic error, which causes the differences of the shown curves with respect to K on the left side of Fig. 8.

Fig. 9 represents the influence of the variance of the involved stochastic parameters on the relative approximation error. The main result of our investigation is that a greater variational coefficient induces a lower determin-

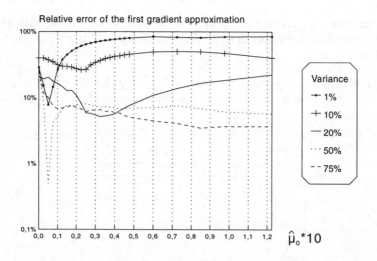

Figure 9: Influence of $\widehat{\mu}_0$ on the approximation error according to different variational coefficients of the stochastic parameters

istic error, which is shown on the right hand side of the picture. This effect surely depends on the special basic problem, which contains a probabilistic objective function.

3.3 The parameter c_0

After the proper determination of the increment for the gradient approximation, the position of the design points used for the Hessian approximation is investigated. Here, the same argument holds concerning a selection of c_n in

(27) relative to the distance between $\underline{x}_l$ and $\underline{x}_u$:

$$c_0 := \widehat{c}_0 \, \|\underline{x}_u - \underline{x}_l\| \quad \text{with } \widehat{c}_0 \approx 0.1 \, . \tag{45}$$

A principal rule for the choice of $\widehat{c}_0$ is that $\widehat{c}_0$ should be greater than $\widehat{\mu}_0$. Before starting the investigation of the proper selection of $\widehat{c}_0$, a measure for the quality of the parameter has to be defined. One possibility is to do one step with a pure quasi–newton–like search direction and then to consider the resulting deviation to the optimum. This was done in Fig. 10, where the influence of the number of realizations K on the selection of $\widehat{c}_0$ is shown. As seen in Fig. 8,

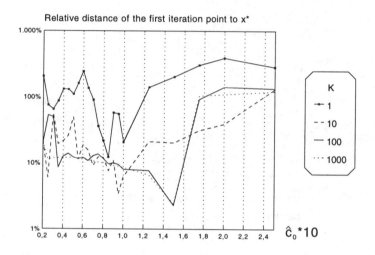

Figure 10: Influence of $\widehat{c}_0$ on the relative distance of the next iteration point to the optimum according to different values of K

we again observe smaller differences on the right hand side of the figure, which represents the deterministic approximation error, and shaky behaviour on the left hand side, caused by greater stochastic errors. Comparing the two figures, Fig. 8 and Fig. 10, we observe a greater sensitivity towards the selection of $\widehat{c}_0$ than towards $\widehat{\mu}_0$.

3.4 Comparison of RSM with other methods

Finally, we compare our algorithm with other well–known methods. The underlying problem is based on a fibre reinforced structure. The deterministic task was to minimize the total thickness under some failure criteria. The

substituting problem, which considers the stochastic character of some parameters, is of the type (1), where the objective f is a weighted sum of the
original objective function and the loss due to a violation of the failure restrictions. The penalty functions taken into account lead to formulations,
which contain a probability function; hence a great number of structural analyses is required to approximate the objective function F of the substituting
problem.

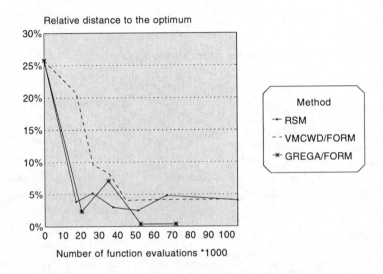

Figure 11: Comparison of different methods for solving stochastic
structural optimization problems

In Fig. 11 the relative distance of the last iteration point to the optimum is
plotted against the maximum allowed number of evaluations of the function f.
Three different methods are compared. We can conclude, that none of these
algorithms is really better than anyone of the others. The advantage of RSM,
however, is that the accuracy of the appproximation becomes better during
the iteration process, while the other methods, approximating the objective
function with a Taylor polynom, produce always produce an unknown break–
off error.

4 Summary

A new stochastic approximation method is presented, which makes use of the response surface technique to estimate the gradient of the objective function. After a short description of an appropriate selection of so–called design points, search directions and step size rules of the algorithm are established, which are based on second order information. The necessary partition of robust algorithms into several phases is explained, and conditions for phase changes are specified. A detailed error analysis and the resulting conditions of convergence lead to explicit formulas for some of the control parameters. Those parameters themselves are based on some tuning parameters, which greatly influence the convergence rate. This is figured out in some diagrams. An appropriate adaptive choice for those parameters is presented. Last but not least, the new implementation is compared to other methods for solving stochastic structural optimization problems. This comparison shows no great differences, what proves, that RSM is an adequate method for solving problems of this type, provided that the involved parameters are selected suitably.

References

[1] Box, G.E.P. and Draper, N.R.: *Empirical Model-Building and Response Surfaces*. Wiley Series in Probability and Mathematical Statistics. John Wiley & Sons, New York, Chichester, Brisbane, Toronto, Singapore, 1987.

[2] Ermoliev, Yu. and Wets, R. J.-B., editors: *Numerical Techniques for Stochastic Optimization*, volume 10 of *Springer Series in Computational Mathemetics*. Springer–Verlag, Heidelberg, 1988.

[3] Marti, K.: Semi–Stochastic Approximation by the Response Surface Methodology (RSM). *Optimization*, 25:209–230, 1992.

[4] Plöchinger, E.: *Realisierung von adaptiven Schrittweiten für stochastische Approximationsverfahren bei unterschiedlichem Varianzverhalten des Schätzfehlers*. Dissertation, Fakultät für Informatik der Universität der Bundeswehr München, Neubiberg, 1991.

[5] Reinhart, J.: *Stochastische Optimierung von Faserkunststoffverbundplatten*. Adaptive stochastische Approximationsverfahren auf der Basis der Response–Surface–Methode. Dissertation, Fakultät für LRT der Universität der Bundeswehr München, Neubiberg, 1996.

Optimization of an Engine Air Intake System for Minimum Noise Tansmission Using Function Approximation Concepts

M.H. van Houten , A.J.G. Schoofs and D.H. van Campen

Department of Mechanical Engineering - Stevin Centre, Eindhoven University of Technology, P.O.Box 513, Eindhoven, The Netherlands

Abstract. Response surface methods are used to construct explicit approximations of objective and constraint functions in optimization problems. Serving as interface between analysis code and optimization algorithm, the function approximations provide a fast but approximate evaluation of response quantities in the optimization process. Application to the quiet design of an engine air intake system is discussed.

Keywords. Response surface, function approximation, acoustic optimization, air intake system

1 Introduction

In structural optimization, often an approximation concept is introduced as interface between structural analysis code and optimization algorithm. Then approximation functions are built of all objective and constraint functions, based upon function values and, if available, sensitivities which follow from structural analysis calculations. In this way, the original optimization problem is replaced by an explicitly known approximate optimization problem.

Local, global and mid-range approximation concepts can be distinguished. Local function approximations, which are most commonly applied, are based upon the function values and sensitivities in a single point of the design space. The approximate objective function and constraints then define an optimization subproblem. Since local approximations are only valid in a limited area of the design space, a new cycle of approximation and optimization can be started at the optimum of the subproblem. This procedure is repeated until an acceptable optimum is achieved, resulting in the popular process which is indicated as sequential approximate optimization. For some optimum design applications, it is profitable to build global or mid-range approximations, which are based on structural analysis results in multiple design points. Global approximations are valid in the whole design space, whereas mid-range approximations are valid in a substantially smaller part of the design space, but this part is expected to be larger than in the case of local approximations. In the optimization process global and mid-range approximations are used in a different way. After the global approximate model has been built, it is passed once to the optimization module to try and find an optimum design. Mid-range approximations are used in a sequential approximate optimization process, in a similar way as described above for local approximations. This paper will be restricted to global and mid-range approximation models.

For building the function approximation models several methods are available. In chapter 2 the so-called response-surface method will be explained. The application of this method to construct global and mid-range approximations will be illustrated with an example of the optimization of an engine air intake system.

2 Function approximation concepts

2.1 Global function approximation concept

Response-surface modelling is a powerful tool to build global approximate models. These strategies were originally developed for the model fitting of physical experiments [1] but can be applied successfully in multidisciplinary optimization.

Construction of response-surface models is an iterative process. One starts with postulating the approximate model functions. For this, the designer must know to some extent which variables play a role and which form is suitable to describe the relation between design variables and responses. A parametric study in the initial design phase may be helpful for deriving this relation. Usually one starts with simple function models to reduce initial computation costs, e.g. first or second order polynomials.

Once the model functions are chosen, an experimental design is determined, containing the points in the design space for which numerical experiments must be carried out. In principle all real design variable values within the design variable bounds are allowed to be chosen. For the purpose of efficiency, however, only a very limited number of discrete values, called levels, of every design variable are chosen. Many experimental design methods exist in literature, of which factorial designs are most often used.

In a full factorial design on each possible combination of levels of design variables a design point is placed (Fig. 1). The choice of the number of levels for a certain design variable depends on the order of the variable in the assumed approximation model. A linear effect can be estimated by

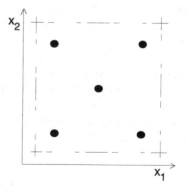

Fig. 1. Full factorial design for 2 design variables with centre point.

means of at least two levels, a quadratic effect needs at least three levels, etc.. Therefore increasing the number of levels or variables, the total number of analyses can become excessively large and unmanageable. Full factorial designs are applicable when the response can be modelled by a simple approximation model or when a systematic search of the design region is needed. To limit the number of analyses, one useful approach is the addition of centre points. With only one extra design point three levels can be estimated.

If the number of design variables becomes large, the desired information can often be obtained by using a fraction of a complete full factorial design, called a fractional factorial design. However, not all combinations of variables can be estimated in this way. For example, for four design variables a fraction can be obtained by constructing a complete full factorial design for three variables. Values of the fourth variable are generated using a combination of the previous three variables. As a consequence of this, the fourth variable and the combination can not be estimated independently and one of them should be removed from the function model. As a result of applying fractional designs, two or more fractions (i.e. lists of design points) exist which form together a complete full factorial design. In the first run, one can use for example the first fraction and if one is not satisfied with the obtained approximation models, a second fraction can be analyzed, and so on.

Each experimental design method requires different knowledge about the problem. In any experimental design problem, a critical decision is the choice of sample size, that is, the number of analyses to be run. Generally, if one is interested in detecting local behaviour in a large part of the design space, more analyses are required than if one is only interested in the global behaviour. Also the complexity of the posed model functions and the number of design variables play an important role in selecting a suitable experimental design method. At least, with data obtained in the design points one must be able to estimate the unknown parameters of the approximation models, hence the number of design points should be equal or greater than the number of unknowns. It is advised to start with a simple model with a moderate number of parameters and requiring a moderate number of design points. In subsequent steps, gaining more knowledge about the behaviour of the responses to be approximated, extra design points can be added and more complex models can be used. It is important to note that all analyses remain valuable during this process.

After determining the design points in the design space, analyses are performed to gather data, i.e. response values and/or sensitivity values. This offers possibilities for parallel computing. When all information has been collected, the model functions are fitted to the results of the numerical experiments. In matrix notation:

$$y(x) = F\beta + \varepsilon \tag{1}$$

with responses:

$$y(x) = [y_1(x), \ldots, y_N(x)]^T \tag{2}$$

the model functions:

$$F = [f_1(x), \ldots, f_k(x)]^T \tag{3}$$

and unknown model parameters:

$$\beta = [\beta_1, \ldots, \beta_m]^T \tag{4}$$

and ε are the errors between calculated and approximated function values. N is the total number of responses, k is the number of function models and m the number of unknown parameters. The estimated model parameters β are calculated using a least squares approach:

$$\beta = (F^T F)^{-1} F^T y \tag{5}$$

The number of constraints in the problem can be large if for each constraint the same function type is used: the system matrix is decomposed only once and the unknown parameters are calculated by a multiplication with the right-hand-side containing the response values. Using more different functions types, computation time for the unknown parameters increases only slightly and will be in general much less than one analysis with an analysis code.

To reduce the number of analyses, sensitivity data may be used in the model fitting process. However, this sensitivity data is not always available or not always available at low costs. It even can disturb the global behaviour if the response is non-smooth or contains discontinuities and should not be used in this case. Since response quantities and their sensitivities are two entities with different (physical) dimensions, weighing factors must be used to express the importance of sensitivity data with respect to response data.

Evaluation of the approximation models is required to check their validity. One possibility is to look at the absolute and relative errors between the exact analyses and the responses according to the approximation models. Also statistical measures can be computed to get an indication of the validity of the models. But remember, in this context experiments are computer experiments and therefore errors in the response-surface models cannot be supposed to be random. Validation of the functions should occur with design points not included in the fit.

Statistician have developed a number of variable selection procedures to find the best subset among a large number of variables or combination of variables specified. These selection procedures give useful information about the relation between variables and responses and can serve as a guide in the function building process. One of the selection methods is called stepwise regression [2]. At each step, before determination of the next variable to be added, the statistics for significance F of the already chosen coefficients are examined to see if a variable elimination is applicable. If the variance contribution of a variable in the regression is insignificant at a specified F-level, this variable is removed from the regression. If no variable is to be removed, the procedure looks whether the variance reduction obtained by adding a variable to the regression is significant at a specified F-level. If so, this variable is entered into the regression. An important property of the stepwise regression procedure is that a variable which entered the regression, may be removed in a later stage when it becomes insignificant, therefore only significant variables are included in the final regression. The procedure stops when, for specified significance levels neither a selection nor a elimination of a variable is indicated. The final results of the procedure are not unique, as they are dependent on the choices of significance levels for addition and deletion.

If further model improvement is necessary, one performs another model building cycle consisting of design data collection, model fitting and testing. This is clearly not an automated process since the decisions depend on the users knowledge and his desire to control the process. If the resulting models describe the response behaviour accurate enough, they can be used as explicit problem functions in the optimization.

The response-surface method is suitable in an early design stage where the number of design variables is relatively small (i.e. 10 to 15) and for problems with non-smooth response behaviour. However, global approximation concepts are not only valuable for the preliminary design investigation. After an optimum has been found, global approximation models build around the optimum can be used to investigate changes in the optimization problem specifications like changes in design variables or constraint bounds without the need to run the analysis code once again. Another important feature is that noise and other irregularities can be averaged out through the smoothing capabilities of functions, avoiding multiple local minima and preventing premature convergence of the optimization algorithm.

2.2 Mid-range function approximation concept

Mid-range approximations [3] are designed to be valid in a smaller region than the region for global approximations. Such a region is bounded by movelimits. As a consequence, simpler model functions and fewer design points can be used in comparison with the global method. Also, contrary to the global method the mid-range approximations are constructed in a sequential way comparable to local methods. The differences with local methods however are that the initial movelimits are larger and previous analyses are used as much as possible to construct and enhance the approximations.

The process of building mid-range approximation functions uses basically the same scheme as building globally valid function models. First, model functions must be selected for each response which can be, for example, simple linear or reciprocal. A restricted number of design points is available for function construction, in principal only the design points within the movelimits. Hence the number of parameters must be kept small, preferably in the order of the number of design variables. In general, model functions should be able to describe curvature well in order to speed up convergence.

Two strategies can be distinguished for the choice of the design points the models are fitted to (Fig. 2). The first strategy uses data from single points along

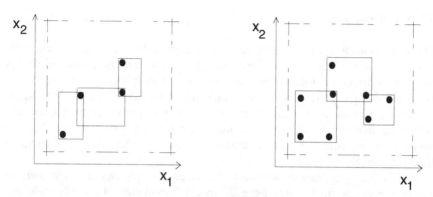

Fig. 2. Design point placement in mid-range SPP (left) and MPP (right) method.

the iteration path in the design space, so called single-point-path (SPP) methods. The design points are the solutions of the sequential optimization subproblems. All points within the movelimits are used to build the approximation. As the optimization progresses, more design points become available to fit the models to, and the approximations are improved in each optimization cycle. The number of design variables can be as large as for a local method.

Approximations derived from data computed in clusters of design points along the optimization path are called multi-point-path (MPP) methods. Around each solution of the optimization subproblem one or more extra points are generated by placing points according to simple experimental designs as discussed in section 2.1. This approach is valuable if no sensitivity data can be used, for instance if sensitivity data is not accurate enough or if it can not be computed, and is preferable above the calculation of finite differences. However, a restriction is placed on the number of design variables to be used, since for each additional design variable extra design points must be analyzed. The number of design variables must be limited to keep the problem manageable.

Both mid-range methods require a robust movelimit strategy: the design region should be large enough to use multiple points but should be not too large to worsen the accuracy of the models. The choice of the size of the design sub-region is based on a mixture of accuracy of the approximation functions, constraint violation and convergence behaviour of the objective function.

After the unknown parameters in the function models are estimated, the approximate optimization problem is solved by an optimization algorithm. A large number of constraints can be handled and their number is only limited by computer resources. Next, around this optimum a new approximation is made based on either a SPP or MPP strategy and the process is repeated until convergence occurs.

3 Example: Acoustic optimization

3.1 Introduction

The main objective of structural-acoustic design is to create quiet structures. Generally, such structures were designed on a trial and error basis and any design better than the original design was accepted as a satisfying, not necessarily the optimum, design. During the past decade attention has shifted to the use of optimization methods, leading to a more structured design approach. Serious difficulties that remain are the derivation of a correct analysis model to describe the physical (acoustic) behaviour of the structure and the inherent long computation time for acoustic analyses.

As in most optimization problems, the success of the optimization and the acceptance of the final design depends on the formulation of the objective and constraint functions. In [4] various objective function formulations for structural-acoustic plate design were considered. These formulations are valid for acoustic optimization problems in general. Sound power radiated from vibrating plates was optimized in [5]. The optimization problem was solved using a gradient based optimization routine. Since the problem had multiple minima, only local minima could be found and the possibility of using global optimization methods was suggested. In reference [6] an engine structure is optimized for minimum noise transmission. Response surface methods were used to construct explicit relations between design variables and responses. Closed form expressions between the transmission loss of sandwich panels and the design variables were derived in [7]. The authors stated that methods which spread the search of the optimum over the entire design space have more chance of finding the global optimum than methods that rely on perturbations near the current design.

The acoustic optimization problem that will be considered here is described in [8]. It comprises most of the aspects encountered in acoustic optimization problems: no sensitivity data readily available and a noisy objective function behaviour. On the other hand, the mathematical model of the air intake system describes the physical model sufficiently accurate and computation time for the acoustic analysis is relatively short.

3.2 Acoustic model of engine air intake system.

In Fig. 3 a four-cylinder engine air intake system is schematically represented by straight tubes. It consists of four air intake runners, one for each of the four cylinders of the engine. The air cleaner is connected to the intake runners by a crossover tube and a throttle body. To the air cleaner an inlet snorkel is mounted.

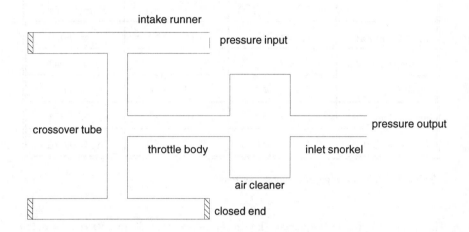

Fig. 3. Model of four-cylinder engine air intake system [8].

The noise produced by the system results from the opening and closing of the inlet valves of the engine cylinders. If one of the inlet valves opens, the pressure in the cylinder is above atmospheric pressure and a positive pulse sets the air in the inlet system into oscillation. The output pressure at the inlet snorkel is a function of the frequency at which the valves open and close. This frequency is directly related to the number of revolutions at which the engine is running. At some specific number of revolutions the natural frequencies of the inlet system are matched and the manifold transmits noise very efficiently.

Plane wave theory (see [9]) is used to derive the relation between the input pressure at the intake runner and output pressure at the inlet snorkel. Part of the equations involved are derived in the appendix. A fluctuating pressure of unit one is assumed at input while the other three valves are closed. The resulting set of complex equations is solved by the NAG-routine F04ADF.

Experiments on the real air intake system showed a good resemblance with the analytical model [8]. The small differences can be explained by the non-smooth junctions and the neglect of mean flow in the tubes. Also, acoustic damping and wall absorption is not taken into account. A comparable system was optimized in [10] using the finite element method. By adding volumes to the initial system, the pressure amplitudes decreased but more eigenfrequencies were introduced.

3.3 Optimization problem formulation

Table 3.1 shows the initial lengths and diameters of the tubes. Only the lengths of four of the tubes are taken as design variables in this study and vary between the bounds given.

	initial length (m)	bounds (m)	diameter (m)
intake runners	$L_a = 0.216$	0.15 - 0.51	$S_a = 0.034$
crossover tube	$L_b = 0.152$		$S_b = 0.041$
throttle body	$L_c = 0.063$	0.05 - 0.50	$S_c = 0.044$
air cleaner	$L_d = 0.152$	0.10 - 0.30	$S_d = 0.251$
inlet snorkel	$L_e = 1.016$	0.51 - 1.50	$S_e = 0.076$

Table 3.1. Initial design of air intake system.

The most undesirable frequencies lay in the range from 50 to 250 Hz. To obtain a low noise transmission in this range, the average output pressure between 50 and 250 Hz is taken as the object function U:

$$U = \frac{1}{n} \sum_{i=1}^{n} P_i(f) \tag{6}$$

where n is the number of frequency intervals used in the summation. In [8] a frequency step of 5 Hz was used to obtain reasonable computation times but caused the objective function to be highly non-smooth. A smaller step size decreases the non-smoothness of the function. Replacing the summation by an integration (e.g. Runge-Kutta integration with variable step size control) even removes the non-smoothness completely. To maintain the noisy behaviour of the objective function and to represent the actual response as close as possible, a step size of 1 Hz is chosen. The objective function value for the initial design is 0.040. Sensitivity data is not directly available and, as will be seen in section 3.4, it is not wise to include it in the optimization process. No further constraints are imposed on the problem.

3.4 Optimization results

To gain some insight in the objective function behaviour, the optimization problem is first solved with the global function approximation method. A full factorial design on two levels results in a total of 16 design points, positioned at the bounds of the design space. On the results of these 16 design points a polynomial with only first order main terms is fitted. Maximum difference between exact and approximated values is 66%. Since the approximation forms a plane in the four dimensional design space, the optimum is found in a corner of the design space, $\mathbf{x} = (0.51, 0.05, 0.30, 0.51)$ where the approximated objective function value is 0.0074 and the exact value is 0.0058. Of course, one should at least add quadratic terms to the function model and perform some additional analyses to obtain a better explicit approximation and a more accurate optimum.

To be able to visualize the design space, the optimization process is limited to two design variables. Construction of a normal probability plot [6] from the 16 design points and a full polynomial with all possible interactions between variables reveals that design variables L_a and L_c are of main importance. In Fig. 4 the objective function is shown, based on a grid of 20 by 20 design points.

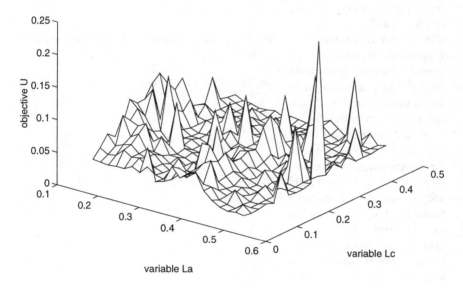

Fig. 4. Design space for design variables L_a and L_c.

The design space is highly non-smooth with many local minima. A direct coupling of the analysis code with an optimization algorithm would almost certainly lead to premature convergence of the algorithm. The use of sensitivity data, if it was available, could even worsen the process of approximation and optimization.

Searching for the global optimum design in the above grid yields the design L_a = 0.49, L_c = 0.22 and the objective function value U = 0.0092. This point will serve as a reference for further optimization.

Next, the mid-range function approximation method is applied to the two-dimensional problem. Linear models are constructed in each optimization cycle. The initial move-limit size is chosen to be 0.4 for each design variable direction, i.e. around the initial design an area of 0.4 times the distance between the bounds on the variable is chosen, and the move-limits are allowed to change in each cycle and for each design variable separately. Choosing a smaller initial step size causes the optimizer to get trapped in a local optimum. Additional design points are chosen along each variable axis. Convergence is achieved after 9 optimization steps and 25 objective function evaluations (Fig. 5). The path followed by the optimizer in the design space is shown in Fig. 6.

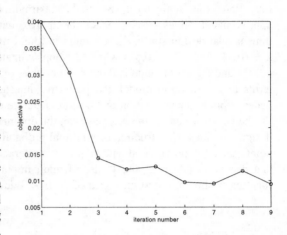

Fig. 5. Iteration history mid-range approximation method.

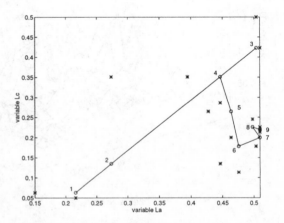

Fig. 6. Optimization path through design space ('o' = optimum point, '*' = additional point).

The optimum design found is $L_a = 0.51$, $L_c = 0.22$ where the approximate objective function value $U = 0.0093$ and is close to the optimum found from searching the 20 by 20 grid. Starting the optimization from different starting points resulted in some cases in local minima but a restart requires only a moderate number of analyses. Use of a larger initial move-limit does not automatically lead to a faster convergence with lesser analyses but, what is more important, the optimum lays in these cases close to the global optimum. A more detailed study with all four design variables was carried out and showed the optimum to be $x = (0.51, 0.05, 0.30, 0.51)$, i.e. in a corner of the design space, and $U = 0.0058$. Design variables L_d and L_e appear to have no a great effect on the final objective function value. A large reduction in noise transmission can be achieved compared to the initial design of the air intake system.

4 Conclusion

Function approximation concepts based on response surface methods are discussed. Response data from multiple design points are used in constructing the explicit approximations which serve as an interface between analysis code and optimization algorithm. The multi-point approximations appear to be useful if no sensitivity data is available or cannot be used due to non-smooth response behaviour.

Application to a four-cylinder engine air intake system for minimum noise transmission is shown. Non-smooth response behaviour is caused by numerical noise which is effectively dealt with in the approximation process.

Acknowledgement
This research was sponsored by the CEC-BRITE-EURAM contract BRE2 CT92-0141, Project No. 5083 - OPTIM

References

[1]   Box, G.E.P. and Draper, N.R. (1987): Empirical Model Building and Response Surfaces. John Wiley, New York.

[2]   Efroymsom, M.A. (1960): Multiple Regression Analysis. In: Ralston, A., Wilf, H.S.: Mathematical methods for Digital Computers. New York, John Wiley.

[3]   Toropov, V.V., Filatov, A.A. and Polynkin, A.A. (1993): Multiparameter structural optimization using FEM and multipoint explicit approximations. Struct. Opt. 6, 7-14.

[4]   Lamancusa, J.S. (1993): Numerical optimization techniques for structural-acoustic design of rectangular panels. Comp. & Struct. 48(4), 661-675.

[5]   Belegundu, A.D., R.R. Salagame and G.H. Koopman (1994): A general optimization strategy for sound power minimization. Struct. Opt. 8, 113-119.

[6]   Milsted, M.G., T. Zhang and R.A. Hall (1993): A numerical method for noise optimization of engine structures. Proc. Instn. Mech. Engrs. 207, 135-143.

[7]   Makris, S.E., C.L. Dym and J.M. Smith (1986): Transmission loss optimization in acoustic sandwich panels. J. Acoust. Soc. Am. 79(6), 1833-1843.

[8]   Lamancusa, J.S. (1988): Geometric optimization of internal combustion engine induction systems for minimum noise transmission. J. Sound and Vibration 127(2), 303-318.

[9]   Munjal, M.L. (1987): Acoustics of ducts and mufflers with application to exhaust and ventilation system design. Wiley & Sons, New York.

[10] Hackenbroich, D. (1988): Reduktion des Innengeräusches bei Nutzfahrzeugen durch rechnerische Optimierung des Mündungsgeräusches von Motoransaug-anlagen. VDI Berichte 669, 631-654.

Appendix

For the derivation of the system of equations, plane wave theory is used. The basic assumption is that the diameters of the tubes are small compared to the acoustic wavelength. Then the acoustic variables are functions of only one coordinate and the wave equation is:

$$\frac{\partial^2 p}{\partial x^2} = \frac{1}{c^2}\frac{\partial^2 p}{\partial t^2} \tag{A.1}$$

where p is the pressure and c is the speed of sound. A good resemblance between measurements and numerical modelling using plane wave theory was shown in [8]. A part of the complete air intake system is shown in figure 1.

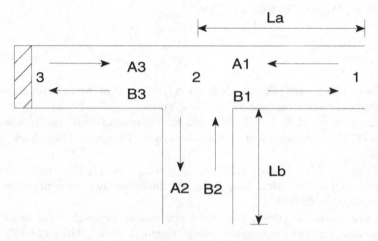

Fig. 7. Part of plane wave model of four-cylinder induction system [8].

The general solution of the wave equation is:

$$p = A\,e^{i(\omega t - kx)} + B\,e^{i(\omega t + kx)} \tag{A.2}$$

where A is the amplitude of the forward travelling pressure wave, B is the amplitude of the backward travelling wave, ω is the frequency (rad/s), k is the wavenumber defined by $k = \omega/c - i\alpha$ (α is the absorption coefficient). Assuming at each junction continuity of pressure and mass conservation, a set of 18 complex equations can be derived describing the relation between input and output pressure as a function of frequency ω, where A_j and B_j ($j = 1, ..., 9$, the number of junctions) are the 18 unknowns to be solved. At position 1, $x_1 = 0$, and a fluctuating input pressure of magnitude 1, $p_1 = e^{i\omega t}$, resulting in:

$$A_1 + B_1 = 1 \tag{A.3}$$

Applying continuity of pressure at junction 2:

$$p_1(x_1 = L_a) = p_2(x_2 = 0) \Rightarrow A_1 e^{-ikL_a} + B_1 e^{ikL_a} = A_2 + B_2 \tag{A.4}$$

$$p_3(x_3 = L_a) = p_2(x_2 = 0) \Rightarrow A_3 e^{-ikL_a} + B_3 e^{ikL_a} = A_2 + B_2 \tag{A.5}$$

And mass conservation at junction 2:

$$S_a u_1 (x_1 = L_a) + S_a u_3 (x_3 = L_a) = S_b u_2 (x_2 = 0) \tag{A.6}$$

with the velocity:

$$u_j = \frac{1}{\rho_0 c}(A_j e^{i(\omega t - kx)} - B_j e^{i(\omega t + kx)}) \tag{A.7}$$

yields:

$$S_a (A_1 + A_3) e^{-ikL_a} - S_a(B_1 - B_3) e^{ikL_a} = S_b(A_2 - B_2) \tag{A.8}$$

At closed end 3 (valve closed) the volume velocity is zero, resulting in:

$$A_3 - B_3 = 0 \tag{A.9}$$

Other junctions are equally treated. At the open end of the inlet snorkel (junction 10) a radiation impedance Z exists since the tube radiates sounds into the surrounding medium. Assuming a flanged pipe:

$$Z = \frac{(ka)^2}{2} + \frac{(ka)^4}{2} + i[\frac{8ka}{3\pi} - \frac{32(ka)^3}{45\pi}] \tag{A.10}$$

where a is the radius of the snorkel end. Since $Z = p/u$:

$$A_9 (1 - Z) e^{-ikL_e} + B_9 (1 - Z) e^{ikL_e} \tag{A.11}$$

and the exit pressure becomes:

$$P_{exit} = A_9 e^{-ikL_e} + B_9 e^{ikl_e} \tag{A.12}$$

Stochastic Structural Optimization of Powertrain Mounting Systems with Dynamic Constraints

Rolf Deges and Thomas Vietor

Ford Werke AG, Vehicle NVH, D-50725 Köln

Abstract. NVH (Noise, Vibration, Harshness) is one of the main attributes in a passenger car. One of the key systems responsible for the overall NVH behaviour is the powertrain mounting system. The optimal layout of this system leads to the definition of a stochastic optimization problem. In this paper the background of the complexity of the powertrain mounting system is highlighted. In a first approach the optimization problem is solved with Design of Experiments (DOE) methods.

Keywords. DOE, Powertrain Mounting System, NVH, Scattering Mechanical Quantities, Vehicle Attributes

1. Introduction

The development of a passenger car is a multidisziplinary task. The vehicle has to fulfill demands out of different attributes like vehicle dynamics, driveability, acoustics, thermal and heat management, safety, crash, economics. This paper concentrates to acoustics or in general Noise, Vibration and Harshness (NVH) of the vehicle. This area is currently one of the main attributes defining overall performance and customer perception of a passenger car. Very often demands out of this area are conflicting with other areas. A main problem is the variability of mechanical quantities describing the NVH performance of a car. To overcome this, the extension of the conventional deterministic optimization problem to a stochastic optimization problem is necessary. The powertrain mounting system, which is the physical link between powertrain and vehicle, is one of the most critical systems with respect to NVH. So this paper concentrates on aspects for the optimal layout of the powertrain mounting system. Because of the complexity of this single task in a first approach a sensitivity calculation of the system is performed with Design of Experiments (DOE) methods.

NVH Vehicle system concepts (e.g. body structure, front- and rear suspension, powertrain mounting systems, etc.), which are selected in an early program phase, have significant influence on NVH performance of the vehicle. It is almost impossible to solve NVH concerns resulting from selection of poor concepts in a later program phase. A good understanding of dynamics of powertrain and its interaction within the vehicle is needed to design and realize mounting systems that allow the company to reach its NVH leadership goals. The selection of the powertrain mounting concept and the design of powertrain mounting system is a highly complex task which requires involvement of several areas.

In the following, demands for powertrain mounting systems, theoretical background, an overview on the design process as well as some practical details of the development process will be given.

2. Demands for Powertrain Mounting Systems

The powertrain mounts are the main links between powertrain and body structure. The main functional demands for the powertrain mounting system are:

- Support of the powertrain under all load cases, e.g.
    - gravity weight
    - drive torque
    - tip in / back out[1] (torque change)
    - switch on / switch off
    - acceleration, cornering
    - road impacts

- Isolation of vibration due to
    - Engine excitation [1], e.g. idle, acceleration, cruise, overrun
    - Driving maneuvers, e.g. torque change, switch on / switch off
    - Road and wheel excitation

The selection of the powertrain mounting system is heavily restricted by both program assumptions as well as corporate demands :
- Powertrain concept or concepts
    - Front-wheel-drive, rear-wheel-drive and/or all-wheel-drive
    - North-South (N-S) or East-West (E-W)[2] powertrain installation

---

[1] Short form for pressing or releasing the gas pedal

[2] Indicates orientation of the powertrain in the vehicle. N-S is along the vehicle x-axis which is in driving direction.

- Engine architecture, e.g. I4, V6, I3, I5, etc.
- Gas- or Diesel Engine
- Manual or automatic transmission
- Available package space[3] (rock angle)
- Crash requirements
- Cost requirements / Investments
- Feasibility and manufacturing demands
- Durability
- Company cross-carline strategy
  - reduced complexity
  - unique parts

All demands for powertrain mounting systems as given above have to be taken into account for selection of the mounting concepts. It is obvious that some of the demands are contradictory.

3. Effects and Vehicle Systems Influenced by the Powertrain Mounting System

A series of phenomena within the vehicle are mainly influenced by the powertrain mounting system. The main NVH effects are given in Tab. 1

Powertrain mounts have effects over a wide frequency range, excitation amplitudes vary from several millimeters in the low frequency range to micrometer in the high frequency range. As already stated above, the selection of the engine mounting system has severe influences on several other vehicle systems.

4. Systems Engineering Approach

Customer wants and customer satisfaction are the key points during development process of a new vehicle [2]. Extensive market research activities including benchmarking, customer drives and translation from customer wording to objective measurables (quality function deployment QFD) result in target values for vehicle attribute performance. With respect to powertrain NVH, these total vehicle targets are e.g. interior noise level at specified driving conditions or idle vibration of the seat track. All vehicle development work, e.g. vehicle system selection, vehicle system optimization and component design is based on these target values, which become program objectives during the development process after confirmation of vehicle system selection.

---

[3] Package indicates the amount of available volume to arrange engine, transmission and all other subsystems.

NVH Effect	Frequency Range
Drive noise and vibration	20 - 500 Hz
Idle shake, vibration and boom	5 - 50 Hz
Road induced shake	10 - 15 Hz
Take-off judder	10 - 30 Hz
Drive-away harshness	20 - 100 Hz
Switch on/off vibration	5 - 20 Hz
Tip in / back out	3 - 10 Hz
Steering column shake	25 - 40 Hz
Powertrain boom and harshness	50 - 500 Hz

Table 1. NVH Effects Influenced by Powertrain Mounting System.

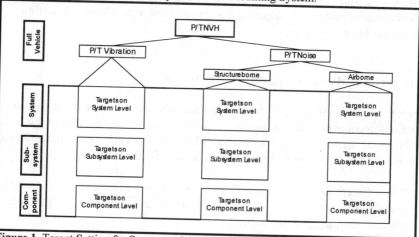

Figure 1. Target Setting for System and Components - Systems Engineering
(P/T = Powertrain).

Target cascading, which is mainly supported and performed by CAE techniques, is the process of deriving system and subsystem targets based on total vehicle targets. One aspect of target cascading for drive noise involving powertrain mounting system characteristics can be seen from Fig. 1. This graph shows some aspects of source identification, source ranking and transfer path analysis. The final outcome of these exercises are targets for excitation levels and transfer characteristics as well as targets and design specifications for components.

It is obvious, that target cascading is performed for all vehicle attributes. If system and component targets derived for different aspects are contradictory, trade-offs have to be performed already in an early program state to ensure high overall customer satisfaction. Tab. 2 as well as Fig. 1 should point out that specifications for each vehicle system can be derived only taking into account total vehicle performance. One main objective during NVH development work is to avoid coincident resonances within the vehicle. It is well known, that best isolation of vibration can be achieved if excitation frequencies are well above resonance frequencies. With respect to powertrain mounting system characteristics, rigid body modes of the powertrain should be as low as possible, but above 1/2 engine order frequency and should be well separated from overall vehicle system resonances. Fig. 2 shows characteristic resonance and excitation frequencies of the main vehicle systems. There is a small frequency window available only for rigid powertrain modes.

It is well known that it is not sufficient for competitive vehicle NVH performance to design a powertrain mounting system which has just all rigid powertrain modes within the frequency gap marked in Fig. 2. Experience showed that mode shape characteristics and frequencies of modes play an important role for the quality of a powertrain mounting system.

5. Powertrain Mounting System Design Process

Several departments have to be incorporated during the powertrain mounting system selection and design process. This process is heavily driven by CAE analyses [3], as there is no hardware available in this early development stage. The powertrain mounting system design process is illustrated in Fig. 3.

- The different powertrains are defined in program assumptions, full vehicle performance requirements, e. g. NVH targets, are derived from the benchmarking and target setting process.

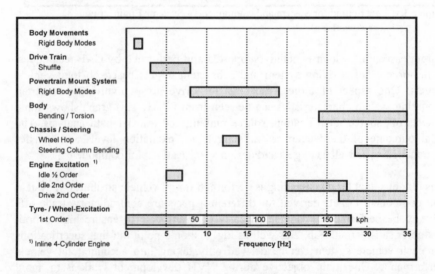

Figure 2. Typical Excitation- and Resonance Frequencies of a Vehicle.

- Based on body, powertrain and chassis design, available package as well as experience from former vehicle programs (bookshelf knowledge), an initial proposal for the powertrain mount positions as well as powertrain mount stiffnesses for idle load is made. These proposals have to be confirmed by CAE analyses. As in this early stage of the development program, no detailed body information is available, first CAE calculations are performed on a linear, 'grounded' powertrain CAE model. The first set of design variables is optimized to fulfill modal demands.

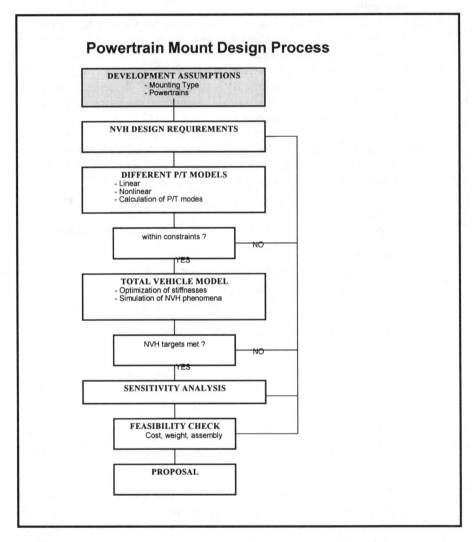

Figure 3. Simplified CAE based optimization of Engine Mounting Systems.

- Linear mount characteristics are not sufficient because of package constraints (e.g. clearance for maximum rock angle). Therefore, progressive mounts characteristics have to be designed. Non-linear, progressive mount stiffnesses with "zero-load" stiffness as confirmed above are selected, which should ensure the following criteria:
  - smooth transition from linear to progressive characteristics range,
  - drive in non-linear characteristics range in low gears only,
  - mount stiffness for drive as low as possible,
  - use of maximum feasible deflection at maximum engine load only,
  - feasibility.

CAE checks of maximum rock angle and powertrain mount forces of non-linear mounts can be performed with non-linear models only, which are mainly grounded powertrain models.

- If all targets and requirements are met, a 'total vehicle system CAE mode' has to be used to check, if full vehicle NVH targets for all phenomena influenced by powertrain mounts are met as well as to optimize mount characteristics as well as body and chassis performance.

If no detailed body information is available at the first total vehicle calculations, CAE runs can be performed using surrogate body models or simplified body CAE models. The following CAE tools are widely used to support powertrain mount design process:

ABAQUS :
Standard Finite Element Program for linear and nonlinear calculation.

ADAMS:
Standard tool for Vehicle Dynamic applications. For NVH the limited upper frequency limit constraints the use rigidly.

NASTRAN:
Standard Finite Element Program for all kinds of calculations.

MOTRAN:
Set of programs developed by FORD, which are used for NVH analysis of vehicles. It allows combination of modal models with simple elements, e.g. linear springs. Massive reduction of computing time.

HYBRID TECHNIQUES:
Techniques, which allow combination of calculated (CAE) and measured (test) data.

It is urgently necessary to use total vehicle CAE models from this step on as rigid body modes of grounded powertrains can significantly differ from modes of the same powertrain with identical mounts in a vehicle. Therefore, a post-optimization of the above developed mount characteristics is necessary. Idle as well as drive vibrations can be calculated with help of standard dynamic FE packages. One sample result of such a calculation of steering wheel idle vibrations is given in Fig. 4. It shows vibrations for the same load for two different sets of engine mounts.

Estimates for total vehicle NVH performance above 80 Hz are not possible with standard FE tools, at least not at this program state. The only way to verify system performance in a higher frequency range is to use hybrid techniques for so-called 'Noise Path Analysis' tools.

• If all NVH targets are met, a sensitivity analysis follows, which identifies main influence factors and enables definition of feasible tolerances or requires a complete redesign to achieve a more robust powertrain mounting system. After a feasibility check, the final proposal will be realized as hardware and will be tested in prototypes or demonstration vehicles.

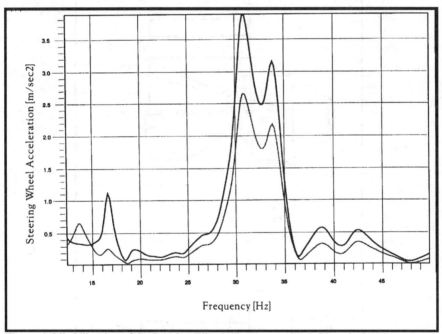

Figure 4. Idle Vibration of the Steering Wheel. The two curves define calculated accelerations in z-direction of the steering wheel.

• The described development process leads to a fast possibility to find an optimal layout of the engine mounting system from NVH point of view. During the design process, it is possible and often necessary to jump back to an arbitrary earlier step. In each step and as soon as available, consideration of requirements from other areas like Vehicle Dynamics, Driveability, component areas and the supplier is necessary. This rises the need for parallel optimization for several attributes. It is impossible to solve this multi-criteria optimization problem purely by CAE. So a well defined process must be established with participation of all related areas including the manufacturer directly from the beginning. By this way only, necessary trade-off decisions are possible in an early stage.

6. Torque Roll Axis System

Since introduction of E-W installed powertrains in front-wheel driven vehicles, a series of mounting systems has been developed. Design of the first systems has been mainly dominated by package reasons whereas in the last years a general trend goes to so-called 'Torque Roll' axis (TRA) systems (Fig 5). These are powertrain mounting systems mainly based on inertia characteristics of the powertrain.

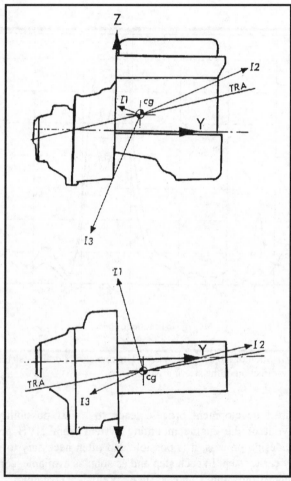

Figure 5. Torque Roll Axis and Principal Axis.

The 'Torque Roll axis' (TRA) is the theoretical axis on which a free powertrain rotates if subjected to torque fluctuations about the crankshaft. The orientation of this torque roll axis is defined only by inertia properties of the powertrain, the center of gravity is located on this axis. The roll axis, which is the axis of rotation of the mounted powertrain, if it is subjected to torque fluctuations, is not identical with the torque roll axis. The roll axis is dependent on powertrain inertia, mount stiffness and excitation frequency, whereas the (static) rock axis depends only on the mount stiffness. The idea of so-called TRA mounting systems is to minimize the difference between roll axis and TRA axis in order to achieve minimum dynamic mount forces in idle, which are mainly caused by combustion pressure torque fluctuations.

7. Scattering Variables

It is necessary for customer satisfaction to design and manufacture a robust vehicle. A great scattering of different parameters describing the overall NVH performance of the vehicle is definitely not tolerable. Identified main parameters responsible for the scattering of the interior noise and vibrations at certain contact points in the vehicle are:

- Engine mount idle stiffnesses. Here a certain tolerance level is realistic for production parts. The exact value is depending from the kind of mount (rubber or hydraulic) and the manufacturing process.

- Engine mount high frequency dynamic stiffnesses. In this paper not included.

- Maximum preload values due to weight of the powertrain and/or engine torque. A deviation of ± (5...10) % seems to be a realistic lower tolerance bound. This scattering is critical because of large deviations of the dynamic engine mount stiffnesses with scattering preloads.

- Load variation in idle with electrical consumers on/off.

- Mounting positions.

Fig. 4 shows calculated accelerations at the steering wheel in z-direction for two different engine mount idle stiffness sets. The original set with nominal stiffnesses shows accelerations at a lower level than the second set. Stiffness data of the second set are varied within the production tolerances. The necessary calculations are performed for a total vehicle CAE model which includes a detailed modal model of the flexible body.

8. Formulation of an Optimization Problem

For the optimization of the engine mounting system an intuitive approach is unsatisfying. So the formulation of an optimization problem is necessary and the use of an optimization procedure for the solution of the problem.

8.1 Definition of a Stochastic Optimization Problem [5]

A simplified definition of the continuous, stochastic optimization problem is given in the following.

$$"Min" \, f_A(Z)$$
$$x \in D$$
(8.1)

with

$$f_{A_i}(Z) = k_1 \, E\big(f_i(Z)\big) + k_2 \, V\big(f_i(Z)\big), \quad Z^T = (X^T, P^T)$$
(8.2)

$$D = \Big\{ x \in R^n \mid h_i = 0 \, \forall i = 1, \ldots, m_{st}; P_{fk} = P\big[g_k(Z) < 0\big] < P_{k_{max}}$$
(8.3)

$$\forall k = 1, \ldots, n_g, \, x_{kl} \leq x_k \leq x_{ku} \, \forall k = 1, \ldots, n \Big\}$$

the vector of inequality constraints $g^T = (g_1, g_2, \ldots, g_{n_g})$
(8.4)

and

f_{A_i} augmented objective function, here interior noise or vibrations,

g_k inequality constraints as a function of stochastic variables, here frequency conditions,

h_i equality constraints $h_i = h_i(\mu)$, here the system equations,

Z vector of the stochastic variables,

X vector of the stochastic design variables, here mount stiffnesses and/or coordinates,

P vector of the stochastic parameters,

$E(f(Z))$ expected value of the objective function,

$V(f(Z))$ variance of the objective function,

k_1, k_2 weighting factors,

D feasible design space,

P_{fk} failure probability of the k-th inequality constraint,

$P_{k_{max}}$ feasible value of the failure probability,

n_g number of stochastic inequality constraints,

x_{kl}, x_{ku} lower and upper bounds of the design variables, respectively.

Here $\mathbf{X}$ ≡ $\mathbf{P}$ is assumed.

8.2 Stochastic Optimization Procedures

The optimization problem defined in equation (8.1) can be solved by means of different procedures. By integrating a stochastic optimization procedure into the optimization procedure SAPOP [4], a complete and extensive optimization environment is available that additionally allows to use further optimization strategies in combination with stochastic optimization [5]. Thus, the stochastic optimization problem in (8.1) is transformed into a quasi-deterministic optimization problem with reliability constraints. Here, fulfillment of constraints is determined by stating probabilities, which are considered during optimization. On the other hand, the deviation of the objective function owing to the stochastic distribution of the design variables and constraints is neglected.

8.3 Practical Solution of the Stochastic Optimization Problem

For the solution of the mentioned optimization problem it it necessary to have a structural model of the total vehicle. This includes

- detailed Finite Element (FE) body model,
- concept FE chassis model,
- detailed FE powertrain model for interior noise or concept FE model for vibration calculations,
- cavity FE or Boundary Element model (only for interior noise calculations).

The CPU time for **one** structural analysis is of the magnitude of several hours on a supercomputer. So it is obvious that a stochastic optimization is only possible

- with the help of simplified models where sufficient,
- with the use of model reduction techniques like superelements or modal models,
- the limitation to a number of stochastic variables < 10.

9. Robust Design with DOE Methods

With CAE methods it is possible in a very early stage of the program to investigate the robustness of the total vehicle and different subsystems, e.g the powertrain mounting system. One of the methods is the Design of Experiments (DOE) [6, 7] methodology. In combination with Response Surface Methodologies (RSM), which can be used with CAE models and tests very effectively, also an optimization is possible. Two different approaches are:
- Design of a robust powertrain mounting system. This is limited to the concept phase of the program.

- Investigate the robustness of a given system, identification of sensitive parameters and formulation of feasible tolerances for the mechanical quantities of the powertrain mounting system like stiffnesses and the manufacturing process.

The first approach is for sure the best but often not possible. The second one is applicable in a running development with a given concept but can only improve a given system. DOE allows to quantify the influence of varying dynamic stiffness of each powertrain mount for each direction on both powertrain forces as well as vehicle NVH performance. Changing stiffness of one mount in one direction only can influence forces of other mounts drastically.

Fig. 6 shows adjusted response curves as a result of a typical DOE study performed with the software package RS [7]. The structural model contains a full vehicle detailed model but with a modal representation of the body. This reduces the structural analysis time by a factor of about 20. In a first step no interactions between the variables is assumed. The meaning of the variables is as follows for a system with three engine mounts: $x_1, \ldots, x_8$ stiffnesses and locations of the engine mounts, for each direction of the mount one indenpendent stiffness is assumed, constraint position variables are not included.

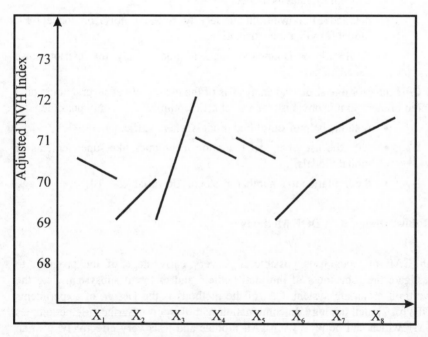

Figure 6. Adjusted Response Curves for the DOE Study.
The adjusted response surface (NVH index)

$$NVHindex(\mathbf{Z}) = f(X_1, X_2, \ldots, X_n) \tag{9.1}$$

is a normalized scalar measure for the vibrations at certain points in the vehicle. A greater value indicates smaller amplitudes of the vibrations. For the vehicle development the target is to maximize this value. From Fig. 6 it is obvious that the influence of the three adjusted variables x_2, x_3, x_6 is dominating the other variables. In a further step a new DOE is necessary including interactions but limited to the main identified variables. With this result the RSM is used to find an optimal design.

10. Conclusion

The powertrain mounting system is one of the key systems for the overall NVH behavior of the vehicle. This paper presents in a short form the mechanical background of this system and explains the stochastic character of the optimization problem. Because of the complex structural model only a simplified approach for the solution is applicable. With the proposed procedure the identification of the main sensitive parameters is possible. In the later development the optimization of these parameters is necessary. Currently the investigation of all parameters is state of the art. With the given methods in future a drastical reduction of the effort is possible.

References

[1] Mass, H., Klier, H.: Kräfte, Momente und deren Ausgleich in der Verbrennungskraftmaschine, Springer-Verlag, Wien, 1981.

[2] Eichhorn, U., Sauerwein, D., Schmitz, T., de Vlugt, A., Teubner, H.-J.: Kundenorientierte Entwicklung am neuen Ford Fiesta, ATZ 97 (1995) 9, S. 522 -531.

[3] Bürger, K.-H.: Analysen des Schwingungsverhaltens des Gesamtfahrzeuges mit Konzeptmodellen, in 'Computergestützte Berechnungsverfahren in der Fahrzeugdynamik', VDI, 1991.

[4] Eschenauer, H.A.; Koski, J.; Osyczka, A.: Multicriteria Design Optimization. Springer, Berlin, 1990.

[5] Vietor, T.: Stochastic Optimization in Mechanical Engineering. In: Marti, K. (ed.): Sonderheft der Zeitschrift für Operations Research "Structural Reliability and Stochastic Structural Optimization". To be published in 1997.

[6] Bandemer, H.; Bellmann, A.: Statistische Versuchsplanung. Teubner, Leipzig, 1994.

[7] N.N.: RS/DISCOUVER Reference Manual. BBN-Softw. Prod., Cambridge, 1992.